# Nanofabrication Handbook

# Nanofabrication Handbook

Edited by **Lindy Bowman**

NY RESEARCH PRESS

New York

Published by NY Research Press,
23 West, 55th Street, Suite 816,
New York, NY 10019, USA
www.nyresearchpress.com

**Nanofabrication Handbook**
Edited by Lindy Bowman



International Standard Book Number: 978-1-63238-336-5 (Hardback)



Printed in China.

# Contents

# Preface

The world is advancing at a fast pace like never before. Therefore, the need is to keep up with the latest developments. This book was an idea that came to fruition when the specialists in the area realized the need to coordinate together and document essential themes in the subject. That's when I was requested to be the editor. Editing this book has been an honour as it brings together diverse authors researching on different streams of the field. The book collates essential materials contributed by veterans in the area which can be utilized by students and researchers alike.

In the wake of recent advancements, science is one of the most swiftly progressing fields. We deal with a lot of challenges today, such as the growing demand for energy and consumer goods throughout the world. We consider that new developments in science and technology will help resolve many of these troubles. Nanofabrication is a key process required for the development of novel equipment, strategy and systems. Specific management of nanomaterials, nanostructures, nanodevices and their performances is necessary for future development in technology. This book gives the most recent research developments in nanofabrication of natural and non-living substances, bio and hybrid materials. With this book we intend to contribute in creating a brighter future for the next generation.

Each chapter is a sole-standing publication that reflects each author's interpretation. Thus, the book displays a multi-facetted picture of our current understanding of application, resources and aspects of the field. I would like to thank the contributors of this book and my family for their endless support.

**Editor**

# 1

# Nanofabrication for Molecular Scale Devices

Susmit Kumar[1], Shilpi Karmakar[1], Alessandro Bramanti[2], Ross Rinaldi[1] and Giuseppe Maruccio[1,3]

*[1]National Nanotechnology Laboratory, Istituto Nanoscienze-CNR,*
*[2]STMicroelectronics Srl,*
*[3]Department of Physics, University of Salento, Lecce, Italy*

## 1. Introduction

The predicted 22-nm barrier which is seemingly going to put a final stop to Moore's law (Fig. 1) is essentially related to the resolution limit of lithography. Consequently, finding suitable methods for fabricating and patterning nanodevices is the true challenge of tomorrow's electronics. However, the pure matter of moulding devices and interconnections is interwoven with research on new materials, as well as architectural and computational paradigms. In fact, while the performance of any fabrication process is obviously related to the characteristic of the materials used, a particular fabrication technique can put constraints on the definable geometries and interconnection patterns, thus somehow biasing the upper levels of the computing machine. Further, novel technologies will have to account for heat dissipation, a particularly tricky problem at the nanoscale, which could in fact prevent the most performing nanodevice from being practically employed in complex networks. Finally, production costs - exponentially growing in the present Moore rush - will be a key factor in evaluating the feasibility of tomorrow technologies.

The possible approaches to nanofabrication are commonly classified into top-down and bottom-up. The former involves carving small features into a suitable bulk material; in the latter, small objects assemble to form more complex and articulated structures. While the present technology of silicon has a chiefly top-down approach, bottom-up approaches are typical of the nanoscale world, being directly inspired by nature where molecules are assembled into supramolecular structures, up to tissues and organs. As top-down approaches are resolution-limited, boosting bottom-up approaches seems to be a good strategy to future nanoelectronics; however, it is highly unlikely that no patterning will be required at all, since even with molecular-scale technologies there is the need of electrically contacting the single elements and this most often happens through patterned metal contacts, although all-molecular devices were also proposed. Here, we will give some insight into both top-down and bottom-up without the intention to be exhaustive, because of space limitations.

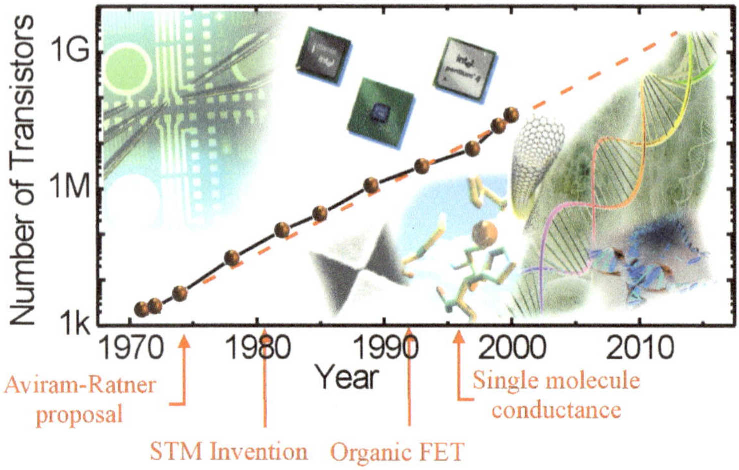


Fig. 1. Moore's law, number of transistors as a function of time (points refer to the various processors introduced by Intel: 4004, 8008, 8080, 8086, 286, 386, 486, Pentium, Pentium II, Pentium III and Pentium 4, respectively). Some related milestones of nano/molecular electronics are also indicated (reprinted with permission from G.Maruccio et al., J. Mat. Chem., 14, 542 (2004). Copyright 2004 Royal Society of Chemistry).

### 2.2 Top-down processes

The top-down approach in building integrated circuits (ICs) usually involves some form of lithography to define the required pattern of features. Many other kinds of processes are required to build an IC, such as ion-implantation, metal evaporation and etching, which are used in several nanoscale devices too; however, here we will concentrate on the problems of lithography, since that is the side of technology directly related to resolution and, consequently, most innovative (and troublesome) in nanodevices. According to a common distinction, we will classify lithographic methods into hard and soft, where the former (often referred to simply as "lithography") are the improved version of those already employed for present ICs, whereas the latter have been purposely developed for nanoscale fabrication (in particular when dealing with fragile molecules).

#### 2.2.1 (Hard) lithography

By 'hard lithography' we mean all those techniques using resist layers and optical masks, derived from standard photolithography. The core-idea is to expose a sensitive material (the 'resist') to an activating agent (light or a particle beam) in order to harden (or soften) it in targeted areas. Subsequently, a developer solution will wash away the soft parts of the resist layer selectively, while the remaining hard parts mask the sample for subsequent processing (see the schematic in Figure 2a). Commonly, either the material is etched where unmasked by the layer (and the remaining resist is taken away after the etching); or a new layer of material is grown (e.g. metal by evaporation) so that it builds up upon the patterned resist, where this is present, or upon the underlying material in the resist's voids; finally, the hardened resist will be removed, carrying away the material grown atop (lift-off) and leaving the material inside the patterned voids untouched.

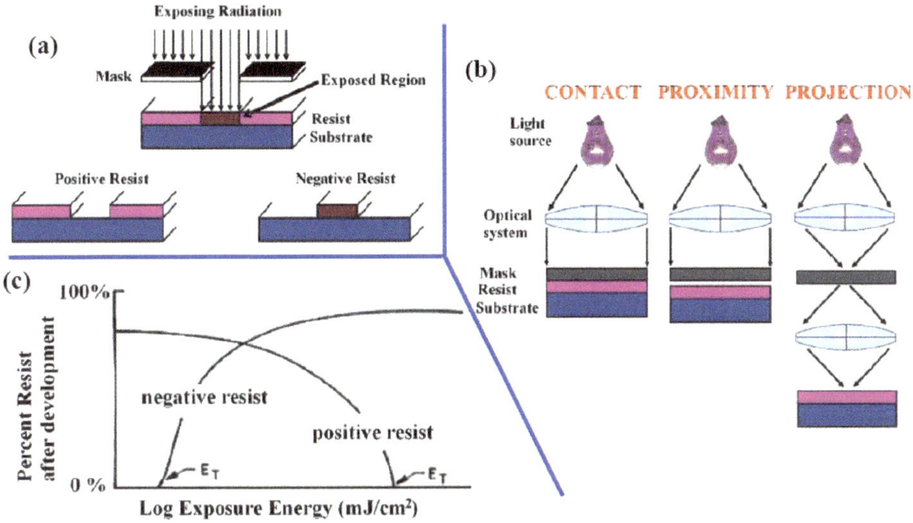


Fig. 2. (a) Positive and negative photoresists; (b) Exposure methods: contact, proximity and projection; (c) Response curves for negative and positive photoresist after exposure and development. A threshold energy $E_T$ characterizes both the positive/negative resist solubilities in their developers.

The key issue is how, and how precisely the pattern can be drawn into the resist. Standard photolithography exposes a suitable type of resist to UV light passing through a mask featuring transparent and opaque zones, but is not suitable to fabricate nanoscale devices because of limited resolution. Actually, the minimum definable feature size, or half-pitch $\mathbf{p}_{1/2}$ is related to the radiation wavelength by the Rayleigh criterion stating that

$$p_{1/2} \approx \kappa\lambda / NA \tag{1}$$

where $\lambda$ is the wavelength itself, ***NA*** is the numerical aperture of the optical system and $\kappa$ summarizes other typical parameters of the system employed (such as the non-point-shaped light source) and typically lies between 0.25 and 1; at the UV wavelengths of photosensitive resists then the half-pitch is typically limited to hundreds of nanometres.

In order to realize sub-100 structures, a gradual shift from the more traditional mercury-based G- or I-Hg ultraviolet lamps as sources working at 436nm or 365nm respectively was deemed necessary. The requirement of the semiconductor industry for denser and faster chipsets in the 1980s drove the search for alternative sources able to meet this demand. In this respect, excimer lasers based on a combination of noble gases and reactive gases (mainly KrF 248 nm and ArF 193 nm) were the system of choice as deep-ultraviolet sources (Figure 3) (Jain et al., 1982).

Present technological trend spearheaded by technology behemoths like Intel, IBM and AMD have then seen a further gradual shift of focus from deep-ultraviolet sources for photolithography to manipulation of the wavevectors of the incident interfering laser light using novel forms of interferometric lithography (Figure 3) (Pfeiffer et al., 1999). Such novel forms take advantage by realizing an optically denser medium such as an index-matching liquid between the excimer source and the chemically amplified photoresist (immersion lithography for 22nm half-pitch). Extreme-ultraviolet lithography (EUVL) technology for now is prohibitively expensive due to cost of processing per wafer. However, extensive research is being carried out on EUVL-based technology as sub-22nm half-pitch structures would require choice of stable sub-193nm source, novel chemically amplified photoresists and understanding of photomask for EUV technology.

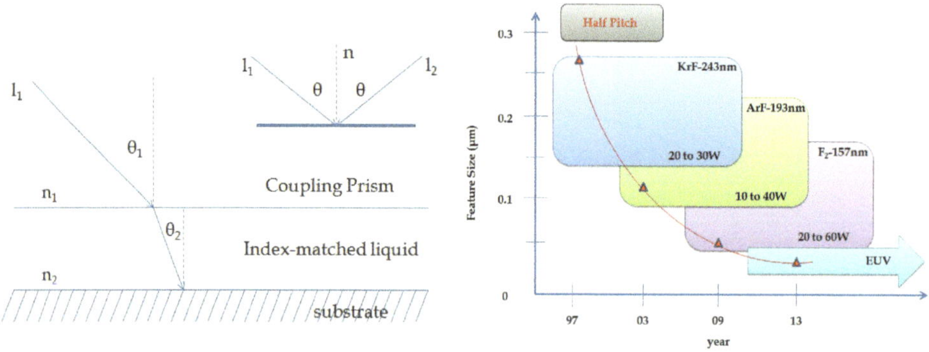


Fig. 3. LEFT Immersion lithography uses beam splitters to split the laser beam in two which are then made incident on a beam coupling prism (inset) which focuses them. Once the beam leaves the coupling prism it enters a denser refractive-index matched liquid at an angle $\theta_2$ providing finer structures (for simplicity one of incident light beam is shown). RIGHT Over the years reduction in feature size has been possible due to the development of excimer laser sources. In order to realize sub-22nm half-pitch structures the use of extreme-ultraviolet sources is predicted to be necessary.

As **Next Generation lithography (NGL)**, IBM is now focussing on using a modified version of immersion lithography technology for fabrication of 22nm nodes called projection reduction exposure with variable axis immersion lenses (Pfeiffer et al., 1999). Whereas, Intel in 2011 has launched its 22nm node transistors called the tri-gate transistor (Chau et al., 2003; Doyle et al., 2003) which uses conventional fabrication tools. Intel believes that it can further reduce the node size to 14nm using existing technology after which they believe that they would reach the Moore's limit for conventional fabrication technologies. Extreme-ultraviolet lithography technology (EUVL) has not found centre stage as of 2011 in next generation lithographic processes due to the extremely high costs. Direct lithography and nanoimprint lithography (NIL) are also being considered as options for NGL techniques.
Alternatively, for resolution enhancement and patterning on the nanometer scale, **electron-beam lithography** (EBL) can be employed due to resolution in the order of tens of nanometers and remarkable versatility especially when combined with other techniques, which make it the most widely employed method for patterning under-100 nm structures. Also, EBL is commonly employed to fabricate masks. At the core of this type of high resolution lithography there is the exposure of a sensitive resist to a thin beam of electrons, which is moved around the layer by deflecting electric fields and turned on and off to design the desired structure. The resist is usually made of polymers in a liquid solvent. High-energy electrons traveling across the resist can either break the main chain bonds (positive resist) thus making the resist removable upon exposition; or, they can promote the

formation of bonds making the resist irremovable (negative resist). A very common positive resist is poly methyl methacrylate (PMMA); an example of negative resist is SU-8.
The usual resolution of EBL (a few tens of nanometers) is larger than the spot of the electron paint-brush, due to the proximity effect, a result of the scattering of electrons passing through the resist up to the substrate. In a classical picture, the electrons experience elastic collisions which make them scatter either preserving the approximate direction of the momentum (forward scattering) or taking up the reverse direction (back-scattering), the latter especially when bouncing on the substrate. Besides, there are inelastic collisions producing secondary, lower energy electrons. The overall exposure results from adding up the three contributions from the forward and backward scattering and from the secondary electrons

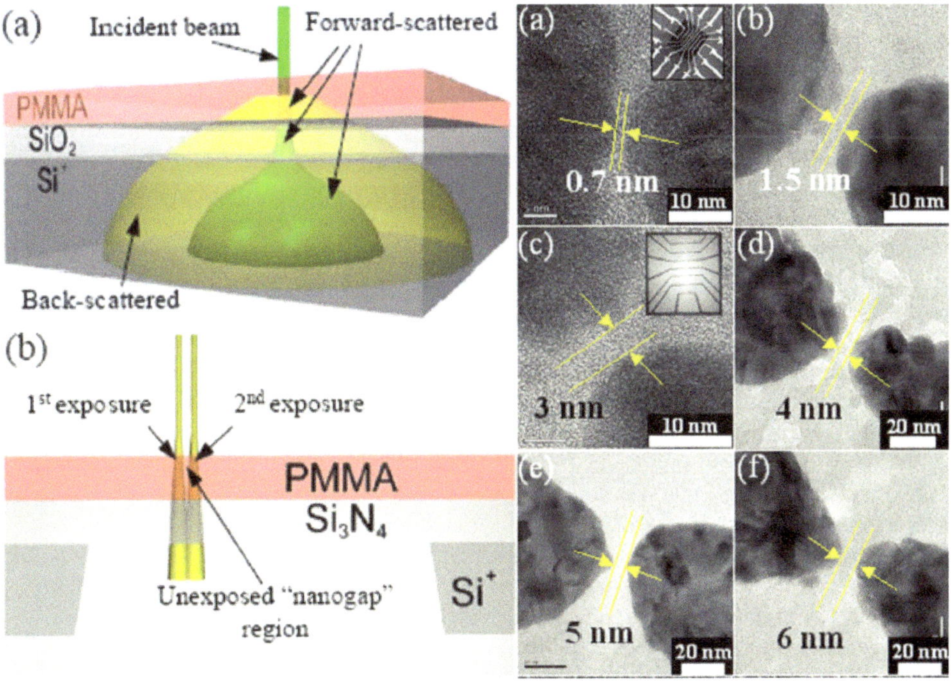


Fig. 4. LEFT (a) In a standard EBL process (e.g. on PMMA-$SiO_2$-$Si^+$ substrates), the electron beam experience both forward scattering (slightly in the PMMA and $SiO_2$ layers) and strong, broadly distributed back scattering (in the $Si^+$) resulting in the exposure of a wider PMMA region and a reduced resolution. On the contrary, using a PMMA-$Si_3N_4$ substrate, the proximity effect can be significantly reduced as illustrated in (b). Right TEM image of nanogaps with sizes 0.7 nm (a), 1.5 nm (b), 3 nm (c), 4 nm (d), 5 nm (e) and 6 nm (f). Insets: SEM and TEM images of electrodes on a suspended $Si_3N_4$ membrane (reprinted with permission from M. D. Fischbein et al., Appl. Phys. Lett. 88 (2006). Copyright 2006 American Institute of Physics).

The second and third terms obviously add supplementary exposure of the resist, causing the images to be larger than desired and very close structures (such as parallel lines) to experience unwanted 'mutual' exposure (the proximity effect). W. Hu et al. (Hu et al., 2005) improved the resolution of EBL by introducing cold development: the chemical developer is kept at a temperature down to 4-8 °C temperature obtaining substantially finer resolution, with definition smaller than 5 nanometres. In the cited work this approach was used to pattern the layout of a QCA adder. On the other hand, Fischbein et al. (Fischbein and Drndic, 2006) demonstrated how resolution close to the spot size can be achieved when EBL is carried out on a membrane due to the limited back scattering (see Figure 4).

If EBL is the most popular advanced lithographic technique, a bunch of other methods should be mentioned. Among them, **ion-beam lithography** (IBL) is the closest to EBL, since it substitutes ions to electrons in the writing beam. The principle and resolution are quite similar to those found in EBL but it presents some advantages for example concerning the backscattering; moreover ions can be used to directly build structures over a substrate.

In **X-ray lithography** (Silverman, 1997) collimated radiation with 0.1-10 nm wavelengths is used to expose a resist in a parallel replication process. The very short wavelength and remarkable penetration capability of X-rays are at the basis of the high resolution achieved. The mask is usually made up of a thin, transparent layer covered with patterned, X-ray-opaque features of metals like gold or tungsten. Several drawbacks have been overcome, notably those concerning the relative weakness of the mask (Peckerar and Maldonado, 1993) which, due to the thinness of the transparent basis, tended to inflect upon stress, e.g. of thermal origin. Another lamented trouble was the lack of demagnification: in optical lithography lenses are commonly used to narrow the light cone by a scaling factor $n$, so that the mask features can be made $n$ times larger than those actually fabricated, easing the mask fabrication process. However, though X-ray lithography makes optical lenses useless, another kind of demagnification can be achieved by means of the so-called sweet-spot method or bias reduction (Vladimirsky et al., 1999). Basically, the metal absorber is narrowed at the sides by a controlled amount to compensate for the effects of diffraction, by which light propagates to some extent under the absorber in the nominally dark region. The entity of the required correction is dependent on the mask gap and exposure dose; in the cited work, demagnifications in the order of 3.5 were shown, comparable to those obtained in optical lithography, thanks to a combined optimization of exposure and diffraction. As of today, however, the most serious drawback to X-ray lithography is the need of a synchrotron as the source of radiation. Research is being carried out on small but reasonably powerful X-ray sources, which might make the difference in the success of this performing processes.

Other available fabrication techniques include:

- **Interference lithography**, exploiting the fringes naturally arising when two or more coherent (laser) light sources interfere (see for example (Solak et al., 2003)). This method is maskless, since the geometry is determined by the interference pattern and a comparatively large area can be exposed simultaneously, unlike in E- or IBL; the obvious drawback is that the patterns - and, so, the defined features - are severely bound as to geometry (typically, they must be symmetric).
- **Shadow-mask** techniques where the deposition of a material on a substrate occurs through a holed mask, kept at little distance from the substrate itself, allowing features in the range of some hundreds of nanometres (see for examples the microtips shown in (Luthi et al., 1999)).

To obtain better resolutions or smaller inter-electrode gaps, it is possible to use **post-processing techniques**, by which large EBL-defined gaps are narrowed, or EBL-defined gapless junction are split. An example of the former technique is shown by Y.V. Kervennic and coworkers (Kervennic et al., 2002). The idea was to EBL-fabricate couples of free-standing contacts and subsequently to narrow the inter-electrode gap by electrodeposition, in a controllable way. In the cited work, platinum was electrodeposited from an aqueous solution of 0.1 mol of $K_2PtCl_4$ and 0.5 mol of $H_2SO_4$. During electrodeposition, the tunneling conductance between the two electrodes was constantly monitored, being obviously dependent on the gap size. The process was interrupted at predefined values, reproducibly obtaining gaps ranging from 20 to 3.5 nm (Figure 5 left).

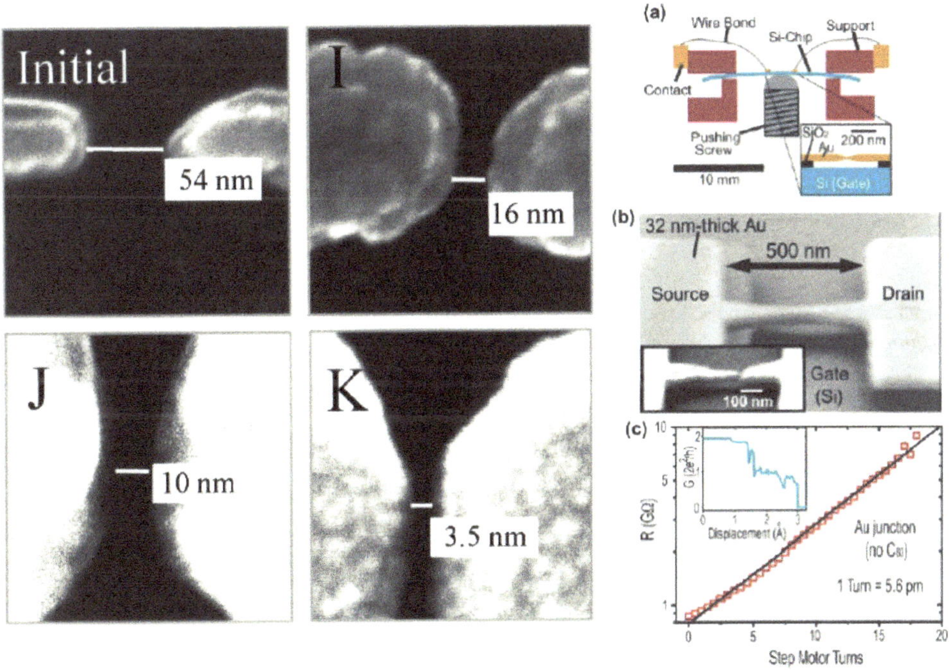


Fig. 5. Left: initial separation of the EBL-defined electrodes; (I-K) final electrode separations (16, 10, and 3.5 nm, respectively) obtained by stopping the electrodeposition process when the monitor current is 30, 90, and 140 nA. (reprinted with permission from Y.V.Kervennic et al., *Appl. Phys. Lett.* 80, 321 (2002). Copyright 2002 American Institute of Physics). Right: (a) Schematic of the mechanically adjustable and electrically gated nanojunction by A.R. Champagne et al. and (b) corresponding SEM-image at a of 78° tilt angle; in the inset of (b), an electromigration-broken nanojunction is shown. (c) Increase in resistance (and drop in the current, inset) as the shape is more and more deformed (reprinted with permission from A. R. Champagne et al., Nano Lett. 5, 305 (2005). Copyright 2005 American Chemical Society).

The inverse pathway is breaking a narrow metal junction defined by EBL to create the gap *a posteriori*. Among the earliest experimented techniques there are electromigration and mechanically controllable break. In the former case (see for example (Park et al., 2002; Heersche et al., 2006; Jo et al., 2006)), current is passed through the electrodes to reach electromigration densities (~$10^6$-$10^7$ J/cm$^2$) in the narrowest spot (Ho and Kwok, 1989). This gets the contact to break, an event easily revealed by the drop in conductivity within the contacts. Such junctions were employed for example for single molecule transistor measurements on magnetic molecules (Heersche et al., 2006; Jo et al., 2006). In mechanically controlled break junctions (Reed et al., 1997; Champagne et al., 2005), the substrate on which the thin junction is fabricated is flexed by the gentle push of a screw underneath, while firmly held at the ends, until the junction cracks (Figure 5 right). Again, the occurrence of this event is checked through constant measurement of the junction's conductance. Obviously, all methods rely on not quite easily controlled nanoscale events - the growing of metal structures from electrodeposition, the dismantling of metal hillocks by electromigration, or the mechanical induction of nanoscale fissures - which can guarantee neither high reproducibility of the results nor their controllability *a priori*. In fact, the typical size of gaps fabricated in these ways has to be checked after the fabrication by scanning electron microscope (SEM) measurements.

Recently, Zandbergen et al. (Zandbergen et al., 2005) introduced a technique for sculpting couples of nanoelectrodes of given gap-size and shape by use of a Transmission Electron Microscope (TEM). A 300kV electron beam with a 2 to 10 nm spot size, carrying a current of 5 nA - corresponding to a flux of ~$10^9$ electrons/s - was scanned on a sample in which a metal bridge had been defined. By modulating the intensity of the beam, the shape and size of this bridge was finely-tuned, compatibly with the characteristics of the metal used for the contacts; to this respect more than one type of Au were tested, obtaining 1 to 5 nm separations. Pt lines were tested too, which, due to the comparatively low mobility of its atoms, required a much longer exposure time than Au (up to one hour vs. a few minutes) but also permitted the smaller gap obtained in this series of experiments (only 0.6 nm).

### 2.2.2 Soft lithography

Instead of creating a pattern by masking and exposure, soft lithography exploits an elastomeric stamp (Xia and Whitesides, 1997) (Figure 6). This approach includes different techniques, sharing this basic concept, which have undergone remarkable development in recent years. Here, we will briefly recall the main two techniques, listing only their main variants (Geissler and Xia, 2004).

In **moulding** techniques, structures are formed inside the voids of a master which is the 3D negative of the desired pattern. With replica moulding (REM), replicas of organic polymers are made against a polydimethylsiloxane (PDMS) mould, reaching feature sizes below 2 nm with some materials. In another variant known as microtransfer moulding (μTM, Figures 6a and 6b) the mould's voids are filled with a liquid prepolymer and the mould is transferred onto the desired substrate. The prepolymer is subsequently cured to a solid and peeled off, exploiting the elasticity of the mould.

The **printing** techniques use the stamp to transfer an ink to the substrate. To this category belongs, among others, the microcontact printing technique (μCP, Figures 6c and 6d), also classified as a relief printing method, by which the stamp is first inked with a solution containing the molecules to be deposited (in the first and most classical example, alkanethiols), then dried and subsequently brought into contact with the surface of the substrate (in the cited case, gold).

Research is now ongoing on novel variants and materials (see for example (Choi and Rogers, 2003)). The main field of application of soft lithography remains microfluidics, particularly in the field of lab on a chip (Becker and Locascio, 2002; Pollack et al., 2002; Erickson and Li, 2004; Stone et al., 2004; Dittrich and Manz, 2006; Primiceri et al., 2010; Chiriacò et al., 2011). However, in a recent report (Briseno et al., 2006), soft lithography was, for example, employed also for the collective fabrication of organic single-crystal transistors, outperforming organic thin film transistors but requiring the growth and placing of one crystal per transistor separately, which is the bottle-neck of this technology. In the cited work, octadecyltriethoxysilane (OTS) films were printed by μCP in patterns on which the crystals were subsequently grown in a way pre-conditioned by the geometry of the soft-lithography-defined sites.

The range of applications of soft-lithography is widening, but it is of particular relevance for biomolecules. J. Damon Hoff et al. (Hoff et al., 2004) demonstrated protein patterning by imprint lithography, exploiting the binding specificity between biotin and streptadivin. J.P. Hulme et al. (Hulme et al., 2006) reported on a replica-moulding derived technique to pattern biological molecules, called biomolecular embossing, by which they were able to replica-mould DNA and then emboss it into poly(ethylene terephtalate). Similar approaches could prove very useful in tomorrow's devices, because of the importance of biomolecules both in fabrication and in functional issues of nanoelectronic devices (hard lithography techniques typically can not be employed due to the fragility of these molecules).

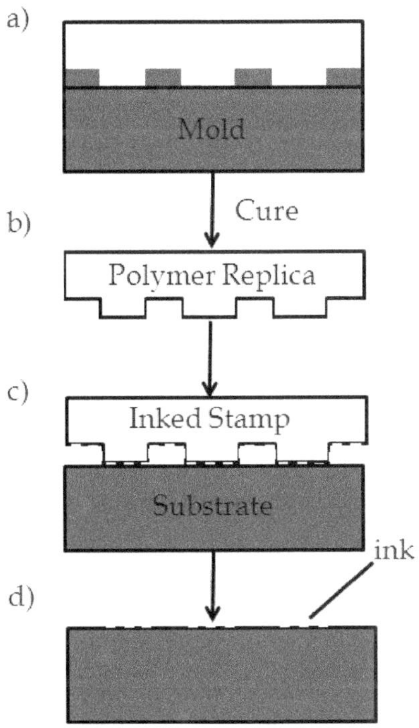


Fig. 6. Schematics of the microtransfer moulding (a and b) and microcontact printing; (c and d).

Finally, we mention an example of a cross-inspired technology, joining ideas from different approaches to overcome their inherent limits. S.Y. Chou et al. (Chou et al., 2002) used a stamp like in soft lithographies, but made of quartz instead of an elastomer, to imprint features on a crystal of silicon – not used with classical soft lithography - after melting its superficial layer by a single excimer laser pulse. The technique, called 'laser-assisted direct imprint' (LADI) was used to impress a 250 nm-periodic grating of parallel silicon wires, which could be used as a template for a crossbar structure. However, there are also 10 nm-wide, 15-nm tall silicon lines on the edges of the bigger wires which are just the negative of the notches created on the mould by the trenching effect during the reactive ion etching, meaning that the resolution of this technique is potentially much better than that of the first stamp.

### 2.2.3 Scanning probe nanolithography

Presently the highest fabrication precision is obtained by writing methods (most of them SPM-based), employing a stylus or pen with variable sharpness. Depending on the resolution of the writing instrument, macroscopic down to atomic-scale features can be defined according to virtually any topology, at the price of a low speed. Beyond the equivalent of a paper and a pen, ink may be required or not, depending whether the writing is additive (deposition of material on the substrate) or subtractive (removal of material from the substrate). Moreover, a local modification of the surface/layer can be achieved by means of an energy beam such as a laser or an electric/magnetic field.

The simplest subtractive solution employs a rigid stylus pressed against the substrate to dig channels that, at their best resolution, can be as slim as 0.1 nm. The choice of the stylus depends of course on the material: good results have been obtained with metals, oxides, and polymers.

Positive writing processes are also called add-on processes. By them, among others, resist-, etching- and development-incompatible materials can be patterned. A popular add-on method is dip-pen nanolithography (DPN) (Piner et al., 1999), developed by the Mirkin group, where an atomic force microscope (AFM) tip is coated with molecules and then drawn onto a surface. The water meniscus formed between the tip and the substrate by the condensation of humidity provides a transfer pathway by which the molecules are transferred from the tip to the substrate (Figure 7a). In (Hyun et al., 2002), the versatility of DPN was exploited in conjunction with other self-assembly methods. Specifically, 16-mercaptohexadecanoic acid was deposited onto gold by DPN and subsequently used as the grafting pattern for an amine-terminated biotin derivative. Subsequently, streptadivin structures were formed upon incubation and biotinylated proteins were deposited, always following the lithographed pattern. Biotin-tagged molecules being very common, this is a powerful molecular patterning method and shows that molecular mediation can be used to make the patterning method less dependent on the chemistry of the substance being patterned. The Mirkin group also used dip-pen nanolithography to construct proteins patterns with features in the range of 100 nm (Lee et al., 2002). Moreover, they also demonstrated a multiple-pen nano-plotter for multiple ink nanolithography in order to pattern different organic molecules down to a 5-nanometer separation (Hong et al., 1999; Hong and Mirkin, 2000).

Once the pattern is fabricated, replica processes can be also exploited to create duplicates as demonstrated by the Stellacci group at MIT (Yu et al., 2005).

The field is of course still in progress: among recent developments, a group at the ETH Zurich fabricated a four-terminal quantum dot and a double quantum dot system with integrated charge readout using a scanning force microscope to achieve a local, direct oxidation in a AlGaAs/GaAs heterostructure containing a two-dimensional electron gas. As a result of the local oxidation, the electron gas is locally depleted in the GaAs surface and the observed Coulomb blockade diamonds demonstrated the high quality of this fabrication process (Dorn et al., 2002). Let us also remember the variant of DPN called electro-pen lithography (Cai and Ocko, 2005) by which a conductive AFM probe coated with ink molecules and biased with a voltage oxidizes the substrate underneath and deposits the ink, all in a sweep. The lines drawn are as large as 50 nm with a writing speed larger than 10 μm/s. For a review on DPN see K. Salaita et al. (Salaita et al., 2007).
Ultradense atomic patterns can be also realized by atomic manipulation using a STM (Manoharan et al., 2000; Agam and Schiller, 2001; Fiete and Heller, 2003; Morr and Stavropoulos, 2004). Despite the advantage given by their ultimate resolutions, however, scanning probe nanolithography techniques are slow and appear too expensive for mass production, though a major performance leap cannot be excluded for these techniques in the future. Even if such dramatic improvement did not happen, however, tomorrow's nanoelectronic circuits may turn to have critical parts requiring precise, very small-scale patterning without the intrinsic faultiness of other techniques (e.g. of self-assembly), for which the writing methods can turn out essential.

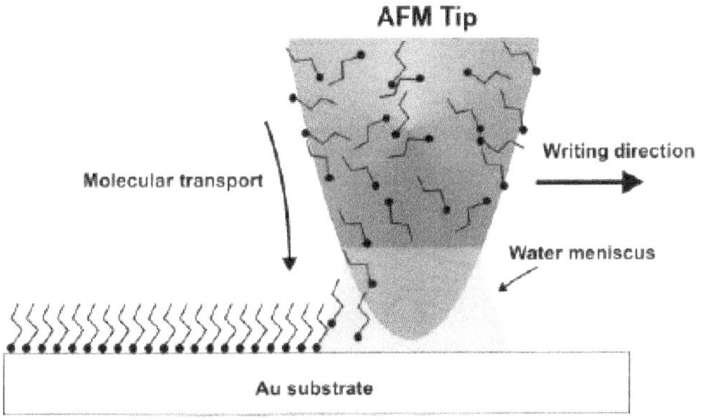


Fig. 7. (a) Schematic representation of dip-pen nanolithography reprinted (with permission from R. D. Piner et al, Science 283, 661 (1999). Copyright 1999 American Association for the Advancement of Science) and (b) stepwise fabrication process to create molecular recognition-mediated protein nanostructures (reprinted with permission from J. Hyun et al, Nano Lett. 2, 1203 (2002). Copyright 2002 American Chemical Society).

### 2.2.4 Other top-down methods

From the previous discussion, it is clear how EBL and other similar methods are neither suitable for mass production of contacts (due to their slowness and related high cost), nor for a reproducible interconnection of nanoscale objects (due to the inevitable changes in the contacts). The task of fabricating reproducible contacts is, in fact, not trivial, because the electrical behavior of nanocontacts is sensitively dependent on their atomic-scale characteristics. Yet, it is a crucial requirement at least for molecular systems, where the performance of the electric contacts can determine the behaviour of the device. Here, we present some non-conventional solutions for creating nanoscopic electrodes, addressing the problems of mass production and/or reproducibility.

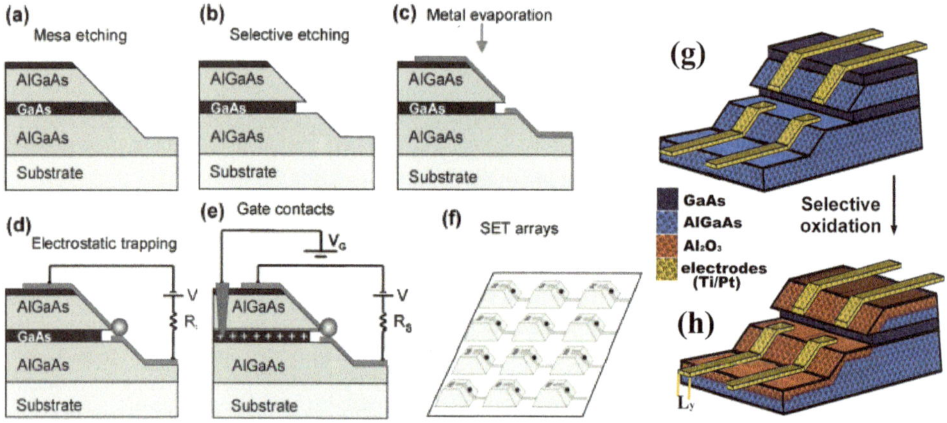


Fig. 8. In the method for fabricating nanodevices proposed by R. Krahne et al. (Krahne et al., 2002) the MESA structure in (a) undergoes a GaAs-selective etching (b), so that the subsequent metal evaporation (c) leaves a gap in which a nanoparticle can be trapped electrostatically (d) and then measured in a three-terminal configuration (e). The method lends itself to one-step mass production of such SET arrays (f). Reprinted with permission from R. Krahne et al., Appl. Phys. Lett. 81, 730 (2002). Copyright 2002 American Institute of Physics. Schematics of the g) non-oxidized and h) oxidized mesa nanojunctions (reprinted with permission from G. Maruccio et al., Small 3, 1184 (2007). Copyright 2007 Wiley-VCH).

In this respect an innovative approach is that proposed by R. Krahne et al. (Krahne et al., 2002) and then improved by G. Maruccio et al. (Maruccio et al., 2007) An AlGaAs/GaAs/AlGaAs quantum well structure is grown by molecular beam epitaxy (MBE) and subsequently carved to a slanted shape by wet-etching (Figure 8a), until reaching the bottom AlGaAs layer. Then the GaAs layer is removed to a few tens of nanometers by selective wet-etching with citric acid and $H_2O_2$ (Figure 8b) and a metal layer is evaporated perpendicular to the substrate to form the contacts (Figure 8c). In such a way, a contacted gap as large as a few nanometers is created where, for example, a gold particle can be immobilized by electrostatic trapping (Figure 8d) and provided with a gate electrode (Figure 8e) to form a SET. In order to reduce the leakage current at room temperature through the underneath semiconductor layer, a selective oxidation was used (Maruccio et al., 2007) to convert the AlGaAs barriers in insulating oxide layers. This approach allowed G.Maruccio et al to investigate electron transfer in individual proteins (Maruccio et al., 2007) and pave the way to applications in molecular electronics. Since only photolithography is used to define the electrode pattern and all processes are carried out at wafer scale, this improved mesa-gap technique enables the simultaneous, economic fabrication of large arrays of nanodevices working at ambient conditions, a crucial advance for the implementation of low-cost mass-production of nanoscale devices and the fabrication of complex circuits consisting of different nanodevices and arrays of sensors at reasonable cost

(Maruccio et al., 2007; Maruccio et al., 2009) (Figure 8f). Similar approaches for parallel fabrication of nanodevices were also developed by other groups. For example Ray et al. reported a CMOS-compatible fabrication of room temperature single-electron devices having source and drain electrodes vertically separated by a thin dielectric film (Ray et al., 2008). Concerning the problem of good and reproducible contacts, T. Dadosh et al. (Dadosh et al., 2005) proposed the use of two gold nanoparticles (NPs) to contact a conductive organic molecule in a controlled way (Figure 9). A NP dimer was assembled using the molecule as a bridge and a device was fabricated, in which the two nanoparticles were also exploited to electrostatically trap the dimers (thanks to their polarizability) within EBL-defined nanojunctions. Their results demonstrated that the transport through a conjugated molecule is quenched by localizing groups placed at its center or near the contacts (Figure 9). Metal nanoparticles were employed as means to probe molecular conduction also by other groups in order to create a bridge between organic monolayers formed on metallic electrodes. Most of these studies focused on conductance through conjugated molecules with thiol terminal groups linked to gold. For example, Amlani et al. (Amlani et al., 2002) demonstrated the particle bridge concept by measuring conductance through a monolayer of (1-nitro- 2,5-diphenylethynyl-4¢-thioacetyl)benzene. Long et al. (Chu et al., 2007) demonstrated magnetic nanoparticle assembly by comparing conductance through undecanethiol, oligo(phenylene ethynylene)- dithiol, and oligo(phenylene vinylene)dithiol.

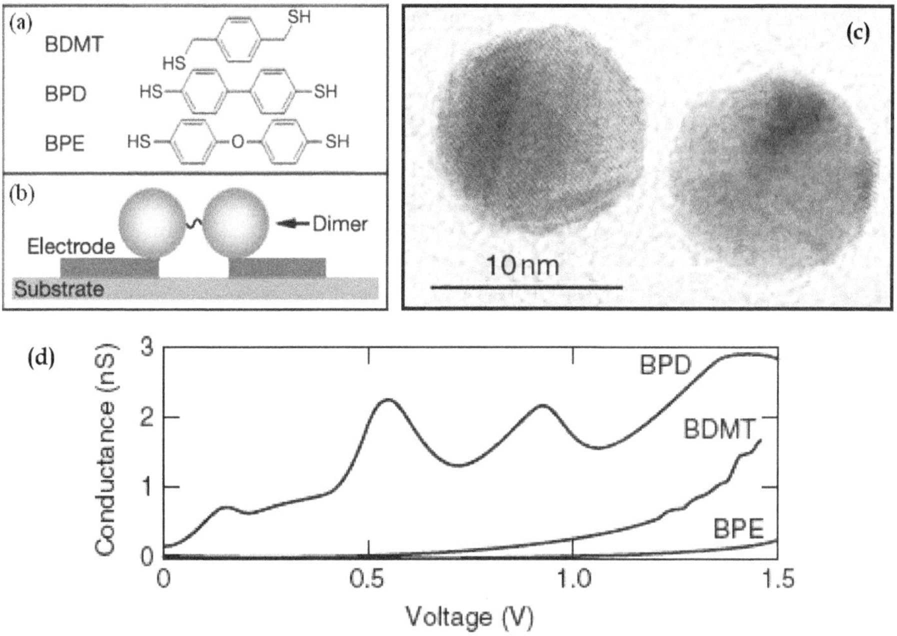


Fig. 9. (a-b) Three organic molecules attached to two gold nanoparticles. (c) high-resolution SEM image of a dimer (d) I-V characteristics of the different molecules. Transport through a conjugated molecule (BPD) is quenched by localizing groups placed at its center (BPE) or near the contacts (BDMT). (reprinted with permission from T. Dadosh et al., Nature 436, 677 (2005). Copyright 2005 Macmillan Publishers Ltd.)

## 2.3 Bottom-up processes

In bottom-up processes, the capability of atomic or molecular building blocks of assembling into ordered, complex arrangements is exploited. Very common in living beings, especially with regeneration purposes, bottom-up processes are made up of the composition of many atomistic processes, on which no punctual control is required, in opposition to top-down processes where the degree of refinement of a defined structure is directly related to the precision employed. What is required in bottom-up processes is instead that the molecular constituents are apt to assemble and that the environment favors the process.

Following (Barth et al., 2005), we classify bottom-up phenomena as 'self-assembly' or 'self-organized growth' processes, where the first term refers to the spontaneous association of molecular constituents into supramolecular structures, while the second term is used to designate autonomous order phenomena guided by mesoscale force fields or kinetics limitations, such as those leading to the deposition of a monolayer of molecules on a substrate. The use of bottom-up processes in nanoelectronics is crucial to the construction of molecular devices, in which the 'active' part is a bunch of molecules (typically a monolayer) or consists of individual conductive molecules.

Bottom-up processes usually involve two main factors, namely a driving force and a recognition mechanism. The driving force is usually exerted by a gradient of concentration, or an electric field, and tends to bring some molecules, usually in solution, in contact with other molecules, often fixed to a substrate or another molecular or supramolecular structure. The recognition mechanism is the necessary chemical affinity between the two molecules, usually provided by two chemical groups playing the role of molecular-scale compatible 'hooks'. Once the suitable conditions and environment have been created, the assembling goes on by itself without need of direct control on each elementary event (which would be impossible, on the other hand).

As in all physical processes and chemical reactions, bottom-up processes are driven by the goal of reaching a minimum of energy, but conditioned to the overcoming of activation energies, i.e. barriers which could get the process trapped into local minima. The accuracy of the final result and the speed of the process depend on the equilibrium between kinetics and thermodynamics.

Let us consider a crystalline substrate on whose surface an organized structure must be grown; for example an ordered monolayer of molecules. The involved atomistic phenomena are the deposition of the molecules from solution onto the surface and their surface diffusion kinetics. The former factor mainly depends on the solution concentration and the applied driving force (possibly just a concentration gradient and the Brownian motion) and can be summarized by the deposition rate **F**. The latter factor is dependent on the molecule-substrate interaction as well as on the reciprocal interaction between the deposited molecules. We then assume the diffusivity **D** of the molecules on the surface as the representative parameter of thermodynamics. If an Arrhenius law is assumed, then

$$\mathbf{D} = \exp(-\mathbf{E_a}/\mathbf{RT}) \tag{2}$$

where $\mathbf{E_a}$ is the activation energy, **T** is the absolute temperature and **R** is the gas constant; now, the thermodynamics-to-kinetics **D/F** ratio summarizes the overall character of the process. After molecules have 'landed' on the substrate, they generally undergo random hopping processes across the lattice, during which they can meet other similar species to enucleate a new one or join an already formed aggregate. The slowest is the deposition rate

**F**, the most likely is the settling of the newly deposited adsorbates onto an equilibrium condition, for they simply have time to move around the lattice until a minimum energy configuration is reached; hence, the largest is **D/F**, the closest is the process to the equilibrium condition. In such condition the individual processes are of little or virtually no importance, since the system evolves towards a minimum quite independently of the random walk followed by each constituent. Conversely, a large deposition rate with respect to diffusivity (small **D/F**) means that kinetics is prevailing on thermodynamics, i.e. the quick deposition of adsorbates on the substrate prevents the movement of the single adsorbate towards the overall minimum energy; in this case, rather, local energy minima are favored corresponding to metastable structures. The effect of the **D/F** ratio is shown pictorially in Figure 10 from (Barth et al., 2005).

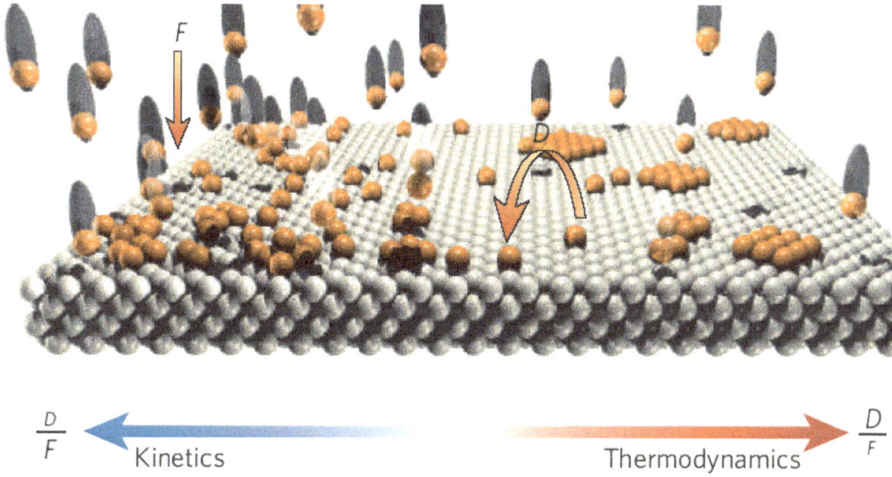


Fig. 10. Atomic-scale view of growth processes at surfaces. The kind of assembled structures is chiefly determined by the D/F diffusivity-to-flux ratio: a small D compared with F favours the aggregation in structures such as metal nanoislands, the reverse situation drives the reaction towards the global energy minimum and is the most suitable for supramolecular assembly (reprinted with permission from J. V. Barth et al., Nature 437, 671 (2005). Copyright 2005 Macmillan Publishers Ltd.)

Tuning the **D/F** through control on the process parameters reflects, in turn, on the average geometric characteristics of the structures to build. For example, metal islands are thermodynamically sub-optimal with respect to a more uniform distribution of metal atoms; consequently, their construction is kinetics-driven, imposing small **D/F**. Intermediate **D/F** values are used for building semiconductors, while supramolecular structures require a high degree of ordering of the constituents, or a close proximity to the equilibrium condition (large **D/F**) because the final, ordered structure just coincides with the global energy minimum.
Basing on the exposed principles one can build also complex multilayered structures (see the review (Hammond, 2004) and references therein) provided that the chemistry of the layers is chosen carefully according to the deposition sequence; a common way is depositing

layers provided with charges of alternated signs (layer-by-layer electrostatic assembly). Varying the constituents during deposition one can tune the variation of the layers' cross profile in a very fine way.
Other techniques can be employed to design the in-plane structures. One widely used is the employment of block copolymers, i.e. at least two variable polymer fragments, immiscible and joined by a covalent bond (Geissler and Xia, 2004). Once deposited, ensembles of block copolymer can form a variety of regular 2D patterns, in turn usable to template other molecules (binding with them) or to mask the deposition of other molecules (by letting them through nanoscale pores).
Before describing some significant examples of bottom-up patterning it is opportune to remind that these processes are intrinsically faulty due to their statistical nature. Their envisaged positive importance in nanoscale fabrication is also the main reason why fault-tolerance is predicted to become a ubiquitous problem in nanotechnology. In what ways and to what extent this will affect future systems depends on both the details of the process and the spatial resolution of the employed devices. In larger-than-molecular-scale devices, where charge transport can be modeled to happen inside a bulk of molecules, a not too high faultiness can be accounted for in statistical terms as a correction in the charge transport performance. Using single-molecule-scaled devices, even a single flawed molecular site - one where the molecule is wrongly oriented, structurally altered or lacking at all - will generally imply alteration of the local and possibly overall processing capabilities, a much more severe problem requiring specific countermeasures like redundancy and error correction mechanisms.

### 2.3.1 Biomolecules for patterning

The world of biological molecules offers a plethora of opportunities from nanoelectronics to biosensing, because of the complexity and richness of structures and functional properties. For space constraints, they can not be discussed here in detail. However, we want to introduce the biomolecules as instruments for patterning and device fabrication. To this aim biological molecules are ranked basing mainly on their mechanical characteristics and manipulability; this is where and why DNA plays a major role.
As is well known, DeoxyriboNucleic Acid has a double-helix structure supported by a sugar backbone and surrounded by counterions yielding global neutrality. The thermodynamical stability of the structure is based on the matching between compatible couples of bases, among four types (adenine, A, thymine T, guanine, G, and cytosine C). Stable links are A-T and C-G; each mismatch is a weak ring in the chain increasing the flexibility of the overall structure (Schallhorn et al., 2005) and decreasing its stability; couples of too mismatched (non-complementary) helices are expected not to bind at all.
Since its discovery in the 1950s by Watson and Crick, the progress in DNA manipulation proceeded at astonishing speed: nowadays it is possible to design and synthesize single or double helices with relative simplicity to create even very complicated structures. The main idea in using DNA as a template is to synthesize single helices and let them self-assemble. Intertwined assembling is possible, since a single helix (say 1) can contain a sequence complementary to that of single helix 2 adjacent to another complementary to 3, and 2 and 3 in turn can be bound to other complementary helices along other spots, and so on. Hence, careful design of the base sequences permits to synthesize true DNA

crystals, with complex patterns and structures (Winfree et al., 1998; Reif, 2002; Chelyapov et al., 2004; He et al., 2005; Liu et al., 2005; Mathieu et al., 2005; Reishus et al., 2005; Paukstelis, 2006; Rothemund, 2006).

Research is therefore ongoing on improving DNA modelling, to tune the base sequence according to the target characteristics, as well as on DNA synthesis. Regarding the latter, for instance, it was shown that the rolling circle amplification technique can be used to synthesize long periodic DNA sequences (Beyer et al., 2005) quickly and reliably, so representing a helpful method to produce relatively large amounts of DNA with repetitive structure to self-assemble into large patterns. J. Zhang et al. (Zhang et al., 2005) assembled straight filaments on DNA on a Si surface by combining the advantages of molecular combing - a technique for stretching and binding molecules on a surface by the force of a receding meniscus of liquid - and microcontact printing. D.C. Chow et al. (Chow et al., 2005) demonstrated a way to grow vertical filaments of DNA beginning from an oligonucleotide initiator nanopatterned on a surface. As a result, an ordinate layer of vertical DNA strands can be grown with controlled and variable lateral structures, to be used as scaffoldings for various types of molecules, of biological origin or not. A.P.R. Johnston et al. (Johnston et al., 2005) demonstrated the assembly of a multilayer structure obtained alternating two-block homopolymeric nucleotides (poly$A_{20}G_{20}/C_{20}T_{20}$); A. Granéli et al. (Graneli et al., 2006) tethered DNA molecules on a lipid bilayers. Since the filamentary shape of DNA suggests that it be used as a template for nanowires, Park et al. (Park et al., 2005a) have proposed that DNA be synthesized in tiles, forming 1D and 2D-lattice structures and subsequently coated with silver to obtain a 20-nm diameter and ohmic current-voltage characteristics. H. Kudo et al. (Kudo and Fujihira, 2006) performed electroless metallization of DNA with copper, upon previous activation with palladium; heights from the substrate as large as some tens of nanometers were shown, depending on the metallization time. Other examples of DNA-based nanowire fabrication are found in G. Braun et al. (Braun et al., 2005) who 'decorated' DNA strands with gold nanoparticles obtaining grainy nanowires of different densities; and in L. Berti et al. (Berti et al., 2005) who photoinduced the reduction of charged silver nanoparticles on DNA and the subsequent formation of chains, possibly usable as seeds to further metallization. Extensive work on the characterization of variously metallised DNA nanowires is found in the literature (see for example (Hosogi et al., 2005)).

In 2002, K. Keren et al. (Keren et al., 2002) demonstrated DNA-based molecular lithography (Figure 11a-e) by which a substrate of possibly patterned single strands of DNA can be selectively metallized or, alternatively, tagged with molecules grafted in specific sites. The two main ideas underlying molecular lithography just point out the analogy to traditional lithographic methods. First, spots of immobilized DNA single strands can be selectively masked by attaching complementary sequences marked with an enzyme, the RecA protein. Upon DNA hybridization (homologous recombination) the proteins polymerize on the substrate making DNA inert, for instance, to metallization (see the sequences in the images of Figure 11b-e and the schematic of Figure 11a). Secondarily, molecules can be tagged with short strands bearing selected base sequences, complementary to deposited spots in specific positions. Unspecific binding is negligible, provided that the reaction is assisted by RecA. In principle, molecular lithography permits remarkable topology flexibility, limited only by the complexity of the patterned structure of DNA and the selectivity of the homologous recombination processes. Selective metallization and coating with selected molecules can prove the basis to build a topologically complex network of 'processing' molecules, interconnected or insulated depending on the metallization (Keren et al., 2003).

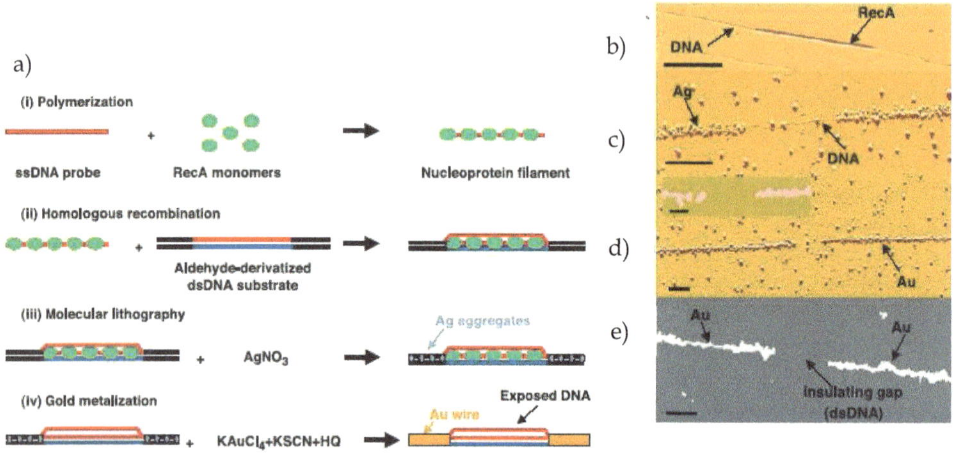


Fig. 11. (a) Molecular lithography permits selective metallisation of DNA with gold after masking some parts of the strand with the RecA protein. The deposition of gold is enhanced by Ag as a catalyst. AFM images of the process are shown, in particular: (b) RecA bound to a sequence of DNA; (c) the sample after exposing to Ag; (d) the metallisation with Au; in the inset, the gap is zoomed in. (e) is a SEM image of the sample in (d). The height of the metallisation is around 50 nm; the scale bars in (b) through (e) are 0.5 µm, except in the inset of (c) where it is 0.25 µm. (Reprinted with permission from K. Keren et al., Science 297, 72 (2002). Copyright 2005 American Association for the Advancement of Science).

Other simpler though less general processes have been devised, such as that by Park et al. (Park et al., 2005b) who built a scaffold assembling two types of DNA tiles (A and B) in a chessboard manner (Figure 12a-f) and got proteins bind at the crossings. Four types of scaffolds (or nanotracks) are shown in the cartoon schematic: the first two form are 2D, i.e. thanks to mutually sticky ends (marked with n and n') they can assemble into planes of indefinite extension. They differ in that in one case only A tiles are modified in the center to carry biotin, by which streptadivin is subsequently bound, while in the other case both A and B are biotinylated. In the other two cases, intentional non-stickiness allows assembling of a 1D nanotrack. The AFM images in Figure 12g-j are significant as to the effectiveness of the method in building large, regular and functionalized structures. To assess the potential importance of such methods, suffice it to think of the possibility of building large, regular component matrixes for memories. It should be pointed out that the DNA-protein interaction may be of general importance in fabricating molecular electronic devices (not just applications like biosensors as might be expected) because of the patterning function briefly illustrated here. Binding and interaction between DNA and proteins are being more and more elucidated by ongoing research (Dixit et al., 2005; Yang and Schepartz, 2005; Hu and Shklovskii, 2006; Sun et al., 2006).

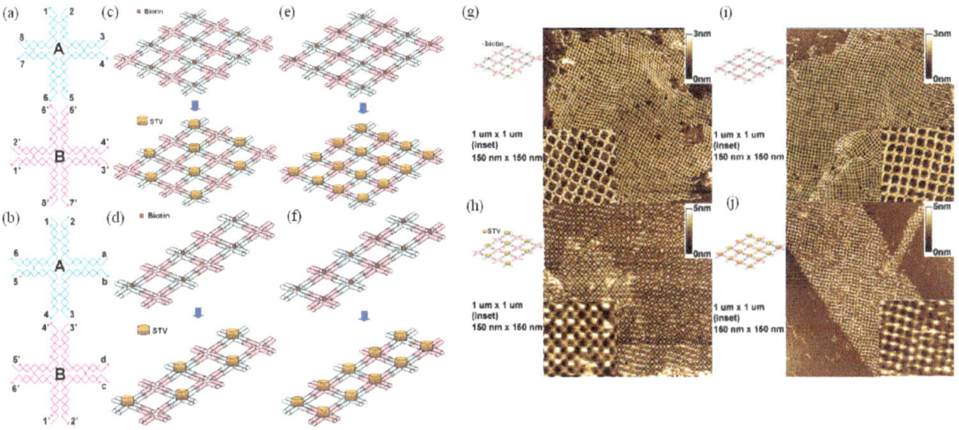


Fig. 12. (a) Geometry of two cross-shaped DNA tiles capable of assembling into lattices; coupled of sticky ends are labelled with *n* and *n'*. (b) These version of the tiles are sticky only along one axis. In (c) and (d) there are the cartoon images of the lattices formed by (a) and (b) respectively, where only tile A is tagged with biotin; due to the differences in stickiness, lattice (c) is 2D while lattice (d) is 1D. In the lattices (e) and (f) both tiles are biotinylated. (g) and (i) are AFM images of lattices (c) and (e) before the attachment of streptadivin; (h) and (i) are the images of the same, respectively, after the streptadivin has been grafted (reprinted with permission from S. H. Park et al., Nano Lett. 5, 729 (2005). Copyright 2005 American Chemical Society).

Application of similar techniques to pattern nanoparticles was also reported, for example by G.H. Woehrle et al. (Woehrle et al., 2004), who attached gold nanoparticles on DNA with fine-controlled separation thanks to molecular lithography, and L. Dillenback et al. (Dillenback et al., 2006) who employed temperature control to direct nanoparticles assembly by tuning the thermal stability of the DNA sequences used as hooks. A somewhat close use of DNA as linker between floating objects - i.e. without needing a substrate - is found in (Goux-Capes et al., 2006) by L. Goux-Capes et al., who linked different single-walled carbon nanotubes coated with streptadivin through biotin- or bis-biotin-terminated DNA links. Future nanoelectronic components might be assembled in similar fashions and then patterned in larger grids or templates.

As another example, we recall the work by Y.Y. Pinto et al. (Pinto et al., 2005) who assembled gold nanoparticles of two different sizes (5 and 10 nm in diameter) on a same self-assembled DNA scaffolding (Figure 13). The DNA tiles were designed in order to let two types of sticky ends prong out of the scaffold in alternate rows. Two groups of nanoparticles where functionalized with thiol-modified DNA, each bearing strands complementary to one type of sticky end; the final result is shown in the AFM image of Figure 13c. Assemblies of photocrosslinked proteins bound to nanoparticles are used in (Hill et al., 2005) to build a scaffolding for a bioelectronic 3D architecture. The recognition properties of DNA can also be used for computation (Adleman, 1994; Braich et al., 2002).

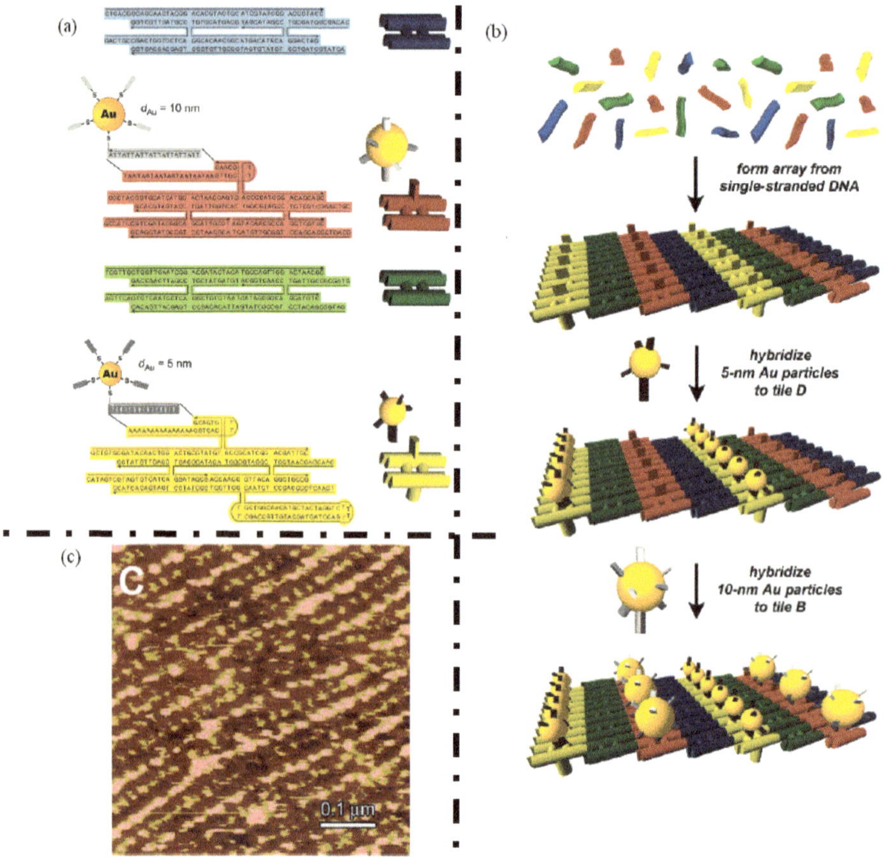


Fig. 13. (a) Three types of DNA tiles; each possesses sticky ends to bind with other tiles; the 'red' and 'yellow' tiles also have other sticky ends pronging outwards and different from each other. In (b) the cartoon image of the assembly sequence: the tiles assemble in a predefined sequence to form a lattice; gold nanoparticles are grafting to the pronging sticky ends, after being functionalized with the complementary strands; thanks to the diversity in the sticky ends, two types of nanoparticles (with diameters 5 and 10 nm) can be made assemble in ordinate, alternate rows. The AFM image in (c) shows the result of this process (reprinted with permission from Y. Y. Pinto et al., Nano Lett. 5, 2399 (2005). Copyright 2005 American Chemical Society).

Besides DNA, other affine molecules can be used like Peptide Nucleic Acid (PNA) (Nielsen et al., 1991). Its main difference to DNA is that it is uncharged, bearing a polyamide backbone. Its binding properties and selectivity are even better than those of DNA, to some respect. K.A. Williams et al. (Williams et al., 2002) coupled carbon nanotubes with single-strand PNA covalently and tested hybridization with single-strand complementary DNA: there resulted that DNA-PNA grafted to nanotubes, especially near the cut ends. The preference for the extremities of the nanotubes ensues from the higher reactivity there, at least in the lack of side-wall defects along the nanotubes themselves, and is a useful feature; moreover, DNA does bind to PNA through specific interaction, rather than with nanotubes through unspecific binding. Such processes could might be envisaged as a future opportunity to exploit molecular lithography for handling nanotubes (as done also by Keren et al. (Keren et al., 2003)), considered among the most promising objects for nanoelectronics, as well as to build more complex objects for computation.

At the end of this overlook of biomolecules in patterning we want to cite the work by L. Riemenschneider et al. (Riemenschneider et al., 2005), a 'hybrid' technique joining the advantages of scanning probe lithography with the use of an enzyme. First, an AFM tip was functionalized with a molecule of substrate of alkaline phosphatase, by a technique ensuring that it was grafted exactly at the apex of the tip. The tip was then moved across a mica substrate in a solution of 5-bromo-4-chloro-3-indolyl phosphate (BCIP) and nitro blue tetrazolium (NBT); since alkaline phosphatase dephosphorylates BCIP, making it precipitate in the presence of NBT, the resulting precipitate layer follows the tips' path. For instance, keeping the tip standing for some tens of seconds produces a spot of precipitate whose large size is proportional to time, and moving it across the substrate at proper speed causes different patterns to form.

In addition to DNA and its derivatives, also other specific interaction (antigen-antibody, ligand-receptor, etc) can be exploited for patterning. Recognition between biotin and streptadivin, as well as between bovine serum albumin (BSA) and the corresponding anti-BSA antibody, were used by Y. Wang et al. (Wang et al., 2005) to assemble CdTe nanowires into branched structures, representing prototypes of nanoscale circuits. Diode-like characteristics were found in the single nanowires, while the junctions were found to be non-conductive, presumably due to the large tunnelling barriers placed by the proteins themselves, creating a ~ 5 nm gap. The authors suggested that the existence of such gap could be viewed as an opportunity to build single-electron devices or, on the other hand, could be reduced by using DNA oligomers in place of proteins. Moreover, scaffolds of nanoparticles were built by McMillan et al. (McMillan et al., 2002) assembling engineered chaperonins (i.e. protein complexes assisting proteins in their folding phase) into regular patterns of units of variable diameter.

### 2.3.2 Other patterning methods based on self assembly

Wiring is essential in any electronic architecture; that is why patterning specific networks of nanowires to connect devices is one of the most debated - and trickiest - tasks in nanoelectronics. In many cases, nanowires are grown from nanoparticles assembled on a suitable pattern by using biomolecules or other methods like those involving force fields (see for instance (Cohen, 2005)). Actually, S.O. Lumsdon et al. (Lumsdon and Scott, 2005) assembled gold and carbon black colloids, as well as carbon nanotubes, into wires thinner than 1 μm. The growth was triggered by ac electric fields exerting electrophoretic forces on the objects and this technique may be helpful in growing nanowires, provided that the electrodes are capable of shaping the electric field with the due precision. Finally, as carbon nanotubes and nanorods become increasingly important, methods are being studied to get them assembled in ordered geometries (Harnack et al., 2003; Gupta et al., 2006; Hu et al., 2006; Ryan et al., 2006; Nobile et al., 2007; Yan et al., 2007).

Another interesting way of patterning might be that of exploiting the natural periodicity of crystals, which can be tuned to some extent by cutting the crystal to expose a particular face.

Let us cite in this respect the work by A. Kirakosian et al.(Kirakosian et al., 2001) who demonstrated an atom-accurate silicon grating with period 5.73 nm, or exactly 17 atoms, by means of a Si(577) surface: notice that a grating of parallel lines is one of the most frequent test for patterning methods and an insidious one for lithography, since chemical etching suffers from capillarity when it comes to penetrating nanometer-narrow channels.
Finally, the self-assembly properties of monodisperse spheres were also exploited to demonstrate shadow nanosphere lithography (NSL) that allow the fabrication of periodic arrays with morphologies ranging from cups to rods and wires by simply changing the substrate position with respect to the evaporation source (Kosiorek et al., 2005; Imperia et al., 2008; Gwinner et al., 2009).

## 3. Conclusions

In conclusion, many nanofabrication methods are today available and the choice among them is typically dictated by the materials employed and the specific application targeted. In general, however, they allow modern researchers and IC companies an unprecedented control on processes and open the way to the fabrication of completely new classes of devices.

## 4. References


Adleman, L. M. (1994). Molecular Computation of Solutions to Combinatorial Problems. *Science*, Vol. 266, No. 5187, pp. 1021-1024.

Agam, O. & Schiller, A. (2001). Projecting the Kondo effect: Theory of the quantum mirage. *Physical Review Letters*, Vol. 86, No. 3, pp. 484-487.

Amlani, I., Rawlett, A. M., Nagahara, L. A. & Tsui, R. K. (2002). An approach to transport measurements of electronic molecules. *Applied Physics Letters*, Vol. 80, No. 15, pp. 2761-2763.

Barth, J. V., Costantini, G. & Kern, K. (2005). Engineering atomic and molecular nanostructures at surfaces. *Nature*, Vol. 437, No. 7059, pp. 671-679.

Becker, H. & Locascio, L. E. (2002). Polymer microfluidic devices. *Talanta*, Vol. 56, No. 2, pp. 267-287, 0039-9140.

Berti, L., Alessandrini, A. & Facci, P. (2005). DNA-templated photoinduced silver deposition. *Journal Of The American Chemical Society*, Vol. 127, No. 32, pp. 11216-11217.

Beyer, S., Nickels, P. & Simmel, F. C. (2005). Periodic DNA nanotemplates synthesized by rolling circle amplification. *Nano Letters*, Vol. 5, No. 4, pp. 719-722.

Braich, R. S., Chelyapov, N., Johnson, C., Rothemund, P. W. K. & Adleman, L. (2002). Solution of a 20-variable 3-SAT problem on a DNA computer. *Science*, Vol. 296, No. 5567, pp. 499-502.

Braun, G., Inagaki, K., Estabrook, R. A., Wood, D. K., Levy, E., Cleland, A. N., Strouse, G. F. & Reich, N. O. (2005). Gold nanoparticle decoration of DNA on silicon. *Langmuir*, Vol. 21, No. 23, pp. 10699-10701.

Briseno, A. L., Mannsfeld, S. C. B., Ling, M. M., Liu, S. H., Tseng, R. J., Reese, C., Roberts, M. E., Yang, Y., Wudl, F. & Bao, Z. N. (2006). Patterning organic single-crystal transistor arrays. *Nature*, Vol. 444, No. 7121, pp. 913-917.

Cai, Y. G. & Ocko, B. M. (2005). Electro Pen Nanolithography. *Journal Of The American Chemical Society*, Vol. 127, No. 46, pp. 16287-16291.

Champagne, A. R., Pasupathy, A. N. & Ralph, D. C. (2005). Mechanically adjustable and electrically gated single-molecule transistors. *Nano Letters*, Vol. 5, No. 2, pp. 305-308.

Chau, R., Boyanov, B., Doyle, B., Doczy, M., Datta, S., Hareland, S., Jin, B., Kavalieros, J. & Metz, M. (2003). Silicon nano-transistors for logic applications. *Physica E: Low-dimensional Systems and Nanostructures*, Vol. 19, No. 1-2, pp. 1-5, 1386-9477.

Chelyapov, N., Brun, Y., Gopalkrishnan, M., Reishus, D., Shaw, B. & Adleman, L. (2004). DNA triangles and self-assembled hexagonal tilings. *Journal Of The American Chemical Society*, Vol. 126, No. 43, pp. 13924-13925.

Chiriacò, M. S., Primiceri, E., D'Amone, E., Ionescu, R. E., Rinaldi, R. & Maruccio, G. (2011). EIS microfluidic chips for flow immunoassay and ultrasensitive cholera toxin detection. *Lab on a Chip*, Vol. 11, No. 4, pp. 658-663, 1473-0197.

Choi, K. M. & Rogers, J. A. (2003). A photocurable poly(dimethylsiloxane) chemistry designed for soft lithographic molding and printing in the nanometer regime. *Journal Of The American Chemical Society*, Vol. 125, No. 14, pp. 4060-4061.

Chou, S. Y., Keimel, C. & Gu, J. (2002). Ultrafast and direct imprint of nanostructures in silicon. *Nature*, Vol. 417, No. 6891, pp. 835-837.

Chow, D. C., Lee, W. K., Zauscher, S. & Chilkoti, A. (2005). Enzymatic fabrication of DNA nanostructures: Extension of a self-assembled oligonucleotide monolayer on gold arrays. *Journal Of The American Chemical Society*, Vol. 127, No. 41, pp. 14122-14123.

Chu, C., Na, J.-S. & Parsons, G. N. (2007). Conductivity in Alkylamine/Gold and Alkanethiol/Gold Molecular Junctions Measured in Molecule/Nanoparticle/ Molecule Bridges and Conducting Probe Structures. *Journal of the American Chemical Society*, Vol. 129, No. 8, pp. 2287-2296, 0002-7863.

Cohen, A. E. (2005). Control of Nanoparticles with Arbitrary Two-Dimensional Force Fields. *Physical Review Letters*, Vol. 94, No., pp.

Dadosh, T., Gordin, Y., Krahne, R., Khivirich, I., Mahalu, D., Frydman, V., Sperling, J., Yacoby, A. & Bar-Joseph, I. (2005). Measurement of the conductance of single conjugated molecules. *Nature*, Vol. 436, No., pp. 677.

Dillenback, L. M., Goodrich, G. P. & Keating, C. D. (2006). Temperature-programmed assembly of DNA : Au nanoparticle bioconjugates. *Nano Letters*, Vol. 6, No. 1, pp. 16-23.

Dittrich, P. S. & Manz, A. (2006). Lab-on-a-chip: microfluidics in drug discovery. *Nature Reviews Drug Discovery*, Vol. 5, No. 3, pp. 210-218, 1474-1776.

Dixit, S., Singh-Zocchi, M., Hanne, J. & Zocchi, G. (2005). Mechanics of binding of a single integration-host-factor protein to DNA. *Physical Review Letters*, Vol. 94, No. 11, pp.

Dorn, A., Sigrist, M., Fuhrer, A., Ihn, T., Heinzel, T., Ensslin, K., Wegscheider, W. & Bichler, M. (2002). Electronic properties of antidot lattices fabricated by atomic force lithography. *Applied Physics Letters*, Vol. 80, No. 2, pp. 252-254.

Doyle, B. S., Datta, S., Doczy, M., Hareland, S., Jin, B., Kavalieros, J., Linton, T., Murthy, A., Rios, R. & Chau, R. (2003). High performance fully-depleted tri-gate CMOS transistors. *Electron Device Letters, IEEE*, Vol. 24, No. 4, pp. 263-265, 0741-3106.

Erickson, D. & Li, D. Q. (2004). Integrated microfluidic devices. *Analytica Chimica Acta*, Vol. 507, No. 1, pp. 11-26, 0003-2670.

Fiete, G. A. & Heller, E. J. (2003). Colloquium: Theory of quantum corrals and quantum mirages. *Reviews Of Modern Physics*, Vol. 75, No. 3, pp. 933-948.

Fischbein, M. D. & Drndic, M. (2006). Nanogaps by direct lithography for high-resolution imaging and electronic characterization of nanostructures. *Applied Physics Letters*, Vol. 88, No. 6, pp. 063116-3

Geissler, M. & Xia, Y. N. (2004). Patterning: Principles and some new developments. *Advanced Materials*, Vol. 16, No. 15, pp. 1249-1269.

Goux-Capes, L., Filoramo, A., Cote, D., Bourgoin, J. P. & Patillon, J. N. (2006). Coupling carbon nanotubes through DNA linker using a biological recognition complex. *Physica Status Solidi A-Applications And Materials Science*, Vol. 203, No. 6, pp. 1132-1136.

Graneli, A., Yeykal, C. C., Prasad, T. K. & Greene, E. C. (2006). Organized arrays of individual DIVA molecules tethered to supported lipid bilayers. *Langmuir*, Vol. 22, No. 1, pp. 292-299.

Gupta, S., Zhang, Q. L., Emrick, T. & Russell, T. P. (2006). "Self-corralling" nanorods under an applied electric field. *Nano Letters*, Vol. 6, No. 9, pp. 2066-2069.

Gwinner, M. C., Koroknay, E., Fu, L. W., Patoka, P., Kandulski, W., Giersig, M. & Giessen, H. (2009). Periodic Large-Area Metallic Split-Ring Resonator Metamaterial Fabrication Based on Shadow Nanosphere Lithography. *Small*, Vol. 5, No. 3, pp. 400-406, 1613-6810.

Hammond, P. T. (2004). Form and Function in Multilayer Assembly: New Applications at the Nanoscale. *Advanced Materials*, Vol. 16, No. 15, pp. 1271-1293.

Harnack, O., Pacholski, C., Weller, H., Yasuda, A. & Wessels, J. M. (2003). Rectifying behavior of electrically aligned ZnO nanorods. *Nano Letters*, Vol. 3, No. 8, pp. 1097-1101.

He, Y., Chen, Y., Liu, H. P., Ribbe, A. E. & Mao, C. D. (2005). Self-assembly of hexagonal DNA two-dimensional (2D) arrays. *Journal Of The American Chemical Society*, Vol. 127, No. 35, pp. 12202-12203.

Heersche, H. B., de Groot, Z., Folk, J. A., van der Zant, H. S. J., Romeike, C., Wegewijs, M. R., Zobbi, L., Barreca, D., Tondello, E. & Cornia, A. (2006). Electron transport through single Mn-12 molecular magnets. *Physical Review Letters*, Vol. 96, No. 20, pp., 0031-9007.

Hill, R. T., Lyon, J. L., Allen, R., Stevenson, K. J. & Shear, J. B. (2005). Microfabrication of three-dimensional bioelectronic architectures. *Journal Of The American Chemical Society*, Vol. 127, No. 30, pp. 10707-10711.

Ho, P. S. & Kwok, T. (1989). Electromigration in metals. *Reports On Progress In Physics*, Vol. 52, No., pp. 310-348.

Hoff, J. D., Cheng, L. J., Meyhofer, E., Guo, L. J. & Hunt, A. J. (2004). Nanoscale protein patterning by imprint lithography. *Nano Letters*, Vol. 4, No. 5, pp. 853-857.

Hong, S. H. & Mirkin, C. A. (2000). A nanoplotter with both parallel and serial writing capabilities. *Science*, Vol. 288, No. 5472, pp. 1808-1811.

Hong, S. H., Zhu, J. & Mirkin, C. A. (1999). Multiple ink nanolithography: Toward a multiple-pen nano-plotter. *Science*, Vol. 286, No. 5439, pp. 523-525.

Hosogi, M., Hashiguchi, G., Haga, M., Yonezawa, T., Kakushima, K. & Fujita, H. (2005). Electrical conductivity of lambda DNA-Pd wire. *Japanese Journal Of Applied Physics Part 2-Letters & Express Letters*, Vol. 44, No. 28-32, pp. L955-L957.

Hu, T. & Shklovskii, B. I. (2006). How does a protein search for the specific site on DNA: The role of disorder. *Physical Review E*, Vol. 74, No. 2, pp.

Hu, W. C., Sarveswaran, K., Lieberman, M. & Bernstein, G. H. (2005). High-resolution electron beam lithography and DNA nano-patterning for molecular QCA. *Ieee Transactions On Nanotechnology*, Vol. 4, No. 3, pp. 312-316.

Hu, Z. H., Fischbein, M. D., Querner, C. & Drndic, M. (2006). Electric-field-driven accumulation and alignment of CdSe and CdTe nanorods in nanoscale devices. *Nano Letters*, Vol. 6, No. 11, pp. 2585-2591.

Hulme, J. P., Gwak, J. & Miyahara, Y. (2006). Biomolecular embossing. *Journal Of The American Chemical Society*, Vol. 128, No. 2, pp. 390-391.

Hyun, J., Ahn, S. J., Lee, W. K., Chilkoti, A. & Zauscher, S. (2002). Molecular recognition-mediated fabrication of protein nanostructures by dip-pen lithography. *Nano Letters*, Vol. 2, No. 11, pp. 1203-1207.

Imperia, P., Kandulski, W., Kosiorek, A., Glaczynska, H., Maletta, H. & Giersig, M. (2008). Magnetic anisotropy study of triangular-shaped Co nanostructures. *Journal Of Magnetism And Magnetic Materials*, Vol. 320, No. 21, pp. 2682-2687, 0304-8853.

Jain, K., Willson, C. G. & Lin, B. J. (1982). Ultrafast deep UV Lithography with excimer lasers. *Electron Device Letters, IEEE*, Vol. 3, No. 3, pp. 53-55, 0741-3106.

Jo, M. H., Grose, J. E., Baheti, K., Deshmukh, M. M., Sokol, J. J., Rumberger, E. M., Hendrickson, D. N., Long, J. R., Park, H. & Ralph, D. C. (2006). Signatures of molecular magnetism in single-molecule transport spectroscopy. *Nano Letters*, Vol. 6, No. 9, pp. 2014-2020, 1530-6984.

Johnston, A. P. R., Read, E. S. & Caruso, F. (2005). DNA multilayer films on planar and colloidal supports: Sequential assembly of like-charged polyelectrolytes. *Nano Letters*, Vol. 5, No. 5, pp. 953-956.

Keren, K., Berman, R. S., Buchstab, E., Sivan, U. & Braun, E. (2003). DNA-templated carbon nanotube field-effect transistor. *Science*, Vol. 302, No. 5649, pp. 1380-1382.

Keren, K., Krueger, M., Gilad, R., Ben-Yoseph, G., Sivan, U. & Braun, E. (2002). Sequence-specific molecular lithography on single DNA molecules. *Science*, Vol. 297, No. 5578, pp. 72-75.

Kervennic, Y. V., Van der Zant, H. S. J., Morpurgo, A. F., Gurevich, L. & Kouwenhoven, L. P. (2002). Nanometer-spaced electrodes with calibrated separation. *Applied Physics Letters*, Vol. 80, No. 2, pp. 321-323.

Kirakosian, A., Bennewitz, R., Crain, J. N., Fauster, T., Lin, J. L., Petrovykh, D. Y. & Himpsel, F. J. (2001). Atomically accurate Si grating with 5.73 nm period. *Applied Physics Letters*, Vol. 79, No. 11, pp. 1608-1610.

Kosiorek, A., Kandulski, W., Glaczynska, H. & Giersig, M. (2005). Fabrication of nanoscale rings, dots, and rods by combining shadow nanosphere lithography and annealed polystyrene nanosphere masks. *Small*, Vol. 1, No. 4, pp. 439-444, 1613-6810.

Krahne, R., Yacoby, A., Shtrikman, H., Bar-Joseph, I., Dadosh, T. & Sperling, J. (2002). Fabrication of nanoscale gaps in integrated circuits. *Applied Physics Letters*, Vol. 81, No. 4, pp. 730-732.

Kudo, H. & Fujihira, M. (2006). DNA-Templated copper nanowire fabrication by a two-step process involving electroless metallization. *Ieee Transactions On Nanotechnology*, Vol. 5, No. 2, pp. 90-92.

Lee, K. B., Park, S. J., Mirkin, C. A., Smith, J. C. & Mrksich, M. (2002). Protein nanoarrays generated by dip-pen nanolithography. *Science*, Vol. 295, No. 5560, pp. 1702-1705.

Liu, Y., Ke, Y. G. & Yan, H. (2005). Self-assembly of symmetric finite-size DNA nanoarrays. *Journal Of The American Chemical Society*, Vol. 127, No. 49, pp. 17140-17141.

Lumsdon, S. O. & Scott, D. M. (2005). Assembly of colloidal particles into microwires using an alternating electric field. *Langmuir*, Vol. 21, No. 11, pp. 4874-4880.

Luthi, R., Schlittler, R. R., Brugger, J., Vettiger, P., Welland, M. E. & Gimzewski, J. K. (1999). Parallel nanodevice fabrication using a combination of shadow mask and scanning probe methods. *Applied Physics Letters*, Vol. 75, No. 9, pp. 1314-1316, 0003-6951.

Manoharan, H. C., Lutz, C. P. & Eigler, D. M. (2000). Quantum mirages formed by coherent projection of electronic structure. *Nature*, Vol. 403, No. 6769, pp. 512-515.

Maruccio, G., Cingolani, R. & Rinaldi, R. (2004). Projecting the nanoworld: Concepts, results and perspectives of molecular electronics. *Journal Of Materials Chemistry*, Vol. 14, No. 4, pp. 542-554.

Maruccio, G., Marzo, P., Krahne, R., Passaseo, A., Cingolani, R. & Rinaldi, R. (2007). Protein conduction and negative differential resistance in large-scale nanojunction arrays. *Small*, Vol. 3, No. 7, pp. 1184-1188, 1613-6810.

Maruccio, G., Primiceri, E., Marzo, P., Arima, V., Della Torre, A., Rinaldi, R., Pellegrino, T., Krahne, R. & Cingolani, R. (2009). A nanobiosensor to detect single hybridization events. *Analyst*, Vol. 134, No. 12, pp. 2458-2461, 0003-2654.

Mathieu, F., Liao, S. P., Kopatscht, J., Wang, T., Mao, C. D. & Seeman, N. C. (2005). Six-helix bundles designed from DNA. *Nano Letters*, Vol. 5, No. 4, pp. 661-665.

McMillan, R. A., Paavola, C. D., Howard, J., Chan, S. L., Zaluzec, N. J. & Trent, J. D. (2002). Ordered nanoparticle arrays formed on engineered chaperonin protein templates. *Nature Materials*, Vol. 1, No. 4, pp. 247-252.

Morr, D. K. & Stavropoulos, N. A. (2004). Quantum corrals, eigenmodes, and quantum mirages in s-wave superconductors. *Physical Review Letters*, Vol. 92, No. 10, pp.

Nielsen, P. E., Egholm, M., Berg, R. H. & Buchardt, O. (1991). SEQUENCE-SELECTIVE RECOGNITION OF DNA BY STRAND DISPLACEMENT WITH A THYMINE-SUBSTITUTED POLYAMIDE. *Science*, Vol. 254, No. 5037, pp. 1497-1500, 0036-8075.

Nobile, C., Fonoberov, V. A., Kudera, S., Della Torre, A., Ruffino, A., Chilla, G., Kipp, T., Heitmann, D., Manna, L., Cingolani, R., Balandin, A. A. & Krahne, R. (2007). Confined optical phonon modes in aligned nanorod arrays detected by resonant inelastic light scattering. *Nano Letters*, Vol. 7, No. 2, pp. 476-479.

Park, J., Pasupathy, A. N., Goldsmith, J. I., Chang, C., Yaish, Y., Petta, J. R., Rinkoski, M., Sethna, J. P., Abruna, H. D., McEuen, P. L. & Ralph, D. C. (2002). Coulomb blockade and the Kondo effect in single-atom transistors. *Nature*, Vol. 417, No. 6890, pp. 722-725.

Park, S. H., Barish, R., Li, H. Y., Reif, J. H., Finkelstein, G., Yan, H. & LaBean, T. H. (2005a). Three-helix bundle DNA tiles self-assemble into 2D lattice or 1D templates for silver nanowires. *Nano Letters*, Vol. 5, No. 4, pp. 693-696.

Park, S. H., Yin, P., Liu, Y., Reif, J. H., LaBean, T. H. & Yan, H. (2005b). Programmable DNA self-assemblies for nanoscale organization of ligands and proteins. *Nano Letters*, Vol. 5, No. 4, pp. 729-733.

Paukstelis, P. J. (2006). Three-Dimensional DNA Crystals as Molecular Sieves. *Journal Of The American Chemical Society*, Vol. 128, No. 31, pp. 6794-6795.

Peckerar, M. C. & Maldonado, J. R. (1993). X-Ray-Lithography - An Overview. *Proceedings Of The Ieee*, Vol. 81, No. 9, pp. 1249-1274.

Pfeiffer, H. C., Dhaliwal, R. S., Golladay, S. D., Doran, S. K., Gordon, M. S., Groves, T. R., Kendall, R. A., Lieberman, J. E., Petric, P. F., Pinckney, D. J., Quickle, R. J., Robinson, C. F., Rockrohr, J. D., Senesi, J. J., Stickel, W., Tressler, E. V., Tanimoto, A., Yamaguchi, T., Okamoto, K., Suzuki, K., Okino, T., Kawata, S., Morita, K., Suziki, S. C., Shimizu, H., Kojima, S., Varnell, G., Novak, W. T., Stumbo, D. P. & Sogard, M. (1999). Projection reduction exposure with variable axis immersion lenses: Next generation lithography. *Journal of Vacuum Science & Technology B: Microelectronics and Nanometer Structures*, Vol. 17, No. 6, pp. 2840-2846, 1071-1023.

Piner, R. D., Zhu, J., Xu, F., Hong, S. H. & Mirkin, C. A. (1999). "Dip-pen" nanolithography. *Science*, Vol. 283, No. 5402, pp. 661-663.

Pinto, Y. Y., Le, J. D., Seeman, N. C., Musier-Forsyth, K., Taton, T. A. & Kiehl, R. A. (2005). Sequence-encoded self-assembly of multiple-nanocomponent arrays by 2D DNA scaffolding. *Nano Letters*, Vol. 5, No. 12, pp. 2399-2402.

Pollack, M. G., Shenderov, A. D. & Fair, R. B. (2002). Electrowetting-based actuation of droplets for integrated microfluidics. *Lab on a Chip*, Vol. 2, No. 2, pp. 96-101, 1473-0189.

Primiceri, E., Chiriacò, M. S., D'Amone, E., Urso, E., Ionescu, R. E., Rizzello, A., Maffia, M., Cingolani, R., Rinaldi, R. & Maruccio, G. (2010). Real-time monitoring of copper ions-induced cytotoxicity by EIS cell chips. *Biosensors and Bioelectronics*, Vol. 25, No. 12, pp. 2711-2716, 0956-5663.

Ray, V., Subramanian, R., Bhadrachalam, P., Ma, L.-C., Kim, C.-U. & Koh, S. J. (2008). CMOS-compatible fabrication of room-temperature single-electron devices. *Nat Nano*, Vol. 3, No. 10, pp. 603-608.

Reed, M. A., Zhou, C., Muller, C. J., Burgin, T. P. & Tour, J. M. (1997). Conductance of a molecular junction. *Science*, Vol. 278, No. 5336, pp. 252-254.

Reif, J. H. (2002). DNA Lattices: A Method for Molecular-Scale Patterning and Computation. *Computing in Science and Engineering*, Vol. 4, No. 1, pp. 32-41.

Reishus, D., Shaw, B., Brun, Y., Chelyapov, N. & Adleman, L. (2005). Self-assembly of DNA double-double crossover complexes into high-density, doubly connected, planar structures. *Journal Of The American Chemical Society*, Vol. 127, No. 50, pp. 17590-17591.

Riemenschneider, L., Blank, S. & Radmacher, M. (2005). Enzyme-assisted nanolithography. *Nano Letters*, Vol. 5, No. 9, pp. 1643-1646.

Rothemund, P. W. K. (2006). Folding DNA to create nanoscale shapes and patterns. *Nature*, Vol. 440, No. 7082, pp. 297-302.

Ryan, K. M., Mastroianni, A., Stancil, K. A., Liu, H. T. & Alivisatos, A. P. (2006). Electric-field-assisted assembly of perpendicularly oriented nanorod superlattices. *Nano Letters*, Vol. 6, No. 7, pp. 1479-1482.

Salaita, K., Wang, Y. H. & Mirkin, C. A. (2007). Applications of dip-pen nanolithography. *Nature Nanotechnology*, Vol. 2, No. 3, pp. 145-155.

Schallhorn, K. A., Freedman, K. O., Moore, J. M., Lin, J. & Ke, P. C. (2005). Single-molecule DNA flexibility in the presence of base-pair mismatch. *Applied Physics Letters*, Vol. 87, No. 3, pp.

Silverman, J. P. (1997). X-ray lithography: status, challenges and outlook for 0.13 um. *Journal of Vacuum Science & Technology B*, Vol. 15, No., pp. 2117.

Solak, H. H., David, C., Gobrecht, J., Golovkina, V., Cerrina, F., Kim, S. O. & Nealey, P. F. (2003). Sub-50 nm period patterns with EUV interference lithography. *Microelectronic Engineering*, Vol. 67-8, No., pp. 56-62.

Stone, H. A., Stroock, A. D. & Ajdari, A. (2004). Engineering flows in small devices: Microfluidics toward a lab-on-a-chip. *Annual Review of Fluid Mechanics*, Vol. 36, No., pp. 381-411, 0066-4189.

Sun, G. X., Fecko, C. J., Nicewonger, R. B., Webb, W. W. & Begley, T. P. (2006). DNA-protein cross-linking: Model systems for pyrimidine-aromatic amino acid cross-linking. *Organic Letters*, Vol. 8, No. 4, pp. 681-683.

Vladimirsky, Y., Bourdillon, A., Vladimirsky, O., Jiang, W. & Leonard, Q. (1999). Demagnification in proximity x-ray lithography and extensibility to 25 nm by optimizing Fresnel diffraction. *Journal Of Physics D-Applied Physics*, Vol. 32, No. 22, pp. L114-L118.

Wang, Y., Tang, Z. Y., Tan, S. S. & Kotov, N. A. (2005). Biological assembly of nanocircuit prototypes from protein-modified CdTe nanowires. *Nano Letters*, Vol. 5, No. 2, pp. 243-248.

Williams, K. A., Veenhuizen, P. T. M., de la Torre, B. G., Eritja, R. & Dekker, C. (2002). Nanotechnology - Carbon nanotubes with DNA recognition. *Nature*, Vol. 420, No. 6917, pp. 761-761.

Winfree, E., Liu, F. R., Wenzler, L. A. & Seeman, N. C. (1998). Design and self-assembly of two-dimensional DNA crystals. *Nature*, Vol. 394, No. 6693, pp. 539-544.

Woehrle, G. H., Warner, M. G. & Hutchison, J. E. (2004). Molecular-level control of feature separation in one-dimensional nanostructure assemblies formed by biomolecular nanolithography. *Langmuir*, Vol. 20, No. 14, pp. 5982-5988, 0743-7463.

Xia, Y. N. & Whitesides, G. M. (1997). Extending microcontact printing as a microlithographic technique. *Langmuir*, Vol. 13, No. 7, pp. 2059-2067.

Yan, Y. H., Chan-Park, M. B. & Zhang, Q. (2007). Advances in carbon-nanotube assembly. *Small*, Vol. 3, No. 1, pp. 24-42, 1613-6810.

Yang, L. & Schepartz, A. (2005). Relationship between folding and function in a sequence-specific miniature DNA-binding protein. *Biochemistry*, Vol. 44, No. 20, pp. 7469-7478.

Yu, A. A., Savas, T. A., Taylor, G. S., Guiseppe-Elie, A., Smith, H. I. & Stellacci, F. (2005). Supramolecular nanostamping: Using DNA as movable type. *Nano Letters*, Vol. 5, No. 6, pp. 1061-1064.

Zandbergen, H. W., van Duuren, R., Alkemade, P. F. A., Lientschnig, G., Vasquez, O., Dekker, C. & Tichelaar, F. D. (2005). Sculpting nanoelectrodes with a transmission electron beam for electrical and geometrical characterization of nanoparticles. *Nano Letters*, Vol. 5, No. 3, pp. 549-553.

Zhang, J. M., Ma, Y. F., Stachura, S. & He, H. X. (2005). Assembly of highly aligned DNA strands onto Si chips. *Langmuir*, Vol. 21, No. 9, pp. 4180-4184.

2

# Nanoscale Architectures for Smart Bio-Interfaces: Advances and Challenges

Serban F. Peteu[1,2], Sabine Szunerits[3], Alina Vasilescu[4] and Wolfgang Knoll[5]
[1]*The National Institute for R&D in Chemistry and Petrochemistry, Bucharest,*
[2]*Department of Chemical Engineering and Materials Science, Michigan State University, East Lansing, Michigan,*
[3]*Institut de Recherche Interdisciplinaire Universite of Lille 1, Villeneuve d'Ascq,*
[4]*International Centre for Biodynamics,*
[5]*Austrian Institute of Technology AIT, Vienna,*
[1,4]*Romania*
[2]*USA*
[3]*France*
[5]*Austria*

## 1. Introduction

*"At the nano level atoms do not belong to any field of science"*, professor Chad Mirkin explained the need to unlock our minds and transform our attitudes, as we continue to live the adventure of nano science and technology. This ably conveys both the uniqueness and diversity of nanotechnology, while stressing the preparation required by those aspiring to it (Papapzoglou & Parthasarathy, 2007). Clearly, the structuring of substance at the molecular scale extend across the entire spectrum of scientific knowledge counting physics, chemistry, medicine, or engineering.

However, why should anyone care about *nanotechnology* or even *nanofabrication*? If we take a step back, we find that the *microfabrication* techniques, such as the conventional lithography, deposition, or etching have enabled micromachining of architectures down to sub-micrometer dimensions, (e.g., 400-900 nm). These techniques have attained an adequate level of maturity and can be found already incorporated commercial MEMS products, like pressure microsensors, micro-accelerometers, or micro-gyroscopes (Arshak, 2005; Cook-Chennault, 2008; Guo, 2009; Liu, 2007; Pal, 2006; Tsai, 2007). Other research-grade, sensitive micro(bio)interfaces (Amatore et al, 2006; Asher et al, 2002; Avramescu et al, 2002 ; Bitziou et al, 2010; Cosnier, 2000; Deo et al, 2003; Grayson, 2004; Huber et al, 2006; Jungblut et al, 2009; Knoll et al, 2006; Lee et al, 2009; Marcon et al, 2010; Peteu et al, 2007; Szunerits & Walt, 2002; Vasilescu et al, 2003 ; Ziaie, 2004a).

More recently, nano-size structures have attracted a colossal interest, due in part to their unique electrical, magnetic, optical, thermal, and mechanical properties. Clearly, once properly developed, these *nano* architectures are expected to lead to a range of electronic, photonic, sensing or actuating devices with superior cost-benefit performances, compared with their *macro* and *micro* counterparts. Herein, the focus is on advances and challenges in nanofabrication of smart bio-interfaces. While staying true to such fascinating thoughts, our efforts aim to achieve the control at molecular scale. By building block-by-block the new nano-architectures, one will not only advance scientific knowledge, but also will design and develop the future nano-parts for the next-generation hybrid smart micro-devices. These will integrate multi-level hierarchically linked nano-parts with electrical, optical, chemical, biological functions. Several nanofabrication methods are more mature, others are currently intensely investigated however these are not *per se* the subject of this chapter. More details can be found in well-documented, recent books, chapters or reviews (Bhushan, 2004; Cui, 2008; Kim et al, 2005; Klauser et al, 2010; Kummar et al, 2005; Mirkin & Rogers, 2001; Wang, Mirkin & Park. 2009; Ziaie, 2004b).

The *nanofabrication* implies making artifacts whose scale is in the nano domain, *1 to 100 nanometers*. As a reminder, 1 nm is one-millionth of 1 mm, 1 nm = $10^{-9}$ mm. To setup stage for nano, **Figure 1** charts a comparison between nano-scale and several familiar components: one human hair, the human cell, an 1980-old transistor, a bacterium, a virus, DNA, etc.

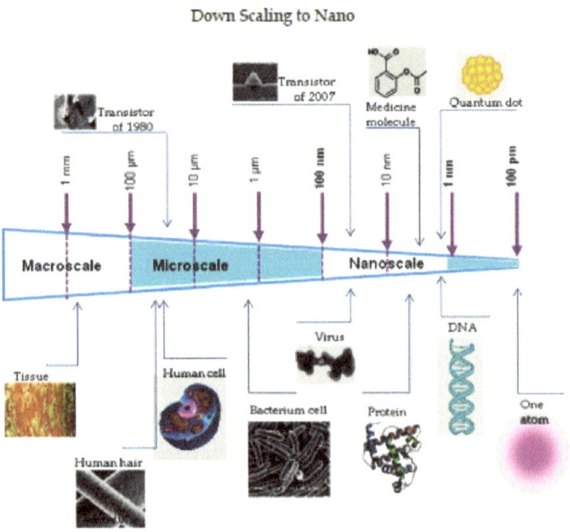


Direct and Indirect Nanofabrication:

| Direct | Resolution |
|---|---|
| Optical lithography | 30 nm |
| Electron beam lithography | 10 nm |
| Nano-imprint lithography | 10 nm |
| Scanning Probe lithography | 20 nm |
| Reactive ion etching | 50 nm |
| Metal lift-off | 10 nm |
| Focused ion beam | 30 nm |
| **Indirect** | |
| Sidewall lithography | 50 nm |
| Dimensional substract.-addition | 50 nm |
| *Self-assembly, DNA directed* | *20 nm* |
| Multi-step processing | 30 nm |
| High resolution nano-patterning | 2 nm |

Fig. 1. The nanofabrication methods /resolution, compared with manmade *vs.* natural micro-nano worlds.

Among nanofabrication methods listed in Figure 1, self-assembly often appears as most promising, due to its low cost and remarkable ability to produce nanostructures at different length scales. Consequently, IBM and Cal Tech teamed up to build *DNA origami* at the sub-22 nm scale, this nano-circuitry using DNA-directed self-assembly, by dry etching the $SiO_2$ with e-beam lithography, to create sites for the DNA to attach & self organize. Thus, DNA becomes a scaffold for nanotubes to self-assemble into sub-22 nm circuits (e.g., Aldaye & Sleiman, 2009; Nanofabrication , 2011 course web).

The strongest driving force for nanofabrication remains perhaps the production of increasingly smaller electronic components, keeping pace with *Moore's Law* forecasting the doubling of IC device density every 18 months (Freebody, 2011; Kahng, 2010). Furthermore, depending on specific applications, additional vectors could be desirable, including: complexity of structures, feature density, and materials generality. In addition, the fidelity/accuracy and scalability (parallel, simple, and cost-effective processing) are also keys to develop new and useful nanofabrication technologies (Wang, Mirkin & Park, 2009). One valued consequence of this drive is providing researchers with new tools and nano-materials to address rewarding topics in materials science and engineering, energy, life sciences, healthcare and more.

Nanofabrication offers two approaches, *top-down* and *bottom-up*. The *top-down* nanofabrication constructs objects from larger entities by removing material, while the *bottom-up* nanofabrication builds devices structured devices *via* the assembly of their molecular parts. While high-resolution lithography methods, like e-beam, can be employed to fabricate nano-size structures, their serial nature and or cost preclude a widespread application. This has forced investigators to explore alternative and potentially superior techniques such as self-assembly or nanoimprint lithography.

This chapter will critically review several specific nanoscale architectures for smart interfaces, selected as being under intense investigation and with exciting advances reported (Ali et al, 2008, 2010; Baca et al, 2011; Chi et al., 2005; Ishihara & Takai, 2009; Iwamoto, Kaneto & Mashiko, 2003; Kim et al., 2009; Knoll et al., 2008; Shvedova et al, 2010; Niedziolka-Jonsson, 2010; You et al., 2006, 2007). The following nano interfaces will be appraised:

1. Nano scale modeling and simulation, enabling nano-fabrication by-design, or "rationally designed materials";
2. Hybrid organic-inorganic nanomaterials, specially conductive polymer hybrids for enhanced sensing and actuation;
3. Nanoplasmonic methods and structures with focus on lamellar plasmonic nano-interfaces for optical sensors;
4. Nanoelectronics and more specifically the controlled molecular functional architectures for thin film transistors.

## 2. Modeling and simulation of nano architectures for rationally designed materials

### 2.1 The need for rational nanomaterials design

The investment in nanoscience has been seen by many countries as building a new chance for the future. The United States has one such team effort, the *Chemical Vision 2020* partnership, a chemical industries lead effort (GE, Dow, Intel, DuPont, Honeywell, Ciba, Rohm and Haas) to accelerate innovation and technology development, to introduce a host of innovative products to revive and energize the economy, solving major societal problems and creating new businesses. There is a resilient confidence that Nanofabrication will catalyze the new manufacturing (www.chemicalvision2020.org).

The traditional, typical methods to discover and develop new nanomaterials, today, are still based on somewhat focused *experimentation* and *scientific inspiration*, rather than on *rigorous engineering design*. A new strategy to achieve rational nanomaterials design (RND) or *nanomaterials by design*, was discussed by many experts (Burello &Worth, 2011; Nicholls et al, 2011). One possible multi-faceted approach is illustrated in **Figure 2**, reflecting an engineering, software-lead, market-oriented mind-set (Mize, 2004).

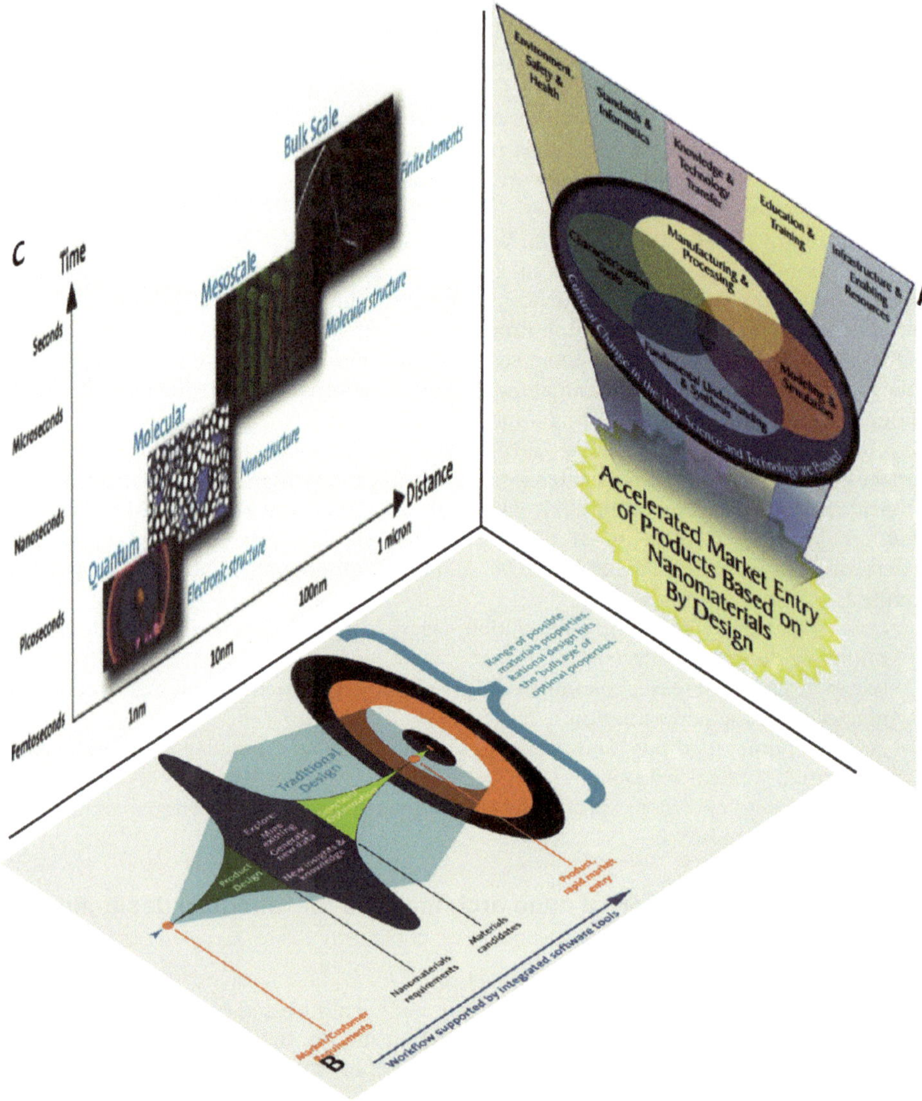


Fig. 2. The rational nanomaterials design (RND) complex process integrates the modeling-simulation methods with the theory, experiment, and converts resulting data into knowledge, then used in nanofabrication (from Mize, 2004).

The concept of *nanomaterials by design*, or *nanomaterials rational design* (NRD) refers to the ability to employ scientific principles in deliberately creating structures with nanoscale features (e.g., size, architecture) that deliver unique functionality and utility for target applications. While advances in technologies including synthesis, manufacturing, and characterization are very important factors in realizing this vision, an overarching strategy and rational design framework is essential. This involves the integration of modeling and simulation methods with theory, experiment, and the transformation of the resulting information into knowledge, which is then applied in processing and manufacturing. Key to this approach is the use of dedicated software – modeling, simulation and informatics (Barnard, 2010; Schommers, 2007).

In Figure 2, each plane-panel is one facet of NRD concept. Fist, the NRD algorithm of *accelerated product market entry* is charted in panel **A**. Next, the traditional manufacturing process is compared with the NRD in panel **B**, using the metaphor of *hitting the bull's eye of a target.* Finally, several modeling and simulation methods in panel **C** are addressing a range of sizes, suited for hierarchical materials (edited from Mize, 2004).

Thus, the panel **A** outlines a "cultural change in the way science and technology are pursued", a system leading to "accelerated market entry of products based on nanomaterials by design ".The goal is creating a nano-material based product able to deliver unique functionality and utility, for some very specific, pre-designed *targeted* applications.

Next, the panel **B** illustrates, in same figure, the process of Rational Nanomaterials Design. Herein, a *traditional* product manufacturing process is compared with a *rational* approach, using the metaphor of preparing to hit a *target,* in this case the *product rapid market entry.* Here, the traditional process, shown in light blue, results in more of a "hit and miss" approach, with more products missing the exact target than compared to the rational approach.. By contrast, the rational approach goes straight to the center of the target, hitting the "bull's eye" of optimal properties (Mize, 2004).

Finally, the panel **C** from figure 2 above, charts the different modeling and simulation methods to address the range of sizes of a complex material, from nano up to micro dimensions. The modeling of molecular systems was lately enabled by the unprecedented capabilities of super computers. This high performance in turn has driven important algorithmic advances, to a point where advanced calculations can be carried out from our desktop computers (Mize, 2004).

Molecular modeling and simulation combines methods that cover a range of size scales (the sub-atomic quantum mechanics; the atomistic level of molecular mechanics methods, the micrometer-scale mesoscale modeling) in order to study material systems. It is extremely expensive in terms of computing power to apply the more fundamental methods. Each step up the length scale offers the ability to model larger and more complex systems, with the tradeoff of a greater level of approximation in property prediction (Nicholls et al, 2011; Thamwattana et al, 2010; Vasiliev et al, 2009)

The modeling and simulation efforts are still exploding in different areas, often far apart, thus somewhat difficult to process, especially by a first-time reader. Thus, for their benefit, we decided to focus our discussion on modeling and simulation, to one major hub, namely the NSF-funded Network for Computational Nanotechnology (NCN), including its web-based interactive educational portal (www.nanoHUB.org).

## 2.2 Modeling and simulation

*Molecular modeling* is the materials representation at atomic- molecular level with 3D computer graphics, *via* graphical mathematical descriptions of the system. This allows scientists to predict fundamental relationships between structure, properties, behavior, composition, and the external stimuli. *Simulation* is the use of a computer to apply these methods to imitate the behavior of a real system, leading to a better understanding of that system. It can allow the prediction of properties of complex systems with many different discrete parts (Barnard, 2010; Batelle Institute, 2007; Mize, 2004).

## 2.3 Case study: The three dimensional nanoelectronic modeling

Device physics and material science meet at the atomic scale of novel nanostructured semiconductors, and the distinction between a new ultra-small material and a device is frequently blurred. The quantum-mechanical effects in the electronic states of the device and also the granular atomistic representation of the underlying material are equally important. In some instances, the approaches based on a continuum representation of the underlying material typically used by device engineers and physicists become invalid. *Ab initio* methods currently used by material scientists typically do not represent the band gaps and masses precisely enough for device design, or they do not scale to realistically large device sizes. The plethora of geometry, material, and doping configurations in semiconductor nano-devices suggest that *a general nano-electronic modeling tool* would be highly desirable (e.g., Qiao et al, 2006; Klimeck et al 2007a; 2007b).

The 3-D NanoElectronic MOdeling (NEMO 3-D) tool was pioneered by Klimeck and others, to address these needs and was recently expanded to NEMO 5 (**Figure 3**). Based on atomistic valence force field and a variety of nearest neighbor tight-binding models (e.g., *s*, *sp*$^3$*s**, and *sp*$^3$*d*$^5$*s**), NEMO 3-D enables the computation of strain and electronic structure for about 64 and 52 million atoms, corresponding to (110 nm)$^3$ and (101 nm)$^3$ volumes, respectively (Klimeck et al, 2007a, b).

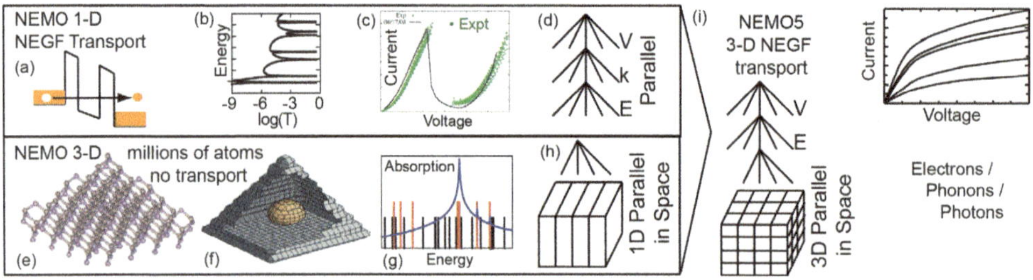


Fig. 3. The NEMO5 3-D simulation charts the electronic properties of individual atoms into realistic structures with millions of atoms, and computes relevant nanostructure properties such as strain relaxation, phonon modes, electronic structure, quantum transport etc (reproduced from Klimeck, 2011a by kind author's permission). More details in text.

The physical problem may involve very large scale computations and NEMO 3-D has been optimized to be scalable from single central processing units to large numbers of processors on supercomputers and clusters. NEMO 5 is a parallel multiscale tool for nanoelectronics, allowing among others, simulation of quantum transport in nanodevices using the non-equilibrium Greens function (NEGF) and open-boundary wave-function formalisms. NEMO 3-D and NEMO5 have been released with an open-source license (in 2003, respectively 2011) and are developed by the NCN interactive educational portal (www.nanoHUB.org). Two interesting examples of theoretical models are briefly discussed below, indicative of essential algorithmic and computational components that have been used in the development and successful deployment of NEMO 3-D.

### 2.3.1 Example: Quantum dots

After much effort in surface chemistry development and optimization by several groups, fluorescent semiconductor nanocrystals probes, also known as quantum dots or Qdots, are now entering the realm of biological applications with much to offer to biologists. The road to success has been paved with hurdles but from these efforts has stemmed a multitude of original surface chemistries that scientists in the biological fields can draw from for their specific biological applications. The ability to easily modulate the chemical nature of Qdot surfaces by employing one or more of the recently developed Qdot coatings, together with their exceptional photophysics have been key elements for Qdots to acquire a status of revolutionary fluorescent bio-probes. Indeed, the unique properties of Qdots not only give biologists the opportunity to explore advanced imaging techniques such as single molecule or lifetime imaging but also to revisit traditional fluorescence imaging methodologies, or based on shifting the diffraction peak, and extract yet unobserved or inaccessible information in vitro or in vivo (Alivisatos, 2004; Asher et al 2002; Cameron, Zhong & Knoll; Cheng et al. 2008; Feng et al, 2007, 2008; Medintz, Mattoussi & Clapp, 2008; Tomczak, et al., 2009).

The improved Qdots semiconductor nanocrystals display longer quenching time compared to conventional fluorescence dyes and size-tunable optical properties, which recommend them for optoelectronic, photonic or bio-labeling uses. The Qdot synthesis typically happens in liquid-phase, the resulting nanomaterial quality being influenced by experimental conditions. Current trends include (i) lowering toxicity for *in vivo* applications by using core-shell architectures (starting with Hines et al, 1996) and (ii) switching to "one pot" green chemistry (Gu, et al, 2004 ; Mekis, et al, 2003). Control over optical properties *via* the size of Qdots is exploited in polymer–Qdots hybrid materials, where designer architectures can be envisioned to match specific applications (Tomczak et al, 2009).

The several still-frames from **Figure 4a** illustrate the simulation and analysis of a pyramid-shaped Qdot using the "Quantum Dot Lab" application. Therein, several powerful analytic features of this tool are demonstrated, including: the visualization of specific 3D wave functions corresponding to discrete energy levels within the quantum dot; rotating the 3D volume of the quantum dot with wave function; scanning through the energy levels of the states inside the Qdot; comparing the absorption curves for different dot sizes (Klimeck & Haley, 2009).

### 2.3.2 Example: Electron density in a circular silicon nanowire transistor

Continued down-scaling of transistors have enabled the tremendous advances in consumer electronics. We are reaching the limits where the individual transistors or on-off electron valves are only a few nanometers in diameters wide. At these atomic length scales the electrons do no longer act like billiard balls but like waves. Sophisticated modeling engines that consider a quantum mechanical description of the electrons, an atomistic description of the material, and non-equilibrium electron distributions are needed for device design and optimization.

The "Nanowire" tool on nanoHUB.org enables modeling and simulation of such device, as illustrated in figure **4b** and enables the visualization of the electron density in such ultra-scaled structures through 3D volume rendering. In each panel, the left and the right regions on the image represent the large electron densities in the source and drain of the transistor. The central section represents the gated region that enables the control of the electron flow through the nano-scale on-off switch (Klimeck & Mehrotra, 2011; Park et al, 2011).

In the following sub-chapter, we will examine the **organic-inorganic hybrid nanomaterials**, specifically conductive polymer hybrids for enhanced sensing and actuation.

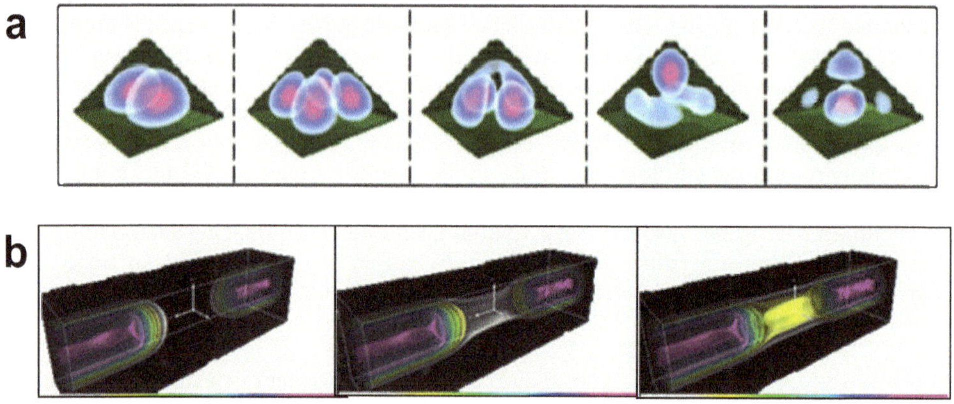

Fig. 4. Examples of using the NEMO 3-D tool. (**a**)Video frames of the simulation and analysis for a pyramid-shaped Qdot using Quantum Dot Lab; (**b**) Electron density in Si nanowire transistor (from Klimeck & Haley, 2009; Klimeck & Mehrotra 2011 reproduced with kind permission from the author). See text for more details.

## 3. Hybrid organic-inorganic material interfaces

Next, we will discuss aspects of organic-inorganic hybrid materials an important class of *functional* nanomaterials. Many of which are designed to be sensitive to the environment, so-called *responsive* or even *intelligent.* Interestingly, these nano-matrices bring a rich blend of complementary attributes, e.g., optical, photocatalytic, electrochemical, mechanical properties, that are tailored for specific applications in nanoelectronics, plasmonics, bio/catalysis, fuel cells, diagnostic imaging, etc. (Dong et al, 2005, 2006, 2009; Feng et al, 2008; Kickelbick, 2007; Gomez-Romero & Sanchez, 2004 ; Ruiz-Hitzky, Ariga & Lvov, 2007; Sanchez, et al, 2011; Vivero-Escoto & Huang, 2011; Willner, Willner & Katz, 2007).

### 3.1 Inorganic and organic molecular components

Hybrid (bio)organic–inorganic materials offer opportunities for both basic research and new exciting applications, *via* their inherent multifunctional properties. Today's design, tailoring of complex hybrid systems, is possible *via* cross-disciplinary, synergistically coupled approaches in biomolecular engineering, smart processing or nanofabrication.

The impressive set of inorganic-organic hybrid nanomaterials span a wide spectrum of properties, yielding innovative applications in areas such as plasmonics, nanoelectronics, health, energy, the environment among others (e.g., Ruiz-Hitzky, Ariga & Lvov, 2007; Knoll et al, 2004; Li et al, 2007; Nakamura, Katagiri & Koumoto, 2010; Peng et al, 2007, 2008; Stemmler et al, 2009; Vivero-Escoto & Huang, 2011).

Today the main hybrid materials that find applications in industry are based mostly on the association between metal oxides or metal–oxo polymers and organic molecules or macro-monomers of all kinds including bio-components. An exhaustive description of all the chemical reactions involved in the construction of organic and inorganic components is beyond of the scope of this work and the reader is referred to several reviews and books (Boissiere, et al., 2007; Cong & Yu, 2009 ; Dong, et al., 2008; Kickelbick, 2007; Romero & Sanchez, 2004 ; Mullen, et al., 2008; Ray, et al., 2005). A brief outline on the synthetic nanofabrication of the *inorganic* and the *organic* components follows.

### 3.1.1 The inorganic component

Typical inorganic partners in the hybrid include noble metal nano-objects (Coffinier et al, 2010; Ghodbane et al, 2010), magnetite ($Fe_3O_4$) nanoparticles, solid or mesoporous silica or quantum dots. The inorganic metal-oxo polymers or metal oxides (Hongqin et al, 2009) are typically produced as an amorphous or nanocrystalline network, or metal-oxo cluster *via* condensation of metal organic precursors or salts. The sol-gel polymerization can be driven through hydrolysis reactions (addition of water to reactive precursors such as alkoxides, or chemical or thermal modification of the pH of aqueous solutions containing metallic salts) to form reactive M-OH species that condense yielding metal-oxo oligomers and polymers assembled via M-O-M and/or M-OH-M bridges. Metal-oxo species can also be generated through thermal elimination of organic moieties. The elimination of ester, ether, and/or alkyl chloride are well known examples of using "sacrificial" routes to produce metal-oxides through thermally induced non hydrolytic sol-gel chemistry (Escribano, et al., 2008 ; Ohara, 2011 ; Pyun, et al., 2001 ; Shen & Shi, 2010 ; ten Elshof, et al., 2010 ; Yamada, 2009; Yao, Gao & Yu, 2010; Yuan & Muller. 2010).

### 3.1.2 The organic component

The organic part is usually a polymer, carbon nanotube, a biomolecule, etc. Organic components can be introduced into an inorganic network in two different ways, as network modifiers (molecules) or network formers (macromolecules). The most commonly used network modifiers or network formers are coupled to inorganic moieties through organo silicon alkoxides or chlorides. The introduction of organic network formers into an inorganic network to form hybrid materials can be performed *via* two main strategies. One method is by using already pre-synthe-sized functional macromonomers that are compatibilized with the inorganic component either via chemical grafting "class II" or through embedding with a growing inorganic network in a common solvent to form "class I" hybrid materials; The second approach is *in situ* generation through photo- or thermo- induced polyadditions in presence of a radical initiator, atomic transfer radical polymerization, chemically or electrochemically driven oxidative polymerisation (polypyrrole, polyaniline, polythiophene etc), or polycondensation (polyesters, polyimine, polyamides formophenolic) (Fabregat-Santiago et al, 2011 ; Hu & Shea, 2011; Ma, et al, 2007; Ohara, 2011 ; Vivero-Escoto & Huang, 2011; Weickert, et al., 2011).

## 3.2 Nanofabrication of organic-inorganic hybrid nanomaterials

In the same line of environmentally friendly, mild synthetic methods for hybrid nanocomposites, a hot research topic in material science is the development of bio-inspired hybrids. Synthetic strategies for functionalized inorganic-organic hybrids based on self-assembly strategies, whether in-situ, template-induced, evaporation-induced, layer-by-layer, etc were recently reviewed (Cong et al, 2009). The "soft chemistry" (*la chimie douce*) approach to synthesize various organic-inorganic nanomaterials was examined (Sanchez et al, 2011) from sol-gel-derived and nano-building blocks-based hybrids, to thin films of nano-structured porous materials and aerosols. The importance of characterizing the hybrid interface was highlighted, by using multiple modern techniques such as diffusion-ordered spectroscopy nuclear magnetic resonance or ellipsometry.

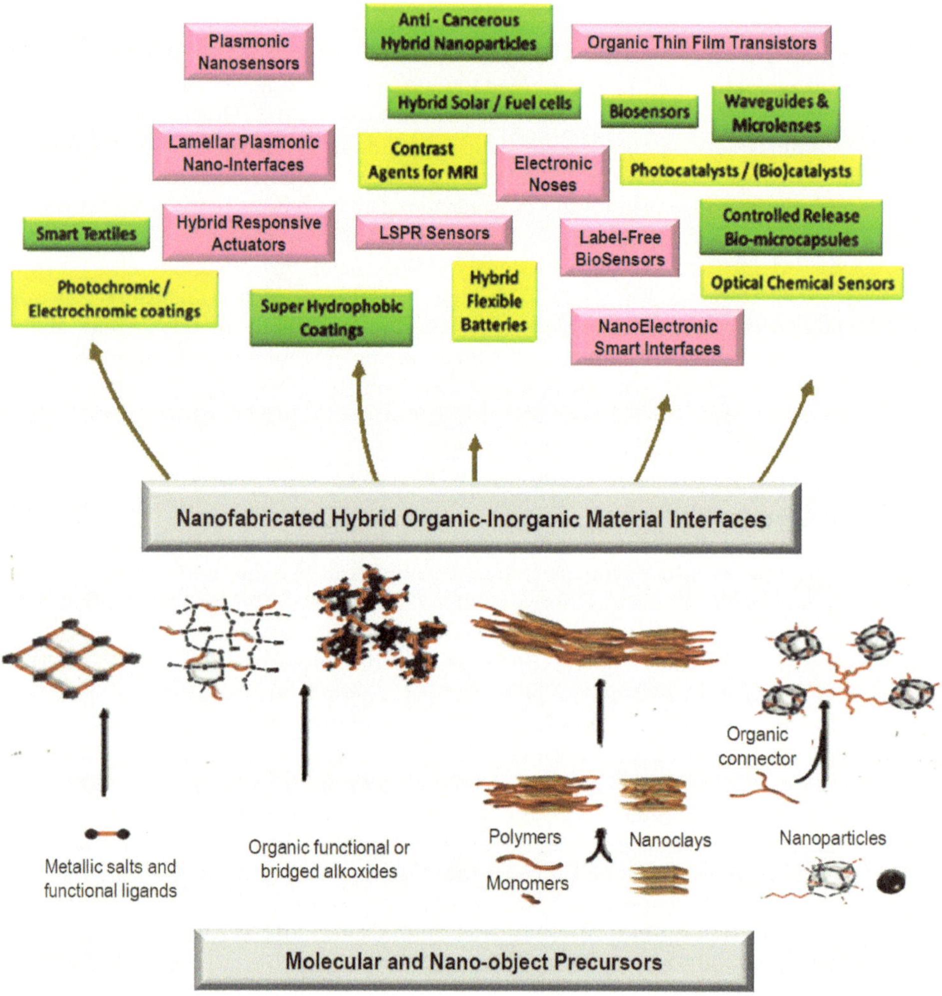


Fig. 5. The process of micro- and nano-fabricating of real life applications starts at the bottom, with the synthesis and self-assembly of molecular and nano- precursors into smart hybrids. Subsequently, these nanofabricated, hierarchically structured, hybrid bio/materials are employed to manufacture real devices by using of complex interactions, primarily from physics, chemistry and bioprocessing (reproduced in part from Sanchez, 2011 by kind permission of the Royal Society of Chemistry).

### 3.3 Hybrids based on nanoscale conductive polymers for enhanced sensing and actuation

The synergistic integration of similar-size biomolecules and nanomaterials resulted recently in novel cross-bred bio-nano-materials displaying an exquisitely unusual set of electronic, photonic, catalytic and recognitive properties and functionalities. More specifically, the electroactive nano-scale polymers have been integrated with receptors, enzymes, antibodies, whole cells, and or nucleic acids in their matrix, leading to uniquely advanced biosensors and bioelectronics.

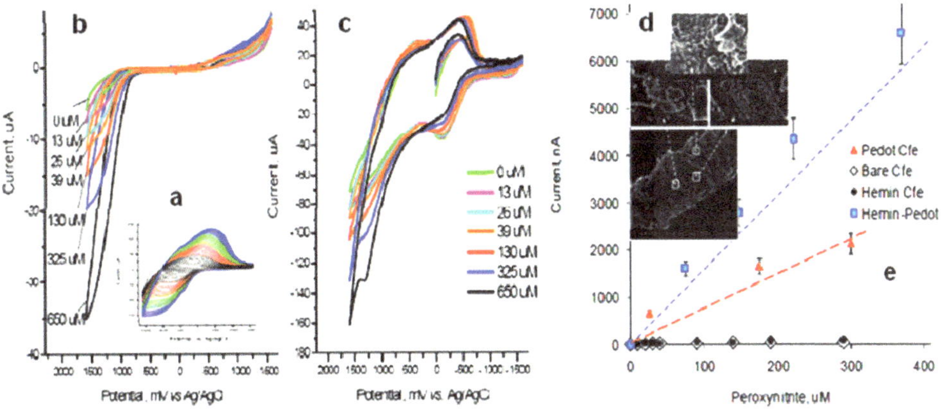

Fig. 6. The PEDOT-nanostructured interface significantly enhanced electrochemical sensing of a reactive nitrogen species as analyte. The panels represent: **(a).** inset showing the catalytic interface electro-deposition from solution; **(b).** Cyclic voltammogram for hemin-modified; **(c).** Same conditions, for hemin-PEDOT-modifed, after addition of different aliquots; **(d).** PEDOT nanostructure as evidenced by SEMs; €. Increased biosensor sensitivity after adding PEDOT to the catalytic interface (reproduced from Peteu et al, 2010 by kind permission of Elsevier).

Moreover, the advanced use of conducting polymers in sensors and actuators has grown over the past decade, also due to their compliance with nano-fabrication. As a result, their reversible, strong biomolecular interactions at nanoscale translated into (i) biomolecular sensing with lower detection limits and enhanced sensitivity and also into (ii) nano-actuating materials (Kulesza, et al, 2006; Lee, et al., 2008).

Thus, electroactive polymers based on polypyrrole or polyethylene dioxythiophene (PEDOT) were employed to prepare electrochemical or optochemical bioanalytical sensors, including at nano scale. These 'synthetic metals' exhibit high conductivity, mediate fast transfer of charge carriers and can be synthesized under mild conditions, through simple deposition onto conductive surfaces from monomer solutions with precise electrochemical control (Liao, Huang &Li. 2009 ; Muller, et al., 2007 ; Rahman, et al., 2008).

More specifically, by comparison with polypyrrole, PEDOT seems to show a better electrochemical stability, with a better conservation of its conductivity and charge. Also, PEDOT has a higher ionization potential, protecting better against the oxidative damage. These soft, synthetic metals were proven to enhance nano-sensing or deliver nano-actuation. Recently, one group reported a 50-100 times increase in sensitivity for peroxynitrite detection, as a result of adding the nano-structured polyethylenedioxythiophene (PEDOT) to a catalytic electrochemical detection matrix (Peteu, et al, 2010), as outlined in **Figure 6.**

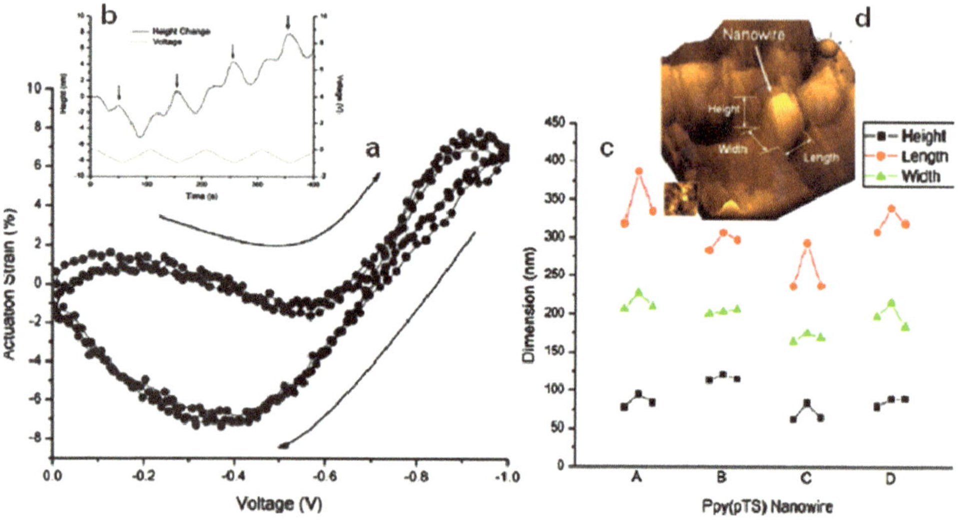


Fig. 7. The actuation of the 50 nm diameter polypyrrole nanowires. **(a)** Real-time actuation analysis for the 50 nm diameter polypyrrole-polytoluenesulphonate nanowire as the voltage is cycled 0 to −1 V. As expected, the nanowire shows cyclical increase and decrease behavior, with the greatest change in height at −1 V; **(b)** Actuation strain plot for 50 nm diameter PPy(pTS) nanowire confined in a polycarbonate membrane as the voltage is cycled from 0 to −1 V; **(c)** Three-dimensional AFM rendering showing a PPy(pTS) nanowire protruding from the surface of the polycarbonate membrane, its surface chemically etched to expose more of the embeded nanowires; **(d)** Analysis of four, 50 nm diameter, PPy(pTS) nanowires. The three points in each data set correspond to the initial oxidizedstate, reduction, then reoxidation of the nanowires (reproduced from Lee et al, 2008, with kind permission by Institute of Physics).

The inherently conductive nanoscale polymers molecules can designer-hybridize with inorganic nano-components leading to tailored, bottom-up nanofabricated, materials with unique properties. Amongst these are nano-scale or nano-structured actuators allowing the conversion of chemical, electrical or thermal energy/ sources into mechanical energy. These so called artificial muscles have found nano-scale applications including for nanorobots or responsive release of countermeasures (Liu, et al, 2010; Tamagawa, et al, 2011).
Additionally, nanoscale actuators are expected to become a major area of development within nanofabrication. They are essential components of the NEMS and nanorobots of the future, and are poised to become a major area of development within nanoscience and nanotechnology. As illustrated in **Figure 7**, our group has reported for the first time the actuation of individual 50 nm diameter polypyrrole nanowires, when triggered electrochemically in solution, by volume change as the result of its oxidation state (Lee et al, 2008).

In the following sub-chapter, we will examine the **plasmonics methods and structures**, with focus on lamellar plasmonic nano-interfaces for optical sensors.

## 4. Nanoplasmonic interfaces

The last decade was marked by exciting discoveries and advances in two young scientific fields that concern optically active nanostructures: *nanophotonics* and *plasmonics*. Nanophotonics represents the study of interactions between light and nanostructured matter, which occur at subwavelength ranges and are determined by the very specific physical, chemical and structural properties of nano- matter. The spotlight herein will be on plasmonics and its applications. However there are several reviews and books available with focus on nanophotonics for those interested. (Eustis, S. & El-Sayed, 2006; Stewart et al, 2008; Willets & Van Duyne, 2007)
Current research in *nanoplasmonics* is directed towards understading through the effects of nanostructures on surface plasmon resonance (SPR), finding new nanomaterials, such as periodically nanostructured thin films or metamaterials and nanoporous silicon layers for increased sensitivity and for supporting excitation using THz frequency range rather than the visible wavelength band. Future research will continue to be directed also towards hyphenation of plasmonic sensors with other techniques such as interferometry for better sensitivity or for obtaining complementary information (Kim, et al., 2008; Sannomiya & Voros, 2011).

### 4.1 Plasmonic metamaterials

The rapid development of synthetic nanofabrication for complex, nanoscale metal structures has led to the emergence of the field of plasmonics, exploring the local and far fields around small metal particles. Additionally, it is concerned with the use of particle morphology as a means to control and tune these fields and with the way energy is dissipated and transported through these structures. It is centered on optical properties of metallic nanostructures and their use for manipulating light at the nanoscale (Ahl et al, 2008; Galopin et al, 2009; Gitsas et al, 2010; Grosserueschkamp et al, 2009; Zhou et al, 2010) and some plasmonic nanomaterials are illutrated in **Figure 8**.
Plasmonic metamaterials, exhibiting simultaneously negative dielectric permittivity and magnetic permeability in a given frequency range, hold the promise for significantly increasing the field depth in imaging applications as well as for a higher sensitivity to changes in the bulk solution. (Mayer & Hafner, 2011; Stewart et al, 2008).

### 4.2 Nanoplasmonic sensitive interfaces

Another interesting recent achievement concerns the nanofabrication of a coherent nano-metallic light source, or surface plasmon amplification by stimulated emission of radiation, or SPASER, consisting in a 44-nm diameter nanoparticle with a Au core surrounded by a shell of dyed silica which acts as a gain medium possible applications including magnetic data-storage industry, nano- lithography, probing and microscopy (Bergman & Stockman, 2003; Noginov, et al., 2009). Some very exciting highlights of nanoplasmonic sensitive interfaces include the plasmonic nanoholes, the surface-enhanced Raman spectroscopy (SERS), the localized surface plasmon resonance, or the localised surface plasmon fluorescence (Galopin et al, 2009a; Tripp, Dluhy & Zhao, 2008). These and others are employed for detection or imaging and will be briefly discusses below.

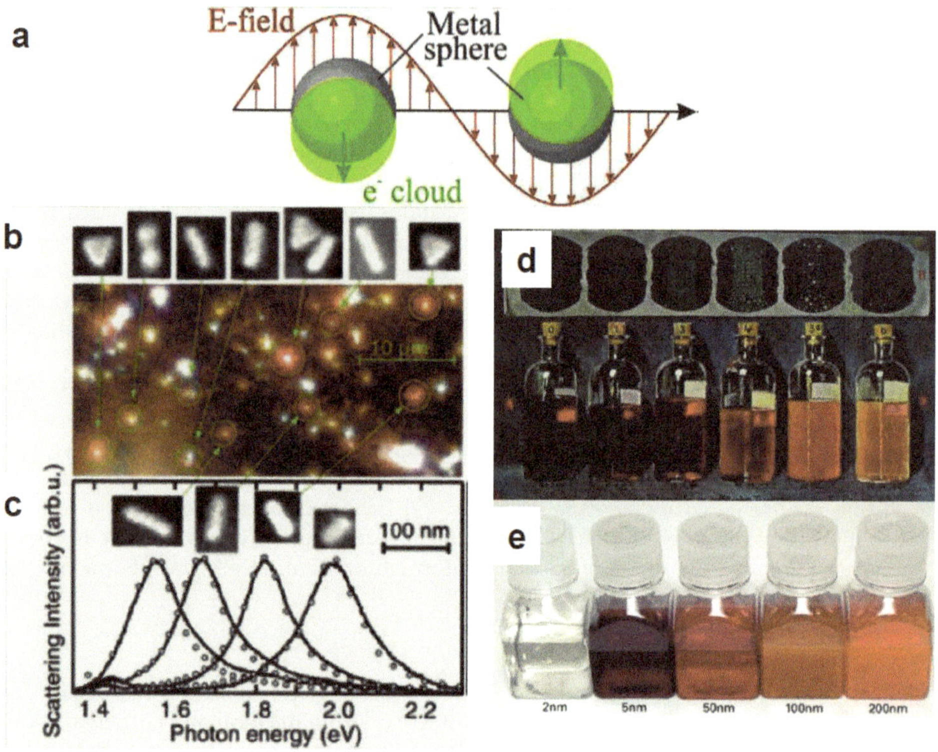


Fig. 8. Plasmonics nanomaterials. (**a**) Localized surface plasmon of a metal sphere showing the displacement of the electron charge cloud relative to the nuclei. (**b**) Dark-field microscopy image and corresponding SEM images; (**c**) light scattering spectra of Au nanocrystals of different shapes; (**d**) illustration of Dark field scattering of Au colloids of various sizes from year 1909); (**e**) actual photo of similar colloid solutions (reprinted from Kelly et al, 2003 with kind permission from American Chemical Society, from Kuwata, et al., 2003 with kind permission American Institute of Physics and from Mayer & Hafner, 2011 with kind permission from the American Chemical Society).

The discovery in 1998 of extraordinary optical transmission (EOT) in arrays of subwavelength nanoholes in a metallic film (Ebbesen, et al, 1998) led to important progress in instrumental design for SPR biosensing, and improvement of sensitivity of both detection and imaging (Lindquist, et al, 2009). Designs based on nanohole arrays served to monitor the binding of organic and biological molecules to the metallic surface or to the spectacular observation, with the naked eye of protein monolayer formation. In an interesting experimental setup, nanoholes served a dual purpose: as the optical active element for analyte detection and as nanochannels facilitating analyte mass transport to the active surface (Eftekhari et al , 2009).

New exciting possibilities for biodetection using plasmonic devices arise from the localized surface plasmon resonance (LSPR) phenomenon displayed by nanomaterials (such as metal nanoparticles, nanoshells,nanodisks, nanowires etc (Lu, et al, 2009; Praig et al, 2009; Zhou et al, 2008) as a result of interaction of light with particles much smaller than the incident wavelength. A result derived from LSPR studies is the advance in highly sensitive surface characterisation methods such as SERS, in particular very sensitive, and producing a specific signal, or fingerprint of certain molecules. For example, the *in situ* cell pH sensor was mapped *in situ* with SERS and gold nanoparticles functionalized with 4-mercapto-benzoic acid. One consequence of LSPR is the significantly higher resulting electric field, which offer opportunities to improve the sensitivity of the detection (McFarland, et al, 2003; Raschke, et al. 2003; Baciu, et al., 2008)

Although LSPR has a short penetration depth of the electric field (around 20 nm), the intensity of SPR signal is strongly dependent on the shape, size, material nature and architecture of nanoplasmonic devices, active research being currently focused on understanding and controlling this relationship. LSPR biosensors include label-free detection of the biomolecular interactions, such as antigen–antibody reactions, DNA–DNA or PNA–DNA hybridizations to monitor cell activity by measuring cell metabolites and sensitive detection small analytes (e.g glucose) or antibodies Besides LSPR, localized surface plasmon coupled fluorescence was also exploited to obtain highly sensitive biosensors (Endo, et al., 2006, 2008; Yamamichi, et al., 2011; Zhou et al, 2009).

### 4.3 Lamellar plasmonic nanointerfaces for optical sensing

Plasmonics applications include high performance near-field optical microscopy (NSOM), high resolution imaging, targeted drug delivery, biosensors, catalysis, solar cells. Also, it is a preferred approach to attain exquisite precision in controlling optical processes. Recent investigations encompass a wide range of examples of interactions between light and nanostructured matter, from guiding the light through metal nanowires below the diffraction limit to optical lenses formed by a thin film of metal, components of metamaterials (artificial materials from nanoscale building blocks), or lamellar plasmonic nanointerfaces for optical sensors (e.g., Galopin, et al., 2010 ; Niedziolka-Jonsson, et al, 2010; Szunerits, et al., 2008a; 2008b; 2010; Touahir et al., 2011)

In particular, the nanostructured noble metals exhibit an intense optical near field due to surface plasmon resonance (SPR), therefore promising widespread applications and being of interest to a broad spectrum of scientists, ranging from physicists, chemists, and materials scientists to biologists. A versatile, highly-sensitive detection of DNA hybridization is described using metal nanostructures-enhanced fluorescence (MEF) emission intensity when fluorescently-labeled DNA oligomers are covalently immobilized on a nanometer-thin amorphous silicon–carbon layer capping the metal nanostructures (Touahir, et al, 2010).

The MEF structures are formed by thermal deposition of silver, gold or silver/gold thin films on glass surfaces and post-annealing at 500 °C. The choice of the metal film allows for tuning the optical properties of the interface. The metallic nanostructures are subsequently coated with an amorphous thin silicon–carbon alloy (a-Si0.80C0.20: H) layer deposited by PECVD. Carboxydecyl groups are attached on these surfaces through hydrosilylation then reacted with amine-terminated single-stranded DNA oligomers, forming a covalent link. The immobilized DNA is hybridized with its complementary strand carrying a fluorescent label. Through optimization of the thickness of the a-Si0.80C0.20: H alloy overlayer and by working close to resonance conditions for plasmon and fluorophore excitation, the hybridization of very dilute oligomers (5 fM) is easily detected, and the hybridization kinetics can be monitored in situ and in real-time as illustrated in **Figure 9**.

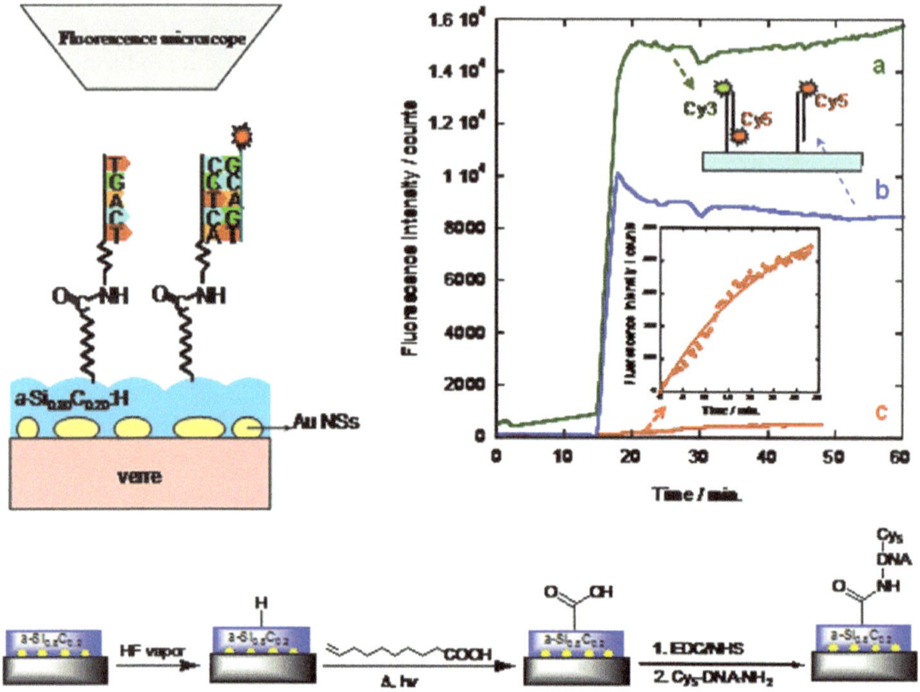


Fig. 9. Lamellar plasmonic sensor (**A**) Schematic of the LSPR-metal enhanced fluorescence structure and set up (**B**) Surface functionalizing reaction scheme. (**C**) Fluorescence intensity of Cy5-ON targets hybridized with immobilized probes (**a**) and probes (**b**). Curve (**c**), enlarged in the inset, correspond to the hybridization of Cy5-ON' targets with immobilized probes. The sensor is obtained from structure 4 coated with 5 nm *a*-$Si_{0.8}C_{0.2}$: H. An exponential fit of curve (c) is shown in the inset (reproduced from Touahir, et al, 2010, with kind permission from Elsevier).

Coating metal nanostructures with a nanometric thin film of amorphous silicon-carbon alloy allows for designing an efficient sensor, which exhibits a good stability in conditions typical of biological assays and with efficient covalent attachment of the biological probes. Such a sensor exhibits a high sensitivity, allowing for the detection of trace amounts of DNA and the investigation of hybridization kinetics in situ and in real-time, in a spatially-resolved, classical geometry of epi-fluorescence.

Such investigations do therefore not require the setting up of custom made, highly accurate instrumentation, which should make this type of measurements easier and help in determining the respective roles of the various factors coming into play for determining hybridization kinetics at solid surfaces. Another attractive characteristic of these substrates is the capability of recording images of labeled fluorescent probes with a high sensitivity, while simultaneously monitoring the interaction of the probes with non-labeled species present in the assay through LSPR imaging. This capability should help in analyzing phenomena like competitive interactions which come into play in realistic diagnostic assays (Barka, et al., 2011; Touahir, et al, 2010; Khor et al, 2007).

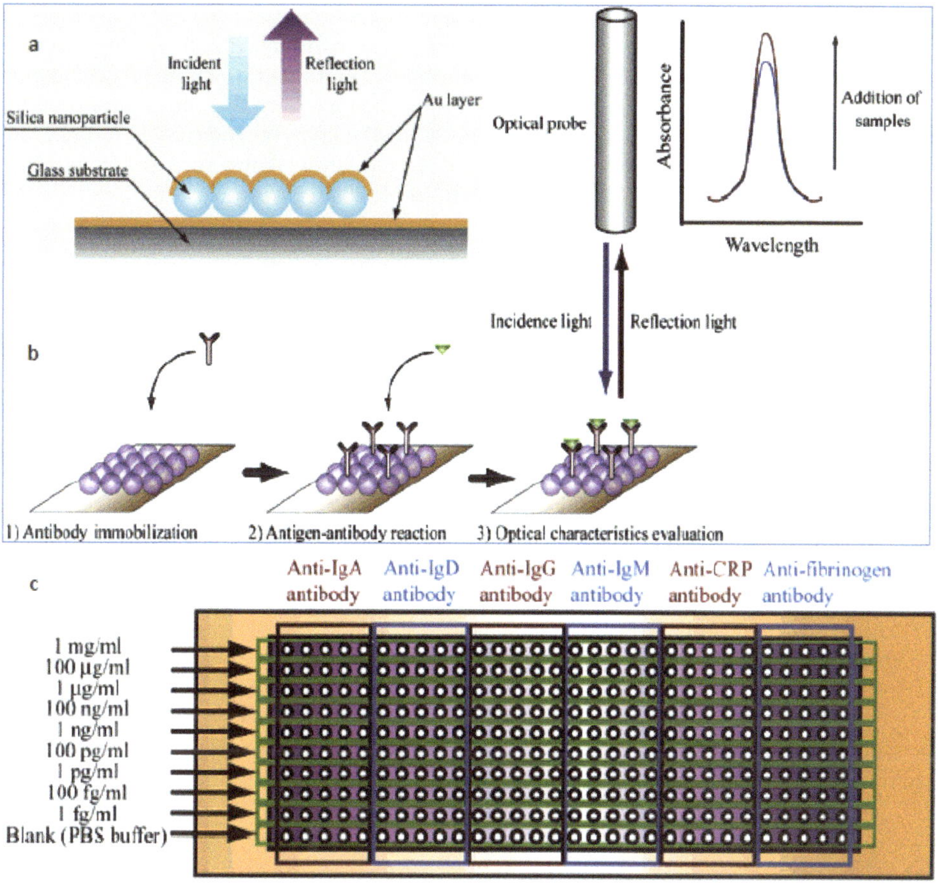


Fig. 10. LSPR-based, core-shell nanoparticle layer device for label-free multi-detection of antigen-antibody reaction. (a) Schematic of the LSPR-based nanochip. The surface modified silica nanoparticles were aligned onto Au deposited glass substrate, with the Au film subsequently deposited onto the silica nanoparticle layer. (b) Experimental setup of the multi-array chip. (c) Chip design illustrating antibody immobilized spots and antigen concentrations. Six kinds of antibodies and antigens were spotted onto the multi=array LSPR-based nanochip surface (reproduced from Endo et al, 2006, with kind permission from the American Chemical Society).

The past decade has witnessed an exciting research effort directed toward the development of hybrid plasmonic interfaces, driven mainly by convincing applications in SPR and LSPR sensors and enabled by rapid progress in nanotechnology and nanoscale science, which in turn facilitates their successful synthesis and characterization.
In the following sub-chapter, we will examine the nanoelectronic interfaces , with focus on organic thin film transistors with molecular scale label free detection.
Before LSPR sensors can become high-throughput laboratory and clinical screening tools to compete with ELISA and other common assays, the technique must be parallelized to

handle large numbers of samples. **Figure 10** illustrates an exciting display of the potential of this technology. An LSPR-based multi-array chip was nano-fabricated by depositing an array of 300 spots of nanoliter volumes of antibody solutions on a film of gold capped nanospheres. A scanning optical probe was employed to measure the optical absorbance of the film, to compare the affinity of each antibody for the analyte. The change in integrated absorbance was measured, rather than an LSPR peak shift. Antigens were detected for concentrations in the region of 100 pg/mL (Endo, et al., 2006).

## 5. Nanoelectronic Interfaces

### 5.1 Nanofabrication and materials for nanoelectronics

*Nanoelectronic* devices are based on structures whose material properties vary on an atomic length scale. Such structures can be created with a variety of experimental methods in a variety of different material systems. Interfaces between different materials can be atomically abrupt. These abrupt interfaces enable device designers to confine electrons quantum mechanically. Interactions/transitions between different man-made quantum mechanical states open a cornucopia of new device applications.

Meanwhile, a model of an organic semiconductor nano-device is illustrated in **Figure 11**. This model from fig. **11a** shows the ultimate miniaturization of nanoscale logic circuits. An electron, entering at the lower left electrode, can be directed with varying degrees of probability, to one of the three output electrodes in the upper part of the picture. An aromatic molecule (naphthalene) bonded to four gold electrodes (green) by sulfur atoms (blue) and polyacetylene wires. An insulator on the surface prevents cross-talk between the electrodes. Parts of the molecule and electrodes are drawn in brighter (darker) colors, to suggest an active (inactive) state during a particular read-out. The X, Y, Z symbols represent three possible logical inputs, which can be operated on in various ways to produce a quasi-digital logical output (Stadler et al, 2003).

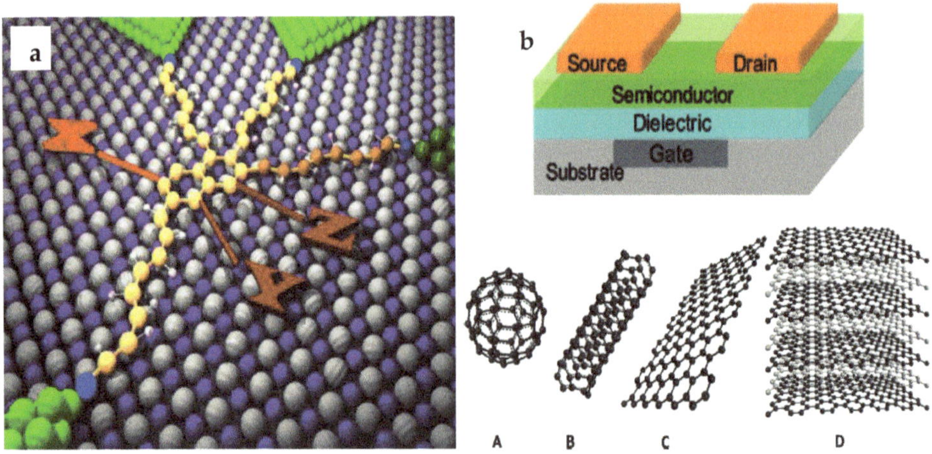


Fig. 11. Organic semiconductor nano-devices. (**a**) Model showing, arguably, the ultimate miniaturization of nanoscale logic circuits (reproduced from Stadler et al, 2003 with kind permission from Institute of Physics); (**b**) A typical organic thin-film transistor (OFET); (**A-D**) The characteristic forms of carbon, and their dimension. See text for more details.

A typical organic thin-film transistor (OFET) is a field-effect transistors (FET) with organic semiconductors in its channel and is made by depositing thin films. The FET has three essential components, source (S), drain (D) and gate (G), as illustrated in Figure **11b**. The gate controls the carrier movement from the source to the drain. Moreover, current OFETs are investigating the use of nano-structured materials such as carbon, to enhance their response. The characteristic forms of carbon, and their dimension as shown in figure 11 are: fullerenes (0D), nanotubes (1D), graphene (2D), graphite (3D).

### 5.2 Organic thin film transistors as biosensors with molecular scale label free detection

Rapid and highly sensitive PNA-DNA hybridization assays have attracted enormous attention for a wide variety of applications ranging from genotyping to molecular diagnosis (Huang et al, 2001; Ito et al, 2007; Nakamura et al, 2006; Kelly et al, 2003). Conventional optical detection systems based on microarrays and real-time PCR involve expensive detection protocols, typically requiring a fluorescent dye and optical sources/detectors; however, this method has become the standard technique for quantifying the extent of hybridization between surface immobilized probes and fluorophore-labeled DNA targets (Yameen et al, 2010; Yin et al, 2011).

The flowchart of fabrication of an organic thin film transistor (OTFT) is illustrated in **Figure 12**. This particular OTFT has been nanofabricated as follows (Khan et al, 2011a). On either n++ silicon or polyimide substrates with a thin Al film, a 25 nm PVP-HDA dielectric layer was spin-coated, followed thermal evaporation of a 25 nm pentacene active layer and a 35 nm CuPc passivation layer. Source-drain (S-D) electrodes with a width (W) of 1000 μm and length (L) of 50 μm were deposited through a shadow mask. The flow cell was laminated on top of OTFTs for operation in buffer media and biosensing demonstrations.

On a similar device (Khan et al, 20111b), charge discrimination experiments were performed for bovine serum albumin (BSA)/anti BSA immunoassay formation in aqueous buffer solutions at different pHs using OTFT and SPR sensors. The solid arrows indicate an injection of anti BSA and open arrows indicate exchange with pure buffer solution. The OTFT current response (IDS/IDS-baseline) with time upon exposure to anti BSA (500 nM) diluted in buffer solution at pH 7 (black curve) and at pH 5 (blue curve) while operating at a constant bias (VG = -5 V, VDS =-2 V), as sown in figure **12a**. The SPR response, as a ratio between the minimum resonance angle shift (RAS) and the RAS baseline with time, upon exposure to anti BSA (500 nM) diluted in buffer solution at pH 7 (black curve) and at pH 5 (blue curve) using SPR sensing platform, is illustrated in figure **12b**.

Recent advances in chemical detection research, in part benefiting from the overwhelming progress made in organic electronics, have shown great promise for a viable, low-cost alternative to current optical detection systems. The utilization of organic transistor technology in chemical sensors is particularly encouraging. This simple platform allows for the fabrication of low-cost, large-area, and flexible devices with air stability, low-power consumption, biocompatibility, and facile surface modification for the detection of a wide range of analyte species (Knoll, et al., 2011a; Khan et al, 2010; Roberts, et al., 2008).

Many examples exist for the detection of analyte vapors using similar organic thin film transistor (OTFT) platforms, with numerous reports addressing the ability to identify particular analytes either through the use of a fingerprint response or by incorporating selective detection layers on functional OTFTs. Few examples of chemical detection in aqueous systems have been demonstrated. However, these devices were not selective toward a particular analyte (Knoll, et al., 2011b; Khan, et al., 2011; Roberts, et al., 2008, 2009). A selective, *in* situ detection with OTFTs requires a versatile method for the immobilization of various selective molecular probes within proximity to the active transport channel.

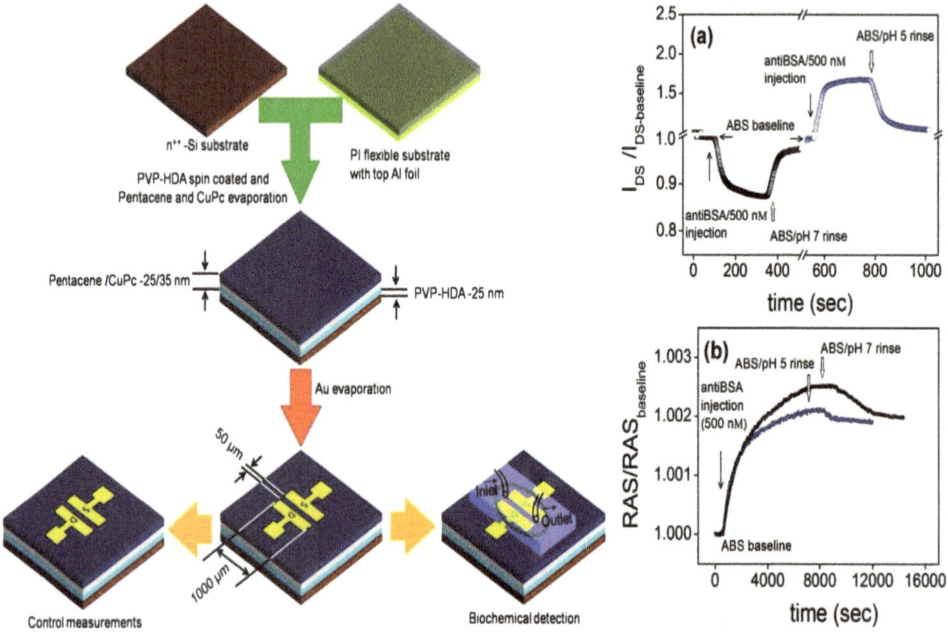


Fig. 12. OTFT sensor fabrication process. Charge discrimination experiments for BSA/antiBSA immunoassay formation in aqueous buffer solutions at different pHs using OTFT and SPR sensors, the solid arrows indicating an injection of antiBSA and open arrows indicate exchange with pure buffer solution, with (**a**) the OTFT current response and (**b**) the minimum resonance angle shift (RAS) response from the SPR sensing platform (reprinted from Khan et al, 2011a and from Khan et al, 2011b with kind permission from the American Chemical Society). See text for more details.

A flowchart of an experimental setup of a carbon nanotube (CNT)-based OTFT sensor and the qualitative, expected detection results are illustrated in **Figure 13**. The use of a target concentration (e.g., 100 nM) of target DNA complement provides excellent discrimination against single/double base mismatches. A Langmuir model can be used to fit the kinetic measurements on the OTFTs sensors based on the PNA-DNA hybridization, which showed that a high affinity constant on the order of $K_A = 4 \times 10^8\ M^{-1}$ can be achieved in these organic transistor-based detection systems. These OTFT sensors would benefit from the transduction of the DNA binding to an easily read electronic signal.

### 5.3 Other nanoelectronic devices

Improvement in fabrication will stem from combining bottom-up with "top-down" and self-assembly strategies. New materials such as graphene nanoribbons, organic-inorganic hybrids, inorganic nanowires need to be integrated in higher-architecture devices. On the other hand, integration of nanostructures in complex systems including micro-fluidics or on Si wafers, etc are extremely important. To this end, the properties at Si-biomolecule interface should be exploited. The nanoelectronics-enabled high computational capacity seems to be critical for future high-throughput screening in clinical medicine or and pharmaceutics. The future nanoelectronic devices should lead to increased functionality and information storage and improve abilities to power and process information from complex in-vivo sensors, drug delivery devices, biomolecules labeling, and protein studies (Yamamoto, et al., 2009).

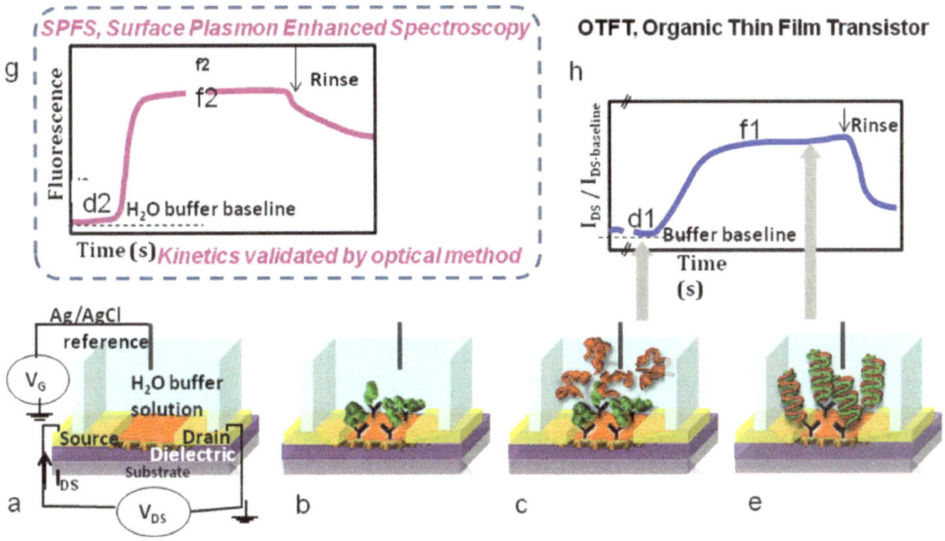


Fig. 13. A model of a carbon nanotube-based, organic thin film transistor sensor. (**a**-**e**) The flowchart schematic of the experimental setup; (**g**-**h**) The expected qualitative detection results for this model.

Several nanoelectronics applications are in the areas of lasers, detectors and/or memory. For *lasers*, the electrons in higher energetic states can emit photons by a transition to lower energetic states. Electrons in *detectors* can be excited by a photon to jump from a lower to a higher energetic state. Furthermore, in the case of *logic memories*, the electrons can transition without loss of energy from one quantum state to the next by quantum mechanical tunneling. The speed of such electronic transport can be extremely high, leading to new low power and high density devices.

The continuing quest for better, faster, inexpensive, smaller devices moved into the nano realm about a decade ago, with spectacular effects, especially visible in the semiconductor industry. This brought along new challenges in *nano-fabrication* including *nano-patterning* methods and other approaches such as *scanning probe lithography*, *nanoimprint lithography* and *self-assembly* show promise for precise manufacturing of nanoelectronic devices.

## 6. Looking ahead, advances and challenges

This chapter has reviewed recent developments in nanoscale architectures for smart bio-interfaces, with focus on four main areas: (i)the modeling and simulation to enable nanofabrication by-design; (ii)the organic-inorganic, hybrid, conductive polymer-based nanomaterials for enhanced sensing, actuation; (iii)the lamellar plasmonic nano-interfaces for optical sensors; (iv)the controlled molecular functional architectures for thin film transistors. The specific nanofabrication methods and results were emphasized in each case and a number of references were selected.

The National Nanotechnology Initiative (NNI) was created in the United States in the year 2000 (Roco, Williams & Alivisatos, 1999). A recent peered-reviewed study, so-called *Nano2 Report,* was prompted by a decade of NNI mandated-efforts (www.wtec.org/nano2). This report "Nanotechnology Research Directions for Societal Needs in 2020"(Roco, Mirkin & Hersam, 2010) aptly highlights the main advances since the year 2000, the fundamental goals by 2020 and the expected challenges to be overcome. Interestingly, the research directions reviewed include "theory, modeling & simulation", "nanoelectronics & nanomagnetics", "nanophotonics & plasmonics", with brief conclusions in an easy to follow, table format. Thus, we would refer the reader to this report, due to text space constraints.

In terms of dealing with risk and benefits of nanofabrication and nanotechnology, there are two extreme views, with everything else in between. While the "optimistic" side emphasizes mostly the colossal potential benefits, the "pessimists" are accentuating the major risks involved, this occasionally heating up as a public policy debate.

In fact that, today, nanofabrication-enabled products are marketed to an estimated *254 thousand million* US dollars (USD) worldwide, with *91 thousand million* USD in the United States alone. These global trends foretell a doubling, every three years, of the number of nanofabricated products and dedicated workers. It seems that the nanotech-fueled optimism of Ratner & Ratner is carried along: experts are now projecting for 2020 a 3 trillion USD market with 6 million workers involved. We would all wish to get some of that, would we not? Perhaps this optimistic view of nanofabrication and nanotechnology would be the best way to conclude.

## 7. Acknowledgements

SFP thanks the Department of Chemical Engineering & Materials Science, Michigan State University for logistical support. SS acknowledges the Agence Nationale de la Recherche (JCJC 2006, Nr. 0149) and the Centre National de la Recherche Scientifique (CNRS) for financial support. AV acknowledges financial support from Marie Curie International Reintegration Grant PIRG08-GA-2010-277126. WK thanks the European Science Foundation (ESF) for project 10-EuroBioSAS-FP-005 awarded to the Austrian Institute of Technology as part of a collaborative research grant.

## 8. References

Ahl, S., Cameron, P. J., Liu, J., Knoll, W., Erlebacher, J., Yu, F. (2008). A comparative plasmonic study of nanoporous and evaporated gold films. *Plasmonics* Vol. 3, No. 1, pp. 13-20. ISSN 1557-1955

Aldaye. A., Sleiman, H.F. (2009). Supramolecular DNA nanotechnology, *Pure & Applied Chemistry*. Vol.81, No.12, pp. 2157-2181. ISSN: 0033-4545

Ali, M., B. Yameen, Neumann, R., Ensinger, W., Knoll, W., Azzaroni, O. (2008). Biosensing and Supramolecular Bioconjugation in Single Conical Polymer Nanochannels. Facile Incorporation of Biorecognition Elements into Nanoconfined Geometries. *Journal of the American Chemical Society* Vol.130, No.48, pp. 16351-16357.

Ali, M., Yameen, B., Cervera, J., Ramírez, P., Neumann, R., Ensinger, W., Knoll, W., Azzaroni, O. (2010). Layer-by-Layer Assembly of Polyelectrolytes into Ionic Current Rectifying Solid-State Nanopores: Insights from Theory and Experiment. *Journal of the American Chemical Society* Vol. 132, No. 24, pp. 8338-8348. ISSN 0002-7863

Alivisatos, P. 2004. The use of nanocrystals in biological detection. *Nature Biotechnology* 22 (1):47-52.

Amatore C; Pebay C; Servant L; Sojic, N., Szunerits, S., Thouin, L. (2006). Mapping electrochemiluminescence asgenerated at double-band microelectrodes by confocal microscopy under steady state. Chemphyschem.,Vol. 7, No. 6,pp. 1322-1327 ISSN : 1439-4235

Avramescu, A., Andreescu, S., Noguer, T., Bala, C., Andreescu, D., Marty, J-L.(2002). Biosensors designed for environmental and food quality control based on screen-printed graphite electrodes with different configurations. Analytical and Bioanalytical Chemistry Vol. 374, No. 1, pp. 25-32. ISSN: 1618-2642

Barnard, A.S. (2010). Modelling of nanoparticles: approaches to morphology and evolution. *Rep. Prog. Phys.* Vol.73, No. 8, pp. 6502. ISSN : 0034-4885

Barras A., Szunerits, S., Marcon, L., Monfilliette-Dupont, N., Boukherroub, R. (2010). Functionalization of Diamond Nanoparticles Using "Click" Chemistry. *Langmuir* Vol. 26, No.16, pp. 13168-13172. ISSN 0743-7463

Battelle Memorial Institute (2007). Productive Nanosystems. A Technology Roadmap. copyright© 2007 Battelle Memorial Institute and Foresight Nanotech Institute, available at <www.productivenanosystems.com/docs>

Bergman, D. J. , Stockman, M. I..(2003) Surface plasmon amplification by stimulated emission ofradiation: Quantum generation of coherent surface plasmons in Nanosystems. *Phys. Rev. Letters* Vol. 90, p. 027402-1–027402-4

Bhushan, B. (editor) (2004). Handbook of Nanotechnology, 2nd Edition, Springer, Berlin Heidelberg New York, ISBN-10:3-540-29855-X.

Bitziou, E., Rudd, N.C.,Szunerits, S., Unwin, Patrick R. (2010). Microelectrode systems for the study of photochemical processes in solution Journal Of Electroanalytical Chemistry. Vol. 646, No. 1, pp. 60-67 ISSN : 1572-6657

Boissiere, M, Allouche, J., Chanéac, C., Brayner, R., Devoisselle, J.-M., Livage, J., Coradin T. (2007). Potentialities of silica/alginate nanoparticles as Hybrid Magnetic Carriers, *International Journal of Pharmaceutics*, Vol. 344, No. 1-2, pp. 128-134, ISSN 0378-5173.

Briseno, A. L.; Mannsfeld, S. C. B.; Reese, C.; Hancock, J. M.; Xiong, Y.; Jenekhe, S. A.; Bao, Z.; Xia, Y.(2007) *Nano Lett.* Vol. 7, No. 325, pp. 2847–2853.

Bruck, R., Melnik, E., Muellner, P., Hainberger, R., Lämmerhofer, M. (2011).Integrated polymer-based Mach-Zehnder interferometer label-free streptavidinbiosensor compatible with injection molding. *Biosensors and Bioelectronics* Vol. 26,pp. 3832–3837. ISSN : 0956-5663.

Burello, E.; Worth, A.P. (2011). QSAR modeling of nanomaterials. Wiley Interdisciplinary Reviews-Nanomedicine And Nanobiotechnology,Vol.3, No. 3, pp. 298-306. ISSN: 1939-5116

Cameron, P. J., Zhong, X. H. , Knoll, W.. (2009). Electrochemically Controlled Surface Plasmon Enhanced Fluorescence Response of Surface Immobilized CdZnSe Quantum Dots. *Journal of Physical Chemistry* Vol.113, No.115, pp.:6003-6008.

Campo, L. F., Rodembusch, F.S., Lerouge, F., Alauzun, J., Cerveau, G., Corriu, R.J.P. (2008). Silica-based organic-inorganic hybrid materials prepared from chiral precursors, *Comptes Rendus Chimie*, Vol.11, No. 10, pp. 1271-1276, ISSN 1631-0748,

*Chemical Vision 2020* partnership (n.d.) (2011), available from <www.chemicalvision 2020.org>

Cheng, Z. Y., S. H. Liu, P. W. Beines, N. Ding, P. Jakubowicz, and W. Knoll. (2008). Rapid and Highly Efficient Preparation of Water-Soluble Luminescent Quantum Dots via Encapsulation by Thermo- and Redox-Responsive Hydrogels. *Chemistry of Materials* Vol. 20, No.23, pp.7215-7219.

Chi, Y. S., Lee, J. K., Lee, K. B., Kim, D. J., Choi, I. S. (2005). Biosurface organic chemistry: Interfacial chemical reactions for applications to nanobiotechnology and biomedical sciences. *Bulletin of the Korean Chemical Society* Vol. 26, No. 3., pp. 361-370. ISSN 0253-2964

Coffinier Y., Galopin, E., Szunerits, S., Boukherroub, R. (2010). Preparation of superhydrophobic and oleophobic diamond nanograss array. *Journal of Materials Chemistry* Vol. 20, No. 47, pp. 10671-10675. ISSN : 1364-5501

Cong, H-P., Yu, S.-H. (2009). Self-assembly of functionalized inorganic-organic hybrids, *Current Opinion in Colloid & Interface Science,* Vol.14, No. 2, pp. 71-80, ISSN 1359-0294

Cook-Chennault, K. A., Thambi, N., Sastry, A. M. . (2008) Powering MEMS portable devices - a review of non-regenerative and regenerative power supply systems with special emphasis on piezoelectric energy harvesting systems *Smart Materials & Structures,* Vol. 17, No. 4. ISSN : 0964-1726

Cosnier, S., Szunerits, S., Marks, R.S., Novoa A., Puech, L., Perez, E., Rico-Lattes, I.(2000) A rapid and easy procedure of biosensor fabrication by microencapsulation of enzyme in hydrophilic synthetic latex films. Application to the amperometric determination of glucose. *Electrochemistry Communications,* Vol.2, No. 12, pp. 851-855 ISSN : 1388-2481

Cui, Z. (2008). *Nanofabrication,* Springer Science Business Media, ISBN 978-0-387-75577-9, New York.

Deo, S., Moschou, E., Peteu, S.(2003). Responsive drug delivery systems. *Analytical Chemistry* Vol. 75, No.9, pp 207A-213A. ISSN : 0003-2700

Dong, L.J., Liu, S.H., Gao, H.T., Ding, N.,Tremel, W., Xiong, C.X., Zhu, Q.M., ,Knoll W. (2009). Self-Assembled FeCo/Gelatin Nanospheres with Rapid Magnetic Response and High Biomolecule-Loading Capacity. *Small* Vol.5, No. 10, pp. 1153-1157. ISSN : 1613-6810

Dong, Q., Huilan, S., Wei, C., Jie, H., Di, Z., Qixin, G. (2008). Biogenic synthesis of hierarchical hybrid nanocomposites and patterning of silver nanoparticles, *Materials Chemistry and Physics*, Vol. 110, No. 1, pp. 160-165, ISSN 0254-0584,

Ebbesen, T.W., Lezec, H.J., Ghaemi, H.F., Thio, T., Wolff, P.A., (1998). Extraordinary optical transmission through sub-wavelength hole arrays *Nature* Vol. 391, No. 6668 , pp. 667–669. ISSN : 0028-0836

Eftekhari, F., Escobedo, C., Ferreira, J., Duan, X., Girotto, E.M., Brolo, A.G., Gordon, R.,Sinton, D. ( 2009). Nanoholes As Nanochannels: Flow-through Plasmonic Sensing. *Analytical Chemistry* Vol. 81, No. 11, pp. 4308–4311. ISSN : 0003-2700

Endo, T, Ikeda, R., Yanagida, Y., Hatsuzawa, T. (2008). Stimuli-responsive hydrogel-silver nanoparticles composite for development of localized surface plasmon resonance-based optical biosensor, *Analytica Chimica Acta*, Vol. 611, No. 2, pp. 205-211, ISSN 0003-2670.

Endo, T, Yamamura., S., Kerman, K., Tamiya, E. (2008). Label-free cell-based assay using localized surface plasmon resonance biosensor, *Analytica Chimica Acta*, Vol. 614, No.2, pp. 182-189, ISSN 0003-2670.

Endo, T., Kerman, K., Nagatani, N., Hiep, H.M., Kim, D.-K., Yonezawa, Y., Nakano, K. , Tamiya, E. (2006). Multiple label-free detection of antigen-antibody reaction using localized surface plasmon resonance-based core-shell structured nanoparticle layer nanochip. Anal. Chem. Vol. 78, No. 18, pp. 6465-6475. ISSN : 0003-2700

Escribano, et al., 2008. Photonic and nanobiophotonic properties of luminescent lanthanide-doped hybrid organic- inorganic materials. *Journal of Materials Chemistry* 18 (1):23-40.

Eustis, S., El-Sayed, M. A. (2006) Why gold nanoparticles are more precious than pretty gold: Noble metal surface plasmon resonance and its enhancement of the radiative and nonradiative properties of nanocrystals of different shapes. *Chem. Soc. Rev.* Vol.35, No.3. pp. 209–217. ISSN : 0306-0012

Fabregat-Santiago, F., Garcia-Belmonte, G. , Mora-Sero, I. Bisquert, J. (2011). Characterization of nanostructured hybrid and organic solar cells by impedance spectroscopy. *Physical Chemistry Chemical Physics* Vol.13, No.20, pp.9083-9118.

Feng, C. L., Zhong, X. H.., Steinhart, M., Caminade, A.M., Majoral, J.P., Knoll, W . (2007). Graded-bandgap quantum-dot-modified nanotubes: A sensitive biosensor for enhanced detection of DNA hybridization. *Advanced Materials* Vol. 19 No.15, pp.1933-1936. ISSN 0935-9648

Feng, C. L., Zhong, X. H.., Steinhart, M., Caminade, A.M., Majoral, J.P., Knoll, W. (2008). Functional quantum-dot/dendrimer nanotubes for sensitive detection of DNA hybridization. *Small* , Vol. 4, No. 5, pp. 566-571. ISSN 1613-6829

Freebody, M. (2011). Preserving Moore's Law Pushes Lithography to its Limits. *Photonics Spectra* Vol. 45, No. 5, pp.45-47.

Galopin, E., Barbillat, J., Coffinier, Y., Szunerits, S., Patriarche, G., Boukherroub, R. (2009a). Silicon nanowires coated with silver nanostructures as ultrasensitive interfaces for surface-enhanced Raman spectroscopy. *ACS Applied Materials & Interfaces* , Vol.1, No.7, pp.1396-1403. ISSN: 1944-8244

Galopin, E., Niedziolka-Jönsson, J., Akjouj, A., Pennec, Y., Djafari-Rouhani,B., Noual, A., Boukherroub,R., Szunerits, S. (2010). Sensitivity of Plasmonic Nanostructures Coated with Thin Oxide Films for Refractive Index Sensing: Experimental and Theoretical Investigations. *Journal of Physical Chemistry C* Vol. 114, No. 27, pp. 11769-11775. ISSN : 1932-7447

Galopin, E.., Noual, A., Niedziolka-Jönsson, J., Jönsson-Niedziolka, M., Akjouj, A., Pennec, Y., Djafari-Rouhani, B., Boukherroub, R., Szunerits, S. (2009b). Short- and Long-Range Sensing Using Plasmonic Nanostrucures: Experimental and Theoretical Studies. *Journal of Physical Chemistry C* Vol.113, No.36, pp. 15921-15927. ISSN : 1932-7447

Ghodbane, S., Haensel, T., Coffinier, Y., Szunerits, S., Steinmüller-Nethl, D., Boukherroub, R., Ahmed Syed I.-U., Schaefer, J. A. (2010). HREELS Investigation of the Surfaces of Nanocrystalline Diamond Films Oxidized by Different Processes. *Langmuir* Vol. 26, No.24, pp. 18798-18805. ISSN : 0743-7463

Ghosh, P.,Han, G., De, M., Kim, CK., Rotello, V.M .(2008). Gold nanoparticles in delivery applications. *Adv. Drug Deliv. Rev.* Vol.60,No.11, pp.1307–1315. ISSN : 0169-409X

Gitsas, A., Yameen, B., Lazzara, T. D., Steinhart, M., Duran, H., Knoll, W. (2010). Polycyanurate Nanorod Arrays for Optical-Waveguide-Based Biosensing. *Nano Letters* Vol.10, No 6, pp. 2173-2177. ISSN : 1530-6984

Gómez-Romero, P., Sanchez, C. (editors). (2004). Functional Hybrid Materials, Wiley-VCH, Weinheim, 2004. ISBN 3-527-30484-3 - Wiley-VCH, Weinheim

Grayson A BioMEMS review: MEMS technology for physiologically integrated devices Author(s): Grayson ACR; Shawgo RS; Johnson AM; et al. Source: PROCEEDINGS OF THE IEEE Volume: 92 Issue: 1 Pages: 6-21 DOI: 10.1109/JPROC.2003.820534 Published: JAN 2004

Grosserueschkamp, M., Nowak, C., Schach, D., Schaertl, W., Knoll, W., Naumann, R. C. . (2009). Silver Surfaces with Optimized Surface Enhancement by Self-Assembly of Silver Nanoparticles for Spectroelectrochemical Applications. *Journal of Physical Chemistry C* Vol. 113, No.41, pp. 17698-17704. ISSN 1932-7447

Gu, H.W., Zheng, R.K., Zhang, X.X., Xu, B. (2004). Facile one-pot synthesis of bifunctional heterodimers of nanoparticles: A conjugate of quantum dot and magnetic nanoparticles. *J. Am. Chem. Soc.* Vol. 126, pp. 5664–5665.

Guo Z., Feng, Z.,Fan, S., Dezhi, Z., Haihan, Z. (2009). Research development of measuring methods on the tribology characters for movable MEMS devices: a review. *Microsystem Technologies-Micro-And Nanosystems-Information Storage & Processing Systems,* Vol. 15, No.3, pp.343-354, ISSN : 0946-7076

Hines, M.A., Guyot-Sionnest, P. (1996).Synthesis and characterization of strongly luminescing ZnS-Capped CdSe nanocrystals. *J. Phys. Chem.* Vol.100, pp. 468–471.

Hongqin, L., Piret, G., Sieber, B., Laureyns, J., Roussel, P., Wenguo, X., Boukherroub, R., Szunerits, S. (2009). Electrochemical impedance spectroscopy of ZnO nanostructures. *Electrochemistry Communications* Vol.11, No.5: pp.945-949. ISSN : 1388-2481

Hu, L. C., Shea, K. J. (2011). Organo-silica hybrid functional nanomaterials: how do organic bridging groups and silsesquioxane moieties work hand-in-hand? *Chemical Society Reviews* Vol.40, No.2, pp.688-695.

Huang, Y., Duan, X., Wei, Q., Lieber, C.M (2001) Directed assembly of one-dimensional nanostructures into functional networks. *Science* Vol. 291, pp. 630–633

Huber, C., Liu, J., Egelseer, E.M., Moll, D., Knoll, W., Sleytr, U.B., Sára, M. (2006). Heterotetramers formed by an S-layer-streptavidin fusion protein and core-streptavidin as a nanoarrayed template for biochip development. *Small* Vol. 2, No. 1, pp. 142-150. ISSN : 1613-6810

Ishihara, K., Takai, M. (2009). Bioinspired interface for nanobiodevices based on phospholipid polymer chemistry. *Journal of the Royal Society Interface,* Vol. 6, pp. S279-S291. ISSN: 1742-5662

Ito, M., Nakamura, F., Baba, A., Tamada, K., Ushijima, H., King, H., Aaron, L., Manna, A., Knoll, W. (2007). Enhancement of surface plasmon resonance signals by gold nanoparticles on high-density DNA microarrays. *Journal of Physical Chemistry C* Vol.111, No.31, pp. 11653-11662. ISSN : 1932-7447

Iwamoto, M., Kaneto, K., Mashiko, S. (2003) *Nanotechnology and Nano-Interface Controlled Devices,* Elsevier Science, ISBN 0-444-51091-5 , Amsterdam, the Netherlands.

Jungblut, M., Knoll, W., Thielemann, C., Pottek, M. (2009).Triangular neuronal networks on microelectrode arrays: an approach to improve the properties of low-density networks for extracellular recording. *Biomedical Microdevices,* Vol. 11, No. 6, pp. 1269-1278 ISSN : 1387-2176

Kahng, A. B. (2010). Scaling: More than Moore's law. *IEEE Design & Test of Computers* Vol. 27, No.3, pp.86-87.

Kelly, K.L., Coronado, E., Zhao, L.L., Schatz, G.C. (2003).The Optical Properties of Metal Nanoparticles: The Influence of Size, Shape, and Dielectric Environment. *J. Phys. Chem. B,* Vol. 107, no. 3, pp 668–677. ISSN : 1520-6106, http://dx.doi.org/10.1021/ jp026731y

Khan, H.U., Roberts, M.E, Johnson, O., Förch,R., Knoll, W., Bao , Z. (2010) In Situ, Label-Free DNA Detection Using Organic Transistor Sensors. *Advanced Materials* Vol. 22, No.40, pp. 4452-4456

Khor, H.L., Kuan, Y., Kukula, H., Tamada, K., Knoll, W., Moeller, M., Hutmacher, D.W. (2007). Response of cells on surface-induced nanopatterns: Fibroblasts and mesenchymal progenitor cells. *Biomacromolecules* Vol. 8, No.5, pp. 1530-1540. ISSN : 1525-7797

Kickelbick, G (editor). (2007). Hybrid Materials: Strategies, Syntheses, Characterization and Applications, Wiley-VCH, Weinheim, ISBN 978-3-527-31299-3.

Kim, D.H., Karan, P., Göring, P., Leclaire, J., Caminade, A.-M., Majoral, J.-P., Gösele, U., Steinhart, M., Knoll, W. (2005). Formation of dendrimer nanotubes by layer-by-layer deposition. *Small* Vol.1, No.1, pp. 99-102. ISSN : 1613-6810

Kim, D-K., Kerman, K., Hiep,H. M., Saito, M., Yamamura, S., Takamura,Y., Kwon, Y.-S., Tamiya, E. (2008). Label-free optical detection of aptamer-protein interactions using gold-capped oxide nanostructures, *Analytical Biochemistry,* Vol. 379, No. 1, pp. 1-7, ISSN 0003-2697,

Kim, F.S., Ren, G., Jenekhe S.A. (2011) One-Dimensional Nanostructures of π-Conjugated Molecular Systems: Assembly, Properties, and Applications from Photovoltaics, Sensors, and Nanophotonics to Nanoelectronics Chem. Mater. 2011, 23, 682–732

Klauser, F., Ghodbane, S., Boukherroub, R., Szunerits, S., Steinmüller-Nethl, D., Bertel, E., Memmel, N. (2010). Comparison of different oxidation techniques on single-crystal and nanocrystalline diamond surfaces. *Diamond and Related Materials* Vol. 19, No. 5-6, pp 474-478. ISSN : 0925-9635

Klimeck, G., 2011a The nanoelectronic modeling group, available at https://engineering.purdue.edu/gekcogrp

Klimeck, G., 2011b https://engineering.purdue.edu/gekcogrp/science-applications/optoelectronics/quantum-dots/

Klimeck, G., Haley, B.P (2009), "Quantum Dot Lab Demonstration: Pyramidal Qdots," http://nanohub.org/resources/6845.

Klimeck, G., Mehrotra, S.R. (2011), "Electron Density in a Nanowire," http://nanohub.org/resources/10666.

Klimeck, G., Ahmed, S. S. , Bae, H. , Kharche, N., Rahman, R., Clark,, S. Haley, B. , Lee, S. H., Naumov, M. Ryu, , H. , Saied, F. , Prada, M. , Korkusinski, M. , Boykin, T. B. (2007a). Atomistic simulation of realistically sized nanodevices using NEMO 3-D - *Part I:* Models and benchmarks. *IEEE Transactions on Electron Devices* Vol.54, No.9, pp.2079-2089.

Klimeck, G., Ahmed, S. S. , Kharche, N. , Korkusinski, M. , Usman, M. , Prada, M. Boykin, T. B.. (2007b). Atomistic simulation of realistically sized nanodevices using NEMO 3-D - *Part II:* Applications. *IEEE Transactions on Electron Devices,* Vol.54, No.9, pp.2090-2099.

Knoll W ; Kasry A ; Yu F ; Wang Y ; Brunsen A ; Dostalek J (2008). New concepts with surface plasmons and nano-biointerfaces. *Journal of Nonlinear Optical Physics & Materials* Vol. 17, No. 2, pp. 121-129. ISSN : 0218-8635

Knoll, W., Caminade, A-M., Char, K., Duran, H., Feng, C.L., Gitsas, A., Kim, D.H., Lau, A., Lazzara, T.D., Majoral, J.P, Steinhart, M., Yameen, B., Zhong, X.H. (2011). Nanostructuring Polymeric Materials by Templating Strategies. *Small* Vol.7, No. 10, pp. 1384-1391.

Knoll, W., Schmid, E.L., Tamada, K., Zhong, X., Han, M.-Y., Li, X., Hernandez-Lopez, J.-L., Manna, A., ,Müllen K., Nakamura, F., Niu, L., Robelek, R. (2004). Nanoscopic building blocks from polymers, metals, and semiconductors for hybrid architectures. *Journal of Nonlinear Optical Physics & Materials* Vol. 13, No. 2, pp. 229-241. ISSN : 0218-8635

Knoll, W., Zhong, X., Stefani, F., Robelek, R., Niu, L., Rochholz, H., Shumaker-Parry, J., Kreiter, M. (2006).Optics with nano-sized structures made from semiconductors and (noble) metals. *Journal of Nonlinear Optical Physics & Materials* Vol.15, No. 3, pp. 355-367. ISSN : 0218-8635

Kulesza, P.J., Skunik, M., Baranowska, B., Miecznikowski, K., Chojak, M., Karnicka, K., Frackowiak, E.,Beguin, F.,Kuhn, A., Delville, M-H., Starobrzynska, B., Ernst, A. (2006). Fabrication of network films of conducting polymer-linked polyoxometallate-stabilized carbon nanostructures *Electrochimica Acta* Vol. 51, No.11, pp. 2373-2379, ISSN 0013-4686

Kumar, B., Tan, H.S, Ramalingam, N., Mhaisalkar, S.G (2009). Integration of ink jet and transfer printing for device fabrication using nanostructured materials. *Carbon* Vol. 47, pp:321–324

Kummar, C.S.S.R., Hormes, J., Leuschner, C. (2005). *Nanofabrication towards Biomedical Applications,*Wiley VCH, ISBN 978-3-527-31115-6, Weinheim.

Kuwata, H., Tamaru, H. , Esumi, K. Miyano, K.. (2003). Resonant light scattering from metal nanoparticles: Practical analysis beyond Rayleigh approximation. *Applied Physics Letters* Vol.83, No.22, pp.4625-4627.

Lee, A. S., Peteu, S. F., Ly, J. V.,. Requicha, A. A. G., Thompson, M. E., Zhou, C. (2008). Actuation of polypyrrole nanowires. *Nanotechnology* Vol.19, No. 16. ISSN : 0957-4484

Lee, B., Roh, S., Park, J. (2009) Current status of micro- and nano-structured optical fiber sensors, *Optical Fiber Technology,* Vol. 15, No. 3, pp. 209-221, ISSN 1068-5200.

Lee,Y. J., Yi, H., Kim, W.-J., Kang, K., Yun, D.S., Strano, M.S., Ceder, G., Belcher, A M. (2009). Genetically Engineered High-Power Lithium-Ion Batteries Using Multiple Virus Genes. *Science,* Vol. 324, pp. 1051-1055.

Li, X., Fu, J., Steinhart, M., Kim, D.H., Knoll, W. (2007). Au/titania composite nanoparticle arrays with controlled size and spacing by organic-inorganic nanohybridization in thin film block copolymer templates. *Bulletin of the Korean Chemical Society* Vol. 28, No. 6, pp. 1015-1020. ISSN : 0253-2964

Liao, Y. Z., Huang, M. R. Li, X. G.. (2009). Biosensors Based on Nanostructured Polypyrroles. *Chinese Journal of Analytical Chemistry* Vol. 37, No. 2, pp. 291-298.

Liu, A. Q., Zhang, X. (2007). A review of MEMS external-cavity tunable lasers. *Journal Of Micromechanics And Microengineering.*Vol. 17, No. 1, pp. R1-R13 ISSN : 0960-1317

Liu, S., Y. Liu, H. Cebeci, R. G. de Villoria, J. H. Lin, B. L. Wardle, and Q. M. Zhang. 2010. High Electromechanical Response of Ionic Polymer Actuators with Controlled-Morphology Aligned Carbon Nanotube/Nafion Nanocomposite Electrodes. *Advanced Functional Materials* 20 (19):3266-3271.

Ma, H., Zin, M. T. , Zareie, M. H. , Kang, M. S. , Kang, S. H. , Kim, K. S. , Reed, B. W. , Behar, C. T. , Sarikaya, M. Jen, A. K. Y. (2007). Assembly of nanomaterials through highly ordered self-assembled monolayers and peptide-organic hybrid conjugates as templates. *Journal of Nanoscience and Nanotechnology* Vol.7, No.8, pp.2549-2566.

Marcon, et al 2010; Cell Adhesion Properties on Chemically Micropatterned Boron-Doped Diamond Surfaces Author(s): Marcon Lionel; Spriet Corentin; Coffinier Yannick; et al. Source: LANGMUIR Volume: 26 Issue: 19 Pages: 15065-15069 DOI: 10.1021/la101757f Published: OCT 5 2010 Times Cited: 1 (from Web of Science)

Marcon, L., Riquet, F., Vicogne, D., Szunerits, S., Bodart, J-F., Boukherroub, R. (2010). Cellular and in vivo toxicity of functionalized nanodiamond in Xenopus embryos. *Journal of Materials Chemistry* Vol. 20, No.37, pp. 8064-8069. ISSN : 0959-9428

Mayer, K. M., J. H. Hafner. 2011. Localized Surface Plasmon Resonance Sensors. *Chemical Reviews* Vol. 111, No.6, pp. 3828-3857. http:\dx.doi.org/10.1021/cr100313v

McFarland, A.D. and Van Duyne, R.P. (2003) Single silver nanoparticles as real-time optical sensors with zeptomole sensitivity. *Nano Letters*. Vol. 3, 1057–1062; ISSN : 1530-6984

Medintz, I. L., Mattoussi, H., Clapp, A.R. (2008). Potential clinical applications of quantum dots. *International Journal of Nanomedicine* Vol.3, No.2, pp151-167. ISSN : 1176-9114

Megouda, N., Cofininier, Y., Szunerits, S., Hadjersi, T., ElKechai, O., Boukherroub, Rabah (2011). Photocatalytic activity of silicon nanowires under UV and visible light irradiation. *Chemical Communications* Vol. 47, No.3, pp. 991-993. ISSN : 1364-548X

Mekis, I., Talapin, D.V., Kornowski, A., Haase, M., Weller, H. (2003). One-pot synthesis of highly luminescent CdSe/CdS core-shell nanocrystals via organometallic and ―greener‖ chemical approaches. *J. Phys. Chem. B* Vol. 107, pp. 7454–7462.

Mirkin, C. A., Rogers, J.A. (2001). Emerging methods for micro- and nanofabrication. *MRS Bulletin* Vol. 26 , No. (7), pp. 506-508.

Mize, A.S. (2004). Toward Nanomaterials by Design: A Rational Approach for Reaping Benefits in the Short and Long Term, available at <*www.scottmize.com/ publications/whitepaper.html*>

Molina, M. A., Rivarola, C. R. , Miras, M. C. , Lescano, D. Barbero, C. A.. (2011). Nanocomposite synthesis by absorption of nanoparticles into macroporous hydrogels. Building a chemomechanical actuator driven by electromagnetic radiation. *Nanotechnology* Vol.22, No.24. ISSN : 0957-4484

Moschou, E. A., Peteu, S. F., Bachas L. G., Madou, M. J., Daunert, S. (2004). Artificial muscle material with fast electroactuation under neutral pH conditions. *Chemistry of Materials* Vol. 16, No.12, pp. 2499-2502. ISSN : 0897-4756

Mullen, T J, Srinivasan, C., Shuster, M.J., Horn, M.W., Andrews. A.M., Weiss. P.S. (2008). Hybrid approaches to nanometer-scale patterning: exploiting tailored intermolecular interactions. *Nanoparticlel Research* Vol. 10, pp. 1231–1240, ISSN : 1388-0764.

Muller, K., Park, M-K., Klapper, M., Knoll, W., Mullen, K., Klapper, M., Ivan, B. (2007). Synthesis and layer-by-layer deposition of spherical poly(3,4-ethylenedioxythiophene) nanoparticles - Toward fast switching times between reduced and oxidized states. *Macromolecular Chemistry and Physics* Vol. 208, No.13, pp. 1394-1401. ISSN : 1022-1352

Nakamura, F., Ito, M., Manna, A., Tamada, K., Hara, M., Knoll, W. (2006). Observation of hybridization on a DNA array by surface plasmon resonace imaging using Au nanoparticles. *Japanese Journal of Applied Physics Part 1-Regular Papers Brief Communications & Review Papers* Vol. 45, No.2A, pp. 1026-1029. ISSN : 0021-4922

Nanofabrication (n.d.) (2011), available from www.physics.uni.edu/chanceyclass/ intermedNano/Nanofab%25207

Nicholls, I.A., Andersson, H.S., Golker, K., Henschel, H., Karlsson, B.C.G., Olsson.G.D., Rosengren, A.M., Shoravi, S., Suriyanarayanan, S., Wiklander, J.G., Wikman, S. (2011). Rational design of biomimetic molecularly imprinted materials: theoretical and computational strategies for guiding nanoscale structured polymer development. *Anal Bioanal Chem* Vol. 400, No. 6, pp.1771–1786. ISSN: 1618-2642.

Nicole, L., Rozes, L., Sanchez, C. (2010). *Adv. Mater.*, Vol. 22, pp. 3208–3214.

Niedziółka-Jönsson, J., Barka, F., Castel, X., Pisarek, M., Bezzi, N., Boukherroub, R., Szunerits, S. (2010). Development of new localized surface plasmon resonance interfaces based on gold nanostructures sandwiched between tin-doped indium oxide films. *Langmuir* Vol. 26, No.6, pp. 4266-4273. ISSN: 0743-7463

Noginov, M. A. , Zhu, G., Belgrave, A. M., Bakker, R., Shalaev, V. M., , Narimanov, E. E., Stout, S., Herz, E., Suteewong, T., Wiesner, U. (2009). Demonstration of a spaser-based nanolaser*Nature,* Vol. 460, No 7259, pp. 1110-1112 ISSN : 0028-0836

Ohara, S., Hatakeyama, Y., Umetsu, M., Tan, Z., Adschiri, T. (2011). Fabrication of Pd-DNA and Pd-CNT hybrid nanostructures for hydrogen sensors, *Advanced Powder Technology,* In Press, Corrected Proof, ISSN 0921-8831,

Pal, P., Kim, Y.J., Chandra, S.(2006). Front-to-back alignment techniques in microelectronics/MEMS fabrication: A review SENSOR LETTERS, Vol. 4, No. , pp. 1-10 ISSN : 1546-198X

Papapzoglou E.S., Parthasarathy, A. (2007). *BioNanotechnology,* Morgan & Claypool, ISBN 1598291386, San Rafael, CA

Park, H-H., Zeng, L., Buresh, M., Wang, S., Klimeck, G., Mehrotra, S.R., Heitzinger, C., Haley, B.P. (2011). Nanowire DOI: 10254/nanohub-r1307.7

Peng, J., Knoll, W., Park, C., Kim, D.H. (2008). Two-dimensional arrays of strings of TiO2 nanoparticles via cooperative block copolymer self-assembly. *Chemistry of Materials* Vol. 20, No.4, pp. 1200-1202. ISSN : 0897-4756

Peng, J., Li, X., Kim, D.H., ; Knoll, W. (2007). Fabrication and photocatalytic activities of morphology-controlled titania nanoobject arrays by block copolymer templates. *Macromolecular Rapid Communications* Vol. 28, No.21, pp. 2055-2061. ISSN : 1022-1336

Peteu, S.F. (2007). Responsive materials configured for micro- and nanoactuation. *Journal Of Intelligent Material Systems And Structures,* Vol. 18, No. 2 , pp. 147-152 ISSN : 1045-389X

Peteu, S. F, Peiris, P., ; Gebremichael E ; Bayachou M (2010). Nanostructured poly(3,4-ethylenedioxythiophene)-metalloporphyrin films: Improved catalytic detection of peroxynitrite. *Biosensors & Bioelectronics* Vol. 25, No.8, pp. 1914-1921. ISSN : 0956-5663

Praig, V. G., Piret, G., Manesse, M., Castel, X., Boukherroub, R., Szunerits, S. (2008). Seed-mediated electrochemical growth of gold nanostructures on indium tin oxide thin films. *Electrochimica Acta* Vol. 53, No. 27, pp.7838-7844. ISSN : 0013-4686

Praig, V.G., McIlwee, H., Schauer, C.L., Boukherroub, R., Szunerits. S (2009). Localized Surface Plasmon Resonance of Gold Nanoparticle-Modified Chitosan Films for Heavy-Metal Ions Sensing. *Journal of Nanoscience and Nanotechnology* Vol. 9, No.1, pp. 350-357. ISSN : 1533-4880

Pyun, J,. Matyjaszewski, K. (2001). Synthesis of nanocomposite organic/inorganic hybrid materials using controlled/"living" radical polymerization. *Chemistry of Materials* Vol.3, No.10, pp.3436-3448.

Qiao W. ; McLennan M. ; Kennell R. ; Ebert D.S. ; Klimeck G. (2006). Hub-based Simulation and Graphics Hardware Accelerated Visualization for Nanotechnology

Applications, *IEEE Transactions on Visualization and Computer Graphics,* Vol. 12, pg. 1061-1068. ISSN 1077-2626

Qin, L., Zou, S., Xue, C., Atkinson, A., Schatz, G.C., Mirkin, C.A. (2006). Designing, fabricating, and imaging Raman hot spots. *Proc Natl Acad Sci U S A.* Vol. 103, No. 36, pp.13300-13303, ISSN : 0027-8424.

Rahman, M. A., P. Kumar, D. S. Park, and Y. B. Shim. 2008. Electrochemical sensors based on organic conjugated polymers. *Sensors* 8 (1):118-141.

Raschke, G. Kowarik, S., Franzl, T., Sönnichsen, C., Klar, T.A., Feldmann, J. Nichtl, A. Kürzinger K. (2003) Biomolecular recognition based on single gold nanoparticle light scattering. *Nano Letters,* Vol.3, No. 7, pp. 935–938, ISSN : 1530-6984

Reimhult, E., Kumar, K., Knoll, W. (2007). Fabrication of nanoporous silicon nitride and silicon oxide films of controlled size and porosity for combined electrochemical and waveguide measurements. *Nanotechnology* Vol. 18, No. 27, pp 275303.1-275303.7, ISSN : 0957-4484.

Roberts, M.E., Mannsfeld, S.C.B., Queraltó, N., Reese, C., Locklin, J., Knoll, W., Bao, Z.N. (2008) Water-stable organic transistors and their application in chemical and biological sensors. *Proc. Natl. Acad. Sci.* vol. 105 no. 34 12134–12139.

Roberts, M.E., Queraltó, N., Mannsfeld, S.C.B., Benjamin N. Reinecke, B.N., Knoll, W., Bao, Z.N. Cross-Linked Polymer Gate Dielectric Films for Low-Voltage Organic Transistors *Chem. Mater.* 2009, *21,* 2292–2299 10.1021/cm900637p

Roco, M.C., R.S. Williams, and P. Alivisatos, P. eds. (1999) Nanotechnology research directions: IWGN workshop report: Vision for nanotechnology R&D in the next decade. Baltimore, Md.: International Technology Research Institute at Loyola College. Available online: http://www.nano.gov/html/res/pubs.html

Roco, M.C., C.A Mirkin, and M. C. Hersam (2010). *Nanotechnology research directions for Societal Needs: Retrospective and Outlook.* Springer. Available online: http://www.wtec.org/nano2/

Ruiz-Hitzky, E. ,Ariga, K., Lvov, Y. M (editors). Bio-inorganic Hybrid Nanomaterials: Strategies, Syntheses, Characterization and Applications. (2007), Wiley-VCH Verlag GmbH & Co. KGaA, Weinheim. ISBN: 978-3-527-31718-9

Sanchez, C., Belleville, P., Popall, M., Nicol L. (2011). Applications of advanced hybrid organic–inorganic nanomaterials: from laboratory to market, *Chem. Soc. Rev,,* vol. 40, pp. 696–753 http://dx.doi.org/10.1039/c0cs00136h:

Sannomiya , T., Voros , J.(2011). Single plasmonic nanoparticles for biosensing : *Trends in Biotechnology,* Vol. 29, No. 7, July 2011, pp.343-351, ISSN :0167-7799

Schmidt, R. C., Healy, K.E. (2009). Controlling biological interfaces on the nanometer length scale. *Journal of Biomedical Materials Research Part A,* Vol. 90A, No. 4, pp.1252-1261. ISSN 1549-3296

Schommers, W. (2007).Directions in theoretical and computational nanoscience, *Journal Of Computational And Theoretical Nanoscience.,*Vol.4, No.4, pp.705-714. ISSN: 1546-1955

Shvedova, A. A., Kagan, V.E., Fadeel, B. (2010). Close Encounters of the Small Kind: Adverse Effects of Man-Made Materials Interfacing with the Nano-Cosmos of Biological Systems. In *Annual Review of Pharmacology and Toxicology.,* Vol. 50, pp.63-88. ISSN: 0362-1642

Sieber, B., Liu, H.Q., Piret, G., Laureyns, J., Roussel, P., Gelloz, B., Szunerits, S., Boukherroub, R. (2009). Synthesis and Luminescence Properties of (N-Doped) ZnO Nanostructures from a Dimethylformamide Aqueous Solution. *Journal of Physical Chemistry C* Vol. 113, No. 31, pp.13643-13650. ISSN : 1932-7447

Stadler, R., Forshaw, M; Joachim, C. (2003).Modulation of electron transmission for molecular data storage. Nanotechnology, Vol. 14, No. 2 , pp. 138-142

Stemmler, M., Stefani, F.D., Ernhardt, S., Bauer, R. E., Kreiter, M., Müllen, K., Knoll, W.. (2009). One-Pot Preparation of Dendrimer-Gold Nanoparticle Hybrids in a Dipolar Aprotic Solvent. *Langmuir* Vol. 25, No. 21, pp. 12425-12428. ISSN : 0743-7463

Stewart, M.E., Anderton, C.R., Thompson, L.B., Maria, J., Gray, S.K., Rogers, J.A., Nuzzo, R.G. (2008). Nanostructured Plasmonic Sensors *Chem. Rev.* vol 108, pp. 494-521, 10.1021/cr068126n

Strachan, A., Klimeck, G., Lundstrom, M.S. 2010b Cyber-enabled simulations in nanoscale science and engineering, *Computing in Science and Engineering,* Vol. 12, number 2, pg: 12-17.

Suh, W.H., Suh, Y.-H., Stucky, G.D. (2009). Multifunctional nanosystems at the interface of physical and life sciences, *Nano Today* Vol. 4, pp. 27—36, ISSN : 1748-0132.

Sun, Z. C., Kim, D.H., Wolkenhauer, M., Bumbu, G. G., Knoll, W., Gutmann, J. (2006). Synthesis and photoluminescence of titania nanoparticle arrays templated by block-copolymer thin films. *Chemphyschem* Vol. 7, No.2, pp. 370-378. ISSN: 1439-4235

Szunerits, S., Walt, D.R. (2002) Fabrication of an optoelectrochemical microring array. Analytical Chemistry,Vol. 74 No. 7, pp.1718-1723 ISSN : 0003-2700

Szunerits, S., Boukherroub, R. (2008). Investigation of the electrocatalytic activity of boron-doped diamond electrodes modified with palladium or gold nanoparticles for oxygen reduction reaction in basic medium. *Comptes Rendus Chimie* Vol. 11, No. 9,pp. 1004-1009. ISSN : 1631-0748

Szunerits, S., Coffinier, Y.,Galopin E., Brenner, J., Boukherroub, R. (2010). Preparation of boron-doped diamond nanowires and their application for sensitive electrochemical detection of tryptophan. *Electrochemistry Communications* Vol. 12, No.3, pp.438-441. ISSN : 1388-2481

Szunerits, S., Das, M.R., Boukherroub, R. (2008). Short- and long-range sensing on gold nanostructures, deposited on glass, coated with silicon oxide films of different thicknesses. *Journal of Physical Chemistry C* Vol.112, No. 22, pp. 8239-8243. ISSN : 1932-7447

Tamagawa, H., Lin W. Y.,, Kikuchi, K. , Sasaki, M.. (2011). Bending control of Nafion-based electroactive polymer actuator coated with multi-walled carbon nanotubes. *Sensors and Actuators B-Chemical* Vol.156, No.1, pp.375-382.

Tan, L. K., Gao, H ; Zong, Y., Knoll, W. (2008). Atomic Layer Deposition of TiO2 to Bond Free-Standing Nanoporous Alumina Templates to Gold-Coated Substrates as Planar Optical Waveguide Sensors. *Journal of Physical Chemistry C* Vol. 112, No. 45, pp. 17576-17580. ISSN : 1932-7447

Tang, Q.; Tong, Y.; Hu, W.; Wan, Q.; Bjørnholm, T. *Adv. Mater.* (2009), Vol. 21,No. 454, pp. 4234–4237.

Tawa, K., Knoll, W. (2004). Mismatching base-pair dependence of the kinetics of DNA-DNA hybridization studied by surface plasmon fluorescence spectroscopy. *Nucleic Acids Research*. Vol.32, No. 8, pp. 2372–2377, ISSN: 0305-1048

Ten Elshof, J. E., Khan, S. U, Gobel, O. F..(2010). Micrometer and nanometer-scale parallel patterning of ceramic and organic-inorganic hybrid materials. *Journal of the European Ceramic Society* Vol.30 , No.7, pp.1555-1577.

Tey, J. N., Wijaya, I. P. M. , Wei, J. ., Rodriguez, I. , Mhaisalkar, S. G.. (2010). Nanotubes-/nanowires-based, microfluidic-integrated transistors for detecting biomolecules. *Microfluidics and Nanofluidics* Vol.9, No.6, pp. 1185-1214.

Thamwattana, N., Hill, J. M., Baowan, D., Cox, B. J. (2010). A Review of Mathematical and Mechanical Modelling in Nanotechnology. *Mathematics And Mechanics Of Solids*.Vol. 15, No. 7, pp. 708-717. ISSN: 1081-2865

Tian, S. J., J. Y. Liu, Tao Z., Knoll, W. (2004). Polyaniline/gold nanoparticle multilayer films: Assembly, properties, and biological applications *Chemistry of Materials* Vol. 16, No. 21, pp. 4103-4108. ISSN : 0897-4756

Tomczak, N, Janczewski, D., Han, M., Vancso, J.G. (2009). Designer polymer-quantum dot architectures, *Progress in Polymer Science*, Vol. 34, No. 5, pp. 393-430, ISSN 0079-6700.

Touahir, L., Galopin, E., Boukherroub, R., Gouget-Laemmel, A.C., Chazalviel, J-N., Ozanam, F., Szunerits, S. (2010). Localized surface plasmon-enhanced fluorescence spectroscopy for highly-sensitive real-time detection of DNA hybridization *Biosensors and Bioelectronics* Vol. 25, no. 12, pp. 2579–2585. ISSN : 0956-5663

Touahir, L., Galopin, E., Boukherroub R., Chantal Gouget-Laemmel A., Chazalviel, J-N., Ozanam, F., ,Saison O., Akjouj, A., Pennec, Y., Djafari-Rouhani, B., Szunerits, S. (2011). Plasmonic properties of silver nanostructures coated with an amorphous silicon-carbon alloy and their applications for sensitive sensing of DNA hybridization. *Analyst* Vol.136, No. 9, pp 1859-1866. ISSN : 1364-5528

Tripp, R.A, Dluhy, R.A., Zhao, Y. (2008). Novel nanostructures for SERS biosensing, *Nanotoday*, 2008, Vol. 3, No. 3-4, pp. 31-37.ISSN: 1748-0132.

Tsai, N.-C. Sue, C-Y. (2007). Review of MEMS-based drug delivery and dosing systems *Sensors And Actuators A-Physical*, Vol. 134, No. 2, pp. 555-564 ISSN : 0924-4247

Vasilescu, A., Andreescu, S., Bala, C., Litescu, S.C., Noguer, T., Marty, J-L. (2003). Screen-printed electrodes with electropolymerized Meldola Blue as versatile detectors in biosensors. Biosensors & Bioelectronics Vol. 18, No. 5-6, pp. 781-790. ISSN : 0956-5663

Vasiliev, I., Lopez del Puerto, M., Jain, M.,Lugo-Solis, A., Chelikowsky, J.R.(2009). Application of time-dependent density-functional theory to molecules and nanostructures, *Journal of Molecular Structure: THEOCHEM*, Vol.914, No.1-3, pp. 115-129, ISSN 0166-1280,

Vivero-Escoto, J. L., Huang. Y.T. (2011). Inorganic-Organic Hybrid Nanomaterials for Therapeutic and Diagnostic Imaging Applications. *International Journal of Molecular Sciences* Vol. 12, No.6, pp.3888-3927. ISSN:1422-0067

Wang, Y. H., Mirkin, C. A., Park, S. J.( 2009). Nanofabrication beyond Electronics. *ACS Nano* Vol.3, No. 5, pp.1049-1056. ISSN:19360851

Wang, Y., Dostalek, J., Brunsen, A., Knoll, W. (2009). ANYL 212-Biosensor based on long-range surface plasmon-enhanced fluorescence spectroscopy and magnetic nanoparticle immunoassay. *Abstracts of Papers of the American Chemical Society*, The 238th ACS National Meeting, Washington, DC, August 16-20, 2009.

Weickert, J., Dunbar, R. B. , Hesse, H. C. , Wiedemann, W. , Schmidt-Mende, L..(2011). Nanostructured Organic and Hybrid Solar Cells. *Advanced Materials* Vol.23, No.16, pp.1810-1828.

Willets, K. A, Van Duyne, R. P.. (2007). Localized surface plasmon resonance spectroscopy and sensing. In *Annual Review of Physical Chemistry*. vol. 58, pp. 267-297. ISSN : 0066-426X

Willner, I, Willner, B., Katz, E. (2007). Biomolecule–nanoparticle hybrid systems for bioelectronic applications . *Bioelectrochemistry* Vol. 70, pp. 2–11, ISSN : 1567-5394

Yamada, M. (2009). Synthesis of Organic Shell-Inorganic Core Hybrid Nanoparticles by Wet Process and Investigation of Their Advanced Functions. *Bulletin of the Chemical Society of Japan* Vol.82, No.2, pp.152-170.

Yamamichi, J., Ojima,T., Yurugi,K., Iida,M., Imamura,T., Ashihara,E., Kimura,S., Maekawa, T. (2011). Single-step, label-free quantification of antibody in human serum for clinical applications based on localized surface plasmon resonance, *Nanomedicine: Nanotechnology, Biology and Medicine*, In Press, ISSN 1549-9634,

Yamamoto, Y., K. Maehashi, Y. Ohno, and K. Matsumoto. 2009. Highly Sensitive Biosensors Based on High-Performance Carbon Nanotube Field-Effect Transistors. *Sensors and Materials* 21 (7):351-361.

Yameen, B., Ali, M., Neumann, R., Ensinger, W., Knoll, W., Azzaroni, O. (2010). Proton-regulated rectified ionic transport through solid-state conical nanopores modified with phosphate-bearing polymer brushes. *Chemical Communications* Vol. 46, No. 11, pp. 1908-1910. ISSN : 1359-7345

Yao, H. B., Gao, M. R. , Yu, S. H. (2010). Small organic molecule templating synthesis of organic-inorganic hybrid materials: their nanostructures and properties. *Nanoscale* Vol.2, No. 3, pp.:323-334.

Yin, M. Z., Feng, C.L., Shen, Jie; S., Yu, Y., Xu, Z., Yang, W., Knoll, W., Müllen, K.. (2011). Dual-Responsive Interaction to Detect DNA on Template-Based Fluorescent Nanotubes. *Small* Vol. 7, No. 12, pp. 1629-1634.

You, C. C., A. Verma, and V. M. Rotello. (2006). Engineering the nanoparticle-biomacromolecule interface. *Soft Matter* Vol. 2, No. 3, pp.190-204.

You, C-C., Chompoosor, A.,. Rotello, V.M. (2007). The biomacromolecule-nanoparticle interface. *Nano Today,* Vol. 2 No.3, pp.34-43. ISSN: 1748-0132

Yuan, J. Y., Muller, A. H. E.. (2010). One-dimensional organic-inorganic hybrid nanomaterials. *Polymer* Vol.51, No.18, pp.4015-4036.

Zhong, X. H., Xie, R. G. ,Ying, Z., Basche , T., Knoll, W. (2005). High-quality violet- to red-emitting ZnSe/CdSe core/shell nanocrystals. *Chemistry of Materials* Vol. 17, No. 16, pp. 4038-4042. ISSN : 0897-4756

Zhou, X. D., Knoll, W., Zhang, N., Liu, H. et al. (2009). Profile calculation of gold nanostructures by dispersed-nanosphere lithography through oblique etching for

LSPR applications. *Journal of Nanoparticle Research* Vol.11, No 5, pp. 1065-1074. ISSN: 1572-896X

Zhou, X. D., Liu, K. Y. , Knoll, W., Quan, C., Zhang, N. (2010). 3D Profile Simulation of Metal Nanostructures Obtained by Closely Packed Nanosphere Lithography. *Plasmonics* Vol. 5, No. 2, pp. 141-148. ISSN : 1557-1955

Zhou, X. D., Virasawmy, S.,Knoll, W., Liu, K.Y., Tse, M.S., Yen, L.W .(2007). Profile simulation and fabrication of gold nanostructures by separated nanospheres with oblique deposition and perpendicular etching. *Plasmonics* Vol. 2, No. 4, pp. 217-230. ISSN: 1557-1955

Ziaie, B. (2004a) Hard and soft micromachining for BioMEMS: review of techniques and examples of applications in microfluidics and drug delivery , *Advanced Drug Delivery Reviews* Vol 56 No 2 Pp: 145-172 DOI: 10.1016

Ziaie, B. (2004b). Introduction to nanofabrication, in *Handbook of Nanotechnology,* edited by Bushan, B., pp147-185, Springer Verlag, ISBN: 3-540-01218-4, Berlin Heidelberg.

# 3

# Nanofabrication of Metal Oxide Patterns Using Self-Assembled Monolayers

Yoshitake Masuda
*National Institute of Advanced Industrial Science and Technology (AIST), Anagahora, Shimoshidami, Moriyama-ku, Nagoya, Japan*

## 1. Introduction

Metal oxides have recently been fabricated in solutions without high-temperature sintering in order to reduce energy consumption and allow application for various substrates having low heat resistance. The fabrication of metal oxide thin films from solutions has been encouraged by the development of environment-friendly chemistry such as Green & sustainable chemistry[1-6], Bioinspired materials chemistry[7], Biomimetic materials chemistry[7], Soft-solution processing[8-10], Soft chemistry ("Chimie douce" in French)[11], Liquid phase deposition[7,12], Chemical bath deposition (CBD)[12,13], Electroless deposition (ED) with catalyst[12,13], Successive ion layer adsorption and reaction (SILAR) [12,13], Sol-gel process[14,15], Hydrothermal reaction[16], Electrodeposition[17,18] and so on. Solution processing of metal oxides allows us to prepare metal oxide thin films on the surface of solids such as substrates, particles, and fibers. Metal oxide nano/microstructures can also be fabricated by applying these solution systems to electronic or photonic devices.
Many kinds of lithography or patterning techniques have been developed to prepare patterns of thin films, for instance, X-ray/electronbeam lithography and photolithography,[19] microcontact printing,[20,21] wet etching,[22] ink-jet printing,[23] embossing,[24,25] slip-pressing,[26] charge-based printing,[27] micromolding,[28] and cold welding[29]. However, etching or lift-off processes are required in many of these methods, which causes degradation of performance, increases waste and energy consumption, and makes the process complicated. Additionally, etching or lift-off processes cannot be applied to corrosion-resistant metal oxides. The deposition of metal oxides only on desired areas of a substrate is thus required for the pattering of metal oxide thin films.
In this section, Liquid Phase Patterning (LPP) of metal oxides was reported [30]. Self-assembled monolayer (SAM), which can modify the surface of solids with various functional groups, was used as the template to enable molecular recognition for LPP. Solution systems were developed and applied to LPP of ceramic thin films on patterned SAMs by the proposed novel LPP processes.

## 2. Liquid phase patterning of metal oxides

### 2.1 SAM preparation for patterning of metal oxides

Self-assembled monolayer (SAM) can modify the surface of solids such as a substrate, particles, or fibers with various functional groups, and the molecular recognition of functional groups of SAM was the key technique for LPP. Patterned SAMs were used as templates in our LPP processes to deposit metal oxide thin films on desired areas of substrates.

An Si wafer (p-type Si [100], NK Platz Co., Ltd.) was sonicated in water, ethanol or acetone for 10 min, respectively, and exposed for 15 min to UV light (184.9 nm and 253.7 nm) (low-pressure mercury lamp 200 W, PL21-200, 15 mW/cm2 for 254 nm, SEN Lights Co.) to clean the surface. UV light (PL21-200) has stronger power than that used in former studies (NL-UV253, Nippon Laser & Electronics Lab.). The OTS(octadecyltrichlorosilane, $C_{18}H_{37}SiCl_3$)-SAM or APTS(aminopropyltrimethoxysilane, $H_2NC_3H_5Si(OCH_3)_3$) were prepared by immersing the Si substrate in an anhydrous toluene (Aldrich Chemical Co., Inc.) solution containing 1 vol% OTS (Acros Organics) for 15 min or APTS (TCI) for 2 h under an $N_2$ atmosphere [31,32](Fig. 1). The substrate with the SAM was baked at 120°C for 5 min to remove residual solvent and promote chemisorption of the SAM. The control of preparation conditions such as humidity is very important to fabricate organically modified surfaces which realize site-selective deposition.

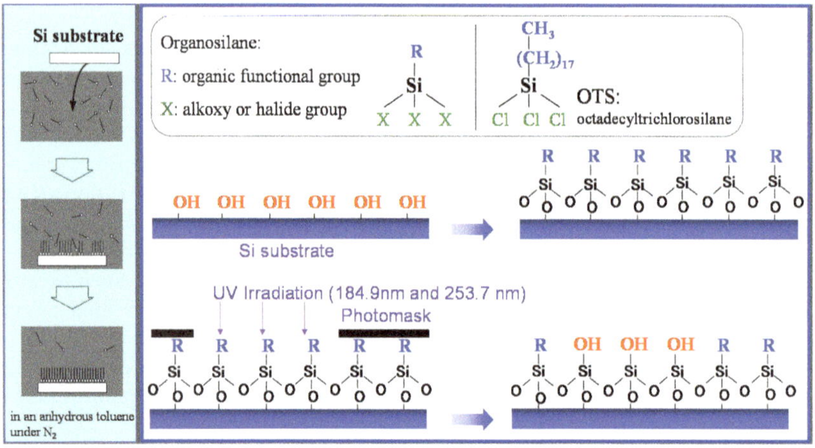


Reprinted with permission from Ref.[31], Masuda, Y., Gao, Y. F., Zhu, P. X., Shirahata, N., Saito, N. and Koumoto, K., 2004, *J. Ceram. Soc. Japan*, 112, 1495. Copyright @ The Ceramic Society of Japan

Fig. 1. Conceptual process for fabrication of a self-assembled monolayer.

The SAMs on the silicon substrates were exposed for 15 min to UV light through a photomask to be used as a template for micropatterning of ZnO crystals. UV-irradiated regions became hydrophilic due to silanol group formation, while the non-irradiated part remained unchanged. Formation of the SAMs and the modification to silanol groups by UV irradiation were verified using the static water drop contact angle ($\theta$w) (a contact angle meter CA-D, Kyowa Interface Science Co., Ltd.) and X-ray photoelectron spectroscopy (XPS) (ESCALAB 210, VG Scientific Ltd.). The X-ray source (MgKα, 1253.6 eV) was operated at 15 kV and 18 mA, and the analysis chamber pressure was 1-3×10-7 Pa. The initially deposited OTS-SAM or APTS-SAM showed a static water contact angle of 105° or 63°, but the UV-irradiated surface of SAM was wetted completely (contact angle < 5°) [33] (Fig. 2). The spectrum peak corresponding to the N 1s binding energy centered at 399.5 eV was observed for the surface of the APTS-treated Si substrate on which APTS-SAM was formed, however, it wasn't detected from the surface after UV irradiation. These experiments show the decomposition and removal of SAMs from the surface of substrates.

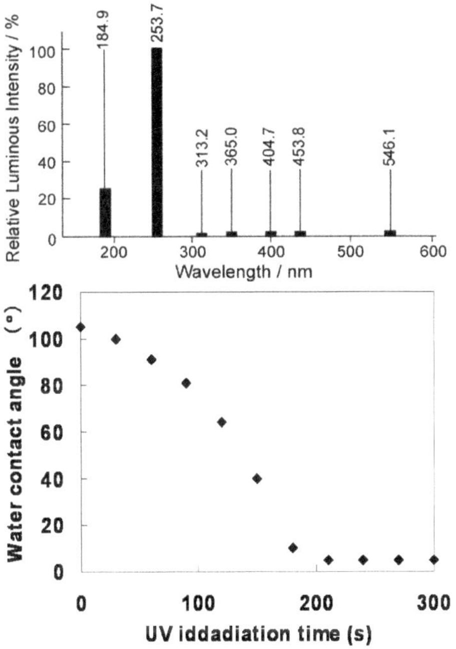


Reprinted with permission from Ref. [33], Masuda, Y., Kinoshita, N., Sato, F. and Koumoto, K., 2006, *Cryst. Growth Des.*, 6, 75. Copyright @ American Chemical Society

Fig. 2. Relative luminous intensity of UV lamp and water drop contact angle of OTS-SAMs as a function of UV-irradiation time.

### 2.2 Liquid phase patterning of amorphous $TiO_2$ thin films[34-37]

Patterned OTS-SAM was immersed in an anhydrous toluene (99.8%, water < 0.002%, Aldrich) solution containing 0.1 M TDD(titanium dichloride diethoxide) for 30 min under an $N_2$ atmosphere using a glove box (Fig. 3) [34-37]. All glassware was dried in a dry box at 50°C before use. The estimated partial pressure of $H_2O$ in an $N_2$ atmosphere is below 0.1 hPa. Chlorine atoms of TDD react with $H_2O$ and change into OH, which further react with silanol groups of SAM resulting in the formation of Ti-O-Si bonds[38]. The ethoxy group, $OC_2H_5$, of TDD is hydrolyzed into hydroxyl groups which are further condensed to form Ti-O-Ti bonds[38]. The thickness of films can be easily controlled by varying the soaking time. After SAM substrates had been rinsed with toluene and preserved in air, thin films appeared on the silanol surfaces of OTS-SAM but were not observed on octadecyl surfaces[34] (Fig. 4). A micropattern of amorphous $TiO_2$ thin films was thus fabricated on a patterned OTS-SAM. Line width measurements at 15 equally spaced points on each line indicated an average printed line width of 23.3 μm. Line edge roughness, as measured by the standard deviation of the line width, was ~0.5 μm, representing a ~2.1% variation (i.e., 0.5/23.2) in the nominal line width[34]. X-ray diffraction measurements (XRD) (Rigaku RU-200) with CuK*α* radiation

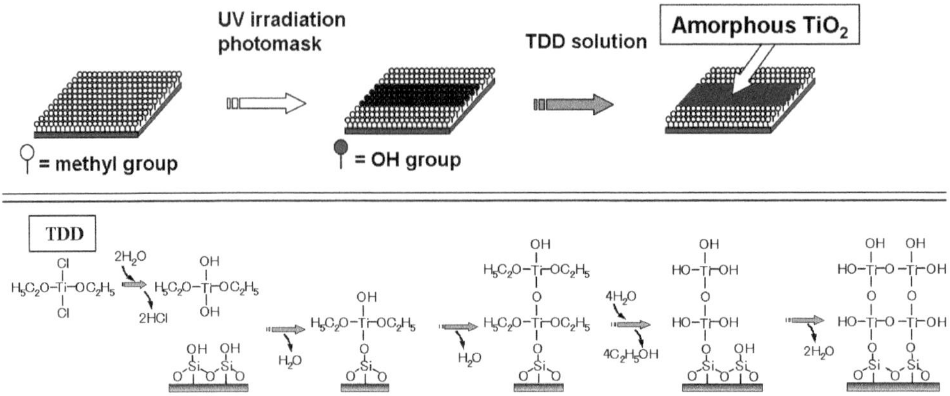


Reprinted with permission from Ref. [34], Masuda, Y., Sugiyama, T., Lin, H., Seo, W. S. and Koumoto, K., 2001, *Thin Solid Films*, 382, 153. Copyright @ Elsevier B.V.

Fig. 3. Conceptual process for selective deposition of amorphous $TiO_2$ thin film using a self-assembled monolayer.

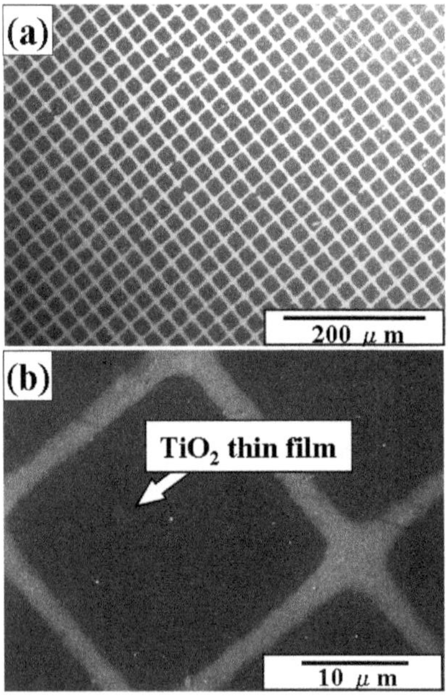


Reprinted with permission from Ref. [34], Masuda, Y., Sugiyama, T., Lin, H., Seo, W. S. and Koumoto, K., 2001, *Thin Solid Films*, 382, 153. Copyright @ Elsevier B.V.

Fig. 4. SEM micrographs of (a) a micropattern of amorphous $TiO_2$ thin films and (b) magnified area of (a).

(40 kV, 30 mA) for as-deposited thin films showed that they were composed of amorphous phases. The ratio of oxygen to titanium was evaluated after 20 min of $Ar^+$ ion sputtering to

avoid the influence of the contaminated layer on the surface. The 1s peak of O can be deconvoluted into two curves (ratio of 529.7 eV (films) and 531.3 eV (silicon oxide) is 1 : 0.22). The ratio of oxygen to titanium was estimated to be 2.2 : 1. Small amounts of chlorine and carbon were also detected (Ti : O : Cl : C = 1 : 2.2 : 0.17 : 0.37)[34].

### 2.3 Liquid phase patterning of crystalline anatase $TiO_2$ thin films using a seed layer[39]

The concept of LPP using a seed layer which accelerates the deposition of thin films was proposed[39]. The deposition process of anatase $TiO_2$ from an aqueous solution was evaluated in detail using a quartz crystal microbalance, and it was found that the nucleation and initial growth of anatase $TiO_2$ were accelerated on amorphous $TiO_2$ thin films compared with silanol, amino, phenyl, or octadecyl groups. Amorphous $TiO_2$ thin films were deposited on silanol regions of a patterned OTS-SAM (Fig. 5(1-a,b)) from a TDD solution (Fig. 5(2-a,b)). This substrate was immersed in an aqueous solution containing a Ti precursor at pH 1.5 for

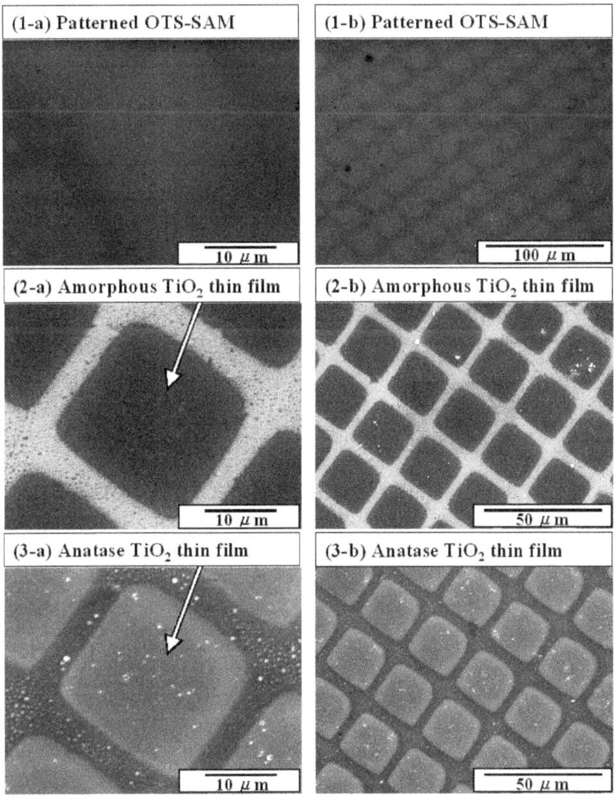


Reprinted with permission from Ref. [39], Masuda, Y., Ieda, S. and Koumoto, K., 2003, *Langmuir*, 19, 4415. Copyright @ American Chemical Society

Fig. 5. SEM micrographs of [(1-a), (1-b)] a patterned OTS-SAM, [(2-a), (2-b)] a micropattern of amorphous $TiO_2$ thin films, and [(3-a), (3-b)] a micropattern of anatase $TiO_2$ thin films deposited at pH 1.5.

1 h to be used as a template for LPP. Anatase $TiO_2$ was selectively deposited on amorphous $TiO_2$ regions to form thin films. Consequently, a micropattern of anatase $TiO_2$ thin film which had high feature edge acuity was successfully fabricated in an aqueous solution (Fig. 5(3-a,b)). The center of the anatase $TiO_2$ thin film region was 61 nm higher than the octadecyl regions, and the thickness of the anatase $TiO_2$ thin film was estimated to be 36 nm considering the thickness of amorphous $TiO_2$ thin film (27 nm)[34] and OTS molecules (2.4 nm)[39]. This result is similar to that estimated by QCM measurement (36 nm). The AFM image showed the film roughness to be 3.7 nm (horizontal distance between measurement points: 6.0 μm), which is less than that of amorphous $TiO_2$ thin film (RMS 9.7 nm, 27 nm thick, horizontal distance between measurement points: 6.0 μm))[34]. Additionally, the roughness of the octadecyl group regions was shown to be 0.63 nm (horizontal distance between measurement points: 1.8 μm)[39]. This study showed the good performance of the LPP process using a seed layer and the importance of quantitative analysis of the deposition process.

### 2.4 Liquid phase patterning of crystalline aanatase $TiO_2$ thin films using site-selective elimination[40]

Ammonium hexafluorotitanate ($[NH_4]_2TiF_6$) (purity 96%, 1.031 g) and boric acid ($H_3BO_3$) (purity 99.5%, 0.932 g) were dissolved separately in deionized water (50°C, 50 ml)[40]. An appropriate amount of HCl was added to the boric acid solution to control pH, and ammonium hexafluorotitanate solution was added. Solutions (100 ml) with 0, 0.1 or 0.6 ml of HCl showed pH 3.8, 2.8 or 1.5, respectively. Supersaturation of solution can be changed by pH value as discussed in ref. 55. $TiO_2$ thin films can be formed fast by the deposition of homogeneously nucleated particles at high pH condition such as pH 3.88, and uniform films can be obtained slowly by heterogeneous nucleation at low pH condition. SAMs were immersed in the solution (100 ml) containing 0.05 M $(NH_4)_2TiF_6$ and 0.15 M ($H_3BO_3$) at pH 1.5, 2.8 or 3.8 and kept at 50°C for 4 h. to deposit anatase $TiO_2$. Ultrasonication was done during immersion period. Deposition of $TiO_2$ proceeded by the following mechanism:

$$TiF_6^{2-} + 2H_2O \leftrightarrow TiO_2 + 4H^+ + 6F^- \quad \text{(a)}$$

$$BO_3^{3-} + 4F^- + 6H^+ \rightarrow BF_4^- + 3H_2O \quad \text{(b)}$$

Equation (a) is described in detail by the following two equations:

$$TiF_6^{2-} \xrightarrow{nOH^-} TiF_{6-n}(OH)_n^{2-} + nF^- \xrightarrow{(6-n)OH^-} Ti(OH)_6^{2-} + 6F^- \quad \text{(c)}$$

$$Ti(OH)_6^{2-} \longrightarrow TiO_2 + 2H_2O + 2OH^- \quad \text{(d)}$$

Fluorinated titanium complex ions gradually change into titanium hydroxide complex ions in an aqueous solution as shown in Eq. (c). Increase of $F^-$ concentration displaces the Eq. (a) and (c) to the left, however, produced $F^-$ can be scavenged by $H_3BO_3$ ($BO_3^{3-}$) as shown in Eq. (b) to displace the Eq. (a) and (c) to the right. Anatase $TiO_2$ was formed from titanium hydroxide complex ions ($Ti(OH)_6^{2-}$) in Eq. (d), and thus the supersaturation degree and the deposition rate of $TiO_2$ depend on the concentration of titanium hydroxide complex ions. The high concentration of $H^+$ displaces the equilibrium to the left in Eq. (a), and the low concentration of $OH^-$, which is replaced with $F^-$ ions, suppresses ligand exchange in Eq. (c) and decreases the concentration of titanium hydroxide complex ions at low pH such as pH 1.5. The solution actually remained clear at pH 1.5, showing its low degree of supersaturation. On the other hand, the solution at high pH such as pH 2.8 or 3.8 became turbid because of homogeneously-nucleated anatase $TiO_2$ particles caused by a high degree of supersaturation. Anatase $TiO_2$ thin film was formed by heterogeneous nucleation in the solution at pH 1.5, while the film was formed by heterogeneous nucleation and deposition of homogeneously nucleated particles at pH 2.8 or 3.8.

After having been immersed in the solution with ultrasonic treatment, the substrates were rinsed with distilled water. Thin films were observed on the silanol group regions to form nano/micro-scaled patterns at pH 3.88, 2.8 or 1.5. The films were deposited for 4 h from the solution without the addition of HCl. Thin films were observed as being dark in an optical micrograph. Separated parallel lines 200–400 nm in width at 100–200 nm intervals were successfully fabricated with this method. The length of the separated parallel lines reached more than 100 $\mu$m. A cross section of the lines was shown as a semicircle, and the thickness of the center of the lines was estimated to be about 100 nm by AFM observation (Fig. 6). Feature edge acuity of the pattern was higher than that of the pattern fabricated by our lift-off process or by the site-selective immersion method.[41] Site-selective deposition was realized at any pH conditions such as pH 3.88, 2.8 or 1.5. Patterns which have higher feature edge acuity can be obtained at low pH conditions because films were formed slowly without the deposition of homogeneously nucleated particles.

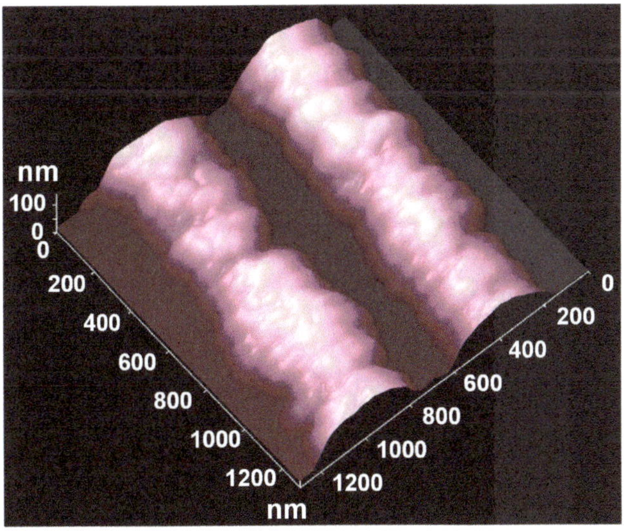


Reprinted with permission from Ref. [40], Masuda, Y., Saito, N., Hoffmann, R., De Guire, M. R. and Koumoto, K., 2003, *Sci. Tech. Adv. Mater.*, 4, 461. Copyright @ American Chemical Society

Fig. 6. AFM image of a nanopattern of anatase $TiO_2$ fabricated by site-selective elimination.

For the adhesion of $TiO_2$ films to silanol groups, the pH of the deposition solution is critical. Pizem *et al.*[13] reported that adherent $TiO_2$ films formed from solutions similar to those used here at pH=3.9, but that the films were less adherent at pH=2.9. They related this difference in adherence to an increased electrostatic attraction of $TiO_2$ to the oxidized surface of silicon at the higher pH. In our experiment, there also were less adherence of $TiO_2$ at pH 2.8 and 1.5 than that at pH 3.5 due to low supersaturaiton shown in Eq. (a) and low electrostatic attraction.[13]

However, some depositions were observed on octadecyl group regions in SEM micrographs. One probable cause is that pinholes and other defects in the films provide at least some degree of access of water to underlying unreacted OH groups in the OTS films. Once exposed to the solution, these sites can act as nucleation points for $TiO_2$ growth. Because the depositions are performed at elevated temperatures, it is likely that pinholes and defects will continually open and close on the OTS film surfaces due to thermal motions of alkyl chains in the films. The $TiO_2$ precursors formed in these defects would act as points for eventual growth of $TiO_2$ over the entire SAM-covered region. This would provide a weakly bound $TiO_2$ film on the OTS film regions due to the limited number of connections to the underlying silanol sites in the film regions. In fact, Sagiv and others have shown that macroscopic defects induced in alkylsiloxane films can readily be accessed by solution species.[42-44] More recently, Dressick and coworkers demonstrated that solvent accessibility to underlying substrates in aromatic siloxane films is also important[45-47] and may be an even greater factor in controlling the properties of those films, which may account for our previous selectivity observation using phenylsiloxane films, as shown in the lift-off process.

In the lift-off process[48], thin films were formed on the entire area of patterned SAM that has silanol group regions and phenyl (or octadecyl) group regions. After being dried, the substrate was sonicated in water to lift off thin films on phenyl (or octadecyl) group regions selectively. Thin films on phenyl (or octadecyl) group regions were peeled off along the cracks that formed during the drying process. Thin films on phenyl (or octadecyl) group regions without cracks were not peeled off because depositions strongly connected to each other to form solid timber (monolith). The lift-off along cracks decreased the feature edge acuity of the pattern in this method. Thin films were formed on silanol group regions selectively and site-selective deposition was realized with our newly developed method. This resulted in high feature edge acuity of the patterns compared to our previous works.[48]

Additionally, the micropattern of thin films was also fabricated by the site-selective immersion method[41]. A solution containing a Ti precursor contacted the hydrophilic regions during the experiment and briefly came into contact with the hydrophobic regions. The solution on the hydrophilic surface was replaced with a fresh solution by continuous movement of bubbles. Thus $TiO_2$ was deposited and a thin film was grown on the hydrophilic regions selectively. This technique can be applied for the formation of many kinds of films from any solution and to fabricate micropatterns for many kinds of thin film because the technique creates the difference in contact time of the solution between hydrophilic regions and hydrophobic regions. However, it is difficult to form a solution layer on nano-scaled hydrophilic regions selectively and replace it with a fresh solution by continuous movement of bubbles while avoiding contact of the solution on hydrophobic regions. This prevents fabrication of nano-scaled pattern with this method. On the other hand, site-selective deposition was realized in the solution with our newly developed method using the difference of adhesive strength of depositions to substrates. Heterogeneously nucleated deposition and homogeneously nucleated particles and/or clusters can be removed from octadecyl group regions even if these regions are designed in nano-scale order in which depositions are smaller. This allowed us to realize high feature edge acuity of the patterns compared to site-selective immersion.[39,41]

The distribution of elements on the surface of the substrates was evaluated by energy dispersive X-ray analysis (EDX; EDAX Falcon, EDAX Co. Ltd.), which is built into SEM. Titanium was detected from thin films selectively and oxygen was detected mainly from silanol group regions by EDX. Other elements, except for silicon from the substrate, were not observed from the thin film and substrate by EDX. Oxygen was detected from not only the deposited thin film but also from the natural oxide layer (amorphous $SiO_2$ layer) formed on all surface areas of a silicon substrate. These observations showed predominant deposition of titanium oxide on silanol group regions.

The deposited thin films were also investigated using an X-ray diffractometer (XRD; RAD-C, Rigaku) with CuKα radiation (40 kV, 30 mA) and Ni filter plus a graphite monochromator. Thin films deposited at pH 3.8 for 4 h showed an XRD pattern of anatase-type $TiO_2$ having orientation similar to that of films deposited in the solution at pH 1.5 or 2.8[41]. The diffraction from parallel to c-plane such as (004) was observed as being strong compared to that of the randomly oriented powder diffraction pattern. Pizem *et al.* postulated that the commonly observed [001] orientation of anatase films could be due to the slight polarity of the planes parallel to the [001] axis, unlike other low-index planes of this structure such as {100}, (110), and (210). The orientation and crystal growth mechanism are further discussed in a separate article.[49]

Thin films were further evaluated by X-ray photoelectron spectroscopy (XPS; ESCA-3200, Shimazu Corporation, $1 \times 10^{-5}$ Pa). The X-ray source (MgKα, 1253.6 eV) was operated at 8 kV and 30 mA. The spectral peaks corresponding to Ti 2p (458.7 eV) were observed from thin films deposited on the silanol region. This binding energy is higher than that of Ti metal (454.0 eV), TiC (454.6 eV), TiO (455.0 eV), TiN (455.7 eV) and $Ti_2O_3$ (456.7 eV), and similar to that of $TiO_2$ (458.4 - 458.7 eV).[50-52]. This suggests that the titanium atoms in thin films are positively charged relative to that of titanium metal by formation of direct bonds with oxygen. On the other hand, this spectrum was not observed from octadecyl group regions. The O 1s spectrum was observed from the silanol regions and divided into O 1s (530.2 eV) and O 1s (532.3 eV). O 1s (532.3 eV) can be assigned to the silicon oxide layer on the surface of the silicon wafer (532.0 eV[51]), whereas the binding energy of O 1s (530.2 eV) is similar to that of $TiO_2$ (529.9 eV[52], 530.1 eV[51]) as observed by Shin *et al.*[53]. This shows that oxygen is negatively charged compared with neutral oxygen molecules (531.0 eV), possibly through the formation of chemical bonds with Ti. The ratio of titanium to oxygen was estimated from the Ti $2p_{3/2}$ (458.7 eV) spectrum and O 1s (530.2 eV) spectrum to be Ti:O = 1:2.0.

## 2.5 Liquid phase patterning of magnetite particulate thin films using pd catalyst[54,55]

Catalyst solution containing $Na_2PdCl_4$ (0.38 mM) and NaCl (0.01 M) in a 0.01 M 2-morpholinoethane sulfonate pH 5 aqueous buffer was prepared[54,55]. The details of preparation of this solution are described in the reference[56]. Hydrolyzed Pd colloids were formed in this solution.[57] The patterned APTS-SAM was immersed into the colloidal dispersion of catalyst at 25 °C for 30 min and catalyzed APTS-SAM was rinsed with water.

Catalyzed SAM was immersed in an aqueous solution containing iron(III) nitrate (0.0025 M) and dimethylamine-borane (DMAB) (0.03 M) and kept at 80 °C using a water bath for 30 min to deposit magnetite particulate thin film.[58] DMAB was used to reduce nitrate ions, giving rise to OH- ions and hence raising the solution pH to precipitate $Fe_3O_4$.

A black colored iron oxide film selectively deposited onto regions of the APTS-SAM that had not been exposed to UV radiation following application of the Pd catalyst dispersion.

Figures 7 show an optical microscope image and a SEM image of as-deposited films, respectively. Black contrast represents a deposited film in an optical microscope image, whereas the white contrast shows deposited films in SEM images. The EDX mapping images shown in Figure 7(c) indicate the films deposited on the amino-surface regions, showing mapping images consisting mainly of iron and oxygen. Thickness of the films was easily controlled in the range from several ten nano meter to several micro meter by change of immersion period. The XRD pattern of the thin film deposited on the whole surface of the APTS-SAM clearly indicates that it is a magnetite ($Fe_3O_4$) film composed of randomly oriented crystallites of about 20 nm in diameter, which was evaluated using the Scherer equation. These evaluations show the successful fabrication of a micropattern of crystalline magnetite films in an aqueous solution using a patterned APTS-SAM and Pd colloid catalysts adsorbed on amino-group (-$NH_2$) regions of a SAM.

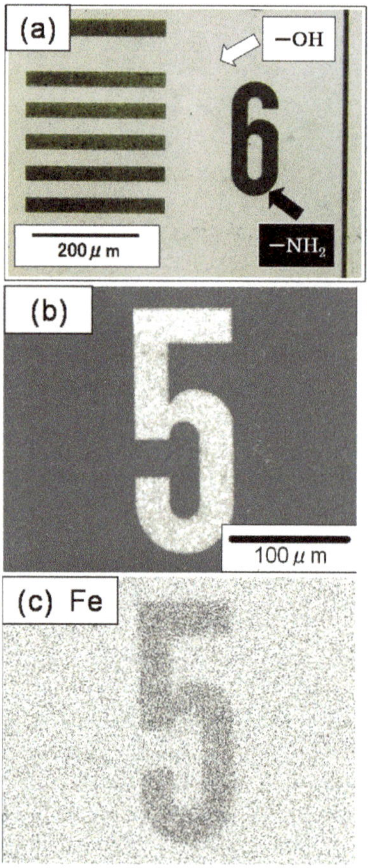


Reprinted with permission from Ref. [54], Nakanishi, T., Masuda, Y. and Koumoto, K., 2004, *Chem. Mater.*, 16, 3484. Copyright @ American Chemical Society

Fig. 7. (a) optical microscope image, (b) SEM image and (c) characteristic X-ray images [Fe] of a micropattern of crystalline Fe3O4.

### 2.6 Liquid phase patterning and morphology control of crystalline ZnO[33]

Zinc acetate ($Zn(CH_3COO)_2$, Kishida Chemical Co., Ltd.) was dissolved into water to be 15 mM at 50°C, and ammonia (28% solution, Kishida) was then added to be 30, 60 or 90 mM ($[NH_3]$ / $[Zn]$ = 2.0, 4.0 or 6.0) with stirring as complexing agent[33]. These solutions showed pH = 7.04, 7.50 or 8.93, respectively. Zinc ions reacted with ammonium ions ($NH_4^+$) formed from ammonia to form tetra amine zinc(II)[59] $[Zn(NH_3)_4]^{2+}$. ZnO was crystallized from the reaction between $[Zn(NH_3)_4]^{2+}$ and $OH^-$. The solution became clouded shortly after adding ammonia due to homogeneous nucleation of ZnO crystals. Morphology of ZnO crystals was controlled by the ratio of ammonia to zinc acetate, i.e., super-saturation degree for crystallization. Patterned OTS-SAMs were immersed downward into the solution containing zinc acetate (15 mM) and ammonia (30 mM) as complexing agent ($[NH_3]$ / $[Zn]$ = 2.0) at 50°C for 3 h (Fig. 8).

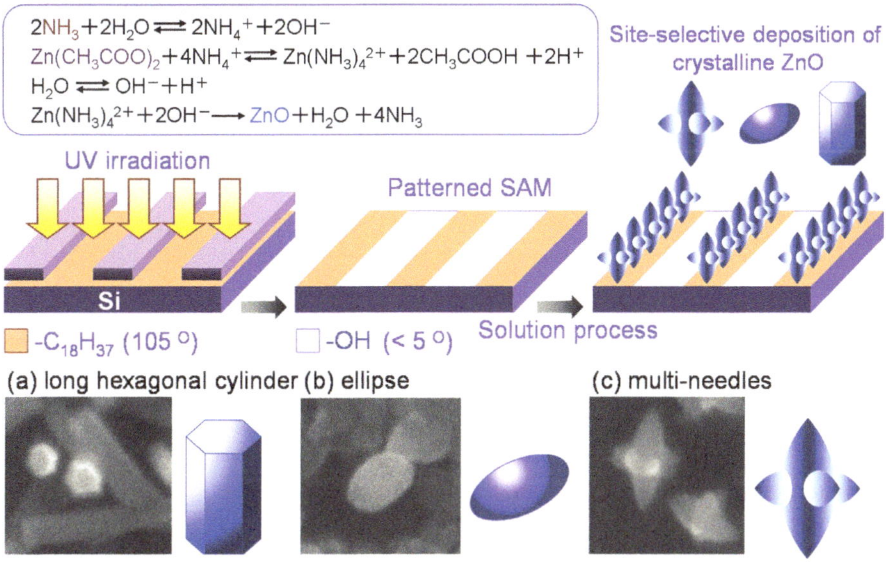


Reprinted with permission from Ref. [33], Masuda, Y., Kinoshita, N., Sato, F. and Koumoto, K., 2006, *Cryst. Growth Des.*, 6, 75. Copyright @ American Chemical Society

Fig. 8. Conceptual process for self-assembly patterning of light-emitting crystalline ZnO nanoparticles in an aqueous solution.

ZnO crystals having long hexagonal cylinder shape were homogeneously nucleated to make the solution turbid shortly after adding ammonia. Crystals showed sharp hexagonal facets of about 100 nm in diameter and larger than 500 nm in length. The morphology indicated high crystallinity of ZnO nanoparticles. The nanoparticles were deposited and further grown on hydrophobic octadecyl group regions of a patterned SAM selectively. Consequently, a micropattern of light-emitting ZnO crystals was successfully fabricated in an aqueous solution without Pd catalyst. ZnO crystals were also deposited on hydrophobic regions of patterned SAMs such as DTS-SAM, HTS-SAM, PTS-SAM, MTS-SAM, PTCS-SAM or APTS-SAM. This showed that the method is highly versatile and offers good potential for the fabrication of devices.

ZnO crystals were deposited on hydrophobic SAM regions such as OTS-, APTS- or other SAMs rather than hydrophilic silanol regions. Zeta potential of ZnO crystals deposited on a silicon substrate was measured to be 10 mV at pH 8.1 and ZnO crystals should thus have positive zeta potential not less than 10 mV in the solution at pH 7.04. SAM of OTS, silanol and APTS showed zeta potential of -3 mV, -38.2 mV or +22.0 mV, respectively. ZnO having positive zeta potential should be deposited on silanol regions having negative zeta potential rather than other SAMs, if the site-selective deposition was caused only by electrostatic interactions. The site-selective deposition of ZnO crystals would be caused by not only electrostatic interactions as shown by the relation of zeta potentials. ZnO crystals having long hexagonal cylinder shape were deposited on a hydrophilic silicon substrate to evaluate the surface of crystals. The substrate covered with many deposited ZnO crystals exhibited high water contact angle (WCA 140°). The deposited ZnO crystals were found from the experiment to have hydrophobic surfaces. Surfaces of naked ZnO crystals would be hydrophilic because of surface hydroxyl groups and they would become hydrophobic by being covered with organic molecules having hydrophobic functional groups. $CH_3COO^-$ ions coming from $Zn(CH_3COO)_2$ might be adsorbed to ZnO crystal surfaces by the interaction between Zn and $-COO^-$ to cover the surface with hydrophobic $-CH_3$ groups and some of $Zn(CH_3COO)_2$ would exist in the surface layer of ZnO crystals. Additionally, deposited ZnO crystals having long hexagonal cylinder shape became hydrophilic (< 10°) and their zeta potential shifted positively by UV irradiation in air. ZnO crystals deposited on a silicon substrate showed zeta potential of 10 mV at pH 8.1, 0 mV at pH 8.8 and -15 mV at pH 9.2, while they shifted to 20 mV at pH 8.1, 10 mV at pH 8.8 and 7 mV at pH 9.2 by UV irradiation. The decomposition of $CH_3COO^-$ ions and the breakage of the bond between $CH_3COO$ and Zn would be caused by light excitation, ozone and active oxygen by UV irradiation in air. This finding suggests that organic molecules, such as $CH_3COO^-$ ions, which show negative zeta potential and can be removed by UV irradiation, would be absorbed onto the surfaces of ZnO crystals. Furthermore, ZnO crystals were confirmed to deposit on a hydrophobic polyethylene terephthalate surface rather than on a hydrophilic polyethylene terephthalate surface modified by UV irradiation in the same solution. Additionally, organic molecule was reported to adsorb to growing ZnO crystals, in which poly (ethylene oxide)-blockpoly (methylacrylic acid) (PEO-b-PMAA) was adsorbed preferentially to {0001} face of ZnO to retard crystal growth perpendicular to this face[60]. Consequently, site-selective deposition was achieved by the effective molecular recognition caused by combination of the forces composed mainly of hydrophobic interactions between functional groups of SAMs and ZnO crystal surfaces.

Patterned SAMs were also immersed into the solution containing zinc acetate (15 mM) and ammonia (60 mM or 90 mM) as complexing agent ($[NH_3]/[Zn] = 4.0$ or 6.0) for 3 h. ZnO crystals having ellipse or multi-needle shape (two large needles and four small needles) were homogeneously nucleated to make the solution turbid shortly after adding ammonia. Nucleation and deposition of ZnO crystals were accelerated by addition of ammonia. Each ZnO crystal was about 500 nm in size. The crystals were deposited and further grown on hydrophobic regions of patterned SAMs selectively. Micropatterns of light-emitting ZnO crystals having ellipse or multi-needle shape were fabricated on patterned SAMs such as OTS-SAM, DTS-SAM, HTS-SAM, PTS-SAM, MTS-SAM, PTCS-SAM or APTS-SAM in aqueous solutions.

XRD spectra of ZnO crystals having ellipse or multi-needle shape showed dominant peaks corresponding to ZnO (0002) planes revealing that ZnO crystals were deposited with a high degree of orientation of their c-axes perpendicular to the substrate. Enhanced (0002) and (10-10) peaks from ZnO crystals having long hexagonal cylinder shape showed that crystals were deposited to make (0002) or (10-10) planes parallel to the substrate. Crystals having high crystallinity and high purity with no additional phase were shown to be prepared in an aqueous solution with precise control of their morphologies without the use of Pd catalyst. The aqueous solution system showed high ability for fabricating nano/micro devices composed of crystalline materials. ZnO crystals are well known to grow along the c-axis. The orientations evaluated from XRD patterns were consistent with SEM observations and were shown to be controlled precisely by the solution conditions.

Photoluminescence properties of ZnO crystal patterns were further evaluated. Micropatterns of ZnO crystals were observed by an optical microscope (Fig. 9) and strong visible-luminescence from ZnO crystals excited by 330-385 nm light was observed by a photoluminescence microscope. ZnO crystals showed strong UV luminescence (around 390 nm) attributed to band-edge luminescence and visible-light luminescence caused from oxygen vacancy (450-600 nm)[61,62]. All of the crystals showed photoluminescencedue to high purity and high crystallinity with optimal oxygen vacancy, and this caused the bright visible-photoluminescence image. Luminescence properties can be controlled by changing the crystalline morphologies. ZnO crystals deposited from an aqueous solution were shown to have high visible-light-emitting properties.

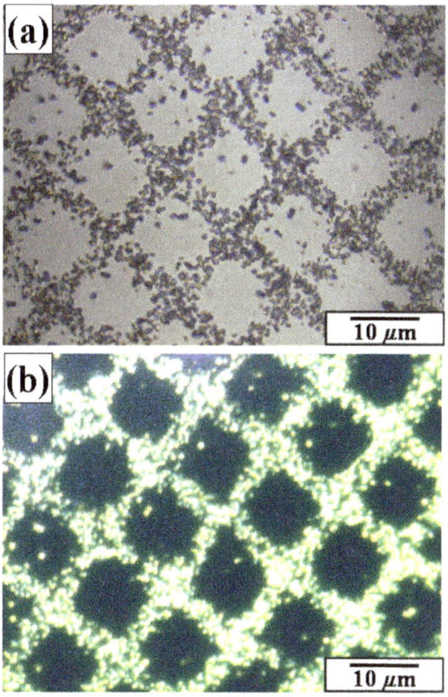


Reprinted with permission from Ref. [33], Masuda, Y., Kinoshita, N., Sato, F. and Koumoto, K., 2006, *Cryst. Growth Des.*, 6, 75. Copyright @ American Chemical Society

Fig. 9. (a) Optical microscope image and (b) photoluminescence image of patterned ZnO particles under white light or UV light (330-385 nm).

## 2.7 Liquid phase patterning of $Y_2O_3$: eu thin films[63]

The patterned APTS-SAM was immersed in an aqueous solution containing $Y(NO_3)_3 \cdot 6H_2O$ (4 mM), $Eu(NO_3)_3 \cdot 6H_2O$ (0.4 mM) and $NH_2CONH_2$ (50 mM) at 25 °C[63]. The solution was heated to 77 °C gradually as shown in Fig. 10 since urea ($NH_2CONH_2$) decomposes to form ammonium ions ($NH_4^+$) above 70 °C (Eq. (a)). The decomposition of urea at elevated temperature plays an essential role in the deposition of yttrium oxide. The aqueous solution of urea yields ammonium ions and cyanate ions ($OCN^-$) at temperatures above 70 °C[64] (Eq. (a)). Cyanate ions react rapidly according to Eq. (b). Yttrium ions are weakly hydrolyzed[65,66] in water to $YOH(H_2O)_n^{2+}$ (Eq. (c)). The resulting release of protons ($H^+$) and/or hydronium ions ($H_3O^+$) accelerates urea decomposition (Eq. (b)). The precipitation of the amorphous basic yttrium carbonate ($Y(OH)CO_3 \cdot xH_2O$, x=1) can take place through the reaction in Eq. (d)[67,68]. The controlled release of cyanate ions by urea decomposition causes deposition of basic yttrium carbonate once the critical supersaturation in terms of reacting component is achieved. Since the decomposition of urea is quite slow, the amount needed to reach supersaturation within a given period of time must be considerably higher than the stoichiometric amount of yttrium ions, as revealed by previous studies of lanthanide compounds[69].

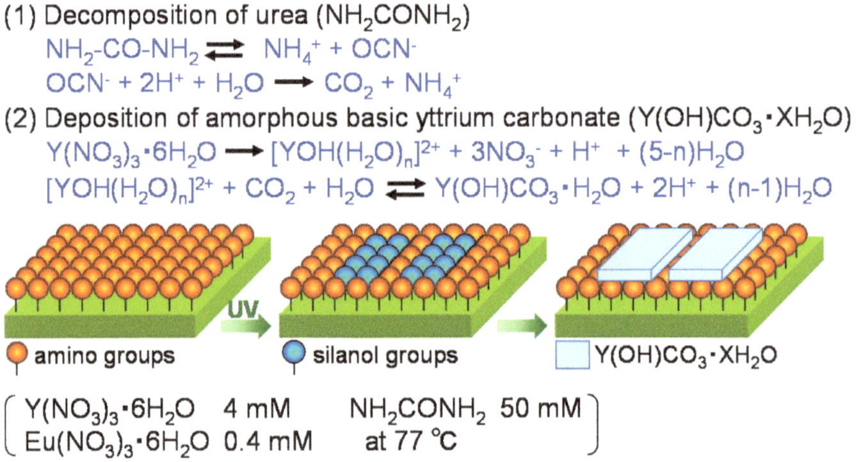


Reprinted with permission from Ref. [63], Masuda, Y., Yamagishi, M. and Koumoto, K., 2007, *Chem. Mater.*, 19, 1002. Copyright @ American Chemical Society

Fig. 10. Conceptual process for site-selective deposition of visible-light emitting Y2O3:Eu thin films using a self-assembled monolayer.

The temperature of the solution increased gradually and reached 77 °C in about 80 min. The solution was kept at ~ 77 °C during deposition. The pH of the solution increased from 5.2 to 5.8 in about 90 min and then gradually decreased to 5.6. Temperature and pH increased for the initial 90 min and became stable after 90 min. The average size of particles homogeneously nucleated in the solution at 100 min was about 227 nm and increased to 262 nm at 150 min, 282 nm at 180 min, 310 nm at 210 min, and 323 nm at 240 min. Particles nucleated and grew after the solution temperature exceeded 70 °C because urea decomposes above 70 °C to form carbonate ions[64] which causes deposition of basic yttrium carbonate[65-68].

The particles grew rapidly at the beginning of the growth period and then their growth rate decreased exponentially. The decrease in growth rate was caused by the decrease of supersaturation degree influenced by a decrease in solution concentration.
Yttrium carbonate films were observed to deposit on amino regions of a patterned SAM after the immersion in an aqueous solution (Fig. 11). Deposits showed white contrast, while silanol regions without deposition showed black contrast in SEM observation. Narrow lines of depositions having 10–50 μm width were successfully fabricated in an aqueous solution. Patterned APTS-SAM showed high ability for site-selective deposition of yttrium carbonate in solution systems.

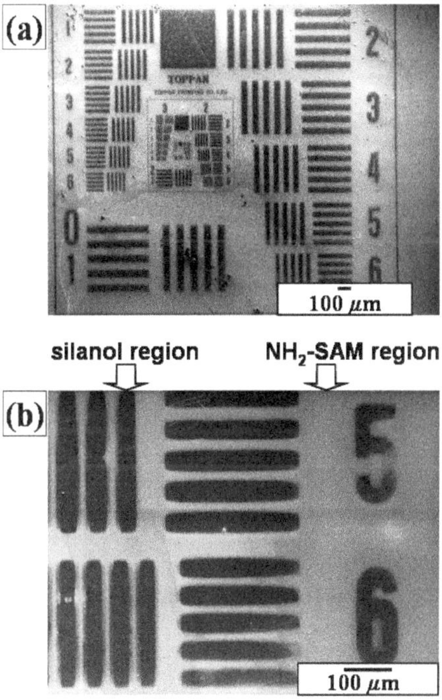


Reprinted with permission from Ref. [63], Masuda, Y., Yamagishi, M. and Koumoto, K., 2007, *Chem. Mater.*, 19, 1002. Copyright @ American Chemical Society

Fig. 11. (a) SEM micrograph of patterned $Y_2O_3$:Eu thin films and (b) magnified area of (a).

Yttrium carbonate films were also deposited on the hydrophobic octadecyl surface of OTS(octadecyltrichlorosilane)-SAM having water contact angle (WCA) of 116 ° and as-purchased silicon wafer having WCA of about 20-50 ° which was kept in a plastic case in air. On the other hand, the films were not deposited on UV irradiated silicon wafer having WCA < 5 °. The super hydrophilic surface of WCA < 5 ° suppressed film deposition, whereas the hydrophobic surface and medium surface of WCA > 20-30 ° accelerated film deposition possibly because of hydrophobic interaction between deposition and substrate surface. This is consistent with a former study[32]. Yttrium carbonate was deposited both on bare single crystal Si wafers, and on Si wafers coated with sulfonate-functionalized organic self-assembled monolayers.

Yttrium, europium, oxygen and carbon were observed from as-deposited thin films on amino regions, while silicon and oxygen were detected from non-covered silanol regions by EDX. The molecular ratio of yttrium to europium was determined to be 100 : 8. It was close to that of $Y(NO_3)_3 \cdot 6H_2O$ to $Eu(NO_3)_3 \cdot 6H_2O$, i.e., 100 : 10, in the solution because the chemistry of $Eu(NO_3)_3$ is similar to that of $Y(NO_3)_3$ to incorporate europium in the precipitation. The content of europium was in the range we had expected. $Y_2O_3$:Eu with atomic ratio Y : Eu = 100 : ~ 8 was reported to have strong photoluminescence[70,71]. Carbon was detected from yttrium carbonate. Silicon and oxygen were detected from silicon wafer covered with a natural oxide layer (amorphous $SiO_2$).

Amino regions were covered with thin films composed of many large particles (about 100–300 nm in diameter) and very high roughness (RMS 25.6 nm). Silanol regions, on the other hand, showed only nano-sized small particles (about 10–50 nm in diameter) and very low roughness (RMS 1.7 nm). The high site-selectivity of deposition and the big difference in surface morphology and roughness were clearly shown by AFM observation. The thickness of the films was estimated from AFM scans across deposited and undeposited regions of the substrate. It increased with immersion time after 45 min (0 nm at 45 min, 60 nm at 70 min and 100 nm at 90 min). The average growth rate (70 nm/h = 100 / 90 min) was higher than that previously reported (2 nm/h = 35 nm / 15 h)[68]. An amorphous yttrium basic carbonate film was deposited at 80 °C from aqueous solutions of $YNO_3 \bullet 5H_2O$ and urea on Si wafers coated with sulfonate-functionalized organic self-assembled monolayers in previous studies. The thickness was then evaluated by TEM after the treatment with ultrasonication for half an hour in distilled water. The difference of growth rate was caused mainly by the difference of the substrate treatment by ultrasonication. Additionally, the thickness of our film was smaller than the particle size in the solution (227 nm at 100 min). Heterogeneous nucleation and attachment of initial particles of yttrium carbonate occurred without the attachment of aggregated large particles. The yttrium carbonate was then grown on the substrate to form a film of 100 nm thickness after immersion for 90 min. The particles of about 100 nm in height were removed by ultrasonication for 30 min and the film of several nm in height remained as reported[68].

Yttrium was not detected by XPS from the substrate immersed for 45 min, however, it was clearly observed from that immersed for 90 min. This indicates that the deposition began between 45 and 90 min after immersion. The solution temperature reached 70 °C in ~ 45 min and then the solution began to decompose and release carbonate ions, causing the deposition of basic yttrium carbonate. The deposition mechanism evaluated by XPS is consistent with the change of solution temperature, decomposition temperature of urea and chemical reaction of this system. The binding energy of Y 3d5/2 spectrum from the deposition (158.2 eV) was higher than that of metal yttrium (155.8 eV)[72]. The spectrum shifted to lower binding energy (156.7 eV) after annealing at 800 °C in air for 1 h and is similar to that of $Y_2O_3$ (157.0 eV)[73]. The binding energies of Y 3d5/2 spectra in as-deposited films and annealed films were higher than that of metal yttrium possibly due to the chemical bonds formed between yttrium ions and oxygen ions. The chemical shift of Y 3d5/2 binding energy by annealing is consistent with crystallization of as-deposited films to crystalline $Y_2O_3$. C 1s spectra were detected at 289.7 eV and 284.6 eV from as-deposited films. The C 1s spectrum at 289.7 eV then disappeared by the annealing. C 1s at 284.6 eV was assigned to surface contamination and C 1s at 289.7 eV was detected from as-deposited yttrium carbonate. The disappearance of C 1s at 289.7 eV is consistent with the phase transition from yttrium carbonate to $Y_2O_3$.

As-deposited film was shown to be an amorphous phase by XRD measurement. The film showed no diffraction peak after annealing at 400 °C for 1 h, however, it showed 222, 400 and 440 diffraction peaks of crystalline cubic $Y_2O_3$[74] without any additional phase after annealing at 600 °C for 1 h and the intensities of diffraction peaks increased further by annealing at 800 °C for 1 h. The film was shown to be a polycrystalline $Y_2O_3$ film constructed from randomly deposited $Y_2O_3$ particles without crystal-axis orientation. The crystal structure model and diffraction pattern of $Y_2O_3$ were calculated from the crystal structure data of ICSD #23811. The crystallization by annealing confirmed from XRD measurement is consistent with XPS evaluation.

$Y_2O_3$ films were attemped to remove from the silicon substrate by debonding with scotch tape or by ultrasonication for 5 min in water. However, the films maintained their bonds with the substrate, indicating that strong adhesion had formed between films and substrate.

The thin film annealed at 800 °C for 1 h, i.e., crystalline $Y_2O_3$:Eu thin film, was shown to be excited by 230–250 nm (center: 243 nm) and emit red light photoluminescence centered at 611 nm in the fluorescence excitation spectrum (Fig. 12a). Neither the as-deposited film nor the film annealed at 400 °C for 1 h showed photoluminescence, on the other hand, the films annealed at 600 °C or 800 °C for 1 h emitted light centered at 617 nm by 250 nm in fluorescence emission spectra (Fig. 12b). The fluorescence intensity of the film annealed at 800 °C was stronger than that of the film annealed at 600 °C. Fluorescence intensity increased by the phase transformation from amorphous yttrium carbonate to yttrium oxide and crystal growth by the heat treatments, and is consistent with the crystallization observed by XRD. The spectra are described by the well-known $^5D_0$–$^7F_J$ line emissions (J = 0, 1, 2, ...) of the $Eu^{3+}$ ion with the strongest emission for J = 2 at 612 nm. The thin film annealed at 800 °C produced visible red light photoluminescence by excitation from Nd: YAG laser (266 nm) (Fig. 12, inset). The white square shows the edges of the $Y_2O_3$:Eu thin film and the red color shows visible red emission from the irradiated area on the substrate.

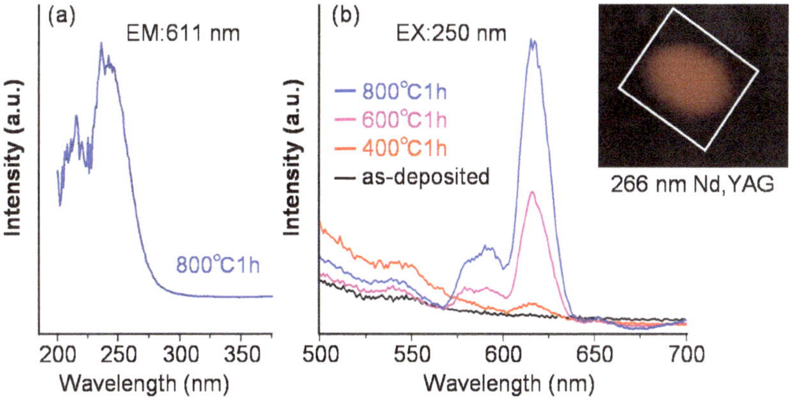


Reprinted with permission from Ref. [63], Masuda, Y., Yamagishi, M. and Koumoto, K., 2007, *Chem. Mater.*, 19, 1002. Copyright @ American Chemical Society

Fig. 12. (a) Fluorescence excitation spectrum (emission: 611 nm) for $Y_2O_3$:Eu thin film after annealing at 800 °C for 1 h. (b) Fluorescence emission spectra (excitation: 250 nm) for $Y_2O_3$:Eu thin films before and after annealing at 400, 600 or 800 °C for 1 h. Inset: Photoluminescence image for $Y_2O_3$:Eu thin film annealed at 800 °C for 1 h (excitation: 266 nm).

### 2.8 Liquid phase patterning of $In_2O_3$ thin films[75]

**Synthesis and patterning of Pd nanoparticles:** A catalyst dispersion[57,76] containing $Na_2PdCl_4$ (0.38 mM) and NaCl (0.01 M) in a 0.01 M 2-morpholinoethane sulfonate pH 5 aqueous buffer was prepared (Fig. 13) as described in references[56,57,76]. Hydrolyzed Pd colloids were formed in this solution[57]. Light-scattering measurements indicated that the catalyst dispersion contained colloid particles of about 30 nm in diameter. Pd nanoparticles showed negative zeta potential (-30.5 eV) at pH 5. APTS-SAM showed positive zeta potential[77] at pH 5 because of protonation of the amino group ($-NH_2$) to $-NH_3^+$. Silanol groups of UV irradiated APTS-SAM, on the other hand, showed negative zeta potential at pH 5 caused by deprotonation of the silanol group (-Si-OH) to $-Si-O^-$. The patterned APTS-SAM was immersed in the colloidal dispersion of catalyst at 25 °C for 30 min and the catalyzed APTS-SAM was rinsed with water. Pd colloids adsorbed on amine groups of APTS-SAM by electrostatic interactions between the negative surface charge of Pd colloids and positive surface charge of APTS-SAM[54] and formed covalent bonds[57], while electrostatic repulsion force kept Pd catalyst particles away from silanol group regions having negative zeta potential.

Pd nanoparticles deposited on amino group regions of a patterned SAM had a diameter of about 30 nm and surface roughness (RMS) of about 1 nm as shown by AFM observation (Fig. 13). Pd was found to be adsorbed on amino group regions uniformly to form a thin catalytic layer with small surface roughness. Even application, small thickness and small surface roughness of the Pd layer are significant for deposition of a uniform transparent indium oxide layer.

Furthermore, site-selective adsorption of Pd was clearly observed by TOF-SIMS mapping[54]. Bright regions due to Pd (m/z = 104, 105, 106, 108, and 110) were observed on the APTS-SAM surface, while no Pd signal was seen on the silanol group surface. This result clearly indicates that site-selective adsorption of Pd catalyst occurred on the APTS-SAM surface.

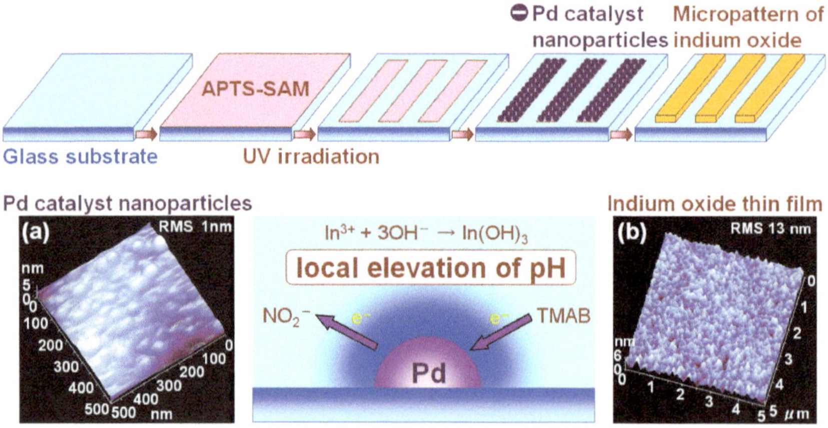


Reprinted with permission from Ref.[75], Masuda, Y.; Kondo, M.; Koumoto, K., 2009, *Cryst. Growth Des.*, 9, 555. Copyright @ American Chemical Society

Fig. 13. Conceptual process for micropatterning of indium oxide thin films on a glass substrate. AFM images of (a) Pd catalyst nanoparticles and (b) indium oxide thin films (lower stand).

**Deposition control of $In(OH)_3$ thin films by Pd catalyst:** The patterned SAM having Pd catalytic nanoparticles on amino group regions was immersed in an aqueous solution containing $In(NO_3)_3$ and TMAB at 65 °C for 1 h (Fig. 13).
Nitrate ions were generated by dissolution of indium nitrate in water according to:

$$In(NO_3)_3 \rightarrow In^{3+} + 3NO_3^- \quad \text{(a)}$$

Pd catalyst is indispensable for Eq. (b) and (c). Oxidation of reducing agent, TMAB, is promoted by Pd catalyst to generate electrons in Eq. (b). Nitrate ions are reduced to nitrite ions by receiving electrons according to Eq. (c). Hydroxide ions are generated by oxidation-reduction reactions near Pd catalyst according to Eq. (c). Local elevation of pH thus occurs near Pd catalysts.

$$(CH_3)_3NBH_3 + 2H_2O \rightarrow BO_2^- + (CH_3)_3N + 7H^+ + 6e^- \quad \text{(b)}$$

$$NO_3^- + H_2O + 2e^- \rightarrow NO_2^- + 2OH^- \quad \text{(c)}$$

$In(OH)_3$ nucleates and grows at high pH according to Eq. (d).
$In(OH)_3$ is thus deposited in Pd-adsorbed regions of patterned SAM (Fig. 13).

$$In^{3+} + 3OH^- \rightarrow In(OH)_3 \quad \text{(d)}$$

Transparent conducting $In_2O_3$ thin film is fabricated by annealing at 300 °C in a reduced atmosphere (3% - $H_2$ / $N_2$) for 1 h according to Eq. (e).

$$2In(OH)_3 \rightarrow In_2O_3 + 3H_2O \quad \text{(e)}$$

**Liquid phase patterning of $In(OH)_3$ thin films:** After having been immersed in the solution containing $In(NO_3)_3$ and TMAB, the patterned SAM having Pd catalytic nanoparticles on amino regions was rinsed with distilled water and dried in air. The thin film was clearly shown by SEM observation to deposit on amino group regions selectively (Fig. 14). Silanol group regions and amino group regions of SAM were shown to be black or white, respectively. Magnified SEM micrograph (b) shows the surface morphology of deposited thin films. The thin films were continuous films without micrometer-scale cracks.
The distribution of elements on the surface of the substrates was evaluated by EDX. Indium was detected from thin films on amino group regions selectively and appeared black in EDX mapping images (Fig. 14). On the other hand, silicon was detected mainly from silanol group regions which were not covered with depositions and exposed bare silicon substrate (Fig. 14). These observations showed the site-selective deposition of thin films containing indium on amino group regions.
Surface morphology was further evaluated by AFM conducted at room temperature under ambient air. Thin films were observed on the amino group regions selectively to form micro-scale patterns. The surface of the thin films showed a uniform morphology and low surface roughness RMS = 13 nm (Fig. 13). The thin films deposited on a glass substrate were found to be transparent, which would be caused by the low surface roughness which reduces diffuse reflection. In-plane particle size was estimated to about 10 - 25 nm in diameter. Film thickness at the edge of the thin film was estimated to be 84 nm.

As-deposited thin film was shown to be crystalline $In(OH)_3$ (JCPDS No. 16-0161) with no additional phase by XRD evaluation. 002 and 004 diffraction peaks of $In(OH)_3$ only were detected. $In(OH)_3$ thin film was shown to have high c-axis orientation. Crystalline size was estimated to be 17.4 nm by using the 002 diffraction peak. This was consistent with the in-plane particle size estimated by AFM observation. Thus, each particle comprising the thin film would be a single crystal.

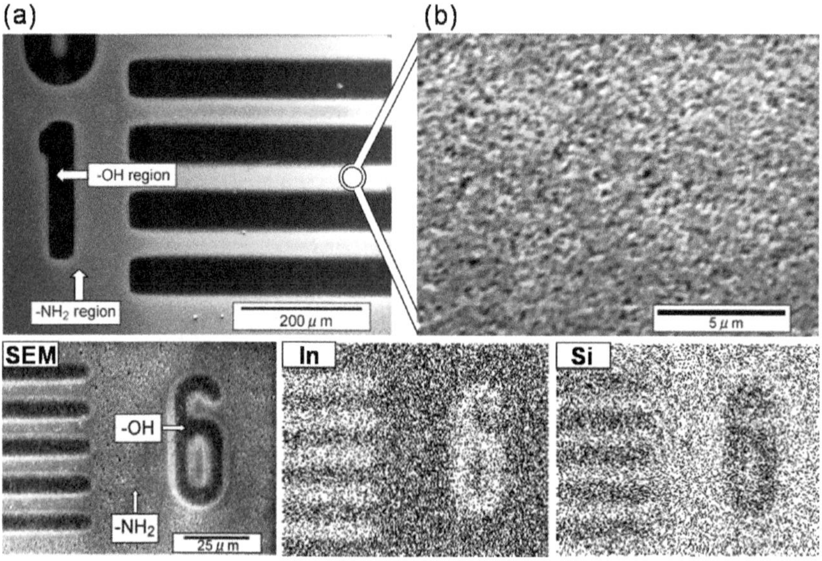


Reprinted with permission from Ref.[75], Masuda, Y.; Kondo, M.; Koumoto, K., 2009, *Cryst. Growth Des.*, 9, 555. Copyright @ American Chemical Society

Fig. 14. SEM micrographs of (a) micropattern of indium oxide thin films and (b) magnified area of (a) (upper stand). SEM micrograph and EDX images for In and Si (lower stand).

**Micropatterning of $In_2O_3$ thin films and its optical properties:** $In(OH)_3$ thin film was annealed at 200, 250 and 300 °C in air. $In(OH)_3$ thin films transformed into single-phase crystalline $In_2O_3$ above 250 °C. The film annealed at 300 °C showed 222, 400, 332, 431 and 440 diffraction peaks of $In_2O_3$ (JCPDS No. 44-1087) with no additional phase. Crystalline size was estimated to be 6.8 nm using the 222 diffraction peak, which was about 0.4 times smaller than that of $In(OH)_3$.

Thin films maintained their uniform surface morphology in the annealing at 250 °C in air and showed a low surface roughness RMS = 11 nm. In-plane particle size was estimated to be about 10 - 25 nm in diameter. Film thickness on the edge of the thin film was estimated to be 45 nm. Shrinkage of film thickness would be caused by volume decrease during crystallization to $In_2O_3$.

The thin films deposited on a glass substrate showed transparency of 60 - 70% in the visible light region and this would be caused by the low surface roughness which reduced diffuse reflection to 5 - 15%.

The optical band gap energy for direct transition in $In_2O_3$ thin films was estimated to be 3.7 eV assuming that all of the interband transition was direct transition.

**Electrical property of $In_2O_3$ thin films:** $In(OH)_3$ thin film was annealed at 300 °C in a reduced atmosphere (3% - $H_2$ / $N_2$) for 1 h instead of atmospheric heating to induce oxygen vacancies to increase the carrier concentration. 222, 400, 332, 431 and 440 diffraction peaks of $In_2O_3$ (JCPDS No. 44-1087) were observed from the thin film after annealing with no additional phase. Crystalline size was estimated to be 9.8 nm after annealing at 300 °C in a reduced atmosphere using the 222 diffraction peak, which was about 0.6 times smaller than that of $In(OH)_3$ and slightly larger than that annealed at 300 °C in air.

Carrier concentration and Hall mobility were evaluated to be $2.1\times10^{19}$ $cm^{-3}$ and 5.2 $cm^2$ $V^{-1}$ $s^{-1}$, respectively by Hall effect measurement. Specific resistance was evaluated to be $5.8\times10^{-2}$ Ωcm by the Van der Pauw method. These electrical properties are similar to those of $In_2O_3$ thin films prepared by the sol-gel method[78] (carrier concentration: $1.7\times10^{19}$ $cm^{-3}$, Hall mobility: 5.9 $cm^2$ $V^{-1}$ $s^{-1}$, specific resistance: $6.1\times10^{-2}$ Ωcm). Electrons would be scattered by grain boundaries in thin films and this would decrease Hall mobility and increase specific resistance. Increase of carrier concentration and decrease of grain boundaries by optimization of the reduction conditions and film formation process would allow us to obtain higher Hall mobility and lower specific resistance, thus improving the electrical properties.

**XPS analysis: (a) XPS analysis for patterning of Pd nanoparticles:** Pd was detected from amino group regions of a patterned SAM by XPS. Detection of Pd from amino group regions is consistent with AFM observation and the deposit observed in the AFM image was shown to be Pd colloids particles. On the other hand, Pd was not observed in silanol regions, i.e., UV-irradiated regions of APTS-SAM, even with XPS which is a highly surface sensitive analysis method. Pd adsorption on silanol regions would be less than the detection limit of XPS. High site-selectivity of Pd deposition only on amino group regions of a patterned SAM was shown in XPS analysis.

Pd 3d $_{5/2}$ and 3d $_{3/2}$ were observed for APTS-SAM at 337.2 eV, 342.5 eV, respectively. The binding energy of Pd 3d $_{5/2}$ observed is higher than that of Pd metal (334.6 eV[79], 335.1 eV[80-82]). Referring to an earlier report[57], the spectrum of Pd 3d $_{5/2}$ can be deconvoluted into 3 peaks with peak positions corresponding to Pd-N (338.7 eV)[83], Pd-Cl (337.8 eV)[84-86] and Pd-O (336.9 eV)[87] to be Pd-N : Pd-Cl : Pd-O = 0.01 : 0.11 : 0.88[54] (peak ratio). Pd on the APTS-SAM was mainly combined with O as Pd-O (336.9 eV). This result indicates that the surface of the Pd colloid layer on APTS-SAM consists of Pd-O or Pd-OH as well as a small amount of Pd-Cl. Although the Calvert group reported that the hydrolyzed Pd particles form covalent bonds with other SAMs which have amine groups[57], the Pd-N bond was not observed in our XPS experiment because of the relatively low depth analyzed by the method (the escape depth of the photoelectrons at the binding energy corresponding to Pd 3d 5/2 is few nanometers), but does not exclude that Pd-N bonds might have been present also in our case.

N 1s spectra of amino group surfaces were detected before and after the immersion into Pd nanoparticle solution, though the spectrum intensities were very low. N1s binding energy of the amino group surface covered with Pd nanoparticles (400.4 eV) was higher than that before immersion (399.6 eV). The positive shift of N1s was about 0.8 eV and is similar to that observed between Pd nanoparticles and the amino group of 3-aminopropyltriethoxysilane-SAM (about 0.8 eV)[88]. The shift of N 1s would be caused by the decrease of electron cloud density around nitrogen atoms and suggests the formation of chemical bonds between nitrogen atoms and Pd ions. The amino group is a strong electron donor and can coordinate to transition metal ions due to the lone pair electrons of nitrogen atoms. On the other hand, the outermost electron of soft metal ion Pd (II) is constructed from $4d^85s^05p^0$ and has an empty lower energy orbit that can accept electrons. Thus, Pd would form strong bonds with nitrogen rather than oxygen and chloride[88].

**(b) XPS analysis for patterning of $In(OH)_3$ thin films:** XPS spectral peaks corresponding to In $3d_{5/2}$ (445.1 eV), In $3d_{3/2}$ (452.8 eV) and O 1s were observed from $In(OH)_3$ thin films deposited on the amino group regions.

The binding energy of In $3d_{5/2}$ is higher than that of In metal (443.1[89], 443.6[90,91], 443.8[92], 444.3 eV[93]) and $In_2O_3$ (444.5[94,95], 444.6[96], 444.7[95,97], 444.8[91], 444.9[98] eV), and similar to that of $In(OH)_3$ (445.0[80], 445.2 eV[99]). This suggests that the indium atoms in thin films are positively charged relative to that of indium metal by formation of direct bonds with oxygen. The binding energy of In $3d_{5/2}$, which is similar to that of $In(OH)_3$ rather than $In_2O_3$, is consistent with XRD evaluation. On the other hand, this spectrum was not observed from silanol group regions, revealing site-selective deposition of $In(OH)_3$ thin film.

**(c) XPS analysis for patterning of $In_2O_3$ thin films:** The spectral peak corresponding to In $3d_{5/2}$ was shifted to a lower binding energy, 444.9 eV, by annealing at 250 °C in air. This was within the range of that of $In_2O_3$ and consistent with crystallization into $In_2O_3$ revealed by XRD evaluation.

Additionally, $In(OH)_3$ thin film was annealed at 300 °C in a reduced atmosphere (3% - $H_2$ / $N_2$) for 1 h instead of atmospheric heating. In $3d_{5/2}$ was shifted to a lower binding energy, 445.0 eV. This was similar to biding energy of $3d_{5/2}$ in $In_2O_3$ and indicated crystallization of $In(OH)_3$ into $In_2O_3$.

### 2.9 Liquid phase patterning of crystalline anatase $TiO_2$ using a superhydrophilic surface[100]

A glass substrate coated with an F doped $SnO_2$ transparent conductive film (FTO, $SnO_2$: F, Asahi Glass Co., Ltd., 9.3-9.7 Ω/□, 26 × 50 × 1.1 mm) showed a water contact angle of 96°. The UV-irradiated surface was, however, wetted completely (contact angle 0–1°). The contact angle decreased with irradiation time (96°, 70°, 54°, 35°, 14°, 5° and 0° for 0 min, 0.5 min, 1 min, 2 min, 3 min, 4 min and 5 min, respectively). This suggests that a small amount of adsorbed molecules on the $SnO_2$: F substrate was removed completely by UV irradiation. The surface of the $SnO_2$: F substrate would be covered by hydrophilic OH groups after irradiation. Consequently, the $SnO_2$: F substrate was modified to have a patterned surface with hydrophobic regions and super-hydrophilic regions.

Aqueous solutions containg ammonium hexafluorotitanate ($[NH_4]_2TiF_6$) and boric acid ($H_3BO_3$) were kept at 50°C for 25h. The substrates were immersed into the solutions at 50°C for 2 h to form a micropattern of $TiO_2$.

After having been immersed in the solution, the substrate was rinsed with distilled water and dried in air. The initial FTO surface appeared to be blue-green under white light due to light diffracted from the FTO layer. On the other hand, $TiO_2$ films deposited on the super-hydrophilic surface appeared to be yellow-green. The color change would be caused by deposition of transparent $TiO_2$ film which influenced the wavelength of the diffracted light.

The micropattern of $TiO_2$ was shown by SEM evaluation to be successfully fabricated (Fig. 13). $TiO_2$ deposited on super-hydrophilic regions showed black contrast, while the initial FTO regions without deposition showed white contrast. The average line width is 55 μm. Line edge roughness[34], as measured by the standard deviation of the line width, is ~2.8 μm.

This represents a ~5% variation (i.e., 2.8/55) in the nominal line width, similar to the usual 5% variation afforded by current electronics design rules. The minimum line width of the pattern depends on the resolution of the photomask and wavelength of irradiated light (184.9 nm). It would be improved to ~ 1 μm by using a high-resolution photomask.

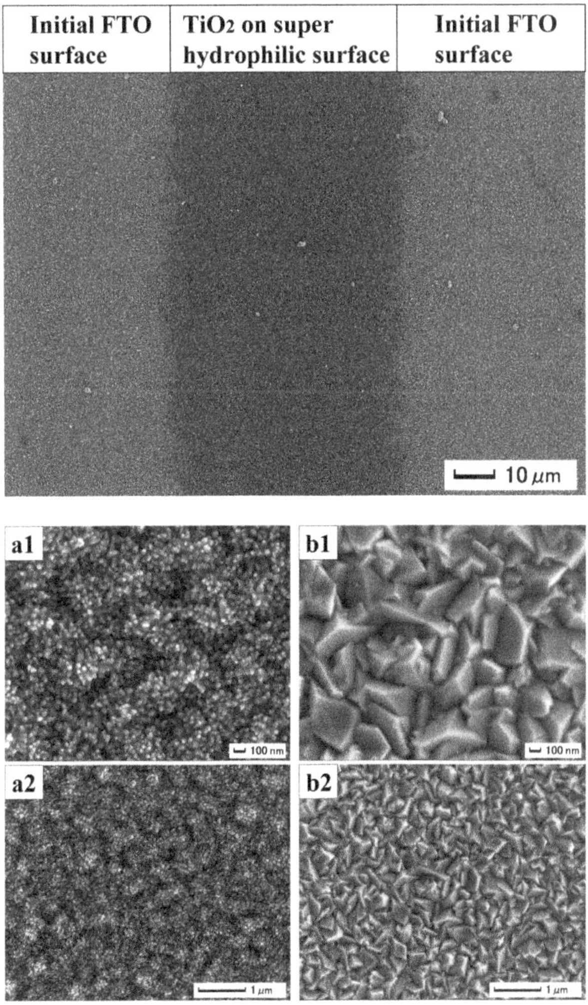


Reprinted with permission from Ref.[100], Masuda, Y. and Kato, K., 2008, *Chem. Mater.*, 20, 1057. Copyright @ American Chemical Society

Fig. 15. SEM micrograph of a micropattern of anatase $TiO_2$ films on $SnO_2$: F substrates. (a1) Surface of anatase $TiO_2$ films deposited on super-hydrophilic region. $TiO_2$ was formed on super-hydrophilic region which was cleaned by UV irradiation before the immersion. (a2) Magnified area of (a1) showing surface morphology of anatase $TiO_2$ film. (b1) Surface of $SnO_2$: F substrate without $TiO_2$ deposition. $TiO_2$ was not formed on non-cleaned region. (b2) Magnified area of (b1) showing surface morphology of $SnO_2$: F substrate.

The FTO layer was a particulate film having a rough surface (Fig. 15-b1, b2). Edged particles of 100 - 500 nm in diameter were observed on the surface. The micropattern of $TiO_2$ thin film was covered by an assembly of nano crystals of 10 - 30 nm in diameter (Fig. 15-a1, a2). The nano crystals would be anatase $TiO_2$ which grew anisotropically. The $TiO_2$ film also had large structural relief of 100 - 500 nm in diameter. As the thin $TiO_2$ film was deposited on the edged particulate surface of the FTO layer, the surface of $TiO_2$ had large structural relief.

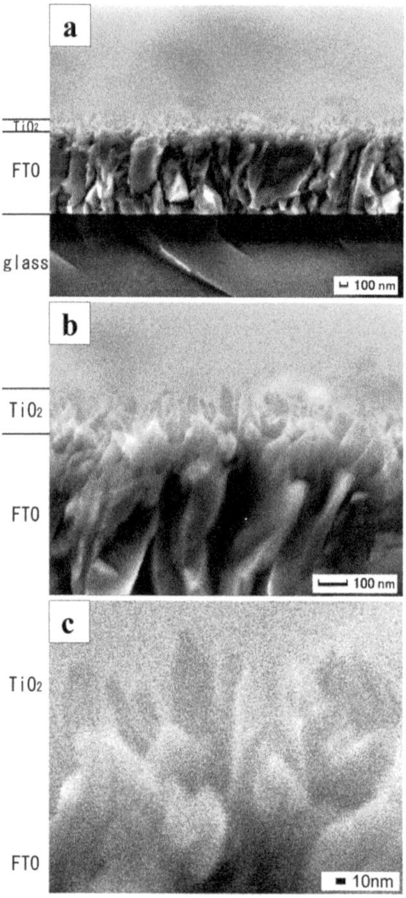


Reprinted with permission from Ref.[100], Masuda, Y. and Kato, K., 2008, *Chem. Mater.*, 20, 1057. Copyright @ American Chemical Society

Fig. 16. SEM micrographs of anatase $TiO_2$ films on $SnO_2$: F substrates. (a) Fracture cross section of $TiO_2$ films. (b, c) Magnified area of (a) showing morphology of nano $TiO_2$ crystals.

The morphology of the $TiO_2$ layer and FTO layer was further observed by fracture cross section profiles (Fig. 16). The polycrystalline FTO layer prepared on a flat glass substrate was shown to have a thickness of ~ 900 nm, and a high roughness of 100 - 200 nm on the surface (Fig. 16a). Nano $TiO_2$ crystals were deposited on the super-hydrophilic FTO surface (Fig. 16a), whereas no deposition was observed on the initial FTO surface. The super-hydrophilic FTO surface was covered with an array of nano $TiO_2$ crystals (Fig. 16b, c), which had a long shape of ~ 150 nm in length and ~ 20 nm in diameter. These observations were consistent with TEM and XRD evaluations[100]. Nano $TiO_2$ crystals would grow along the c-axis and thus enhance the 004 X-ray diffraction peak and 004 electron diffraction peak. They formed a long shape having a high aspect ratio of 7.5 (150 nm in length / 20 nm in diameter) as shown in the SEM fracture cross section profile (Fig. 16b, c) and TEM micrograph[100]. The orientation of nano $TiO_2$ crystals with their long axis perpendicular to the FTO layer (Fig. 16b, c) would also enhance the 004 diffraction peak.

The film deposited on the substrate was evaluated by XRD analysis. Strong X-ray diffractions were observed for films deposited on FTO substrates and assigned to $SnO_2$ of FTO films. The 004 diffraction peak of anatase $TiO_2$ was not observed clearly for $TiO_2$ film on FTO substrates because both of the weak 004 diffraction peak of $TiO_2$ and the strong diffraction peak of FTO were observed at the same angle. Glass substrates with no FTO coating were immersed in the solution. Weak X-ray diffraction peaks were observed at $2\theta$ = 25.3, 37.7, 48.0, 53.9, 55.1 and 62.7° for the films deposited on glass substrates. They were assigned to 101, 004, 200, 105, 211 and 204 diffraction peaks of anatase $TiO_2$ (ICSD No. 9852). A broad diffraction peak from the glass substrate was also observed at about $2\theta$ = 25°.

The intensity of the 004 diffraction peak was stronger than that of the 101 diffraction peak for the film obtained by the liquid phase crystal deposition method, though the intensity of 101 was stronger than that of 004 for anatase $TiO_2$ powders with no orientation (ICSD No. 9852). The integral intensity or peak height of 004 was 2.6 times or 2.2 times that of 101, respectively, suggesting high c-axis orientation of anatase $TiO_2$ crystals. Crystallite size perpendicular to the 101 or 004 planes was estimated from the full-width half-maximum of the 101 or 004 peak to be 9 nm or 17 nm, respectively. Elongation of crystals in the c-axis direction was also suggested by the difference of crystallite size. These evaluations were consistent with high c-axis orientation observed by TEM and electron diffraction[100]. Crystallite size estimated by XRD was similar to that in $TiO_2$ under layer rather than that of acicular crystals observed by TEM. $TiO_2$ thin film prepared on a glass would be constructed of not acicular crystals but polycrystals in under layer.

## 2.10 Liquid phase patterning of tin oxide nanosheets on ITO/PET films[101]

Tin oxide have been widely used in gas sensors[102], optical devices[103], lithium batteries[104], bio-sensors[105], catalyst for chemical reaction(anisole to 2,6-xylenol, etc.)[106] and so on. Various kinds of tin oxide nano-structures such as nano-fibers[107], nano-wires[108,109], nano-belts[110], nano-tubes[111], nano-rods[112,113], spirals, nano-rings[114], zigzag nano-belts[115], grains[116], flakes[116], plates[116], meshes[117] and columnar thin films[118] were reported. Surface coating of substrates with tin oxide nano-structures are strongly required for the future devices. Especially, 2D patterns of tin oxide nano-structures on flexible polymer films are required for light-weight flexible sensors and solar cells. 2D patterning of nanomaterials[119,120] such as metal oxides[33,54,63,75,121,122], Au clusters[123], Pt nanoparticles[124], organic monolayers[125], bio molecular layers[126], DNA[127], etc. is expected to lead to a large step forward in materials science research and the development of new devices. Recently, syntheses of tin oxide nano-structures in aqueous solutions were developed. They were based on precise crystal growth control of tin oxide. Additionally, 2D patterning of metal oxides including amorphous tin oxide films[121] was realized with the use of self-assembled monolayers (SAMs)[33,54,63,75,122]. Molecular recognition, organic-inorganic interaction, surface nucleation energy control, static interaction between SAMs and nano-particles, etc. were effectively utilized for site-selective deposition of metal oxides. More recently, super-hydrophilic surface was applied for 2D patterning of $TiO_2$ crystals[100]. FTO substrates were modified to have super-hydrophilic/hydrophobic patterns. $TiO_2$ nano-crystals were site-selectively formed on the super-hydrophilic area of the substrates. It realized 2D patterning of $TiO_2$ without use of self-assembled monolayers. In this study, a novel process was reported to realize micro-patterning of tin oxide nano-sheets on flexible polymer films.

**Surface modification of ITO/PET films and deposition of tin oxide**

Flexible poly(ethylene terephthalate) (PET) Films were coated with transparent conductive indium tin oxide (ITO) layers. They were exposed to a vacuum ultraviolet light through a photomask for 10 min (VUV, low-pressure mercury lamp PL16-110, air flow, 100 V, 200 W, SEN Lights Co., 14 $mW/cm^2$ for 184.9 nm at distance from lamp 10mm, 18 $mW/cm^2$ for 253.7 nm at distance from lamp 10mm). The initial substrates had hydrophobic surface showing a water contact angle of 111°. The VUV-irradiated surfaces however, were wetted completely (contact angle 0–5°).

Recently, light irradiation has been used for surface modification and material syntheses. Light irradiation was, for instance, used to clean $TiO_2$ surfaces and to connect oxide nanoparticles without heat treatment. UV light (Philips HPL-N lamp, 125 W) was irradiated to $TiO_2$ nanocrystals for 3 h to yield $TiO_2$ porous electrodes[128]. Medium pressure Hg lamp (125W, HPK Cathodeon) was used for $TiO_2$ porous films for dye-sensitized solar cell[129]. UV-$O_3$ treatment of the ITO-PEN film before $TiO_2$ paste application slightly improved the cell efficiency of dye-sensitized solar cell[130]. Light irradiation was effective strategy for nano-structure formation and device application. The irradiation had the advantage of simple, low cost and ordinary temperature process. It is suitable for surface treatment of low heat resistant polymer films.

$SnF_2$ (Wako Pure Chemical Industries, Ltd., No. 202-05485, FW: 156.71, purity 90.0%) was used as received. Distilled water in polypropylene vessels (200 mL) were capped with polymer films and kept at 90°C. $SnF_2$ (870.6 mg) was added and dissolved in the distilled water at 90°C to be 5 mM[131]. The substrates were immersed in the middle of the solutions with the bottom up at an angle. They were tilted at 15 degrees to the upright. The solutions were kept at 90°C using a drying oven (Yamato Scientific Co., Ltd., DKN402) for 2 h with no stirring. The substrates were rinsed under running water and dried by a strong air spray.

**Chemistry and the surface chemistry**

$SnO_2$ and SnO are formed in the aqueous solutions as follows:

$$SnF_2 \xrightarrow{OH^-} SnF(OH) + F^- \xrightarrow{OH^-} Sn(OH)_2 + F^- \quad (a)$$

$$Sn(OH)_2 \longrightarrow SnO + H_2O \quad (b)$$

$$Sn(OH)_2 \longrightarrow SnO_2 + H_2 \quad (c)$$

$$SnO + H_2O \longrightarrow SnO_2 + H_2 \quad (d)$$ [132,133]

Generation of gas was observed immediately after the addition of $SnF_2$ in water at 90°C. Ion concentration, valance of ions and pH changed during synthesis period. They affected formation of SnO and $SnO_2$. After having been immersed in the solutions, the substrates covered with nanocrystals were rinsed under running water and dried by a strong air spray. The nanocrystals were not removed under running water or by air spray to keep their adhesion. Direct crystallization of tin oxide on the films caused a high adhesion strength, which is required for sensors or solar cells. Some of fluorine ions were doped in tin oxide. Fluorine doping was known to be effective for performance advances of sensors and solar cells[134]. Residual fluorine ions were remained in the solutions.

Contact angle, wettability, surface tension strongly relate to nucleation of tin oxide. Function $f(\theta)$ and critical energy decreases from 1 to 0 as decrease of contact angle (θ) (Supporting Information in Ref. 101, Fig. S1). Critical energy decreases with decrease of contact angle (θ) and nucleation occurs easily on hydrophilic surface. The phenomenon is explained with knowledge of crystallography.

$$\gamma_{ls} = \gamma_{cs} + \gamma_{lc} \cos\theta \qquad \text{(a)(Young equation)}$$

$$\Delta G_{hetero} = \Delta G_{homo} f(\theta) = (\Delta G_v 4\pi r^3/3 + 4\pi r^2 \gamma_{lc})(2 - 3\cos\theta + \cos^3\theta)/4 \qquad \text{(b)}$$

$\gamma$: interfacial surface energy, θ: contact angle, l: liquid, s: substrate, c: crystal, r: radius,
Additionally, hydrophilic surface of ITO was covered with hydroxyl groups (-OH). Density of them was much higher than that on hydrophobic surface of ITO before the irradiation. Initial ITO surface was covered with adsorbed organic molecules. They covered pure surface of $In_2O_3$ crystals and increased water contact angle. Hydroxyl groups formed chemical bonds with hydroxyl groups on surface of tin oxide nanocrystals, clusters and/or related ions. Formation of them accelerates nucleation and growth of tin oxide on the surface. Therefore, hydrophilic surface was effective for tin oxide deposition compared to hydrophobic surface.

**Patterning of tin oxide on ITO/PET films**

Super-hydrophilic area of the ITO/PET films was successfully coated with tin oxide nano-sheets in the aqueous solutions (Fig. 17). On the other hand, hydrophobic area suppressed deposition of tin oxide. Super-hydrophilic/hydrophobic surface modification realized 2D patterning of tin oxide nano-sheets on flexible ITO/PET films. Super-hydrophilic area was covered with the nano-sheets uniformly (Fig. 18a). They were 100-300 nm of in-plane size and 5-10 nm in thickness (Fig. 18b). Surface observation (Fig. 18a, 18b) and fractured cross section image (Fig. 18c) revealed morphology of the nano-sheets clearly.

ITO/PET film with tin oxide nano-sheets were evaluated with XRD. It was similar to XRD pattern from ITO/PET film without deposition. Tin oxide surface coating was very thin and strong x-day diffraction peaks were not observed with XRD analyses. Crystal phase of surface coatings were evaluated with electron diffraction patterns which was built into TEM equipment as mentioned below.

The nano-sheets were formed on ITO layers of PET films directly, indicating that super-hydrophilic ITO surface accelerated crystal growth of tin oxide. Surface of ITO layer was covered with nano-sheets of 5-10 nm in size. Some of them further grew to be 100-300 nm of in-plane size. Anisotropic crystal growth of tin oxide realized formation of sheet structure. Lattice fringes were observed from all area of the sheets. Electron diffraction patterns revealed that the nano-sheets in area of circle 1 and 2 were single crystal of SnO. Diffraction spots were clearly observed and assigned to SnO. Estimated lattice spacings matched that of SnO. Meanwhile, nano-sheets in area 3 were consisted of SnO and $SnO_2$. Small amount of $SnO_2$ crystallized at initial stage of the immersion period.

Depth profile of XPS revealed chemical composition of the nano-sheets. Chemical composition on the topmost surface was estimated to Sn : O = 1 : 1.87 which is similar to $SnO_2$. Oxygen content would include slight oxygen in surface contamination. It changed to Sn : O = 1 : 0.75-0.93 by the sputtering. They were similar to that of SnO. Chemical composition of oxygen to tin was known to decrease with the sputtering. Electron diffraction patterns indicated that the structures were mixture of SnO and $SnO_2$ as mentioned above. These effected variation of chemical composition.

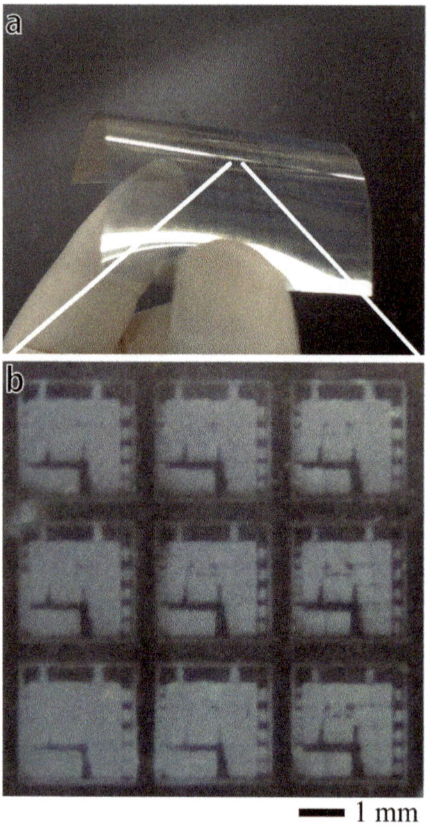


Reprinted with permission from Ref.[101], Masuda, Y.; Ohji, T.; Kato, K., 2011, *Eur. J. Inorg. Chem.*, 18, 2819-2825.. Copyright @ Wiley.

Fig. 17. (a) Photograph of 2D patterns of tin oxide nano-sheets on a flexible ITO/PET film. (b) Magnified area of (a) showing details of 2D patterns.

XPS analyses also indicated the followings. Indium and oxygen were observed from ITO layers after 12 seconds etching. Carbon was detected from PET films after 30 seconds etching. The nano-sheets were composed of tin, oxygen and fluorine, without contaminant.

Spectral peak corresponding to Sn $3d_{5/2}$ was observed at 486.4 eV (Fig. 7b). It was similar to that of $Sn^{4+}$ in $SnO_2$ (486.6 eV[135]) and higher than that of $Sn^{2+}$ site in SnO (485.9 eV[135]) or Sn metal (484.8 eV[136], 484.85 eV[80], 484.87 eV[137], 484.9 eV[95], 485.0 eV[93]). It suggested that tin atoms were positively charged by forming direct bonds with oxygen.

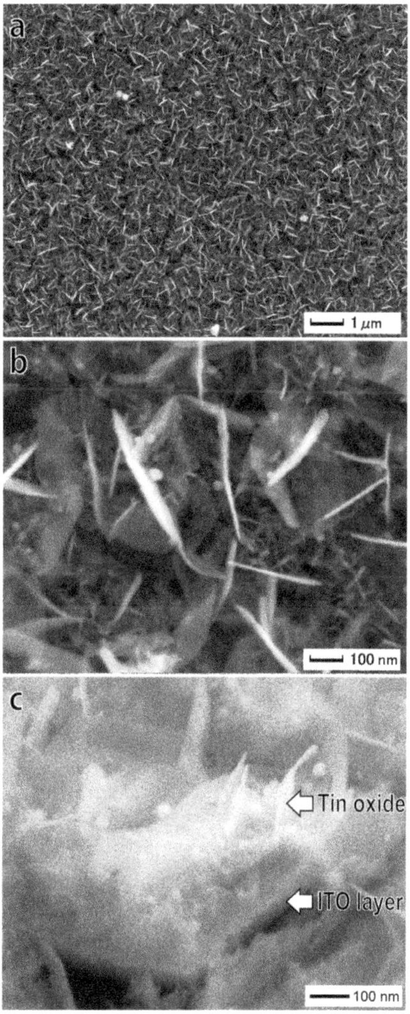


Reprinted with permission from Ref.[101], Masuda, Y.; Ohji, T.; Kato, K., 2011, *Eur. J. Inorg. Chem.*, 18, 2819-2825. Copyright @ Wiley.

Fig. 18. (a) FE-SEM image of tin oxide nano-sheets on super-hydrophilic area of the film. (b) Magnified area of (a) showing surface morphology of tin oxide nano-sheets. (c) Tilted image of fracture cross-section of (b) showing sheet morphology of tin oxide nano-sheets.

O 1s spectrum at 530.4 eV corresponded to that of SnO. It indicated that oxygen atoms were negatively charged compared to neutral oxygen molecules (531.0 eV) possibly through the

formation of direct bonds with tin atoms. O1s peak was rather broad with a shoulder at high energies that confirmed the presence of several contributions ($Sn^{2+}$-O, $Sn^{4+}$-O, -OH, C-O, etc.). It was consistent with the presence of a mixture of SnO and $SnO_2$.
Chemical composition on the topmost surface was estimated to F : Sn = 0.082. It would include fluorine in surface contamination. Because $SnF_2$ was used as precursor and it was difficult to be removed totally under running water. Chemical composition decreased by the sputtering and it was maintained relatively constant to F : Sn = 0.047-0.052. Fluorine remained even after the sputtering. F 1s spectrum was observed at 684.5 eV. The binding energy was similar to that of fluorine atoms which were doped in tin oxide (684.4 eV[138]), suggesting fluorine doping in the nano-sheets. Fluorine doping in tin oxide is well known to be suitable for sensors and solar cells[134]. A key factor for these applications is the electronic conductivity. It increases with fluorine doping. The 2D patterns developed in this study can be applied for the devices.
For comparison, high photocurrent was obtained from fluorine-doped tin oxide nano-structures on FTO (F-doped $SnO_2$ transparent conductive film) substrates under photo irradiation. They were modified with dye-labeled protein or dye-labeled DNA. High electronic conductivity of the structures contributed to high photocurrent. Direct evaluation of electronic conductivity of nano-structures is technically difficult, but it has intrinsic value and is now in progress.

## 3. Summary

A novel concept "Liquid Phase Patterning" was proposed based on scientific knowledge obtained from investigations of interactions and chemical reactions between functional groups of SAMs and ions, clusters and homogeneously nucleated particles in solutions. Nano/micropatterns of metal oxides such as $TiO_2$, $Fe_3O_4$, ZnO, etc. were successfully fabricated on a patterned SAM. Mechanisms were further discussed. These studies showed high performance and high potential of solution chemistry for inorganic materials. The novel concepts and technologies in these studies would open new doors for next generation ceramic science.

## 4. References

[1] http://www.epa.gov/greenchemistry/whats_gc.html.
[2] Anastas, P. T.; Warner, J. C. *Green Chemistry:Theory and Practice*; Oxford University Press: New York, 1998.
[3] http://www.meti.go.jp/policy/chemistry/.
[4] *Green chemistry workshop, Ministry of Economy, Trade and Industry* Japan, 1999.
[5] http://www.rsc.org/is/journals/current/green/greenpub.htm.
[6] http://www.gdch.de/.
[7] Mann, S. *Biomimetic Materials Chemistry*; VCH Publishers: Weinheim, 1996.
[8] Yoshimura, M.; Livage, J. *MRS Bulletin* 2000, *25*, 12-13.
[9] Yoshimura, M.; Suchanek, W. L.; Byrappa, K. *MRS Bull.* 2000, *25*, 17-25.
[10] Yoshimura, M. *J. Mater. Res.* 1998, *13*, 796-802.
[11] Figlarz, M. *Chem. Scr.*, 1988, *28*, 3-7.
[12] Niesen, T. P.; DeGuire, M. R. *J. Electroceram.* 2001, *6*, 169-207.

[13] Pizem, H.; Sukenik, C. N.; Sampathkumaran, U.; McIlwain, A. K.; De Guire, M. R. *Chem. Mater.* 2002, *14*, 2476-2485.
[14] Lev, O.; Wu, Z.; Bharathi, S.; Glezer, V.; Modestov, A.; Gun, J.; Rabinovich, L.; Sampath, S. *Chem. Mater.* 1997, *9*, 2354-2375.
[15] Klein, L. *Sol-Gel Optics: Processing and Applications*; Kluwer Acadmic Publishers: Boston, 1994.
[16] Byrappa, K.; Yoshimura, M. *Handbook of Hydrothermal Technology*; LLC/Noyes Publications: Park Ridge, NJ, 2000.
[17] Yoshimura, M.; Suchanek, W. *Solid State Ionics* 1997, *98*, 197-208.
[18] Yoshimura, M.; Suchanek, W.; Han, K. S. *J. Mater. Chem.* 1999, *9*, 77-82.
[19] Cho, Y. R.; Lee, J. H.; Song, Y. H.; Kang, S. Y.; Hwang, C. S.; Jung, M. Y.; Kim, D. H.; Lee, S. K.; Uhm, H. S.; Cho, K. I. *Mater. Sci. Eng. B* 2001, *79*, 128-132.
[20] Jeon, N. L.; Clem, P. G.; Nuzzo, R. G.; Payne, D. A. *Journal of Materials Research* 1995, *10*, 2996-2999.
[21] Aizenberg, J.; Braun, P. V.; Wiltzius, P. *Phys. Rev. Lett.* 2000, *84*, 2997-3000.
[22] Nashimoto, K.; Haga, K.; Watanabe, M.; Nakamura, S.; Osakabe, E. *Appl. Phys. Lett.* 1999, *75*, 1054-1056.
[23] Mott, M.; Song, J. H.; Evans, J. R. G. *J. Am. Ceram. Soc.* 1999, *82*, 1653-1658.
[24] Stutzmann, N.; Tervoort, T. A.; Bastiaansen, C. W. M.; Feldman, K.; Smith, P. *Adv. Mater.* 2000, *12*, 557-562.
[25] Matsuda, A.; Matsuno, Y.; Tatsumisago, M.; Minami, T. *J. Am. Ceram. Soc.* 1998, *81*, 2849-2852.
[26] Bauer, W.; Ritzhaupt-Kleissl, H. J.; Hausselt, J. *Ceram. Inter.* 1999, *25*, 201-205.
[27] Jacobs, H. O.; Whitesides, G. M. *Science* 2001, *291*, 1763-1766.
[28] Kim, E.; Xia, Y. N.; Whitesides, G. M. *Nature* 1995, *376*, 581-584.
[29] Kim, C.; Burrows, P. E.; Forrest, S. R. *Science* 2000, *288*, 831-833.
[30] Masuda, Y. *http://staff.aist.go.jp/masuda-y/index.html.*
[31] Masuda, Y.; Gao, Y. F.; Zhu, P. X.; Shirahata, N.; Saito, N.; Koumoto, K. *J. Ceram. Soc. Japan* 2004, *112*, 1495-1505.
[32] Masuda, Y. *J. Jpn. Soc. Powder Powder Metallurgy* 2007, *54*, 854-862.
[33] Masuda, Y.; Kinoshita, N.; Sato, F.; Koumoto, K. *Cryst. Growth Des.* 2006, *6*, 75-78.
[34] Masuda, Y.; Sugiyama, T.; Lin, H.; Seo, W. S.; Koumoto, K. *Thin Solid Films* 2001, *382*, 153-157.
[35] Masuda, Y.; Jinbo, Y.; Yonezawa, T.; Koumoto, K. *Chem. Mater.* 2002, *14*, 1236-1241.
[36] Masuda, Y.; Wang, D. J.; Yonezawa, T.; Koumoto, K. *Key Eng. Mater.* 2002, *228-229*, 125-130.
[37] Wang, D. J.; Masuda, Y.; Seo, W. S.; Koumoto, K. *Key Eng. Mater.* 2002, *214*, 163-168.
[38] Lee, L. H. *Journal of Colloid and Interface Science* 1968, *27*, 751-760.
[39] Masuda, Y.; Ieda, S.; Koumoto, K. *Langmuir* 2003, *19*, 4415-4419.
[40] Masuda, Y.; Saito, N.; Hoffmann, R.; De Guire, M. R.; Koumoto, K. *Sci. Tech. Adv. Mater.* 2003, *4*, 461-467.
[41] Masuda, Y.; Sugiyama, T.; Koumoto, K. *J. Mater. Chem.* 2002, *12*, 2643-2647.
[42] Gun, J.; Sagiv, J. *J. Colloid Interface Science* 1986, *112*, 457-472.
[43] McGovern, M. E.; Kallury, K. M. R.; Thompson, M. *Langmuir 1994* 1994, *10*, 3607-3614.
[44] Duchet, J.; Chabert, B.; Chapel, J. P.; Gerard, J. F.; Chovelon, J. M.; Jaffrezic-Renault, N. *Langmuir* 1997, *13*, 2271-2278.

[45] Dressick, W. J.; Chen, M.-S.; Brandow, S. L. *J. Amer. Chem. Soc.* 2000, *122*, 982-983.
[46] Dressick, W. J.; Chen, M.-S.; Brandow, S. L.; Rhee, K. W.; Shirey, L. M.; Perkins, F. K. *Appl. Phys. Lett.* 2001, *78*, 676-678.
[47] Dressick, W. J.; Nealey, P. F.; Brandow, S. L. *Proc. SPIE 2001* 2001, *4343*, 294-305.
[48] Koumoto, K.; Seo, S.; Sugiyama, T.; Seo, W. S.; Dressick, W. J. *Chem. Mater.* 1999, *11*, 2305.
[49] Masuda, Y.; Sugiyama, T.; Seo, W. S.; Koumoto, K. *Chem. Mater.* 2003, *15*, 2469-2476.
[50] Collins, R. J.; Shin, H.; De Guire, M. R.; Heuer, A. H.; Sukenik, C. N. *Appl. Phys. Lett.* 1996, *69*, 860-862.
[51] Huang, D.; Xiao, Z. D.; Gu, J. H.; Huang, N. P.; Tuan, C.-W. *Thin Solid Films* 1997, *305*, 110-115.
[52] Zhang, F.; Mao, Y.; Zheng, Z.; Chen, Y.; Liu, X.; Jin, S. *Thin Solid Films* 1997, *310*, 29-33.
[53] Shin, H.; Collins, R. J.; DeGuire, M. R.; Heuer, A. H.; Sukenik, C. N. *J. Mater. Res.* 1995, *10*, 699-703.
[54] Nakanishi, T.; Masuda, Y.; Koumoto, K. *Chem. Mater.* 2004, *16*, 3484-3488.
[55] Nakanishi, T.; Masuda, Y.; Koumoto, K. *J. Cryst. Growth* 2005, *284*, 176-183.
[56] Brandow, S. L.; Dressick, W. J.; Marrian, C. R. K.; Chow, G. M.; Calvert, J. M. *J. Electrochem. Soc.* 1995, *142*, 2233-2243.
[57] Dressick, W. J.; Dulcey, C. S.; Georger, J. H.; Calabrese, G. S.; Calvert, J. M. *J. Electrochem. Soc.* 1994, *141*, 210-220.
[58] Izaki, M.; Shinoura, O. *Adv. Mater.* 2001, *13*, 142-145.
[59] Call, R. L.; Jaber, N. K.; Seshan, K.; Whyte, J. R. *Solar Energy Mater.* 1980, *2*, 373-380.
[60] Oner, M.; Norwig, J.; Meyer, W. H.; Wegner, G. *Chem. Mater.* 1998, *10*, 460-463.
[61] Wu, X. L.; Siu, G. G.; Fu, C. L.; Ong, H. C. *Appl. Phys. Lett.* 2001, *78*, 2285-2287.
[62] Kang, J. S.; Kang, H. S.; Pang, S. S.; Shim, E. S.; Lee, S. Y. *Thin Solid Films* 2003, *443*, 5-8.
[63] Masuda, Y.; Yamagishi, M.; Koumoto, K. *Chem. Mater.* 2007, *19*, 1002-1008.
[64] Shaw, W. H. R.; Bordeaux, J. J. *J. Am. Chem. Soc.* 1955, *77*, 4729-4733.
[65] Ryabchikov, D. E.; Ryabukhin, V. A. *Analytical Chemistry of Yttrium and the Lanthanide Elements*; Humphrey Science: Ann Arbor, MI, 1970.
[66] Baes, C. F.; Mesmer, R. E. *The Hydrolysis of Captions*; Wiley: New York, 1976.
[67] Aiken, B.; Hsu, W. P.; Matijevic, E. *J. Am. Ceram. Soc.* 1988, *71*, 845-853.
[68] Agarwal, M.; DeGuire, M. R.; Heuer, A. H. *Appl. Phys Lett.* 1997, *71*, 891-893.
[69] Matijevic, E.; Hsu, W. P. *J. Colloid Interface Sci.* 1987, *118*, 506-523.
[70] Sharma, P. K.; Jilavi, M. H.; Nass, R.; Schmidt, H. *J. Lumin.* 1999, *82*, 187-193.
[71] Kwaka, M. G.; Parkb, J. H.; Shon, S. H. *Solid State Commun.* 2004, *130*, 199-201.
[72] Fuggle, J. C.; Martensson, N. *J. Electron Spectrosc. Relat. Phenom.* 1980, *21*, 275.
[73] Wagner, C. D. *Practical Surface Analysis*; 2 ed.; John Wiley, 1990; Vol. 1.
[74] Paton, M. G.; Maslen, E. N. *Acta Crystallographica* 1967, *1*, 1948-23.
[75] Masuda, Y.; Kondo, M.; Koumoto, K. *Cryst. Growth Des.* 2009, *9*, 555-561.
[76] Dressick, W. J.; Kondracki, L. M.; Chen, M. S.; Brandow, S. L.; Matijevic, E.; Calvert, J. M. *Colloids and Surfaces A-Physicochemical and Engineering Aspects* 1996, *108*, 101-111.
[77] Masuda, Y.; Koumura, T.; Okawa, T.; Koumoto, K. *J. Colloid Interface Sci.* 2003, *263*, 190-195.
[78] Tahar, R. B. H.; Ban, T.; Ohya, Y.; Takahashi, Y. *J. Appl. Phys.* 1997, *82*, 865-870.
[79] Hilaire, L.; Legare, P.; Holl, Y.; Maire, G. *Solid State Commun.* 1979, *32*, 157-160.

[80] Wagner, C. D.; Riggs, W. M.; Davis, L. E.; Moulder, J. F.; Muilenberg, G. E. *Handbook of X-ray Photoelectron Spectroscopy*; Perkin-Elmer Corp., Physical Electronics Div., Eden Prairie: Minnesota, 1979.
[81] Weightman, P.; Andrews, P. T. *J. Phys. C-Solid State Phys.* 1980, *13*, L815-L819.
[82] Weightman, P.; Andrews, P. T. *J. Phys. C-Solid State Phys.* 1980, *13*, L821-L825.
[83] Nefedov, V. I.; Zakharova, I. A.; Moiseev, I. I.; Porai-koshits, M. A.; Vargoftik, M. N.; Belov, A. P. *Zh. Neorg. Khim.* 1973, *18*, 3264-3268.
[84] Nefedov, V. I.; Kokunov, Y. V.; Buslaev, Y. A.; Poraikos.Ma; Gustyako.Mp; Ilin, E. G. *Zhurnal Neorganicheskoi Khimii* 1973, *18*, 931-934.
[85] Choudary, B. M.; Kumar, K. R.; Jamil, Z.; Thyagarajan, G. *Journal of the Chemical Society-Chemical Communications* 1985, 931-932.
[86] Sakurada, O.; Takahashi, H.; Taga, M. *Bunseki Kagaku* 1989, *38*, 407-412.
[87] Datye, A. K.; Bravo, J.; Nelson, T. R.; Atanasova, P.; Lyubovsky, M.; Pfefferle, L. *Applied Catalysis a-General* 2000, *198*, 179-196.
[88] Bazzicalupi, C.; Bencini, A.; Bianchi, A.; Giorgi, C.; Valtancoli, B. *Coordination Chemistry Reviews* 1999, *184*, 243-270.
[89] Ouchene M., S. C., Belin E., Gheorghiu A., Theye M. *J. Non-Cryst. Solids* 1983, *59&60*, 625-628.
[90] Bertrand, P. A. *J. Vac. Sci. Technol.* 1981, *18*, 28-33.
[91] Kazmerski, L. L.; Jamjoum, O.; Ireland, P. J.; Deb, S. K.; Mickelsen, R. A.; Chen, W. *J. Vac. Sci. Technol.* 1981, *19*, 467-471.
[92] Sen, P.; Hegde, M. S.; Rao, C. N. *Appl. Surf. Sci.* 1982, *10*, 63.
[93] Wagner, C. D. *Discuss. Faraday Soc.* 1975, *60*,, 291-318.
[94] Fan, J. C. C.; Goodenough, J. B. *J. Appl. Phys.* 1977, *48*, 3524-3531.
[95] Lin, A. W. C.; Armstrong, N. R.; Kuwana, T. *Anal. Chem.* 1977, *49*, 1228-1235.
[96] Clark, D. T.; Fok, T.; Roberts, G. G.; Sykes, R. W. *Thin Solid Films* 1980, *70*, 261-283.
[97] Cahen, D.; Ireland, P. J.; Kazmerski, L. L.; Thiel, F. A. *J. Appl. Phys.* 1985, *57*, 4761-4771.
[98] Hewitt, R. W.; Winograd, N. *J. Appl. Phys.* 1980, *51*, 2620-2624.
[99] Faur, M.; Faur, M.; Jayne, D. T.; Goradia, M.; Goradia, C. *Surf. Interface Anal.* 1990, *15*, 641-650.
[100] Masuda, Y.; Kato, K. *Chem. Mater.* 2008, *20*, 1057-1063.
[101] Masuda, Y.; Ohji, T.; Kato, K. *Eur. J. Inorg. Chem.* 2011, *18*, 2819-2825.
[102] Huang, J.; Matsunaga, N.; Shimanoe, K.; Yamazoe, N.; Kunitake, T. *Chem. Mater.* 2005, *17*, 3513-3518.
[103] Ginley, D. S.; Bright, C. *MRS Bull.* 2000, *25*, 15-21.
[104] Idota, Y.; Kubota, T.; Matsufuji, A.; Maekawa, Y.; Miyasaka, T. *Science* 1997, *276*, 1395 - 1397.
[105] Tokudome, H.; Yamada, Y.; Sonezaki, S.; Ishikawa, H.; Bekki, M.; Kanehira, K.; Miyauchi, M. *Appl. Phys. Lett.* 2005, *87*, 213901-213903.
[106] Jyothi, T. M.; Sugunan, S.; Rao, B. S. *Green Chemistry* 2000, *2*, 269-271.
[107] Yuan, R. S.; Lin, C.; Wu, B. C.; Fu, X. Z. *European Journal of Inorganic Chemistry* 2009, 3537-3540.
[108] Mathur, S.; Barth, S. *Small* 2007, *3*, 2070-2075.
[109] Mathur, S.; Barth, S.; Shen, H.; Pyun, J. C.; Werner, U. *Small* 2005, *1*, 713-717.
[110] Dai, Z. R.; Gole, J. L.; Stout, J. D.; Wang, Z. L. *J. Phys. Chem. B* 2002, *106*, 1274 -1279.
[111] Wang, Y.; Lee, J. Y.; Zeng, H. C. *Chem. Mater.* 2005, *17*, 3899-3903.

[112] Liu, Y. K.; Zheng, C. L.; Wang, W. Z.; Yin, C. R.; Wang, G. H. *Adv. Mater.* 2001, *13*, 1883 - 1887.

[113] Sun, J. Q.; Wang, J. S.; Wu, X. C.; Zhang, G. S.; Wei, J. Y.; Zhang, S. Q.; Li, H.; Chen, D. R. *Cryst. Growth Des.* 2006, *6*, 1584-1587.

[114] Yang, R. S.; Wang, Z. L. *J. Am. Chem. Soc.* 2006, *128*, 1466-1467.

[115] Duan, J. H.; Yang, S. G.; Liu, H. W.; Gong, J. F.; Huang, H. B.; Zhao, X. N.; Zhang, R.; Du, Y. W. *J. Am. Chem. Soc.* 2005, *127*, 6180 -6181.

[116] Ohgi, H.; Maeda, T.; Hosono, E.; Fujihara, S.; Imai, H. *Cryst. Growth Des.* 2005, *5*, 1079-1083.

[117] Uchiyama, H.; Imai, H. *Cryst. Growth Des.* 2007, *7*, 841-843.

[118] Zhang, Y. L.; Liu, Y.; Liu, M. L. *Chem. Mater.* 2006, *18*, 4643-4646.

[119] Ding, L.; Li, C. Q.; Zhou, W. W.; Chu, H. B.; Sun, X. A.; Cao, Z.; Yang, Z. H.; Yan, C. H.; Li, Y. *European Journal of Inorganic Chemistry*, 4357-4362.

[120] Wu, X. C.; Chi, L. F.; Fuchs, H. *European Journal of Inorganic Chemistry* 2005, 3729-3733.

[121] Shirahata, N.; Shin, W.; Murayama, N.; Hozumi, A.; Yokogawa, Y.; Kameyama, T.; Masuda, Y.; Koumoto, K. *Adv. Func. Mater.* 2004, *14*, 580-588.

[122] Masuda, Y. *J. Ceram. Soc. Japan* 2007, *115*, 101-109.

[123] Shibu, E. S.; Radha, B.; Verma, P. K.; Bhyrappa, P.; Kulkarni, G. U.; Pal, S. K.; Pradeep, T. *Acs Applied Materials & Interfaces* 2009, *1*, 2199-2210.

[124] Acevedo, D. F.; Salavagione, H. J.; Lasagni, A. F.; Morallon, E.; Mucklich, F.; Barbero, C. *Acs Applied Materials & Interfaces* 2009, *1*, 549-551.

[125] Mizuno, H.; Buriak, J. M. *Acs Applied Materials & Interfaces* 2009, *1*, 2711-2720.

[126] Li, B.; Franking, R.; Landis, E. C.; Kim, H.; Hamers, R. J. *Acs Applied Materials & Interfaces* 2009, *1*, 1013-1022.

[127] Geissler, M.; Roy, E.; Diaz-Quijada, G. A.; Galas, J. C.; Veres, T. *Acs Applied Materials & Interfaces* 2009, *1*, 1387-1395.

[128] Tebby, Z.; Babot, O.; Toupance, T.; Park, D. H.; Campet, G.; Delville, M. H. *Chem. Mater.* 2008, *20*, 7260-7267.

[129] Gutierrez-Tauste, D.; Zumeta, I.; Vigil, E.; Hernandez-Fenollosa, M. A.; Domenech, X.; Ayllon, J. A. *Journal of Photochemistry and Photobiology a-Chemistry* 2005, *175*, 165-171.

[130] Yamaguchi, T.; Tobe, N.; Matsumoto, D.; Arakawa, H. *Chem. Comm.* **2007**, 4767-4769.

[131] Masuda, Y.; Kato, K. *J. Cryst. Growth* 2009, *311*, 593-596.

[132] Baes, C. F.; Mesiner, R. E. *The Hydrolysis of Cations*; John Wiley & Sons, Inc., Wiley-Interscience: New York, 1976.

[133] Ararat Ibarguena, C.; Mosqueraa, A.; Parrab, R.; Castrob, M. S.; Rodríguez-Páeza, J. E. *Mater. Chem. Phys.* 2007, *101*, 433-440.

[134] Han, C. H.; Han, S. D.; Singh, I.; Toupance, T. *Sensors and Actuators B-Chemical* 2005, *109*, 264-269.

[135] Kwoka, M.; Ottaviano, L.; Passacantando, M.; Santucci, S.; Czempik, G.; Szuber, J. *Thin Solid Films* 2005, *490*, 36-42.

[136] Parry-Jones, A. C.; Weightman, P.; Andrews, P. T. *J. Phys. C. Sol. State Phys.* 1979, *12*, 1587-1600.

[137] Pessa, M.; Vuoristo, A.; Vulli, M.; Aksela, S.; Väyrynen, J.; Rantala, T.; Aksela, H. *Phys. Rev. B* 1979, *20*, 3115-3123.

[138] Martinez, A. I.; Huerta, L.; de Leon, J.; Acosta, D.; Malik, O.; Aguilar, M. *Journal of Physics D-Applied Physics* 2006, *39*, 5091-5096.

# 4

# Nanofabrication of Metal Oxide Nanostructures in Aqueous Solutions

Yoshitake Masuda
*National Institute of Advanced Industrial Science and Technology (AIST), Anagahora, Shimoshidami, Moriyama-ku, Nagoya, Japan*

## 1. Introduction

Metal oxides have been widely used in electro devices, optical devices, etc. Recently, liquid phase syntheses of them attract much attention as future technology and novel academic field. Especially, liquid phase syntheses of anisotropic particles or films are expected for next generation metal oxide devices. This section describes liquid phase morphology control of anisotropic metal oxide. They were realized by precise control of nucleation and crystal growth. They showed high performance of solution systems for future metal oxide devices. Liquid phase morphology control of anisotropic metal oxide particles and films would contribute to development of metal oxide science and technology.

## 2. Morphology control of acicular $BaTiO_3$ particles[1]

Acicular $BaTiO_3$ particles were developed using solution systems. The morphology of $BaC_2O_4 \cdot 0.5H_2O$ was controlled to acicular shape. Its phase transition to $BaTiO_3$ was realized by introducing Ti ions from the coprecipitated amorphous phase. Acicular $BaTiO_3$ particles have an aspect ratio as high as 18 and the particle size can be controlled by varying the growth period of $BaC_2O_4 \cdot 0.5H_2O$ which governs the size of $BaC_2O_4 \cdot 0.5H_2O$ particles. Acicular particles of crystalline $BaTiO_3$ can be used for ultra-thin multilayer ceramic capacitors.

Multilayer ceramic capacitors (MLCC) are indispensable electronic components for advanced electronic technology[2-12], but larger capacity and smaller size are needed for future electronic devices. To meet these needs, $BaTiO_3$ particles were downsized, but ferroelectric ceramics lose their ferroelectricity when their particle size is decreased and lose ferroelectricity entirely at a critical size[2-11]. This is known as the size effect and it impedes the progress of MLCC, so a novel solution has been eagerly anticipated.

Here, we propose MLCC using acicular $BaTiO_3$ particles[13]. An ultra-thin ferroelectric layer and high capacity can be realized by acicular particles having a high aspect ratio. The short side provides an ultra-thin ferroelectric layer and the large volume caused by the long side avoids the loss of ferroelectricity at the critical size. Anisotropic $BaTiO_3$ particles are thus a candidate for MLCC. $BaTiO_3$ has, however, an isotropic cubic or tetragonal structure, and its morphology is extremely difficult to control due to its isotropic crystal faces. We focused on triclinic $BaC_2O_4 \cdot 0.5H_2O$ which has an anisotropic crystal structure and controlled the morphology of these particles by precisely controlling crystal growth. We also achieved phase transition of $BaC_2O_4 \cdot 0.5H_2O$ to crystalline $BaTiO_3$ by introducing Ti ions from the coprecipitated amorphous phase. Having developed several key technologies, we were successfully able to produce anisotropic acicular $BaTiO_3$ particles.

Morphology control of $BaTiO_3$ to rod-shape was reported previously. Additionally, metal oxalates ($MC_2O_4$) have been used for synthesis of rod-shaped oxides or hydroxides. Y. Hayashi *et al.* reported preparation of rod-shaped $BaTiO_3$ from rod-shaped $TiO_2$-$nH_2O$ and $BaCO_3$ in molten chloride at high temperature[14]. Li *et al.* reported preparation of nanoflakes and nanorods of $Ni(OH)_2$, $Co(OH)_2$ and $Fe_3O_4$ by hydrothermal conversion at 160 °C for 12h from $MC_2O_4{\cdot}2H_2O$ in NaOH solutions[15]. Sun *et al.* prepared flowerlike $SnC_2O_4$ submicrotubes in ethanol solutions containing $SnCl_2$ and oxalic acid. They were annealed at 500 °C for 2 h in ambient atmosphere to obtain flowerlike $SnO_2$ submicrotubes[16].

Oxalic acid (252 mg) was dissolved into isopropyl alcohol (4 ml) (Fig. 1)[1]. Butyl titanate monomer (0.122 ml) was mixed with the oxalic acid solution, and the solution was then mixed with distilled water (100 ml). The pH of the solution was increased to pH = 7 by adding NaOH (1 M) and distilled water, while the volume of the solution was adjusted to 150 ml by these additions. The aqueous solution (50 ml) with barium acetate (39.3 mg) was mixed with the oxalic acid solution. The mixed solution containing barium acetate (0.77 mM), butyl titanate monomer (2 mM) and oxalic acid (10 mM) was kept at room temperature for several hours with no stirring, and the solution gradually became cloudy. Stirring causes the collision of homogeneously nucleated particles and destruction of large grown particles, and so was avoided in this process. The size of the precipitate was easily controlled from nanometer order to micrometer order by changing the growth period. Large particles were grown by immersion for several hours to evaluate the morphology and crystallinity in detail.

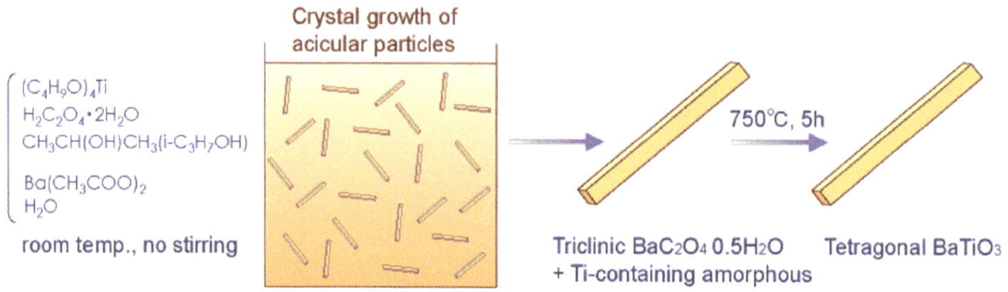


Fig. 1. Conceptual process for fabricating acicular $BaTiO_3$ particles. Morphology control of $BaC_2O_4 \bullet 0.5H_2O$ particles and phase transition to $BaTiO_3$. Reprinted with permission from Ref.[1], Masuda, Y., Yamada, T. and Koumoto, K., 2008, *Cryst. Growth Des.*, 8, 169. Copyright @American Chemical Society

Oxalate ions ($C_2O_4^{2-}$) react with barium ions ($Ba^{2+}$) to form barium oxalate ($BaC_2O_4 \cdot 0.5H_2O$). $BaC_2O_4 \cdot 0.5H_2O$ is dissolved in weak acetate acid provided by barium acetate ($(CH_3COO)_2Ba$), however, it can be deposited at pH 7 which is adjusted by adding NaOH. $BaC_2O_4 \cdot 0.5H_2O$ was thus successfully precipitated from the solution.

Acicular particles were homogeneously nucleated and precipitated from the solution. They were on average 23 μm (ranging from 19 to 27 μm) in width and 167 μm (ranging from 144 to 189 μm) in length, giving a high aspect ratio of 7.2 (Fig. 2). They had sharp edges and clear crystal faces, indicating high crystallinity. A gel-like solid was also coprecipitated from the solution as a second phase.

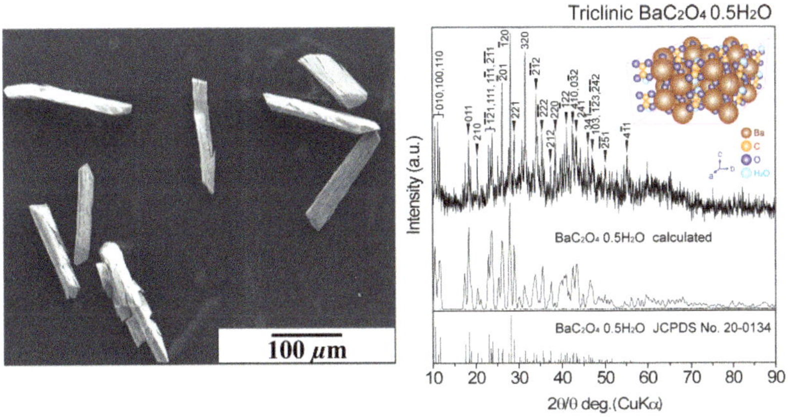


Fig. 2. SEM micrograph and XRD diffraction pattern of acicular $BaC_2O_4 \bullet 0.5H_2O$ particles precipitated from an aqueous solution at pH = 7. XRD diffraction measurement data (first step), XRD pattern calculated from crystal structure data16 (second step) and XRD pattern of JCPDS No. 20-134 (third step) are shown for triclinic $BaC_2O_4 \bullet 0.5H_2O$. Reprinted with permission from Ref.[1], Masuda, Y., Yamada, T. and Koumoto, K., 2008, *Cryst. Growth Des.*, 8, 169. Copyright @American Chemical Society

XRD diffraction patterns for the mixture of acicular particles and gel-like solid showed sharp diffraction peaks of crystalline $BaC_2O_4 \cdot 0.5H_2O$ with no additional phase. Acicular particles were crystalline $BaC_2O_4 \cdot 0.5H_2O$ and the gel-like solid would be an amorphous phase.

Fortunately, $BaC_2O_4 \cdot 0.5H_2O$ has a triclinic crystal structure as shown by the model calculated from structure data[17] (Fig. 2 XRD first step) and thus anisotropic crystal growth was allowed to proceed to produce an acicular shape. Each crystal face has a different surface energy and surface nature such as zeta potential and surface groups. Anisotropic crystal growth is induced by minimizing the total surface energy in ideal crystal growth. Additionally, site-selective adsorption of ions or molecules on specific crystal faces suppresses crystal growth perpendicular to the faces and so induces anisotropic crystal growth. These factors would cause anisotropic crystal growth of $BaC_2O_4 \cdot 0.5H_2O$ and hence allow us to control morphology and fabricate acicular $BaC_2O_4 \cdot 0.5H_2O$ particles. The positions of diffraction peaks corresponded with that of JCPDS No. 20-0134 (Fig. 2 XRD third step) and that calculated from crystal structure data[17] (Fig. 2 XRD second step), however, several diffraction peaks, especially 320 and 201, were enhanced strongly compared to their relative intensity. The enhancement of diffraction intensity from specific crystal faces would be related to anisotropic crystal growth; a large crystal size in a specific crystal orientation increases the x-ray diffraction intensity for the crystal face perpendicular to the crystal orientation.

EDX elemental analysis indicated the chemical ratio of the precipitate, which included acicular particles and gel-like solid, to be about Ba / Ti = 1 to 1.5. The chemical ratio indicated that the coprecipitated amorphous gel contained Ti ions. Additional Ba ions can be transformed into $BaCO_3$ by annealing and removed by HCl treatment in the next step. The ratio was thus controlled to slightly above Ba / Ti = 1 by adjusting the volume ratio of acicular particles and gel-like solid. Consequently, acicular particles of crystalline $BaC_2O_4 \cdot 0.5H_2O$ with Ti-containing gel-like solid were successfully fabricated in an aqueous solution process.

In comparison, isotropic particles of barium titanyl oxalate ($BaTiO(C_2O_4)_2 \cdot 4H_2O$) were precipitated at pH 2. $TiOC_2O_4$ was formed by the following reaction in which the reaction of oxalic acid ($H_2C_2O_4 \cdot 2H_2O$) with butyl titanate monomer ($(C_4H_9O)_4Ti$) and hydrolysis can take place simultaneously[18].

$$(C_4H_9O)_4Ti + H_2C_2O_4 \cdot 2H_2O \rightarrow TiOC_2O_4 + 4C_4H_9OH + H_2O \quad \text{(a)}$$

$TiO(C_2O_4)$ was then converted to oxalotitanic acid ($H_2TiO(C_2O_4)_2$) by the reaction:

$$TiO(C_2O_4) + H_2C_2O_4 \cdot 2H_2O \rightarrow H_2TiO(C_2O_4)_2 + 2H_2O \quad \text{(b)}$$

Alcoholic solution containing oxalotitanic acid ($H_2TiO(C_2O_4)_2$) formed by reaction (b) was subjected to the following cation exchange reaction by rapidly adding an aqueous solution of barium acetate at room temperature:

$$H_2TiO(C_2O_4)_2 + Ba(CH_3COO)_2 \rightarrow BaTiO(C_2O_4)_2 \downarrow + 2CH_3COOH \quad \text{(c)}$$

$BaTiO(C_2O_4)_2$ isotropic particles were formed by reaction (c).

On the other hand, neither $BaC_2O_4 \cdot 0.5H_2O$ nor $BaTiO(C_2O_4)_2$ was precipitated at pH 3 to pH 6. Gel-like solid was formed in the solution and their XRD spectra showed no diffraction peaks. The amorphous gel that precipitated at pH = 3 to 6 would be the same as the amorphous gel coprecipitated at pH 7.

These comparisons show that the crystal growth and morphology control of $BaC_2O_4 \cdot 0.5H_2O$ are sensitive to the solution conditions.

The precipitate was annealed at 750 °C for 5 h in air. Acicular $BaC_2O_4 \cdot 0.5H_2O$ particles were reacted with Ti-containing amorphous gel to introduce Ti ions to transform into crystalline $BaTiO_3$. X-ray diffraction of the annealed precipitate showed crystalline $BaTiO_3$ and an additional barium carbonate phase ($BaCO_3$). Excess precipitation of $BaC_2O_4 \cdot 0.5H_2O$ caused the generation of barium carbonate phase ($BaCO_3$) as expected.

The annealed precipitate was further immersed in HCl solution (1 M) to dissolve barium carbonate ($BaCO_3$). Acicular particles of crystalline $BaTiO_3$ were successfully fabricated with no additional phase. Particles showed acicular shape with 2.8×10×50 μm and x-ray diffraction of single-phase crystalline $BaTiO_3$ (Fig. 3). The high aspect ratio of the particles (17.8 = 50 / 2.8) would be provided by that of $BaC_2O_4 \cdot 0.5H_2O$ particles. The particle size of acicular $BaTiO_3$ can be easily controlled by the growth period and solution concentration for $BaC_2O_4 \cdot 0.5H_2O$ precipitation which decides the particle size of $BaC_2O_4 \cdot 0.5H_2O$.

$BaTiO_3$ has a cubic crystal structure at high temperature above phase transition and has a tetragonal crystal structure at room temperature. The cubic crystal structure is completely isotropic and the tetragonal crystal structure results from stretching a cubic lattice along one of its lattice vectors. For both of the crystal structures it is difficult to control anisotropic crystal growth, however, with our newly developed process we could successfully control the morphology and fabricate acicular particles. This was achieved by controlling the morphology of triclinic $BaC_2O_4 \cdot 0.5H_2O$ to acicular shape and the phase transition to $BaTiO_3$

by introducing Ti ions from the coprecipitated amorphous phase. The novel concept can be applied to a wide variety of morphology control and crystal growth control for advanced electronic devices composed of crystalline materials.

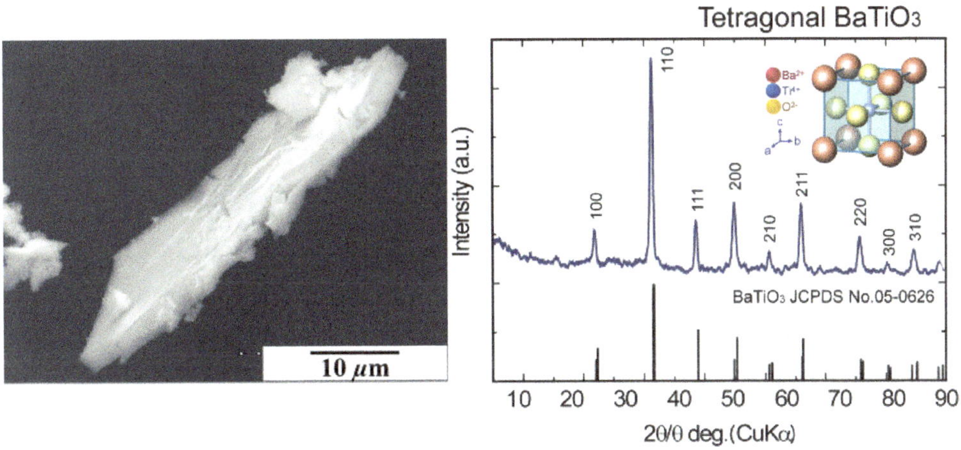


Fig. 3. SEM micrograph and XRD diffraction pattern of acicular $BaTiO_3$ particles after annealing at 750 °C for 5 h and HCl treatment. XRD diffraction measurement data (first step) and XRD pattern of JCPDS No. 05-0626 (second step) are shown for tetragonal $BaTiO_3$. Reprinted with permission from Ref.[1], Masuda, Y., Yamada, T. and Koumoto, K., 2008, *Cryst. Growth Des.*, 8, 169. Copyright @American Chemical Society

## 3. Morphology control of ZnO particles[19,20]

Ethylenediamine ($H_2N$-$CH_2CH_2$-$NH_2$, 15–45 mM, Sigma-Aldrich) was added to the zinc acetate aqueous solution ($Zn(CH_3COO)_2$, 15 mM, Kishida Chemical Co., Ltd.) to promote deposition of ZnO[19]. Zinc chelate ($Zn(H_2N\text{-}CH_2CH_2\text{-}NH_2)^{2+}$) was formed from zinc acetate and ethylenediamine in reaction (a). ZnO was crystallized from zinc chelate and hydroxide ion ($OH^-$) in reaction (c).

$$Zn(CH_3COO)_2 + 2H_2NCH_2CH_2NH_2 \rightleftharpoons Zn(H_2NCH_2CH_2NH_2)_2^{2+} + 2CH_3COO^- \quad \text{(a)}$$

$$H_2O \rightleftharpoons OH^- + H^+ \quad \text{(b)}$$

$$Zn(H_2NCH_2CH_2NH_2)_2^{2+} + 2OH^- \longrightarrow ZnO + 2H_2NCH_2CH_2NH_2 + H_2O \quad \text{(c)}$$

The solution became turbid shortly after adding ethylenediamine. The molar ratio of ethylenediamine to Zn was [ethylenediamine] / [Zn] = (a) 1 : 1, (b) 2 : 1 or (c) 3 : 1. pH of the solutions were (a) pH=7.3, (b) pH=8.0 or (c) pH=8.7, respectively. Crystal growth rate and deposition of ZnO were attemped to control to change particle morphology. Si substrate (Newwingo Co., Ltd.) was immersed to evaluate deposited ZnO particles and particulate films. The solution in a glass beaker was kept at 60 °C for 3 h using a water bath. The silicon substrate was cleaned before immersion as described in references. The substrate was rinsed with distilled water after immersion.

ZnO particles having hexagonal cylinder shape were homogeneously nucleated and deposited in the aqueous solution containing 15 mM ethylenediamine ([ethylenediamine] / [Zn] = (a) 1 : 1) (Fig. 4a). X-ray diffraction patterns showed the deposition to be well crystallized ZnO (Fig. 5a). The relative intensity of (10-10) and (0002) is similar to that of randomly deposited ZnO particles, indicating the random orientation of deposited ZnO hexagonal cylinders, which is consistent with SEM observations. Crystals showed hexagonal facets of about 100–200 nm in diameter and about 500 nm in length. ZnO has a hexagonal crystal structure and thus hexagonal cylinders can be obtained by sufficiently slow crystal growth. A low crystal growth rate allows enough ions to diffuse to form a complete crystal structure.

Fig. 4. SEM micrographs of (a) ZnO hexagonal cylinder particles, (b) ZnO long ellipse particles, (c1) ZnO hexagonal symmetry radial whiskers and (c2) magnified area of (c1). Reprinted with permission from Ref.[19], Y. Masuda, N. Kinoshita and K. Koumoto, *J. Nanosci. Nanotechnol.*, in press. Copyright @American Scientific Publisher

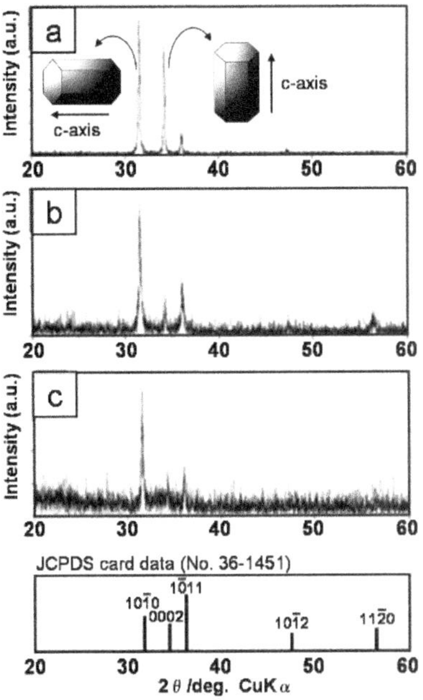


Fig. 5. XRD patterns of (a) ZnO hexagonal cylinder particles, (b) ZnO long ellipse particles, (c) ZnO hexagonal symmetry radial whiskers and ZnO (JCPDS No. 36-1451). Reprinted with permission from Ref.[19], Y. Masuda, N. Kinoshita and K. Koumoto, *J. Nanosci. Nanotechnol.*, in press. Copyright @American Scientific Publisher

ZnO having hexagonal cylinder shape showed strong photoluminescence intensity in the UV region at about 370–400 nm and weak intensity in the visible light region at about 530–550 nm by 350 nm excitation light which appears in spectra (Fig. 6a). ZnO crystals were reported to show UV luminescence (around 390 nm) attributed to band-edge luminescence and visible-light luminescence caused from oxygen vacancy (450–600 nm)[21,22]. Oxygen vacancies would be generated in ZnO during crystallization to show visible-light luminescence.

Concentration of ethylenediamine was increased twice to [ethylenediamine] / [Zn] = (b) 2 : 1. ZnO particles with long ellipse shape were deposited homogeneously from the solution (Fig. 4b). ZnO particles were about 100–200 nm in diameter and about 500 nm in length, and were similar to those of hexagonal cylinder shape. XRD showed the deposition to be well crystallized ZnO (Fig. 5b). Relative intensity of (10-10) diffraction is much stronger than that of (0002), indicating that mainly ZnO particles with long ellipse shape were laid on the silicon substrate. This was also observed in SEM micrographs (Fig. 4b). Deposition speed of ZnO with long ellipse shape was slightly faster than that of ZnO with hexagonal cylinder shape because of the high concentration of ethylenediamine. Ethylenediamine accelerates the crystallization of ZnO. In other words, supersaturation degree of the solution was increased by increasing the ethylenediamine concentration. As a result, ZnO grew slightly faster and formed not sharp hexagonal facets but rounded hexagonal cylinders, i.e., long ellipse shape. The photoluminescence spectrum of ZnO with long ellipse shape (Fig. 6b) was similar to that of ZnO with hexagonal cylinder shape (Fig. 6a). The luminescence property was clearly shown to be stable and not influenced by the synthesis conditions in the range of [ethylenediamine] / [Zn] = 1 : 1 to 2 : 1. High repeatability and stability of the photoluminescence property without being affected by the deposition conditions are major advantages of this system for large-scale production.

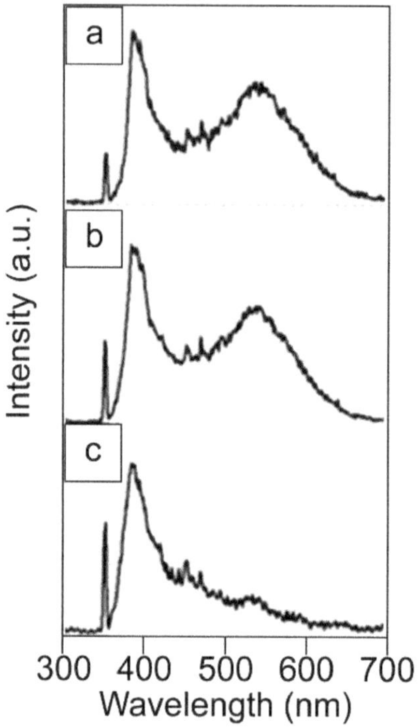


Fig. 6. Photoluminescence spectra of (a) ZnO hexagonal cylinder particles, (b) ZnO long ellipse particles and (c) ZnO hexagonal symmetry radial whiskers. Reprinted with permission from Ref.[19], Y. Masuda, N. Kinoshita and K. Koumoto, *J. Nanosci. Nanotechnol.*, in press. Copyright @American Scientific Publisher

Further control the morphology was attempted by increasing the supersaturation degree. Ethylenediamine was added to the solution to be [ethylenediamine] / [Zn] = (c) 3 : 1 in the deposition process. ZnO whiskers were successfully grown and deposited from the solution. The morphology was drastically changed by the precise investigation and control of solution conditions. The whiskers were about 10–100 nm in diameter and about 1000–2000 nm in length (Fig. 4c1). A high aspect ratio was realized by high crystal growth rate of end faces to elongate the whiskers. Details of morphology were further evaluated. They were dispersed in water and dropped on a silicon substrate. Whiskers were connected to form hexagonal symmetry radial whiskers (Fig. 4c2). Six whiskers connected to form one particle. They had hexagonal symmetry. Tips of whiskers were finer than that at center of the particles. XRD showed the whiskers to be well crystallized ZnO (Fig. 5c). The relative intensity of (10-10) diffraction is much stronger than that of (0002), showing that mainly ZnO whiskers were laid on the substrate. This was also observed in SEM micrographs (Fig 4 c). Deposition of ZnO whiskers having high aspect ratio on the substrate would provide a ZnO network film having a high specific surface area. These whisker films can be applied to gas sensors[23,24]or solar cells[25] which require high specific surface area. The whisker films also have high conductivity per unit volume[26] compared to conventional particulate films or mesoporous materials because the whiskers carry an electric current for a long distance without grain boundaries. Photoluminescence intensity in the visible light region was quite different from that of ZnO with hexagonal cylinder shape or long ellipse shape (Fig. 5c). The large change of morphology indicates that the crystal growth mechanism differs greatly between ZnO particles having hexagonal cylinder or long ellipse shape and ZnO whiskers. Basically, ethylenediamine increased crystal growth rate to generate oxygen vacancies. The oxygen vacancies increased photoluminescence intensity in visible light region. However, hexagonal symmetry radial whiskers prepared in the solution of [ethylenediamine] / [Zn] = (c) 3 : 1 showed very weak photoluminescence intensity in visible light region compared to hexagonal cylinders or long ellipse particles. Excess ethylenediamine would decrease photoluminescence intensity in visible light region. Ion concentration of the solution of [ethylenediamine] / [Zn] = (c) 3 : 1 would drastically change at the initial stage. The solutions of [ethylenediamine] / [Zn] = (a) 1 : 1 and (b) 2 : 1 became turbid shortly after adding ethylenediamine and they changed to transparent gradually. The solutions of [ethylenediamine] / [Zn] = (c) 3 : 1 also became turbid shortly after adding ethylenediamine, however, they changed to transparent rapidly compared to that of [ethylenediamine] / [Zn] = (a) 1 : 1 and (b) 2 : 1. It indicated that ions were consumed to form particles rapidly in the solution of [ethylenediamine] / [Zn] = (c) 3 : 1 and ion concentration would decrease drastically with color change. It can be assumed that high concentration of ethylenediamine increased crystal growth rate to form ZnO particles at the first stage. The particles generated in the first stage would not be whiskers but small ZnO particles. Ion concentration decreased rapidly to make solution transparent by formation of small particles because Zn ions were consumed in formation of the particles. ZnO whiskers would grow slowly on the small particles in dilute solutions at second stage. Consequently, hexagonal symmetry radial whiskers were formed in the solution of [ethylenediamine] / [Zn] = (c) 3 : 1. This phenomenon was consistent with reported ZnO whiskers which had high crystallinity, high photoluminescence intensity in UV region and low photoluminescence intensity in visible light region. Novel properties such as unique morphology, high specific surface area, high conductivity per unit volume, low photoluminescence intensity in visible light region and high photoluminescence intensity in UV region may pave the way to a new age of ZnO devices. Furthermore, they can be fabricated on low heat resistant materials such as polymers, paper or organic materials for flexible devices.

## 4. Morphology control of multi-needle ZnO particles and their particulate films [27]

ZnO has attracted much attention as a next-generation gas sensor for CO[28-30], $NH_3$[31], $NO_2$[32], $H_2S$[33], $H_2$[29,34], ethanol[34,35], $SF_6$[34], $C_4H_{10}$[34] or gasoline[34] and dye-sensitized solar cells[26,36-39]. Sensitivity directly depends on the specific surface area of the sensing material. ZnO particles, particulate films or mesoporous material having high specific surface area were thus strongly required.

ZnO has been crystallized to a hexagonal cylinder shape for gas sensors or solar cells in many studies[26] by using the hexagonal crystal structure of ZnO at low supersaturation degree. However, strategic morphology design and precise morphology control for high specific surface area should be developed to improve the properties. ZnO particles should be controlled to have multi-needles or high surface asperity to increase the specific surface area by the crystallization at high supersaturation degree.

Recently, morphology control[20,40-42] and nano/micro manufacturing[43-46] of oxide materials were proposed in solution systems. Solution systems have the advantage of adjustment of supersaturation degree and high uniformity in the system for particle morphology control. However, many factors affect the system compared to gas phase systems or solid state reactions. Solution chemistry for oxide materials is therefore being developed and many areas remain to be explored.

Morphology control of ZnO particles to hexagonal cylinder shape, ellipse shape and multi-needle shape was recently developed[20]. Photoluminescence property was improved by changing the morphology and oxygen vacancy volume in this system. Morphology control of ZnO has also been proposed based on control of crystal growth[47-50]. W. Q. Peng et al. reported flower-like bunches synthesized on indium-doped tin oxide glass substrates through a chemical bath deposition process[47]. Y. R. Lin et al. fabricated nanowires on a ZnO-buffered silicon substrate by a hydrothermal method[48]. H. Zhang et al. prepared flowerlike, disklike, and dumbbell-like ZnO microcrystals by a capping-molecule-assisted hydrothermal process[49]. K. H. Liu et al. reported a hierarchical polygon prismatic Zn-ZnO core-shell structure grown on silicon by combining liquid-solution colloidals together with the vapor-gas growth process[50]. These studies showed high morphology controllability of ZnO, however, morphology should be optimized to have high specific surface area to apply to solar cells or gas sensors.

In this study, the morphology design of ZnO particles was proposed for solar cells or gas sensors in which high specific surface area, high electrical conductivity and high mechanical strength are required. Multi-needle ZnO particles having ultrafine surface relief structure, as well as particulate films constructed from multi-needle particles and thin sheets were fabricated[27]. Morphology control was realized based on a new idea inspired from the morphology change in our former study[20]. High supersaturation degree of the solution was used for fast crystal growth which induces the formation of multi-needle particles and low super saturation was used for the formation of ZnO thin sheets.

Zinc nitrate hexahydrate ($Zn(NO_3)_2 \cdot 6H_2O$, > 99.0%, MW 297.49, Kanto Chemical Co., Inc.) and ethylenediamine ($H_2NCH_2CH_2NH_2$, > 99.0%, MW 60.10, Kanto Chemical Co., Inc.) were used as received. Glass (S-1225, Matsunami Glass Ind., Ltd.) was used as a substrate. Zinc nitrate hexahydrate (15 mM) was dissolved in distilled water at 60 °C and ethylenediamine (15 mM) was added to the solution to induce the formation of ZnO. Glass substrate was immersed in the middle of the solution at an angle and the solution was kept at 60 °C using a water bath for 80 min with no stirring. The solution became clouded shortly after the addition of ethylenediamine. Ethylenediamine plays an essential role in the formation of crystalline ZnO. ZnO was homogeneously nucleated and grown to form a large amount of particles to make the solution clouded. ZnO particles were gradually deposited and further grown on a substrate. Homogeneously nucleated particles precipitated gradually and the solution became light white after 80 min. The supersaturation degree of the solution was high at the initial stage of the reaction for the first 1 h and decreased as the color of the solution changed.

ZnO particulate films constructed from ZnO particles and thin sheets were fabricated by immersion for 48 h. The glass substrate was immersed in the middle of the solution at an angle and the solution was kept at 60 °C using a water bath for 6 h with no stirring. The solution was then left to cool for 42 h in the bath. The solution became clouded shortly after the addition of ethylenediamine and clear after 6 h. The bottom of the solution was covered with white precipitate after 6 h. The supersaturation degree of the solution was high at the initial stage of the reaction for the first 1 h and then decreased as the color of the solution changed.

Morphology of ZnO particles and particulate films was observed by a field emission scanning electron microscope (FE-SEM; JSM-6335FM, JEOL Ltd.) after heating at 150 °C for 30 min in vacuum for drying of carbon paste (Vacuum oven, VOS-201SD, EYELA, Tokyo Rikakikai. Co., Ltd.) and Pt coating for 3 nm (Quick cool coater, SC-701MCY, Sanyu Electronic Company). Crystal phases were evaluated by an X-ray diffractometer (XRD; RINT-2100V, Rigaku) with CuKα radiation (40 kV, 40 mA).

**Morphology design of ZnO particles and particulate films.** The morphology of ZnO particles was designed to increase the specific surface area, electrical conductivity and mechanical strength of the base material of solar cells and sensors (Fig. 7). Typical ZnO particles grown at low supersaturation degree are shown in Fig. 7a[20]. The particles show edged hexagonal faces and elongate parallel to the *c*-axis. ZnO particles grow to have a hexagonal cylinder shape by slow crystal growth due to the hexagonal crystal structure of ZnO. The morphology of ZnO particles was controlled to have a multi-needle shape in an aqueous solution (Fig. 7b)[20]. Multi-needle particles have a high specific surface area compared to hexagonal cylinder particles, but particulate films constructed from small particles have many grain boundaries which reduce the electrical conductivity (Fig. 7c). Particles should thus have large grain size to decrease grain boundaries and increase electrical conductivity (Fig. 7d). Furthermore, the specific surface area of ZnO particles should be increased to improve the sensing performance of sensors or generating efficiency of solar cells (Fig. 7e). An ultrafine surface relief structure on ZnO particles is a candidate

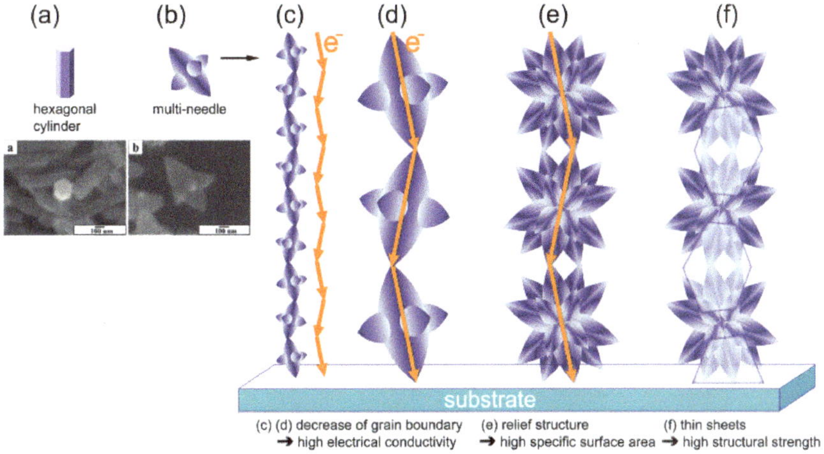


Fig. 7. Design and morphology control of ZnO particles and particulate films for high specific surface area, high electrical conductivity and high mechanical strength. Reprinted with permission from Ref.[27], Masuda, Y. and Kato, K., 2008, *Cryst. Growth Des.*, 8, 2633. Copyright @American Chemical Society

morphology for increasing the specific surface area. Additionally, particles should be connected to each other and with a substrate by a combined member such as ZnO thin sheets (Fig. 7f). The thin sheets increase the mechanical strength of particulate films and help raise the electrical conductivity and specific surface area. Morphology control of ZnO particles was attempted based on the strategic morphology design for application to sensors.

**Morphology control of ZnO particles.** After having been immersed in the solution for 80 min, the substrate covered with ZnO particles was evaluated by SEM and XRD. ZnO particles were shown to be multi-needle shape in which many needles were grown from the center of the particles (Fig. 8). The particles have more needles compared with the particles previously reported which were constructed from two large needles and several small needles[20]. The size of particles was in the range from 1 - 5 μm which is larger than the particles prepared previously[20] (Fig. 7b). Needles were constructed from an assembly of narrow acicular crystals and thus the side surfaces of needles were covered with arrays of pleats. The tips of the needles were rounded V-shape with many asperities. Edged hexagonal shapes were observed at the tips of needles, thus clearly showing high crystallinity and the direction of the *c*-axis. The *c*-axis would be the long direction of multi-needles and narrow acicular crystals. Elongation of the *c*-axis observed by SEM is consistent with high diffraction intensity of 0002 (Fig. 9). The 0002 diffraction intensity of multi-needle ZnO particles was much stronger than $10\bar{1}0$ or $10\bar{1}1$ peaks though 0002 diffraction is weaker than $10\bar{1}0$ or $10\bar{1}1$ diffractions in randomly orientated ZnO particles (JCPDS No. 36-1451). High diffraction intensity from (0002) planes which are perpendicular to the *c*-axis would be caused from the crystalline ZnO particles which grew to elongate the *c*-axis. The particles have more stacks of (0002) crystal planes compared to that of $(10\bar{1}0)$ planes which are parallel to the *c*-axis or $(10\bar{1}1)$ planes and hence the intensity from (0002) planes was stronger than that from $(10\bar{1}0)$ or $(10\bar{1}1)$ planes.

Fig. 8. SEM micrograph (a-d) of multi-needle ZnO particles having ultrafine surface relief structure. Reprinted with permission from Ref.[27], Masuda, Y. and Kato, K., 2008, *Cryst. Growth Des.*, 8, 2633. Copyright @American Chemical Society

ZnO grows to a hexagonal cylinder shape at low supersaturation degree because of its hexagonal crystal structure, however, ZnO grows to a multi-needle shape at high supersaturation degree which induces fast crystal growth. ZnO was thus grown to a multi-needle shape in our solution in spite of its hexagonal crystal structure. The growth of ZnO was halted by a rapid decrease of supersaturation degree and removal of particles from the solution to obtain ZnO multi-needle particles having an ultrafine surface relief structure. The morphology of the ZnO particles was controlled by the fast crystal growth due to high supersaturation degree and by the suppression of crystal growth due to the rapid decrease of supersaturation degree and removal of particles from the solution.

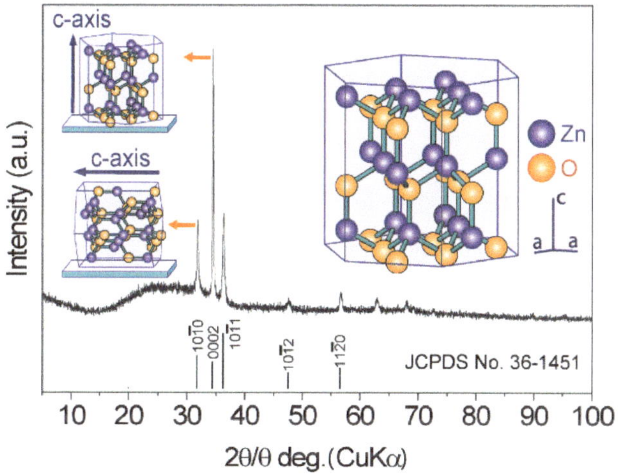


Fig. 9. XRD diffraction pattern of multi-needle ZnO particles having ultrafine surface relief structure. Reprinted with permission from Ref.[27], Masuda, Y. and Kato, K., 2008, *Cryst. Growth Des.*, 8, 2633. Copyright @American Chemical Society

**Morphology control of ZnO particulate films.** ZnO particulate films showed a multi-needle shape and were connected to each other by thin sheets (Fig. 10a-c). The morphology of the particles was similar to that of the particles prepared by immersion for 80 min to have a high specific surface area. Thin sheets had a thickness of 10 - 50 nm and width of 1 - 10 μm and were connected to particles closely with no clearance. The particulate films had continuous open pores ranging from several nm to 10 μm in diameter. The particulate films showed x-ray diffraction patterns of ZnO crystal with no additional phase (Fig. 11). Diffraction peaks were very sharp, showing high crystallinity of the particulate films. The high intensity of 0002 would be caused by elongation of multi-needle particles in the *c*-axis direction which increases the stacks of (0002) crystal planes.

ZnO multi-needle particles having an ultrafine surface relief structure were prepared at 60 °C in the white solution during the initial 80 min (Fig. 7e). The supersaturation degree was high at the initial stage of the reaction due to the high concentration of ions. ZnO particles were then precipitated, making the bottom of the solution white and the solution itself clear. Ions were consumed to form ZnO particles and thus the ion concentration of the solution decreased rapidly. Thin sheets were formed at 25 °C in the clear solution after the formation of multi-needle particles (Fig. 7f). Solution temperature and supersaturation degree would influence on precipitates. Consequently, the particulate films constructed from multi-needle particles and thin sheets were successfully fabricated by the two-step growth (Fig. 7f).

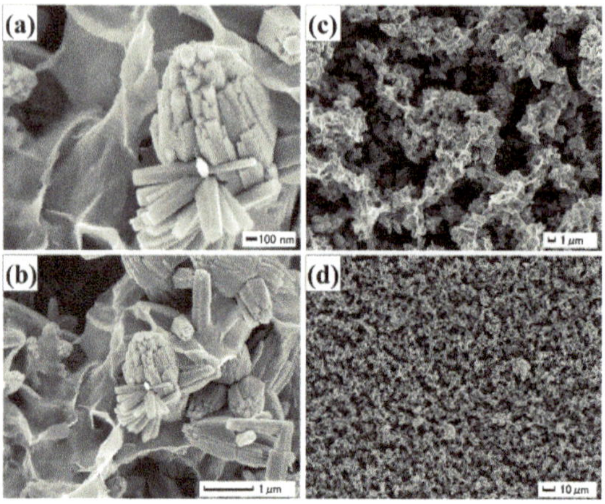


Fig. 10. SEM micrograph (a-d) of ZnO particulate films constructed from ZnO multi-needle particles and thin sheets. Reprinted with permission from Ref.[27], Masuda, Y. and Kato, K., 2008, *Cryst. Growth Des.*, 8, 2633. Copyright @American Chemical Society

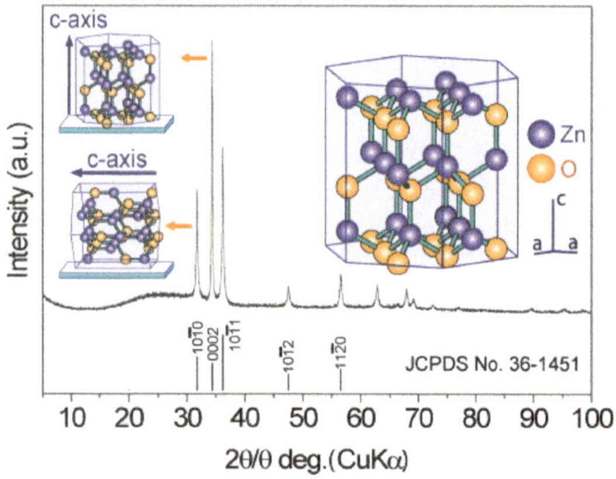


Fig. 11. XRD diffraction pattern of ZnO particulate films constructed from ZnO multi-needle particles and thin sheets. Reprinted with permission from Ref.[27], Masuda, Y. and Kato, K., 2008, *Cryst. Growth Des.*, 8, 2633. Copyright @American Chemical Society

For comparison, assemblies of thin sheets were prepared at air-liquid interfaces of the same solution we used in this study[51]. XRD patterns of the sheets were assigned to ZnO. The sheets had *c*-axis orientation parallel to the sheets, i.e., in-plane *c*-axis orientation. TEM observations showed the sheets were dense polycrystals consisted of nano-sized ZnO crystals. Electron diffraction pattern showed strong isotropic diffraction ring from (0002) planes. It suggested in-plane *c*-axis orientation of ZnO crystals which was consistent with XRD evaluations. Mechanical strength and electrical property would be affected by crystal orientation and microstructures. The sheets would have stronger mechanical strength compared to porous sheets because of their dense structure. They would have different electrical properties from randomly oriented sheets due to in-plane *c*-axis orientation, because ZnO has anisotropic electrical properties caused from anisotropic hexagonal crystal structure. The sheets prepared at air-liquid interfaces[51] would be similar to that prepared in the solutions (Fig. 10) because both of them were prepared from the same solution and showed XRD patterns assigned to ZnO[51] (Fig. 11). The sheets prepared in this study (Fig. 4, 5) would have similar mechanical and electrical properties to the sheets prepared at air-liquid interfaces[51].

Thin sheets were transformed to particles and porous particulate films by annealing at 500 °C for 1 h in air. The sheets did not maintain their thin sheet shape due to high slimness and/or phase transformation. Thin sheets would be inorganic films containing Zn ions such as crystalline ZnO, amorphous ZnO or zinc hydroxide and were transformed to porous ZnO particulate films by the annealing. Further investigation of the thin sheets would contribute to more precise morphology control of ZnO structure and further improvement of specific surface area. Additionally, precise evaluation of mechanical strength and electrical properties should be performed to clarify the potential of ZnO particulate films for sensors or solar cells and to produce guidelines for improving their properties.

Multi-needle ZnO particles having an ultrafine surface relief structure were successfully fabricated by the precise control of crystal growth in an aqueous solution. The morphology of ZnO was further controlled for ZnO particulate films constructed from ZnO multi-needle particles and thin sheets. The thin sheets connected particles to each other and with a substrate. The morphology design and morphology control described here will facilitate the progress of crystal science for developing future advanced materials and devices.

## 5. Morphology control of high *c*-axis oriented stand-alone ZnO self-assembled film[51]

Recently, crystalline ZnO[20,26,47-50,52-59] have been synthesized to utilize the high potential of the solution process for future devices and to realize green chemistry for a sustainable society. For instance, ZnO nanowire arrays have been synthesized using seed layers in aqueous solutions for dye-sensitized solar cell[26,52]. Full sun efficiency of 1.5% is demonstrated in this study. O'Brien et al. prepared specular ZnO films consisting of clumps of elongated triangular crystals[53], small ZnO spherical clumps consisting of particles of ca. 100 nm[53], ZnO films consisting of randomly rod-shaped particles of up to 1000 nm in length[53], ZnO films consisting of very thin random rod-shaped particles of ca. 1000 nm[53] and specular films consisting of flowers with well-formed triangle features[59] in aqueous solutions.

However, ZnO films have usually been prepared on substrates[54,57,60-76], and in particular crystalline ZnO films having high *c*-axis orientation require expensive substrates such as single crystals or highly-functional substrates. A simple and low-cost process for self-supporting crystalline ZnO films is expected to be used for a wide range of applications such as windows of optical devices or low-value-added products. Self-supporting crystalline ZnO films can also be applied by being pasting on a desired substrate such as low heat-resistant polymer films, glasses, metals or papers.

In this section, high *c*-axis oriented stand-alone ZnO self-assembled films was fabricated using an air-liquid interface[51]. ZnO was crystallized from an aqueous solution without heat

treatment or catalyst. The ZnO film was fabricated at the air-liquid interface without using ammonia vapor.
Zinc nitrate hexahydrate ($Zn(NO_3)_2 \cdot 6H_2O$, > 99.0%, MW 297.49, Kanto Chemical Co., Inc.) and ethylenediamine ($H_2NCH_2CH_2NH_2$, > 99.0%, MW 60.10, Kanto Chemical Co., Inc.) were used as received. Zinc nitrate hexahydrate (15 mM) was dissolved in distilled water at 60°C and ethylenediamine (15 mM) was added to the solution to induce the formation of ZnO. The solution was kept at 60°C using a water bath for 6 h with no stirring. The solution was then left to cool for 42 h in the bath. Polyethylene terephthalate (PET) film, glass (S-1225, Matsunami Glass Ind., Ltd.) and an Si wafer (p-type Si [100], NK Platz Co., Ltd.) were used as substrates.
Morphology of ZnO film was observed by a field emission scanning electron microscope (FE-SEM; JSM-6335FM, JEOL Ltd.) and a transmission electron microscope (TEM; H-9000UHR, 300 kV, Hitachi). Crystal phase was evaluated by an X-ray diffractometer (XRD; RINT-2100V, Rigaku) with CuKα radiation (40 kV, 40 mA). Si wafer was used as a substrate for XRD evaluation. The crystal structure model and diffraction pattern of ZnO were calculated from ICSD (Inorganic Crystal Structure Database) data No. 26170 (FIZ Karlsruhe, Germany and NIST, USA) using FindIt and ATOMS (Hulinks Inc.).
The solution became clouded shortly after the addition of ethylenediamine by the homogeneous nucleation and growth of ZnO particles. ZnO particles were gradually deposited to cover the bottom of the vessel, and the solution became light white after 1 h and clear after 6 h. The supersaturation degree of the solution was high at the initial stage of the reaction for the first 1 h and decreased as the color of the solution changed.
White films were formed at the air-liquid interface and they grew to large films. The films had sufficiently high strength to be obtained as stand-alone films. Additionally, a film was scooped to past onto a desired substrate such PET film, Si wafer, glass plate or paper, and the pasted ZnO film was then dried to bond it to the substrate. Both sides of the film can be pasted on substrate. The film physically adhered to the substrate. The film maintained its adhesion during immersion in lightly ultrasonicated water, however, it can be easily peeled off again by strong ultrasonication. The film can be handled easily from substrate to other substrate. It also can be attached strongly to substrate by annealing or addition of chemical regents such as silane coupling agent to form chemical bonds between the film and the substrate
The film grew to a thickness of about 5 μm after 48 h, i.e., 60°C for 6 h, and was left to cool for 42 h.
The air side of the stand-alone film had a smooth surface over a wide area due to the flat air-liquid interface (Fig. 12-a1), whereas the liquid side of the film had a rough surface (Fig. 12-b1). The films consisted of ZnO nano-sheets were clearly observed from the liquid side (Fig. 12-b2) and the fracture edge-on profile of the film (Fig. 12-c1, 12-c2). The nano-sheets had a thickness of 5-10 nm and were 1-5 μm in size. They mainly grew forward to the bottom of the solution, i.e., perpendicular to the air-liquid interface, such that the sheets stood perpendicular to the air-liquid interface. Thus, the liquid side of the film had many ultra-fine spaces surrounded by nano-sheet and had a high specific surface area. The air side of the film, on the other hand, had a flat surface that followed the flat shape of the air-liquid interface. The air-liquid interface was thus effectively utilized to form the flat surface of the film. This flatness would contribute to the strong adhesion strength to substrates for pasting of the film. The air-side surface prepared for 48 h had holes of 100-500 nm in diameter (Fig. 12-a2), and were hexagonal, rounded hexagonal or round in shape. The air-side surface prepared for 6 h, in contrast, had no holes on the surface. The air-side surface was well

crystallized to form a dense surface and ZnO crystals would partially grow to a hexagonal shape because of the hexagonal crystal structure. Well-crystallized ZnO hexagons were then etched to form holes on the surface by decrease in pH. The growth face of the film would be liquid side. ZnO nano-sheets would grow to form a large ZnO film by Zn ion supply from the aqueous solution. Further investigation of the formation mechanism would contribute to the development of crystallography in the solution system and the creation of novel ZnO fine structures.

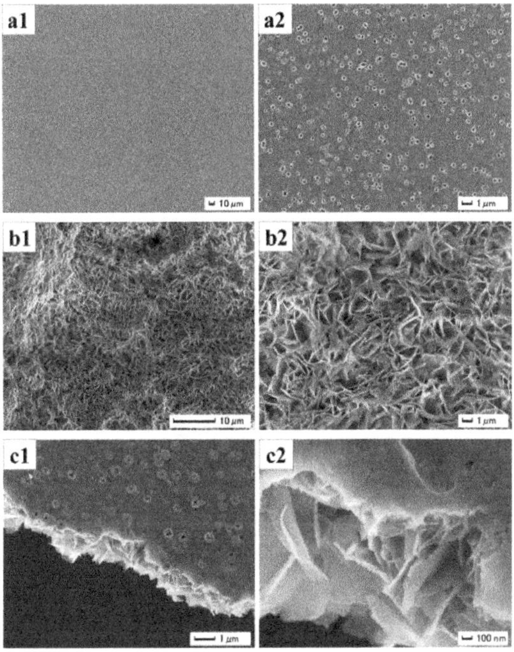


Fig. 12. SEM micrographs of high *c*-axis oriented stand-alone ZnO self-assembled film. (a1) Air-side surface of ZnO film. (a2) Magnified area of (a1). (b1) Liquid-side surface of ZnO film. (b2) Magnified area of (b2). (c1) Fracture cross section of ZnO film from air side. (c2) Magnified area of (c1). Reprinted with permission from Ref.[51], Masuda, Y. and Kato, K., 2008, *Cryst. Growth Des.*, 8, 275. Copyright @American Chemical Society

The film showed a very strong 0002 x-ray diffraction peak of hexagonal ZnO at $2\theta = 34.04°$ and weak 0004 diffraction peak at $2\theta = 72.16°$ with no other diffractions of ZnO (Fig. 13). (0002) planes and (0004) planes were perpendicular to the *c*-axis, and the diffraction peak only from (0002) and (0004) planes indicates high *c*-axis orientation of ZnO film. The inset figure shows that the crystal structure of hexagonal ZnO stands on a substrate to make the *c*-axis perpendicular to the substrate. Crystallite size parallel to (0002) planes was estimated from the half-maximum full-width of the 0002 peak to 43 nm. This is similar to the threshold limit value of our XRD equipment and thus the crystallite size parallel to (0002) planes is estimated to be greater than or equal to 43 nm. Diffraction peaks from a silicon substrate were observed at $2\theta = 68.9°$ and $2\theta = 32.43°$. Weak diffractions at $2\theta = 12.5°$, 24.0°, 27.6°, 30.5°, 32.4° and 57.6° were assigned to co-precipitated zinc carbonate hydroxide ($Zn_5(CO_3)_2(OH)_6$, JCPDS No. 19-1458).

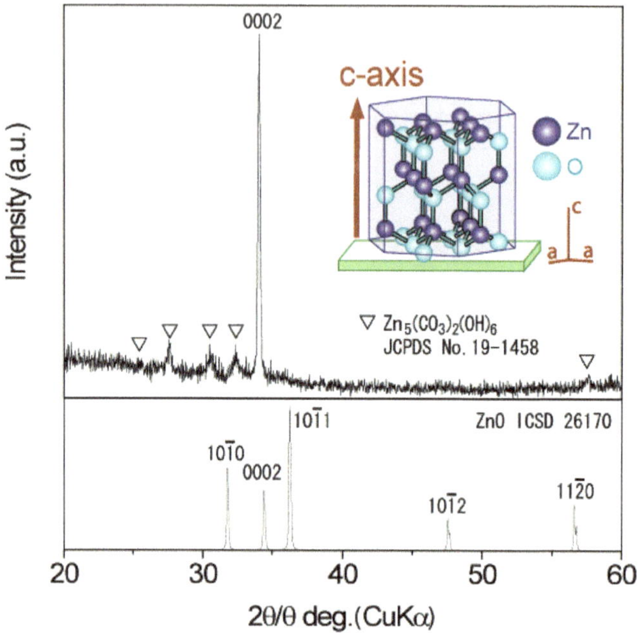


Fig. 13. XRD diffraction pattern of high *c*-axis oriented stand-alone ZnO self-assembled film. Reprinted with permission from Ref.[51], Masuda, Y. and Kato, K., 2008, *Cryst. Growth Des.*, 8, 275. Copyright @American Chemical Society

Stand-alone ZnO film was further evaluated by TEM and electron diffraction. The film was crushed to sheets and dispersed in an acetone. The sheets at the air-liquid interface were skimmed by a cupper grid with a carbon supporting film. The sheets were shown to have uniform thickness (Fig. 14a). They were dense polycrystalline films constructed of ZnO nanoparticles (Fig. 14b). Lattice image was clearly observed to show high crystallinity of the particles. The film was shown to be single phase of ZnO by electron diffraction pattern. These observations were consistent with XRD and SEM evaluations.

The film pasted on a silicon wafer was annealed at 500°C for 1 h in air to evaluate the details of the films. ZnO film maintained its structure during the annealing (Fig. 15). The air side of the film showed a smooth surface (Fig. 15-a1) and the liquid side showed a relief structure having a high specific surface area (Fig. 15-b1, 15-b2). The air side showed the film consisted of dense packing of small ZnO nanosheets and the size of sheets increased toward the liquid-side surface (Fig. 15-a2). ZnO sheets would grow from the air side to the liquid side, i.e., the sheets would nucleate at the liquid-air interface and grow down toward the bottom of the solution by the supply of Zn ions from the solution. Annealed film showed X-ray diffractions of ZnO and Si substrate with no additional phases. As-deposited ZnO nanosheets were shown to be crystalline ZnO because the sheets maintained their fine structure during the annealing without any phase transition. High *c*-axis orientation was also maintained during the annealing, showing a very strong 0002 diffraction peak.

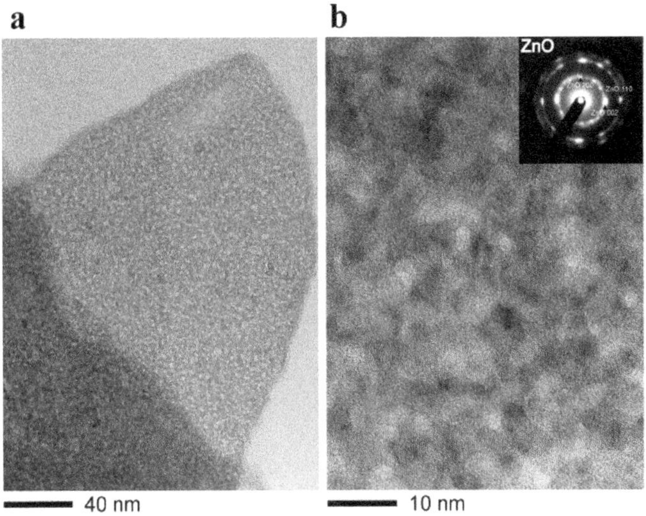


Fig. 14. (a) TEM micrograph of ZnO nano-sheets. (b) Magnified area of (a). (Insertion) Electron diffraction pattern of ZnO. Reprinted with permission from Ref.[51], Masuda, Y. and Kato, K., 2008, *Cryst. Growth Des.*, 8, 275. Copyright @American Chemical Society

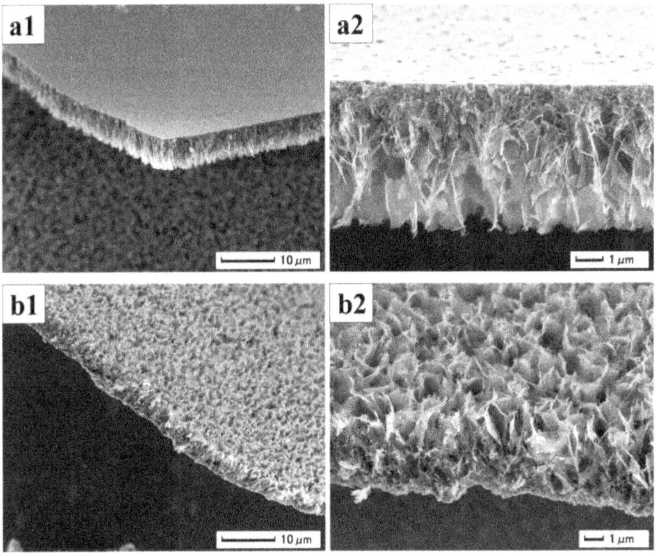


Fig. 15. SEM micrographs of high *c*-axis oriented stand-alone ZnO self-assembled film annealed at 500°C for 1 h in air. (a1) Fracture edge-on profile of ZnO film from air side. (a2) Cross-section profile of ZnO film from air side. (b1) Fracture edge-on profile of ZnO film from liquid side. (b2) Cross-section profile of ZnO film from liquid side. Reprinted with permission from Ref.[51], Masuda, Y. and Kato, K., 2008, *Cryst. Growth Des.*, 8, 275. Copyright @American Chemical Society

The solution was further kept at 25°C for 1 month to evaluate the details of the deposition mechanism. The film prepared at the air-liquid interface for 1 month was not hexagonal ZnO. The film showed strong X-ray diffractions of zinc carbonate hydroxide single phase. ZnO would be dissolved by decrease in pH. ZnO would be crystallized at the initial reaction stage for the first 48 h. ZnO was then gradually etched and dissolved by nitric acid and zinc carbonate hydroxide was crystallized using Zn ions which were supplied by the dissolution of crystalline ZnO.

High *c*-axis oriented stand-alone ZnO self-assembled film was fabricated using a simple solution process. The film consisted of ZnO nanosheets was crystallized at air-liquid interface. The nanosheets grew perpendicular to the film. The film had high *c*-axis orientation and showed a strong 0002 diffraction peak and weak 0004 peak. The air side of the film had a flat surface, whereas the liquid side had a rough surface having many ultra-fine spaces surrounded by ZnO nano-sheets. The rough surface of the liquid side can be utilized for sensors or dye-sensitized solar cells. The ZnO film was also pasted on a desired substrate such as PET films, Si substrate or glass plates. The surface of low heat-resistant flexible polymer film was modified with high *c*-axis oriented crystalline ZnO film without heat treatment. This low-cost, low-temperature technique can be used for a wide range of applications including sensors, solar cells, electrical devices and optical devices using the various properties of high *c*-axis oriented crystalline ZnO.

## 6. Morphology control of nanocrystal assembled $TiO_2$ particles[77]

Anatase $TiO_2$ particles, 100–200 nm in diameter, were developed in aqueous solution at 50°C. The particles were assemblies of nano $TiO_2$ crystals covered with nanorelief surface structures. The crystals grew anisotropically along the *c*-axis to form acicular crystals. The particles showed *c*-axis orientation due to high-intensity X-ray diffraction from the (004) planes. The particles had a BET specific surface area of 270 $m^2/g$. BJH and DFT/Monte-Carlo analysis of adsorption isotherm indicated the existence of pores ~3 nm and ~1 nm in diameter. Crystallization and self-assembly of nano $TiO_2$ were effectively utilized to fabricate nanocrystal assembled $TiO_2$ particles having high surface area and nanorelief surface structure.

Nanoporous $TiO_2$ architecture with micropores (<2 nm), mesopores (2–50 nm) and/or macropores (>50 nm) is of considerable interest for both scientific and technical applications. The latter include cosmetics, catalysts[78], photocatalysts[79-82], gas sensors[83,84], lithium batteries[85-87], biomolecular sensors[88] and dye-sensitized solar cells[89,90]. Crystalline anatase generally exhibits higher properties than rutile in photocatalysts, biomolecular sensors and dye-sensitized solar cells. Electrons are obtained from dyes adsorbed on $TiO_2$ electrodes in sensors and solar cells. Photoelectric conversion efficiency strongly depends on dye adsorption volume and surface area of $TiO_2$. High surface area is required to achieve high efficiency and sensitivity of the devices. Additionally, the surface of $TiO_2$ should be covered with nano/microrelief structures to adsorb large amounts of dye, molecules and DNA for biomolecular sensors and dye-sensitized solar cells.

$TiO_2$ nanoparticles have been prepared by flame synthesis[91,92], ultrasonic irradiation[93,94], chemical vapor synthesis[95], sol-gel methods[79,96-100], sonochemical method[101] and liquid phase deposition of amorphous $TiO_2$[102-105]. High temperature in the treatment processes, however, causes aggregation of nanoparticles and decreased surface area. Formation of nanorelief structures on the surface of the particles is difficult to achieve in these processes.

Highly porous materials have been prepared via template-based methods, including soft templates (surfactants, chelating agents, block polymers, etc.)[106-109] and hard templates (porous anionic alumina, porous silica, polystyrene spheres, carbon nanotubes, etc.)[110,111]. However, the nanostructures of these materials usually change due to amorphous-phase crystallization to anatase $TiO_2$ during annealing. This decreases the surface area and damages the surface nanostructures.

In this section, porous anatase $TiO_2$ particles were developed in aqueous solution. Nano $TiO_2$ was crystallized in the solution to assemble into particles 100–200 nm in diameter. The surface of the particles was covered with nanorelief structures. The particles showed *c*-axis orientation due to anisotropic crystal growth of $TiO_2$ along the *c*-axis. BET surface area of the particles was estimated to be 270 $m^2/g$[112]. BJH and DFT/Monte-Carlo analysis of adsorption isotherm indicated the existence of pores ~2.8 nm and ~3.6 nm in diameter, respectively. The existence of micropores ~1 nm was also indicated. Crystallization and self-assembly of acicular $TiO_2$ were effectively utilized to fabricate nanocrystal assembled $TiO_2$ particles having high surface area and nanorelief surface structure.

Ammonium hexafluorotitanate ($[NH_4]_2TiF_6$) (Morita Chemical Industries Co., Ltd., FW: 197.95, purity 96.0%) and boric acid ($H_3BO_3$) (Kishida Chemical Co., Ltd., FW: 61.83, purity 99.5%) were used as received. Ammonium hexafluorotitanate (12.372 g) and boric acid (11.1852 g) were separately dissolved in deionized water (600 mL) at 50°C. Boric acid solution was added to ammonium hexafluorotitanate solution at a concentration of 0.15 and 0.05 M, respectively.

The solution was kept at 50°C for 30 min using a water bath with no stirring, after which it was centrifuged at 4000 rpm for 10 min (Model 8920, Kubota Corp.). Preparation for centrifugation, centrifugation at 4000 rpm, deceleration from 4000 to 0 rpm and preparation for removal of supernatant solution took 4 min, 10 min, 10 min and 8 min, respectively. The particles contacted with the solution, the temperature of which was gradually lowered for 32 min after maintaining 50°C for 30 min. Precipitates were dried at 60°C for 12 h after removal of supernatant solution.

The crystal phase of the particles was evaluated by X-ray diffractometer (XRD; RINT-2100V, Rigaku) with CuKα radiation (40 kV, 30 mA). Diffraction patterns were evaluated using JCPDS, ICSD (Inorganic Crystal Structure Database) data (FIZ Karlsruhe, Germany and NIST, USA) and FindIt. Morphology of $TiO_2$ was observed by transmission electron microscopy (TEM; JEM2010, 200 kV, JEOL). Zeta potential and particle size distribution were measured by electrophoretic light-scattering spectrophotometer (ELS-Z2, Otsuka Electronics Co., Ltd.) with automatic pH titrator. Samples of 0.01 g were dispersed in distilled water (100 g) and ultrasonicated for 30 min prior to measurement. The pH of colloidal solutions was controlled by the addition of HCl (0.1 M) or NaOH (0.1 M). Zeta potential and particle size distribution were evaluated at 25°C and integrated 5 and 70 times, respectively.

Nitrogen adsorption-desorption isotherms were obtained using Autosorb-1 (Quantachrome Instruments) and samples of 0.137 g were outgassed at 110°C under $10^{-2}$ mmHg for 6 h prior to measurement. Specific surface area was calculated by BET (Brunauer-Emmett-Teller) method using adsorption isotherms. Pore size distribution was calculated by BJH (Barrett-Joyner-Halenda) method using adsorption isotherms because an artificial peak was observed from BJH size distribution calculated from desorption branches. Pore size distribution was further calculated by DFT/Monte-Carlo method ($N_2$ at 77 K on silica (cylinder/sphere, pore, NLDFT ads. model), adsorbent: oxygen) using adsorption branches.

Total pore volume (V) and average pore diameter (4V/A) were estimated from pores smaller than 230 nm (diameter) at P/Po = 0.99 through the use of BET surface area (A). However, adsorption volume increased drastically at high relative pressure in isotherms, indicating a large data error at high relative pressure. Therefore, total pore volume and average pore diameter estimated from pores smaller than 11 nm at P/Po = 0.80 would provide useful information. Relative pressure P/Po = 0.80 was selected for estimations because it was lower than the drastic adsorption increase and higher than adsorption hysteresis. The data can be compared to that of $TiO_2$ having similar morphology with small errors. Total pore volume (cumulative pore volume) was also estimated by BJH method from pores smaller than 154 nm. This usually has a small error compared to that estimated from isotherm data including high relative pressure such as P/Po = 0.99 because estimation by BJH method is not effected by adsorption volume errors at high relative pressure.

**Liquid phase crystal deposition of anatase $TiO_2$.** The solution became clouded about 10 min after mixing ammonium hexafluorotitanate solution and boric acid solution. The particles were homogeneously nucleated in the solution, turning the solution white.

Deposition of anatase $TiO_2$ proceeds by the following mechanisms[43,113,114]:

$$TiF_6^{2-} + 2H_2O \rightleftharpoons TiO_2 + 4H^+ + 6F^- \quad (a)$$

$$BO_3^{3-} + 4F^- + 6H^+ \longrightarrow BF_4^- + 3H_2O \quad (b)$$

Equation (a) is described in detail by the following two equations:

$$TiF_6^{2-} \xrightarrow{nOH^-} TiF_{6-n}(OH)_n^{2-} + nF^- \xrightarrow{(6-n)OH^-} Ti(OH)_6^{2-} + 6F^- \quad (c)$$

$$Ti(OH)_6^{2-} \longrightarrow TiO_2 + 2H_2O + 2OH^- \quad (d)$$

Fluorinated titanium complex ions gradually change into titanium hydroxide complex ions in an aqueous solution, as shown in Eq. (c). The increase of $F^-$ concentration displaces Eqs. (a) and (c) to the left; however, the produced $F^-$ can be scavenged by $H_3BO_3$ ($BO_3^{3-}$) as shown in Eq. (b) to displace Eqs. (a) and (c) to the right. Anatase $TiO_2$ is formed from titanium hydroxide complex ions ($Ti(OH)_6^{2-}$) in Eq. (d).

**Crystal phase of $TiO_2$ particles.** X-ray diffraction peaks for the particles were observed at $2\theta$ = 25.1, 37.9, 47.6, 54.2, 62.4, 69.3, 75.1, 82.5 and 94.0° after evaluation of $N_2$ adsorption. They were assigned to the 101, 004, 200, 105 + 211, 204, 116 + 220, 215, 303 + 224 + 312 and 305 + 321 diffraction peaks of anatase $TiO_2$ (JCPSD No. 21-1272, ICSD No. 9852) (Fig. 16).

The 004 diffraction intensity of randomly oriented particles is usually 0.2 times the 101 diffraction intensity as shown in JCPDS data (No. 21-1272). However, the 004 diffraction intensity of the particles deposited in our process was 0.36 times the 101 diffraction intensity. Additionally, the integral intensity of the 004 diffraction was 0.18 times the 101 diffraction intensity, indicating the *c*-axis orientation of the particles. Particles were not oriented on the glass holder for XRD measurement. Therefore, $TiO_2$ crystals would be an anisotropic shape in which the crystals were elongated along the *c*-axis. The crystals would have a large number of stacks of c planes such as (001) planes compared to stacks of (101) planes. The diffraction intensity from the (004) planes would be enhanced compared to that from the (101) planes.

Crystallite size perpendicular to the (101) or (004) planes was estimated from the full-width half-maximum of the 101 or 004 peak to be 3.9 nm or 6.3 nm, respectively. Elongation of crystals in the *c*-axis direction was also suggested by the difference in crystallite size.

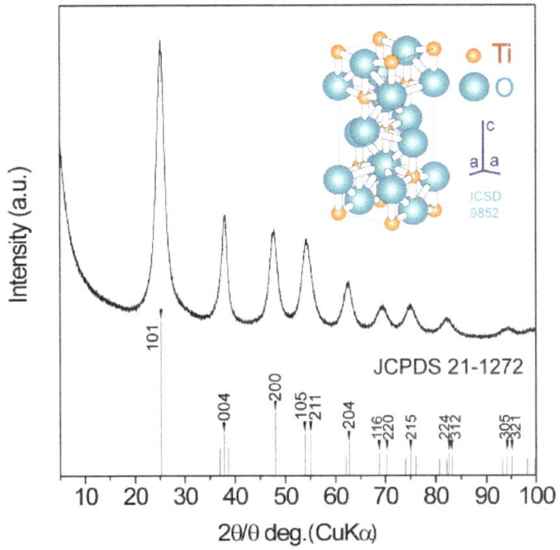


Fig. 16. XRD diffraction pattern of anatase $TiO_2$ particles. Reprinted with permission from Ref.[77], Masuda, Y. and Kato, K., 2008, *Cryst. Growth Des.*, 8, 3213. Copyright @American Chemical Society

**TEM observation of $TiO_2$ particles.** The particles were shown to be assemblies of nano $TiO_2$ crystals (Fig. 17a). Particle diameter was estimated to be 100–200 nm. Relief structures had formed on the surfaces and open pores had formed inside because the particles were porous assemblies of nanocrystals.

Nanocrystals were shown to have acicular shapes (Fig. 17b). They were about 5–10 nm in length. The longer direction of acicular $TiO_2$ is indicated by the black arrow. The inserted FFT image shows the 101 and 004 diffractions of anatase $TiO_2$. Nanocrystals are assigned to the single phase of anatase $TiO_2$. It is notable that the diffraction from the (101) planes has a ring shape due to random orientation but that from the (004) planes was observed only in the upper right region and lower left region in the FFT image. Anisotropic 004 diffractions indicated the direction of the *c*-axis, which was perpendicular to the (004) planes, as shown by the white arrow. It was roughly parallel to the longer direction of acicular $TiO_2$. These results suggest that acicular $TiO_2$ grew along the *c*-axis to enhance the diffraction intensity from the (004) planes. Crystal growth of anatase $TiO_2$ along the *c*-axis was previously observed in $TiO_2$ films[113]. Anisotropic crystal growth is one of the features of liquid phase crystal deposition.

Acicular nanocrystals showed lattice images of anatase $TiO_2$ (Fig. 17c). They were constructed of anatase $TiO_2$ crystals without amorphous or additional phases. Anatase crystals were not covered with amorphous or additional phases even at the tips. Bare anatase crystal with nanosized structure is important to achieve high performance for catalysts and devices.

Crystallization of $TiO_2$ was effectively utilized to form assemblies of acicular nanocrystals in the process. Open pores and surface relief structures were successfully formed on the particles.

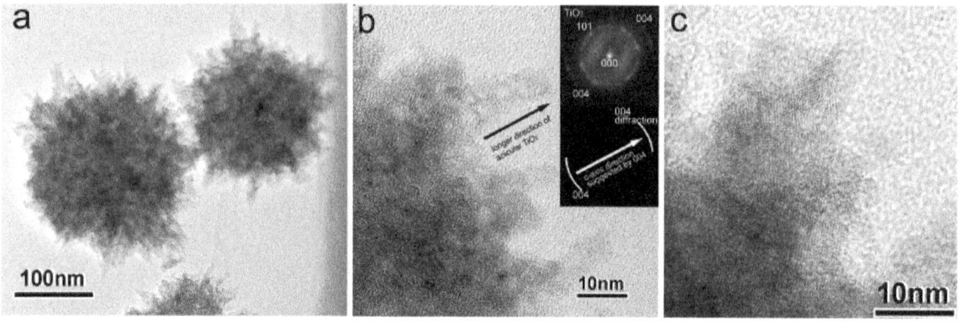


Fig. 17. (a): TEM micrograph of anatase $TiO_2$ particles. (b): Magnified area of (a) showing morphology of acicular crystals. Insertion in (b): FFT image of (b) anatase $TiO_2$. (c): Magnified area of (a) showing lattice images of anatase $TiO_2$.
Reprinted with permission from Ref.[77], Masuda, Y. and Kato, K., 2008, *Cryst. Growth Des.*, 8, 3213. Copyright @American Chemical Society

**Zeta potential and particle size distribution.** The dried particles were dispersed in water to evaluate zeta potential and particle size distribution after evaluation of $N_2$ adsorption. The particles had positive zeta potential of 30.2 mV at pH 3.1, which decreased to 5.0, −0.6, −11.3 and −36.3 mV at pH 5.0, 7.0, 9.0 and 11.1, respectively (Fig. 18). The isoelectric point was estimated to be pH 6.7, slightly higher than that of anatase $TiO_2$ (pH 2.7–6.0)[115]. Zeta potential is very sensitive to the particle surface conditions, ions adsorbed on the particle surfaces, and the kind and concentration of ions in the solution. The variations in zeta potential were likely caused by the difference in the surface conditions of $TiO_2$ particles, affected by the interaction between particles and ions in the solution.

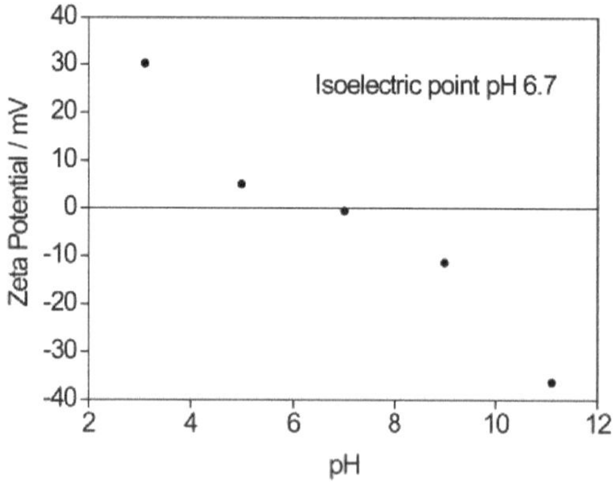


Fig. 18. Zeta potential of anatase $TiO_2$ particles as a function of pH. Reprinted with permission from Ref.[77], Masuda, Y. and Kato, K., 2008, *Cryst. Growth Des.*, 8, 3213. Copyright @American Chemical Society

Mean particle size was estimated to be ~550 nm in diameter with a standard deviation (STD) of 220 nm at pH 3.1 (Fig. 19a). This was larger than that observed by TEM. Slight aggregation occurred at pH 3 because the particles were dried completely prior to measurement. Particle size increased with pH and showed a maximum of near the isoelectric point (550 nm at pH 3.1, 3150 nm at pH 5, 4300 nm at pH 7, 5500 nm at pH 9 or 2400 nm at pH 11.1) (Fig. 19b). Strong aggregation resulted from the lack of repulsion force between particles near the isoelectric point.

The particles were generated in the solution at pH 3.8 in this study. It would be suitable to obtain repulsion force between particles for crystallization without strong aggregation.

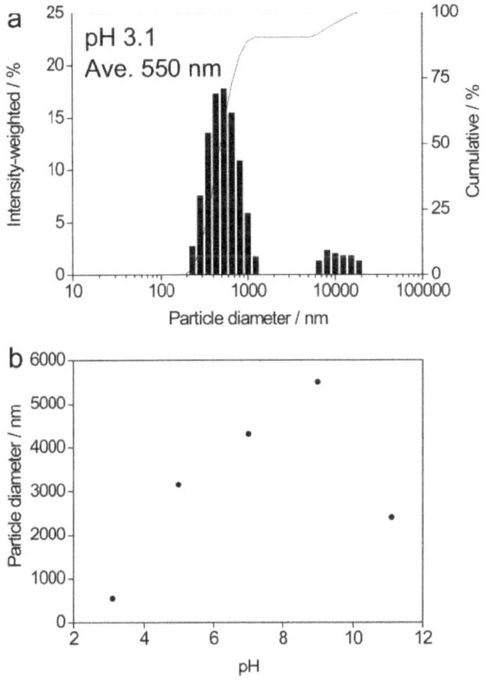


Fig. 19. (a): Particle size distribution of anatase $TiO_2$ particles at pH 3.1. (b): Particle size of anatase $TiO_2$ particles as a function of pH. Reprinted with permission from Ref.[77], Masuda, Y. and Kato, K., 2008, *Cryst. Growth Des.*, 8, 3213. Copyright @American Chemical Society

**$N_2$ adsorption characteristics of $TiO_2$ particles.** $TiO_2$ particles exhibited $N_2$ adsorption-desorption isotherms of Type IV (Fig. 20a). The desorption isotherm differed from adsorption isotherm in the relative pressure ($P/P_0$) range from 0.4 to 0.7, showing mesopores in the particles. BET surface area of the particles was estimated to be 270 $m^2/g$ (Fig. 20b). This is higher than that of $TiO_2$ nanoparticles such as Aeroxide P25 (BET 50 $m^2/g$, 21 nm in diameter, anatase 80% + rutile 20%, Degussa), Aeroxide P90 (BET 90–100 $m^2/g$, 14 nm in diameter, anatase 90% + rutile 10%, Degussa), MT-01 (BET 60 $m^2/g$, 10 nm in diameter, rutile, Tayca Corp.) and Altair TiNano (BET 50 $m^2/g$, 30–50 nm in diameter, Altair Nanotechnologies Inc.)[116]. A high BET surface area cannot be obtained from particles having a smooth surface even if the particle size is less than 100 nm. A high BET surface area would be realized by the unique morphology of $TiO_2$ particles constructed of nanocrystal assemblies.

Total pore volume and average pore diameter were estimated from pores smaller than 230 nm at P/Po = 0.99–0.431 cc/g and 6.4 nm, respectively. They were estimated to be 0.212 cc/g and 3.1 nm, respectively, from pores smaller than 11 nm at P/Po = 0.80. Total pore volume was also estimated by the BJH method from pores smaller than 154 nm to be 0.428 cc/g. Average pore diameter was estimated to be 6.3 nm using BET surface area.

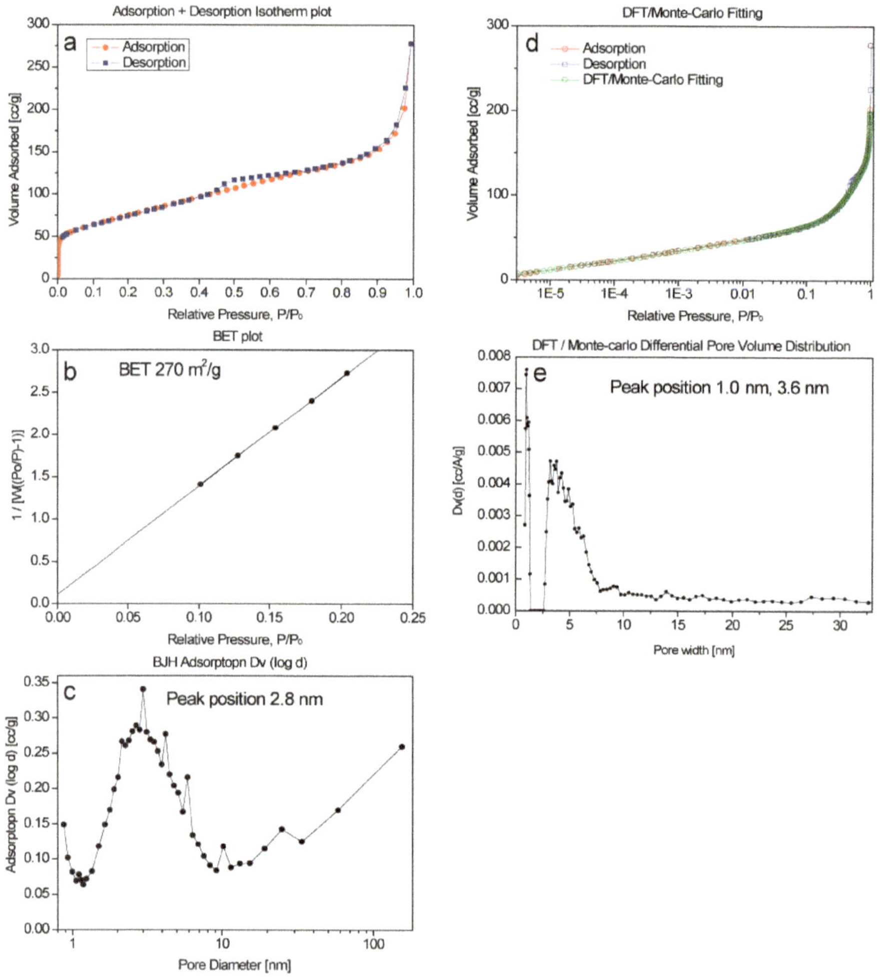


Fig. 20. (a): $N_2$ adsorption-desorption isotherm of anatase $TiO_2$ particles. (b): BET surface area of anatase $TiO_2$ particles. (c): Pore size distribution calculated from $N_2$ adsorption data of anatase $TiO_2$ particles using BJH equation. (d): $N_2$ adsorption-desorption isotherm and DFT/Monte-Carlo fitting curve of anatase $TiO_2$ particles. (e): Pore size distribution calculated from $N_2$ adsorption data of anatase $TiO_2$ particles using DFT/Monte-Carlo equation. Reprinted with permission from Ref.[77], Masuda, Y. and Kato, K., 2008, *Cryst. Growth Des.*, 8, 3213. Copyright @American Chemical Society

Pore size distribution was calculated by the BJH method using adsorption isotherms (Fig. 20c). It showed a pore size distribution curve having a peak at ~2.8 nm and pores larger than 10 nm. $TiO_2$ particles would have mesopores of ~2.8 nm surrounded by nanocrystals. Pores larger than 10 nm are considered to be interparticle spaces. The pore size distribution also suggested the existence of micropores smaller than 1 nm.
Pore size distribution was further calculated by the DFT/Monte-Carlo method. The model was in fair agreement with adsorption isotherms (Fig. 20d). Pore size distribution showed a peak at ~3.6 nm that indicated the existence of mesopores of ~3.6 nm (Fig. 20e). The pore size calculated by the DFT/Monte-Carlo method was slightly larger than that calculated from the BJH method because the latter method is considered to have produced an underestimation[117-119]. The pore size distribution also suggested the existence of micropores of ~1 nm, probably resulting from microspaces surrounded by nanocrystals and the uneven surface structure of nanocrystals.
The particles were shown to have a large surface area as well as micropores of ~1 nm, mesopores of ~2.8-3.6 nm and pores larger than 10 nm, by $N_2$ adsorption characteristics. Assembly of acicular nanocrystals resulted in unique features and high surface area.
$TiO_2$ particles were generated in the solutions at 90°C for 1h using an oil bath with no stirring for comparison. The solutions became clouded after the addition of boric acid solutions into ammonium hexafluorotitanate solutions. High temperature accelerated crystal growth of $TiO_2$. Hydrogen chloride of 0.6 ml was added into the solutions of 200ml to decrease crystallization speed of $TiO_2$. The pH of the solutions was 2.4 one hour after mixing the solutions. BET surface area of the particles was estimated to 18 $m^2/g$. This is much lower than that of the particles prepared at 50°C and slightly lower than that prepared at 90°C for 8 min in our previous work (44 $m^2/g$)[120]. Formation of $TiO_2$ was accelerated at high temperature and it decreased surface area. The particles grew in the solutions to decrease surface area as function of time. Crystallization of $TiO_2$ was shown to be strongly affected by growth conditions such as solution temperature and growth time.
Anatase $TiO_2$ particles, 100-200 nm in diameter, were successfully fabricated in aqueous solution. They were assemblies of nanocrystals 5-10 nm that grew anisotropically along the *c*-axis to form acicular shapes. The particles thus had nanorelief surface structures constructed of acicular crystals. They showed *c*-axis orientation due to high-intensity X-ray diffraction from the (004) crystal planes. The particles had a high BET surface area of 270 $m^2/g$. Total pore volume and average pore diameter were estimated from pores smaller than 230 nm at P/Po = 0.99-0.43 cc/g and 6.4 nm, respectively. They were also estimated from pores smaller than 11 nm at P/Po = 0.80-0.21 cc/g and 3.1 nm, respectively. BJH and DFT/Monte-Carlo analysis of adsorption isotherm indicated the existence of pores ~2.8 and ~3.6 nm, respectively. Additionally, the analyses suggested the existence of micropores of ~1 nm. Crystallization and self-assembly of nano $TiO_2$ were effectively utilized to fabricate nanocrystal assembled $TiO_2$ particles having high surface area and nanorelief surface structure.

## 7. Morphology control of multi-needle $TiO_2$ particles[121]

Flower-like multi-needle anatase $TiO_2$ particles were developed in aqueous solutions. They were pure anatase $TiO_2$ crystals containing no cores, organic binders or solvents. Furthermore, micro-structured silicon wafers were covered with the $TiO_2$ particles uniformly in the solutions. Their unique crystals growth and physicochemical profiles were precisely evaluated and discussed.

**Micro-structured silicon wafers and their surface modification to super-hydrophilic surfaces**

Silicon wafers were modified to have micro-structures on the surfaces with cutting work. They were cut using a precise diamond cutter under running water. Width and height of salient lines were 200 $\mu$m and 150 $\mu$m, respectively. They were formed at 500 $\mu$m intervals. They were blown by air to remove dust and were exposed to vacuum-ultraviolet light (VUV light, low-pressure mercury lamp PL16-110, air flow, 100 V, 200 W, SEN Lights Co., 14 mW/cm$^2$ for 184.9 nm at a distance of 10 mm from the lamp, 18 mW/cm$^2$ for 253.7 nm at a distance of 10 mm from the lamp) for 10 min in air. Bare silicon surfaces were covered with small amount of surface contamination. The VUV irradiation modified them to clean surfaces that showed super hydrophilic surfaces of water contact angle about 0-1 °.

**Morphology control of anatase $TiO_2$**

Ammonium hexafluorotitanate (206.20 mg) and boric acid (186.42 mg) were separately dissolved in deionized hot water (100 mL) at 50°C. Boric acid solution was added to ammonium hexafluorotitanate solution at concentrations of 15 mM and 5 mM, respectively. The silicon wafers having patterned surfaces were immersed in the middle of the solutions with the bottom up at an angle. They were tilted at 15 degrees to the upright. The solutions were kept at 50°C for 19 hours or 7 days using a drying oven (Yamato Scientific Co., Ltd., DKN402) with no stirring. The substrates were washed with running water and dried by air blow. The solutions were centrifuged at 4000 rpm for 10 min (Model 8920, Kubota Corp.). Precipitated particles were dried at 60°C for 12 h after removal of supernatant solutions. The particles were dispersed in distilled water. They were centrifuged and dried again for purification.

1. **Morphology control of multi-needle $TiO_2$ particles**

**TEM observation of multi-needle $TiO_2$ particles**

Flower-like multi-needle $TiO_2$ particles were successfully formed in aqueous solutions (Fig. 21a, b). Needle shaped crystals grew from the center of the particles. Especially, the needles grew parallel to TEM observation direction from the center of the particles as shown in red circles. They clearly showed that needles radiated in all directions to form flower-like morphology. Each particle had about 6-10 taper needles. Width and length of them were about 200 nm and 100 nm, respectively. Aspect ratio was about 2 (200 nm / 100 nm). Width of the needles became smaller as growth direction to have tips. The particles had no core or pore at the center of their bodies. Nucleation and crystal growth were well controlled to have flower-like multi-needle morphology.

Electron diffraction pattern showed that the particles were single phase of anatase $TiO_2$ crystals (Fig. 21c). Interplanar spacing of (004), (200), and (204) planes were estimated to 0.243 nm, 0.201 nm and 0.151 nm. Diffractions from (004), (200), and (204) planes were clear single spots (Fig. 21c). Needle shaped crystal was shown to single crystal of anatase $TiO_2$. Long direction of the needle shaped crystals was perpendicular to (004) crystal faces (Fig. 21c). It indicated that anatase $TiO_2$ crystals grew along *c*-axis to form needle shape morphology. They were thus surrounded by *a*-faces of anatase $TiO_2$ crystals.

Surfaces of the needle crystals were observed carefully (Fig. 21c, d). There were no amorphous layers or second phase layers on the surfaces. The particles had pure and bare anatase $TiO_2$ surfaces.

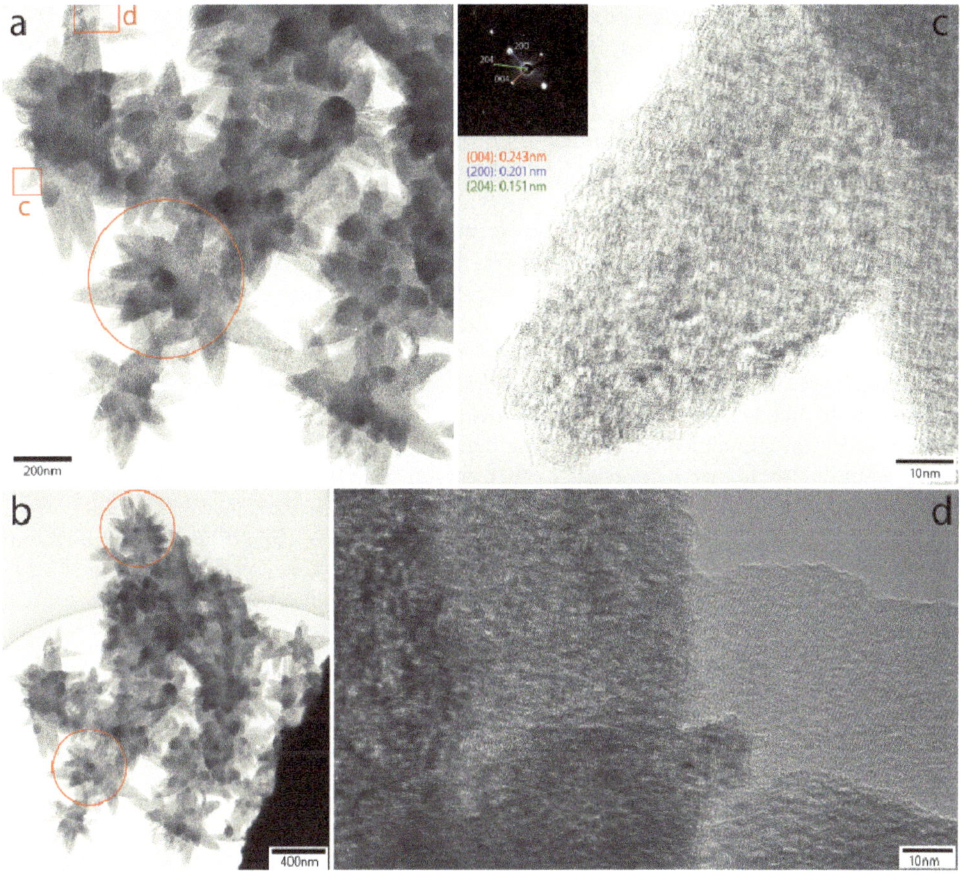


Fig. 21. Transmission electron microscope images of the multi-needle $TiO_2$ particles. (a, b) Low magnification images of the particles. (c) High resolution image of the needles grew from the center of the particle showing taper shape of the needle. (d) High resolution image of surface of the needles showing bare $TiO_2$ surface. (c) and (d) are magnified area of (a) and (b). Insert in (c) is electron diffraction pattern indicating crystal phase and interplanar spacing. Red circles indicate typical images of the multi-needle $TiO_2$ particles. Reprinted with permission from Ref.[121], Masuda, Y., Ohji, T. and Kato, K., 2010, *Cryst. Growth Des.*, 10, 913. Copyright @American Chemical Society

**XRD analysis of multi-needle $TiO_2$ particles**

XRD analysis showed that the particles were single phase of anatse $TiO_2$. X-ray diffraction peaks were observed at $2\theta$ = 25.12, 36.8, 37.7, 47.7, 53.7, 54.7, 62.4, 68.5, 69.8, 74.7, 82.2, 93.7 and 94.4 °. Pure anatase $TiO_2$ particles were obtained from the aqueous solutions in this study. Anatase phase have been prepared using high temperature annealing in many reports. It caused deformation of nano/micro-structures, aggregation of the particles and decrease of surface area. However, crystallization of anatase $TiO_2$ was realized at 50 °C in this study to avoid degradation of the properties.

Crystallite size perpendicular to the (101), (004) or (200) planes was estimated from the full-width half-maximum of the 101, 004 or 200 peak to be 18.2, 29.8 or 19.2 nm, respectively. Crystallite size perpendicular to (004) was larger than that of others. Difference in crystallite size indicated anisotropic crystal growth along *c*-axis.

For comparison, nanocrystal assembled $TiO_2$ particles were prepared in aqueous solutions at 50°C for 30 min in previous report[77]. Crystallite size perpendicular to the (101), (004) or (200) planes was estimated to be 3.9, 6.3 or 4.9 nm, respectively. The crystalline degree of the multi-needle $TiO_2$ nanostructures was much improved compared with that of the previous nanocrystal[77]. The key to achieve highly crystalline degree even at low-temperature, i.e., 50°C, was slow and long-term growth. It was achieved by low super saturation degree of the solutions. Concentration of ammonium hexafluorotitanate and boric acid were one-tenth of the previous nanocrystal[77]. Synthesis parameters of multi-needle $TiO_2$ nanostructures were developed based on previous reports as follows. Acicular nanocrystals were homogeneously formed immediately after mixing of two solutions in previous report[77]. They aggregated into nanocrystal assembled particles. The particles were removed from the solutions 30 minutes after the mixing to prevent further crystal growth. The crystallite size was thus small and it contributed to high specific surface area. On the other hand, $TiO_2$ films were prepared in the same solutions at 50°C for several hours[114,122,123]. The films were consisted of two layers. Under layer was consisted of small nanocrystals. They were formed at an early stage of immersion period in high ion concentration solutions. Ions were consumed gradually to form the crystals. Upper layer of acicular crystal assembly was then formed. They grew in the solutions with low ion concentrations. It indicated that low ion concentration and low super saturation degree realized formation of acicular $TiO_2$ crystals. Additionally, size of acicular crystals was much larger than initially deposited nanocrystals. We tried to form multi-needle particles consisted of large acicular crystals on the basis of these results. Crystal growth in the solution with low ion concentration and low super saturation degree was utilized for anisotropic crystal growth. They were grown to large crystals with long period such as 7 days. Slow growth rate caused formation of euhedral crystals that were affected by crystal structure of tetragonal anatase.

**Raman spectroscopy of multi-needle $TiO_2$ particles**

The particles showed Raman peaks at 157 $cm^{-1}$, 412 $cm^{-1}$, 506 $cm^{-1}$ and 628 $cm^{-1}$. It was typical of anatase $TiO_2$ phase. They were assigned to Eg (v6) mode (157 $cm^{-1}$), B1g mode (412 $cm^{-1}$), doublet of the A1g and B1g modes (506 $cm^{-1}$), and Eg (v1) mode (628 $cm^{-1}$), respectively. Additional peaks indicating rutile $TiO_2$ or other phases were not observed. It was notable that Eg (v6) mode shifted with respect to those of bulk crystals or sintered powders. For the single crystal, Ohsaka et al. determined following allowed bands: 144 $cm^{-1}$ (Eg), 197 $cm^{-1}$ (Eg), 399 $cm^{-1}$ (B1g), 513 $cm^{-1}$ (A1g), 519 $cm^{-1}$ (B1g) and 639 $cm^{-1}$ (Eg)[124]. $TiO_2$ particles (P25, Degussa) showed peaks at 143.3 $cm^{-1}$ (Eg), 196 $cm^{-1}$ (Eg), 396 $cm^{-1}$ (B1g), 516 $cm^{-1}$ (A1g+B1g) and 638 $cm^{-1}$ (Eg). Raman peak of Eg (v6) mode (144 $cm^{-1}$) has been reported to shift due to several factors such as effect of crystalline size (quantum size confinement effect)[125,126], temperature[125] or pressure[127]. Additionally, nitrogen doped anatase $TiO_2$ particles of 8.40-9.80 nm in diameter were reported to have Eg (v6) mode at 151 $cm^{-1}$[126]. The multi-needle particles showed large peak shift of Eg (v6) mode from 144 $cm^{-1}$ to 157 $cm^{-1}$ because of fluorine doping effect and size effect. The large shift of Eg (v6) mode was one of the characteristics of the multi-needle $TiO_2$ particles.

**FT-IR absorbance of multi-needle $TiO_2$ particles**

The particles showed absorption spectra in infrared light region. The absorption bands related to $TiO_2$ were observed at 907 $cm^{-1}$ and 773 $cm^{-1}$. They were assigned to stretching vibrations of Ti=O and -Ti-O-Ti-, respectively. They were consistent with TEM, XRD and Raman analyses. The bands observed in the range of 1400-1750 $cm^{-1}$ were attributed to bending vibrations of O-H[128]. Absorption band in frequency range 600-450 $cm^{-1}$ were reported to be attributed to stretching vibrations of Ti-F bonds in $TiO_2$ lattice[129,130]. The bands were observed at 513, 532, 540, 558 or 568 $cm^{-1}$ from $xTiOF_2 \cdot yBaF_2 \cdot zMnF_2$ glasses[129]. They indicated that F ions were partially replaced to O ions in the lattice[129]. Actually, absorption bands were observed at 455, 488, 505, 520, 552 and 567 $cm^{-1}$ in the spectra of the multi-needle $TiO_2$ particles. Additionally, absorption band at 1080 $cm^{-1}$ was reported to be assigned to surface fluorinated Ti-F species[131]. Absorption band was observed at about 1058 $cm^{-1}$ from the multi-needle $TiO_2$ particles. These analyses suggested that F ions were partially doped into $TiO_2$ crystals to replace O ions.

**UV-Vis spectroscopy and band gap of multi-needle $TiO_2$ particles**

Optical property of the particles was evaluated with UV-Vis spectroscopy. Transparency gradually decreased as decrease of wavelength. The bulk band gap structures are known as direct-transition for anatase and indirect-transition for rutile in titania, respectively. Band gap of the particles was estimated to 3.20 eV (388 nm) by assuming the direct-transition. It was similar to that of anatase $TiO_2$ nanostructures such as anatase $TiO_2$ nanorods (3.2 eV (388 nm))[132], anatase $TiO_2$ nanowalls (3.2 eV (388 nm))[132], anatase $TiO_2$ nanotubes (3.2 eV (388 nm))[133], single-layered $TiO_2$ nano-sheet with a thickness less than 1 nm (3.15 eV (394 nm))[134], stacked $TiO_2$ nano-sheets (3.15 eV (394 nm))[134] and anatase $TiO_2$ films (3.2 eV (388 nm))[135]. It was higher than that of rutile $TiO_2$ films (2.9eV (428 nm))[135], rutile $TiO_2$ nanorods (3.0 eV (414 nm))[132], rutile $TiO_2$ single crystal (3.0 eV (414 nm), SHINKOSHA Co., Ltd.), and lower than that of amorphous $TiO_2$ films (3.5 eV (355 nm))[135].

**$N_2$ adsorption characteristics of multi-needle $TiO_2$ particles**

BET surface area of the particles was estimated to be 178 $m^2/g$ from an adsorption branch in the range of $P/Po = 0.1$-0.29. Average pore diameter was estimated to be 11.1 nm using BET surface area. They were estimated to 166 $m^2/g$ and 11.9 nm, 151 $m^2/g$ and 17.6 nm from $P/Po=0.05$-0.1 or $P/Po=0.02$-0.07, respectively. BET surface area was higher than that of $TiO_2$ nanoparticles such as Aeroxide P25 (BET 50 $m^2/g$, 21 nm in diameter, anatase 80% + rutile 20%, Degussa), Aeroxide P90 (BET 90–100 $m^2/g$, 14 nm in diameter, anatase 90% + rutile 10%, Degussa), MT-01 (BET 60 $m^2/g$, 10 nm in diameter, rutile, Tayca Corp.) and Altair TiNano (BET 50 $m^2/g$, 30–50 nm in diameter, Altair Nanotechnologies Inc.)[116]. Pore size distribution calculated by BJH method indicated that spaces of 2-3 nm were existed in the particles. DFT/Monte-Carlo analysis showed several types of mesopores in the range of 3-10 nm (3.5 nm, 4.9 nm, 6.3 nm and 10 nm) and micropores of ~0.8 nm. Total pore volume and average pore diameter were estimated from pores smaller than 241 nm at $P/Po = 0.99$ to be 0.493 cc/g and 13 nm, respectively. They were estimated to be 0.158 cc/g and 4.2 nm, respectively, from pores smaller than 11 nm at $P/Po = 0.81$. Total pore volume was also estimated by the BJH method from pores smaller than 156 nm to be 0.505 cc/g using desorption isotherm. Total pore volume was also estimated by the BJH method from pores smaller than 160 nm to be 0.628 cc/g using adsorption isotherm.

## 2. Surface coating of micro-structured substrates with multi-needle $TiO_2$ particles

### SEM observation of the surface coatings

Micro-structured silicon wafers were immersed in the aqueous solutions at 50°C for 7 days. The substrates were successfully covered with multi-needle $TiO_2$ particles (Fig. 22-a). Both of salient regions and concave regions were modified with the particles uniformly (Fig. 22-b1, c1). The original micro-structure of silicon wafers were well maintained because the $TiO_2$ surface coatings were uniform thin layers of about 200-600 nm in thickness (Fig. 22-b2, c2). The coating layers were consisted of flower-like multi-needle $TiO_2$ particles of about 200-400 nm in diameter (Fig. 22-b3, c3, red circles). The particles had several needles which grew from the center of the particles. The needles had taper shape along growth direction. The particle in red circle of Fig. 22-b3 clearly indicated that tetragonal crystal phase of anatase $TiO_2$ caused fourfold symmetry crystal growth of four needles parallel to the substrate. The *a*-axis and *c*-axis of the particle were parallel and perpendicular to the substrate, respectively. This shape was basic morphology of the particles in this system. Other needles also grew from the center to form multi-needle shape as shown in red circles in Fig. 22-c3. Salient regions were slightly trapezoidal geometry (Fig. 22-a, b1). $TiO_2$ particles were deposited on walls of them. The area between the salient and concave regions was thus white in SEM images.

For comparison, micro-structured silicon wafers were immersed in the aqueous solutions at 50°C for 19 hours. The multi-needle $TiO_2$ particles were formed on both of salient regions and concave regions (Fig. 23-a). The micro-structures of the substrates were maintained because the surface coatings were uniform thin layers (Fig. 23-b1, c1). However, coverage of the surfaces was different from that immersed for 7 d. The $TiO_2$ particles covered approximately half of the salient regions and one third of concave regions (Fig. 23-b2, c2). Coverage of the salient regions was higher than that of concave regions because concave regions set back far from the salient region surfaces. The particles and ions were not well supplied to concave regions. The particles had several needles grown from the center (Fig. 23-b3, c3). The morphology of them was similar to that immersed for 7d, in contrast, their size was slightly smaller to be about 100-300 nm. These observations indicated growth mechanism as follows. The particles were homogeneously nucleated from ions in the solutions. They were then adhered on the substrates to form surface coatings. The particles further grew to increase number and size of the needles.

### XPS analysis of the surface coatings

Surfaces of the $TiO_2$ coatings were analyzed with XPS. Titanium, oxygen, carbon, silicon and fluorine were observed from the surface before (Fig. 24-1a) and after (Fig. 24-1b) $Ar^+$ sputtering. Ti $2p_{3/2}$ spectrum was observed at 458.6 eV (Fig. 24-2a). The binding energy was higher than that of Ti metal (454.0 eV), TiC (454.6 eV), TiO (455.0 eV), TiN (455.7 eV) and $Ti_2O_3$ (456.7 eV), and similar to that of $TiO_2$ (458.4-458.7 eV)[136-138]. This suggested that the titanium atoms in the particles were positively charged relative to that of titanium metal by formation of direct bonds with oxygen. Ti 2p spectrum changed its shape by the sputtering due to the decrease of cation valence (Fig. 24-2b). The phenomenon was often observed in sputtering of titanium oxide[136]. The spectrum after sputtering was not suitable for estimation of chemical ratio because the ratio was slightly changed by decrease of titanium cation valence. The chemical ratio of Ti to O was thus estimated from the spectra before the sputtering.

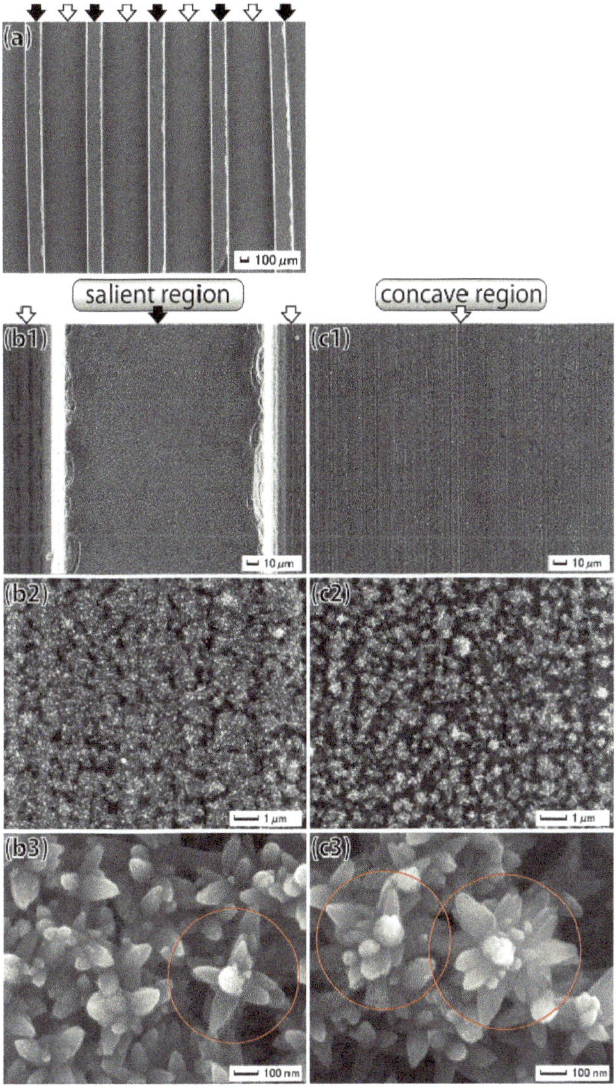


Fig. 22. (a) SEM micrograph of a micro-structured substrate coated with the multi-needle $TiO_2$ particles deposited for 7 d. (b1) Salient region coated with the multi-needle $TiO_2$ particles. (b2) Magnified area of (b1) showing uniform surface coating. (b3) Magnified area of (b2) showing morphology of the multi-needle $TiO_2$ particles. (c1) Concave region coated with the multi-needle $TiO_2$ particles. (c2) Magnified area of (c1) showing uniform surface coating. (c3) Magnified area of (c2) showing morphology of the multi-needle $TiO_2$ particles. Black arrows show salient regions. White arrows show concave regions. Red circles indicate typical images of the multi-needle $TiO_2$ particles. Reprinted with permission from Ref.[121], Masuda, Y., Ohji, T. and Kato, K., 2010, *Cryst. Growth Des.*, 10, 913. Copyright @American Chemical Society

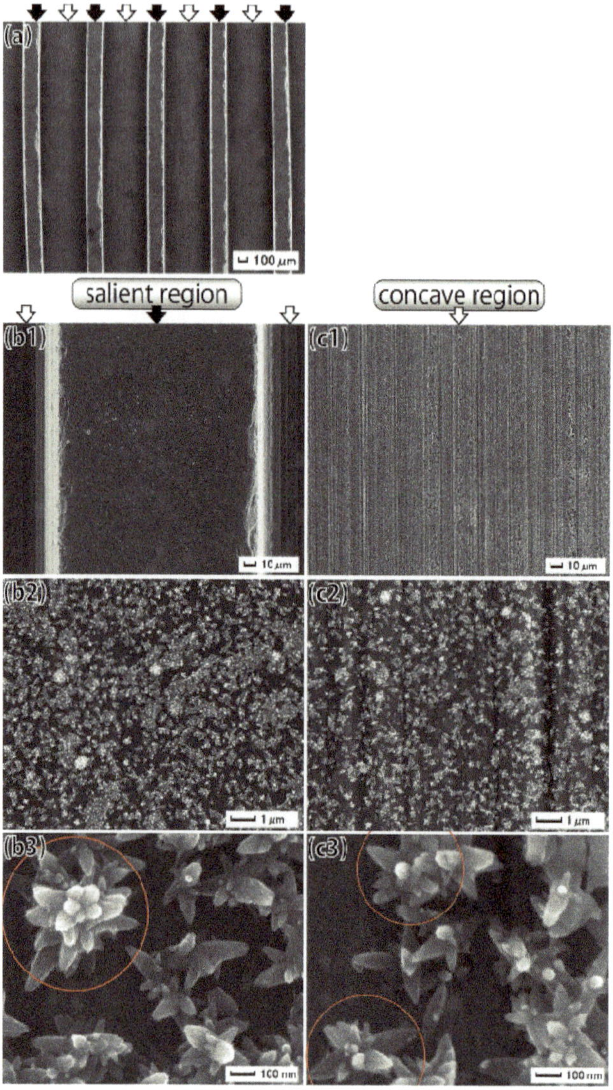


Fig. 23. (a) SEM micrograph of a micro-structured substrate coated with the multi-needle $TiO_2$ particles deposited for 19 h. (b1) Salient region coated with the multi-needle $TiO_2$ particles. (b2) Magnified area of (b1) showing uniform surface coating. (b3) Magnified area of (b2) showing morphology of the multi-needle $TiO_2$ particles. (c1) Concave region coated with the multi-needle $TiO_2$ particles. (c2) Magnified area of (c1) showing uniform surface coating. (c3) Magnified area of (c2) showing morphology of the multi-needle $TiO_2$ particles. Black arrows show salient regions. White arrows show concave regions. Red circles indicate typical images of the multi-needle $TiO_2$ particles. Reprinted with permission from Ref.[121], Masuda, Y., Ohji, T. and Kato, K., 2010, *Cryst. Growth Des.*, 10, 913. Copyright @American Chemical Society

O 1s spectrum was decomposed into two Gaussian curves after removal of background (Fig. 24-3a). The integral intensity ratio of O 1s peak at 532.12 eV to O 1s peak at 529.85 eV was 0.662 : 0.338. The binding energy of O 1s peak at 529.85 eV was similar to that of $TiO_2$ (529.9 eV[138], 530.1 eV[136,137]) showing that oxygen was negatively charged compared to neutral oxygen molecules (531.0 eV) through the formation of direct bonds with Ti. High binding energy component at 532.12 eV was assigned to oxygen atoms combined to carbon atoms as C-O or C=O (532.8 eV). It was decreased by the sputtering for 10 sec (Fig. 24-3b). It was included in surface contaminations. Drastic decreasing of C 1s spectrum by the sputtering supported this ascription (Fig. 24-4a, 4b). Chemical ratio of Ti to O was estimated to 1 : 1.76 using Ti 2p at 458.6 eV (Fig. 24-2a) and O 1s at 529.85 eV (Fig. 24-3a). It was slightly smaller than that expected from $TiO_2$. Oxygen vacancy and doping of fluorine ions would decrease oxygen volume. Fluorine was, in fact, observed from the surfaces at 684.5 eV (Fig. 24-5a). Chemical ratio was estimated to Ti : O : F = 1 : 1.76 : 0.18. Sum of oxygen and fluorine was 1.94 which was similar to 2 expected from $TiO_2$. Fluorine spectrum intensity was not decreased by the sputtering to be Ti : F = 1 : 0.16 (Fig. 24-5b). These indicated that fluorine

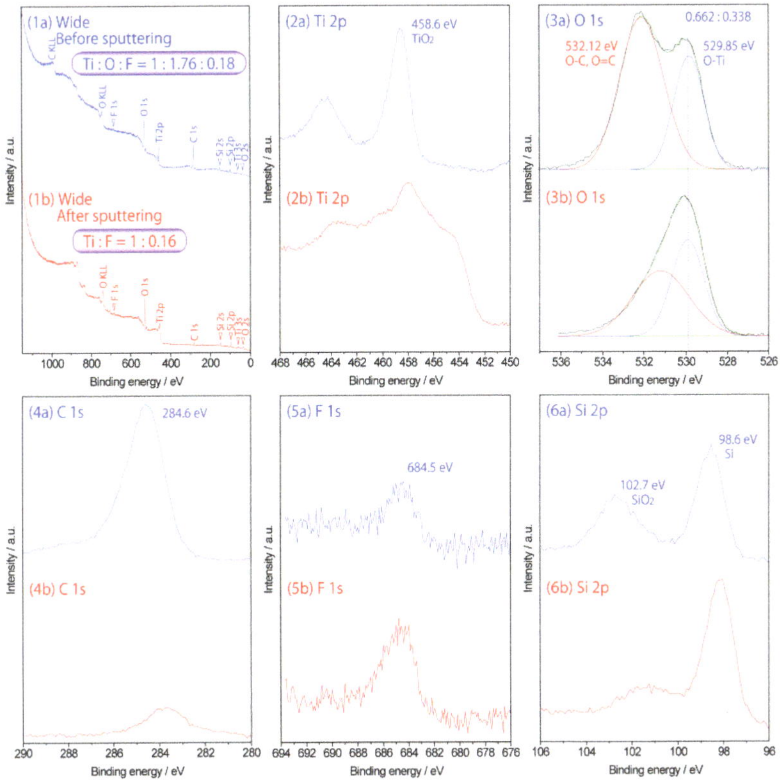


Fig. 24. XPS spectra of (1) wide scan, (2) Ti 2p, (3) O 1s, (4) C 1s, (5) F 1s and (6) Si 2p for a micro-structured substrate coated with the multi-needle $TiO_2$ particles deposited for 7d (a) before and (b) after $Ar^+$ sputtering. (3a) O1s spectra decomposed to two peaks after removal of background. Reprinted with permission from Ref.[121], Masuda, Y., Ohji, T. and Kato, K., 2010, *Cryst. Growth Des.*, 10, 913. Copyright @American Chemical Society

would not be included in surface contaminations but in the $TiO_2$ particles. Fluorine has been reported to improve properties of $TiO_2$. Fluorination of $TiO_2$ increased surface acidity due to the strongest electronegativity of fluorine[139]. Fluorine doping into $TiO_2$ increased the surface OH radicals that were suitable for photocatalytic reactions and improved photocorrosion resistance[140-145]. These properties expand application area of $TiO_2$.

Si $2p_{3/2}$ was observed at 98.6 eV and 102.7 eV (Fig. 24-6a). It was detected from silicon regions uncovered with $TiO_2$. The binding energies were similar to those in silicon wafers (99.6 eV) and $SiO_2$ (103.4 eV). The latter component was decreased by the sputtering because surface native oxide layer of amorphous $SiO_2$ was removed (Fig. 24-6b).

In this section, Multi-needle $TiO_2$ nanostructures having high surface area, fluorine doping and novel physicochemical characteristics were successfully fabricated. Unique aqueous synthesis realized their distinct morphologies and properties. The needle crystals grew along *c*-axis from the center of the particles. They were surrounded by *a*-faces of anatase $TiO_2$ crystals. Width and length of needles were about 200 nm and 100 nm. Aspect ratio was about 2 (200 nm / 100 nm). Diffraction patterns showed that the particles were single phase of anatase $TiO_2$. Interparticle spaces and micro/meso pores of 1-10 nm allowed us to realize high surface area of 178 $m^2/g$. Large Raman peak shift of Eg (v6) mode suggested fluorine doping and size effect. Band gap was estimated to 3.20 eV (388 nm) with UV-Vis. These were characteristics of the multi-needle $TiO_2$ particles. Furthermore, self-assembly surface coating of micro-structured substrates was successfully realized. The coating layers were consisted of multi-needle $TiO_2$ particles of about 200-400 in diameter. XPS analyses indicated chemical bonds between Ti and O. Chemical ratio was estimated to Ti : O: F = 1 : 1.76 : 0.18 suggesting fluorine doping in $TiO_2$. The multi-needle particles and the surface coating with the particles having nano/micro structures would contribute to future metal oxide devices of solar cells, photo catalysts and high sensitive sensors.

## 8. Morphology control of nano-sheet assembled tin oxide[146]

Tin oxide particles were synthesized in aqueous solutions. They were consisted of nanosheets of tin oxide crystals. The sheets were about 50-100 nm in size and 5-10 nm thickness. X-ray diffraction analysis revealed that the particles were crystals of $SnO_2$ and SnO. The particles had BET surface area of 85 $m^2/g$ estimated with $N_2$ adsorption characteristics. BJH analysis indicated that mesopores of 3.9 nm in size contributed to increase surface area.

Tin dioxide ($SnO_2$), with a rutile-type crystalline structure, is an n-type wide band gap (3.5 eV) semiconductor. It is an oxide of great interest for gas sensors[1-3], optical device[147], lithium batteries[148-151], white pigments for conducting coatings, transparent conducting coatings for furnaces and electrodes [9], surge arrestors (varistors)[152,153], catalysts[154,155], opto-conducting coatings for solar cells[156], dye-sensitized molecular sensors, etc. Transparency, semiconductivity and surfaced properties of tin oxide are suitable for these applications.

Nanoparticles of tin dioxide have been synthesized by several methods such as precipitation[151,157], hydrothermal synthesis[158,159], sol–gel[160,161], hydrolytic[162], carbothermal reduction[163] and polymeric precursor[164] methods. A variety of $SnO_2$ nanostructures including nanowires, nanobelts[165], nanotubes[150,166,167], nanorods[168,169], spirals, nanorings[170], zigzag nanobelts[171], grains[172], flakes[172], plates[172], meshes[173] and columnar thin film[149] were synthesized.

Tin chlorides ($SnCl_2$ or $SnCl_4$) were commonly used in many reports. However, chlorine ions were difficult to remove from the systems and it seriously altered superficial and electrical properties. For instance, chlorine ions caused sensitivity degradation of gas sensors[174], aggregation of particles[175] and increasing of sintering temperatures[176]. Chloride problems can be avoided through usage of organic tin compounds, such as alkoxides. However, these reagents were costly which makes industrial syntheses implementation hardly attainable. Additionally, sol-gel processes using alkoxides formed amorphous phases. High temperature annealing which increased cost, energy consumption and $CO_2$ emission was necessary to obtain tin oxide crystals.

Recently, aqueous syntheses of metal oxide crystals including tin oxide were developed[1,20,177]. Morphology control of metal oxide crystals were realized by anisotropic crystal growth. They were induced by crystal growth control or organic additives. Thermodynamically stable crystal faces, for instance, depend on crystal growth conditions. Organic molecules adsorbed on typical crystal faces suppress crystal growth perpendicular to the faces. Morphology control realized improvement of properties of metal oxide devices. In this study, tin oxide particles were prepared in aqueous solutions[146]. They were assembly of nanosheets. Morphology, crystal phases and $N_2$ adsorption characteristics were evaluated. Specific surface area and size distribution of pores were analyzed using isotherms.

**Aqueous synthesis of nano-sheet assembled tin oxide particles.** $SnF_2$ (Wako Pure Chemical Industries, Ltd., No. 202-05485, FW: 156.71, purity 90.0%) was used as received. Distilled water in polypropylene vessels (200 mL) were capped with polymer films and kept at 90°C. $SnF_2$ (870.6 mg) was dissolved in the distilled water at 90°C to be 5 mM. The solutions were kept at 90°C for 30 min using a drying oven (Yamato Scientific Co., Ltd., DKN402) with no stirring. The solutions became clouded shortly after the addition of $SnF_2$. The bottoms of the vessels were covered with white precipitates. The solutions were centrifuged at 4000 rpm for 10 min (Model 8920, Kubota Corp.). Precipitated particles were dried at 60°C for 12 h after removal of supernatant solutions.

The solutions became clouded shortly after the addition of $SnF_2$ because of the homogeneous nucleation and growth of tin oxide particles. $SnO_2$ and SnO are formed in the aqueous solutions as follows:

$$SnF_2 \xrightarrow{OH^-} SnF(OH) + F^- \xrightarrow{OH^-} Sn(OH)_2 + F^- \qquad (a)$$

$$Sn(OH)_2 \longrightarrow SnO + H_2O \qquad (b)$$

$$Sn(OH)_2 \longrightarrow SnO_2 + H_2 \qquad (c)$$

$$SnO + H_2O \longrightarrow SnO_2 + H_2 \qquad [178,179](d)$$

The particles were gradually deposited to cover the bottom of the vessels. The supersaturation degree of the solutions was high at the initial stage and decreased as the color of the solutions changed.

Precipitated particles showed broad X-ray diffraction peaks at $2\theta$ = 27, 34, 38, 52 and 57.5. They were assigned to $SnO_2$ (JDPDS No. 41-1445. The peaks had large width due to small crystallite size of $SnO_2$.

**Morphology of nano-sheet assembled tin oxide particles.** Nano-sheet assembled tin oxide particles were successfully fabricated. Crystallization of tin oxide nanosheets in aqueous solutions allowed us to obtain unique morphology of tin oxide. The particles with 300-800 nm in diameter were observed with FE-SEM (Fig. 25). They were not dense particles but assemblies of tin oxide nanosheets. The sheets were about 50-100 nm in size and 5-10 nm thickness. Aspect ratio was estimated to about 10. They were randomly aggregated to form the particles. The particles had thus continuous open pores inside. It contributed to increase surface area of the particles.

The nanosheets would be generated in the solutions homogeneously. They were aggregated to form particles which made solutions cloudy. They were then precipitated to cover bottoms of vessels. These were consistent with color change observation of solutions.

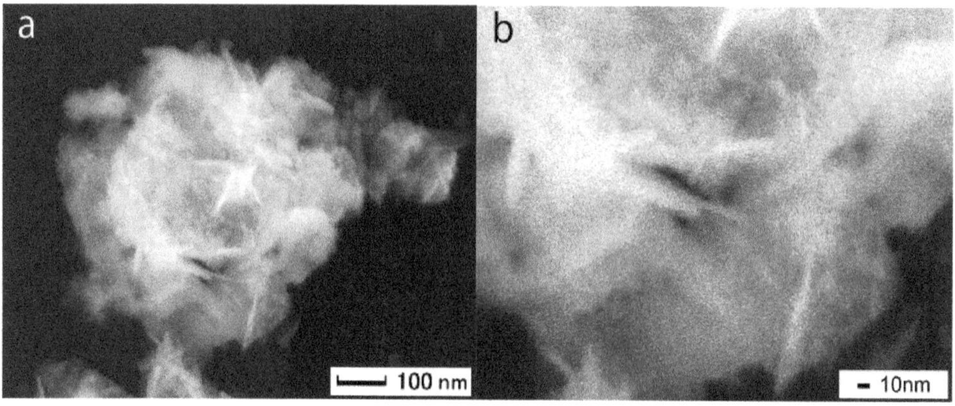


Fig. 25. SEM micrographs of (a) tin oxide particles and (b) magnified area of (a). Reprinted with permission from Ref.[146], Masuda, Y. and Kato, K., 2009, *J. Cryst. Growth*, 311, 593. Copyright @Elsevier

**$N_2$ adsorption characteristics of nano-sheet assembled tin oxide particles.** The particles exhibited $N_2$ adsorption-desorption isotherms of Type IV. The desorption isotherm differed from adsorption isotherm in the relative pressure ($P/P_0$) range from 0.45 to 0.97, showing mesopores in the particles. BET surface area of the particles was estimated to be 85 $m^2/g$. This is higher than that of $SnO_2$ nanoparticles such as $SnO_2$ (BET 47.2 $m^2/g$, 18.3nm in diameter, No. 549657-25G, Aldrich), $SnO_2$ (BET 25.9 $m^2/g$, 34nm in diameter, Yamanaka & Co., Ltd.), $SnO_2$ (BET 23 $m^2/g$, 26nm in diameter, No. 37314-13, NanoTek, C. I. Kasei Co.,Ltd.) and $In_2O_3$-$SnO_2$ (BET 3-6$m^2/g$, 100-300nm in diameter, Sumitomo Chemical Co., Ltd.). The particles were assembly of nanosheets. Unique morphologies of the sheets contributed to increase surface area of the particles.

Total pore volume and average pore diameter were estimated to 0.343 cc/g and 16.1 nm, respectively from pores smaller than 259 nm at P/Po = 0.9925. They were estimated to be 0.088 cc/g and 4.1 nm, respectively, from pores smaller than 10.6 nm at P/Po = 0.7994. Total pore volume was also estimated by the BJH method from pores smaller than 174 nm to be 0.354 cc/g.

Pore size distribution was calculated by the BJH method using desorption braches. It showed a pore size distribution curve having a strong peak at ~3.9 nm and a broad peak from 25 nm to 175 nm. The particles would have a large amount of mesopores with ~3.9 nm in diameter and mesopores with 25-175 nm in diameter. The pores were spaces surrounded by nanosheets and interparticle spaces. Micropores smaller than 2 nm were not suggested by BJH pore size distribution.

In this section, nanosheet assembled tin oxide particles were fabricated by aqueous solution synthesis. Anisotropic crystal growth of tin oxide was effectively utilized to form nanosheet assembled structure. They were crystallized at ordinary temperature without annealing. It allowed us to avoid aggregation of the particles and decrease of surface area. The particles were 300-800 nm in diameter and were crystals of $SnO_2$ and SnO. The sheets were about 50-100 nm in size and 5-10 nm thickness. The particles had BET surface area of 85 $m^2/g$ estimated with $N_2$ adsorption characteristics. BJH analysis indicated that mesopores of 3.9 nm in size contributed to increase surface area. The particles were prepared by environmentally friendly process. The system had advantages on low cost, low energy consumption and low $CO_2$ emission.

## 9. Enhancement of surface area of tin oxide[180]

Tin oxide nano-crystals with high surface area were firstly synthesized in aqueous solutions at 50°C. BET surface area was successfully reached to 194 $m^2/g$[180]. It was much higher than that of $SnO_2$ (BET 47.2 $m^2/g$, Aldrich), $SnO_2$ (BET 25.9 $m^2/g$, Yamanaka & Co., Ltd.), $SnO_2$ (BET 23 $m^2/g$, C. I. Kasei Co.,Ltd.) and $In_2O_3$-$SnO_2$ (BET 3-6$m^2/g$, Sumitomo Chemical Co., Ltd.). $N_2$ adsorption characteristics revealed that they had pores of 1-3 nm which contributed high surface area. TEM, ED and XRD indicated morphology, crystal structure and chemical composition of nano-crystals. Novel process allowed us to avoid sintering and deformation of the crystals and hence, realized high surface area and unique morphology.

In this study, tin oxide nanosheets were formed in the solutions. BET surface area was successfully reached to 194 $m^2/g$. Origin of high surface area was discussed with pore size distribution and morphology observations.

**Synthesis of Tin Oxide Nanosheets.** $SnF_2$ (870.6 mg) was dissolved in distilled water (200 mL) of 50°C to be 5 mM. The solutions were kept at 50°C for 20 min and then at 28°C for 3 days without stirring. The nanocrystals precipitated to cover bottom of the vessels. For comparison, the solutions were centrifuged at 4000 rpm for 10 minafter keeping at 50°C for 20 min. Precipitated particles were dried at 60°C for 12 h after removal of supernatant solutions.

**Morphology and crystal phase of tin oxide nanosheets.**

The nanosheets synthesized at 50°C for 20min and at 28°C for 3 days were mixture of $SnO_2$ main phase and SnO additional phase (Fig. 26a). X-ray diffraction peaks at $2\theta$ = 26.5, 33, 51.4, 62, 64.5, 80 and 89 were assigned to 110, 101, 211, etc. of $SnO_2$ (JDPDS No. 41-1445). 101, 110 and 002 diffraction peaks of SnO (JDPDS No. 06−0395) were over lapped to peaks of $SnO_2$. For comparison, diffraction pattern of the nanosheets synthesized at 50°C for 20min was shown in Fig. 26b. Half maximum full-width of the peaks was smaller than that in Fig. 26a. Lower SnO content resulted sharp peaks of $SnO_2$ in Fig. 26b. These observations indicated that SnO was mainly formed at 28°C rather than at 50°C.

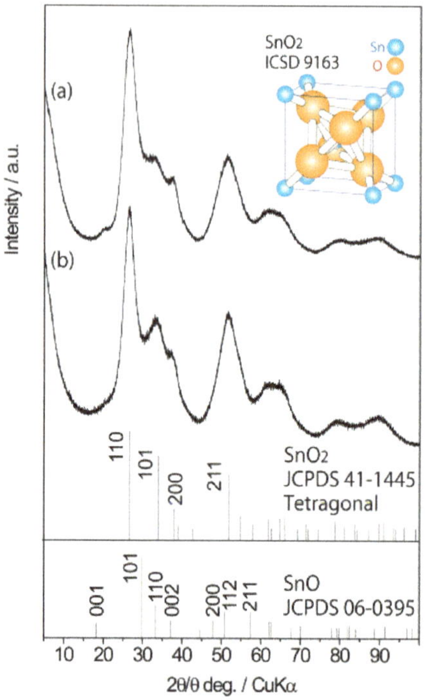


Fig. 26. XRD patterns of (a) tin oxide nanosheets fabricated at 50°C for 20min and at 28°C for 3 days, and (b) tin oxide nanosheets fabricated at 50°C for 20min. Reprinted with permission from Ref.[180], Masuda, Y., Ohji, T. and Kato, K., 2010, *J. Am. Ceram. Soc.*, 93(8), 2140. Copyright @Wiley

The nanosheets were well dispersed in ethanol. The nanosheets in supernatant solutions were skimmed with Cu grids for TEM observations. They were 20-50 nm in diameter having uniform thickness (Fig. 27A). Similar structures were observed from many areas. We observed that they curled up during long-term observation. It indicated that they had sheet structure. Electron beams damaged them to transform the structures. Some of them tightly connected each other (Fig. 27B). They had clear interfaces without pores or small grains. The nanosheet showed electron diffractions (Fig. 27B, Insert). Lattice spacing calculated from spots indicated with a red line and a yellow line were 0.283 nm and 0.154 nm, respectively. They can be assigned to $SnO_2$ and/or SnO. The former one can be assigned to101 crystal plane of $SnO_2$ (0.264 nm) or 110 crystal plane of SnO (0.269 nm). The later one can be assigned to 310 crystal plane of $SnO_2$ (0.149 nm), 221 crystal plane of $SnO_2$ (0.148 nm), 202 crystal plane of SnO (0.149 nm) or 103 crystal plane of SnO (0.148 nm). Additionally, diffraction spots related to lattice spacing of 0.283 nm were observed at upper part. They were shown with two white circles and a red line in Fig. 27B Insert. The double spots indicated that the area shown in a white circle in Fig. 27B was consisted of two crystals, i.e., stacked two nanosheets. High magnification image was also obtained from other observation area (Fig. 27C). The structure was thinner than that in Fig.27A and 27B. Clear image was not obtained in low magnification images due to low contrast. However, it showed clear high magnification image and lattice fringe (Fig. 27C). Electron diffraction patterns showed lattice spacing of 0.277 nm (red line), 0.279 nm (green line) and 0.196 nm (yellow line). They were assigned to 110, 1-10 and 200 crystal planes of SnO, respectively. Lattice spacing and their angles were well matched to that of SnO. Chemical composition was estimated from several points of tightly-packed area. The area included nanosheets and spherical crystals. Chemical composition varied among each positions in the range of Sn : O = 1 : 1.7-2.7 which was similar to that of $SnO_2$ rather than SnO. These observations indicated that the crystals were mixture of $SnO_2$ and SnO.

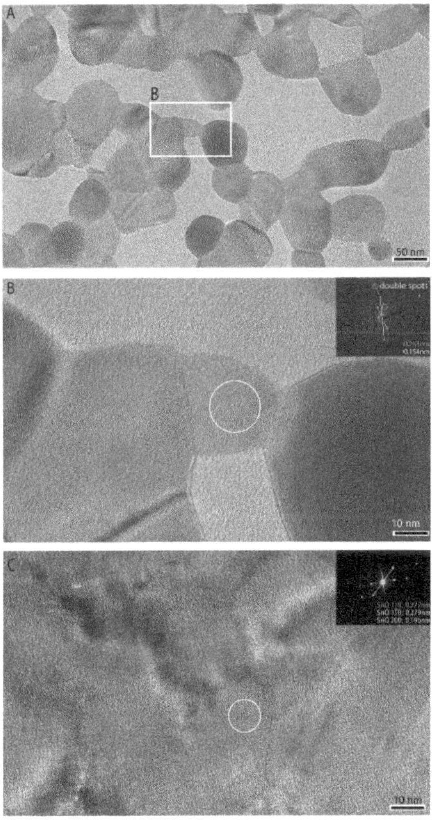


Fig. 27. TEM micrographs and electron diffraction patterns of tin oxide nanosheets. (b): high magnification image of (a). (c): high magnification image of other area. Reprinted with permission from Ref.[180], Masuda, Y., Ohji, T. and Kato, K., 2010, *J. Am. Ceram. Soc.*, 93(8), 2140. Copyright @Wiley

**$N_2$ adsorption characteristics of tin oxide nanosheets.**

BET surface area of the nanosheets was estimated from $N_2$ adsorption isotherm (Fig. 28a). The nanosheets had successfully high surface area of 194 $m^2/g$ (Fig. 28b). It was much higher than that of $SnO_2$ nanoparticles such as $SnO_2$ (BET 47.2 $m^2/g$, 18.3nm in diameter, No. 549657-25G, Aldrich), $SnO_2$ (BET 25.9 $m^2/g$, 34nm in diameter, Yamanaka & Co., Ltd., Osaka, Japan), $SnO_2$ (BET 23 $m^2/g$, 26nm in diameter, No. 37314-13, NanoTek, C. I. Kasei Co.,Ltd., Tokyo, Japan) and $In_2O_3$-$SnO_2$ (BET 3-6$m^2/g$, 100-300nm in diameter, Sumitomo

Chemical Co., Ltd.). Additionally, it was more than double higher than that of previous report (85 $m^2/g$)[181]. Pore size distribution was analyzed with BJH method using adsorption isotherm (Fig. 28c). It indicated that the nanosheets included pores of 1-2 nm. Micropore analysis was performed with DFT/Monte-Carlo Fitting which was completely-consistent with isotherms (Fig. 28d). Pores of 1-3 nm were shown to be in the nanosheets (Fig. 28e). The micro and meso-pores contributed high surface area of 194 $m^2/g$ in this system. For comparison, BET surface area of the nanosheets synthesized at 50°C for 20min was estimated to 146 $m^2/g$. It was also higher than that of nanoparticles in former studies.

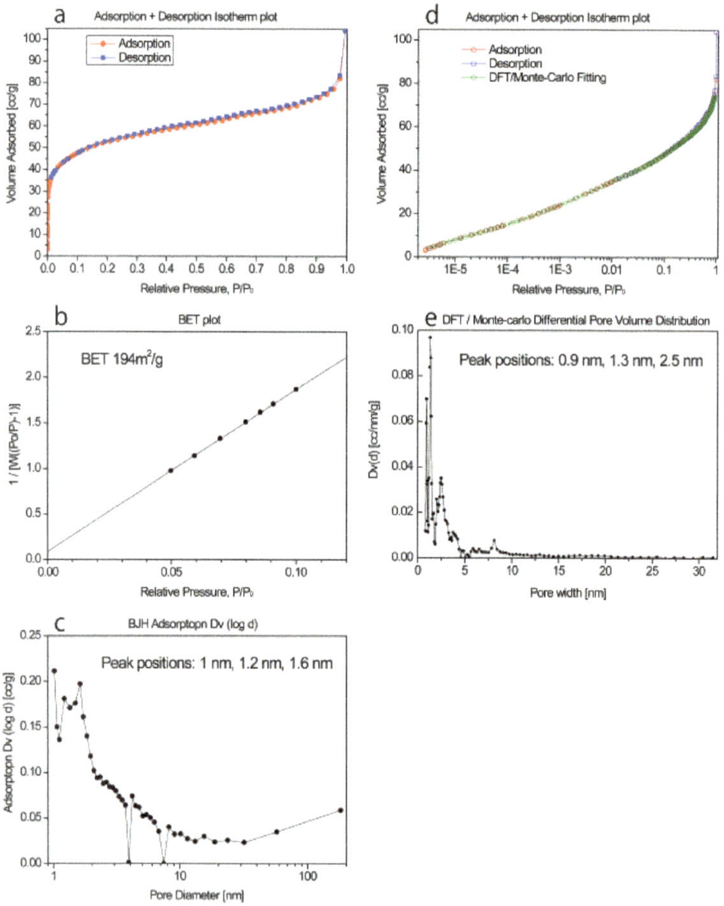


Fig. 28. (a): $N_2$ adsorption-desorption isotherm of tin oxide nanosheets. (b): BET surface area of tin oxide nanosheets. (c): Pore size distribution calculated from $N_2$ adsorption data of tin oxide nanosheets using BJH equation. (d): $N_2$ adsorption-desorption isotherm and DFT/Monte-Carlo fitting curve of tin oxide nanosheets. (e): Pore size distribution calculated from $N_2$ adsorption data of tin oxide nanosheets using DFT/Monte-Carlo equation. Reprinted with permission from Ref.[180], Masuda, Y., Ohji, T. and Kato, K., 2010, *J. Am. Ceram. Soc.*, 93(8), 2140. Copyright @Wiley

In this section, nanosheets of tin oxides were fabricated in aqueous solutions at ordinary temperature. BET surface area successfully reached to 194 $m^2/g$. It was much higher than that of nanoparticles in former studies. 2-dimensional sheet structure was one of ideal structures for high surface area per unit weight. Nano sized thickness directly contributed high surface area. Crystalline nanosheets were prepared without high temperature annealing which degraded surface area and nanostructures. High surface area and unique nanostructures of the sheets can be applied to gas sensors, dye-sensitized solar cells and molecular sensors.

## 10. Tin oxide coating on polytetrafluoroethylene films in aqueous solutions[182]

Tin Polytetrafluoroethylene(PTFE) films were successfully coated with tin oxide in aqueous solutions. Tin oxide was crystallized in the solution and formed nanocrystal coatings on the polymer films. The coatings were consisted of $SnO_2$ and SnO crystals. They were assemblies of tin oxide nanosheet of about 10 nm to 50 nm in size and about 5 nm in thickness. The nanocrystal films can be exfoliated from the PTFE substrates. Tin oxide nanocrystal films had a rough liquid-surface and a dense substrate-side surface. Transparency of PTFE films coated with tin oxide was same to that of bare PTFE films in the range from 400 nm to 800 nm. Tin oxide decreased transparency about 25 % at 320 nm. The PTFE films coated with tin oxide nanocrystals can be pasted on desired substrates.

Organic-inorganic hybrid materials such as metal oxide electronics on polymer flexible films have received considerable attention in recent years for light-weight flexible sensors, displays, dye-sensitized solar cells, etc. The polymer films offer the advantages of flexibility, light-weight, low-cost and impact resistance.

Tin oxide ($SnO_2$) is an important semiconductor with a wide band gap of 3.6 eV at room temperature. They have been widely used in gas sensors[167,183], optical devices[147], lithium batteries[148-150], etc. A novel type of bio-sensor was proposed to detect environmental toxins such as bisphenol-A or dioxin[88,184]. Tin oxide is a candidate material for the sensor because of its suitable band gap, surface characteristics, and high transparency.

Metal oxides including tin oxide have been synthesized with high temperature processes at several hundred degrees for many years. Recently, the aqueous syntheses of metal oxides have attracted much attention as a next generation science and technology[1,51,114,177]. Aqueous systems are environmentally friendly and have advantages of low energy consumption, low cost and an organic solvent free process.

In recent years, metal oxide films and their microstructures were fabricated on organic surfaces such as polyethylene terephthalate (PET) films[185,186] or self-assembled monolayers[20,113,187,188]. However, metal oxide formation on polytetrafluoroethylene (PTFE) was difficult compared to PET films. PTFE films are widely used in electronic applications because of low chemical reactivity, low coefficient of friction, high melting point (327 °C), high-corrosion resistance, a high dielectric strength over many different frequencies, a low dissipation factor, and a high surface resistivity. They were selected as substrates in this study.

In this section, tin oxide nanocrystals were prepared on PTFE films in aqueous solutions[182,184]. They were crystallized in the solution containing tin ions to form films consisted of nanosheets. The process realized tin oxide film formation without high temperature annealing and unique morphology of tin oxide crystals.

**1. Aqueous synthesis of tin oxide nanocrystals on PTFE films**

$SnF_2$ (870.6 mg) was dissolved in the distilled water (200 mL) at 90°C to be 5 mM. PTFE films (thickness: 50 μm, ASF-110, Chukoh) with silicone adhesive (thickness: 30 μm) were pasted on quarts substrates (25 × 50 × 1 mm). They were immersed in the middle of the solutions with the bottom up at an angle or with the top up at an angle. They were tilted at 15 degrees to the upright. The solutions were kept at 90°C using a drying oven (Yamato Scientific Co., Ltd., DKN402) for 2 h with no stirring. The solutions became lightly clouded after 2 h. The as-deposited nanocrystals on substrates were rinsed under running water and dried by a strong air spray. Additionally, the solutions kept at 90°C for 2 h were centrifuged at 4000 rpm for 10 min (Model 8920, Kubota Corp.). Precipitated particles were dried at 20°C for 12 h after the removal of supernatant solutions.

**2. Morphology of tin oxide nanocrystals on PTFE films**

Bare PTFE films pasted on quarts substrates had cracks of about 50 nm to 200 nm in length (Fig. 29). Longer directions of cracks were perpendicular to the extensional direction. They were formed during adhesive processes. The quarts substrates and the PTFE films were transparent or slightly white, respectively. The PTFE films can be pasted on desired substrates such as quarts, metals and polymers with silicone adhesive.

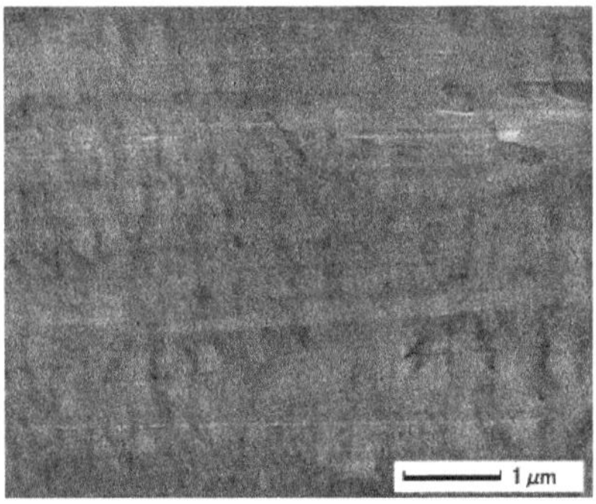


Fig. 29. SEM micrograph of a bare PTFE film. Reprinted with permission from Ref.[182], Masuda, Y. and Kato, K., 2010, *Polym. Adv. Technol.*, 21(3), 211. Copyright @Wiley

The surface of PTFE films was completely covered with assemblies of nanosheets (Fig. 30a). Uniform formation of tin oxide coatings is one of the advantages of solution processes. Large sheets were also observed from the surfaces (Fig. 30b). They were about 200 nm to 300 nm in size and about 10 nm in thickness (Fig. 30b). Some of them stood perpendicular to the PTFE films at an angle. They had an angular outline, which was connected by straight lines. They were caused from anisotropic crystal growth of tin oxide reflected in their crystal structure. The large sheets connected in a cross shape were also observed at the lower right (Fig. 30b). Nanosheets were roughly estimated to about 10 nm to 50 nm in size and about 5 nm in thickness (Fig. 30c). They connected to each other to form continuous films on the PTFE surfaces. The tin oxide films were exfoliated from PTFE films by scratch using a metal spatula. Exfoliated tin oxide films were placed on a substrate for SEM observation. Three exfoliated tin oxide films were observed to be partially-overlapping (Fig. 30d). The top film and bottom film showed their liquid-side surfaces. They were similar to that observed in Fig. 30a-c. The middle film showed a substrate-side surface, which was contacted with PTFE film during the immersion period. Substrate-side surface of the middle film was indicated by a white arrow. It had a dense surface, which consisted of nanocrystals of about 5 nm to 10 nm in size. These observations indicated that dense films consisted of nanocrystals of about 5 nm to 10 nm in size were formed on the PTFE films at first stage of tin oxide film formation. Tin oxide then grew to sheet-shapes to form assemblies of nanosheets. Additionally, large sheet crystals grew on the tin oxide dense films.

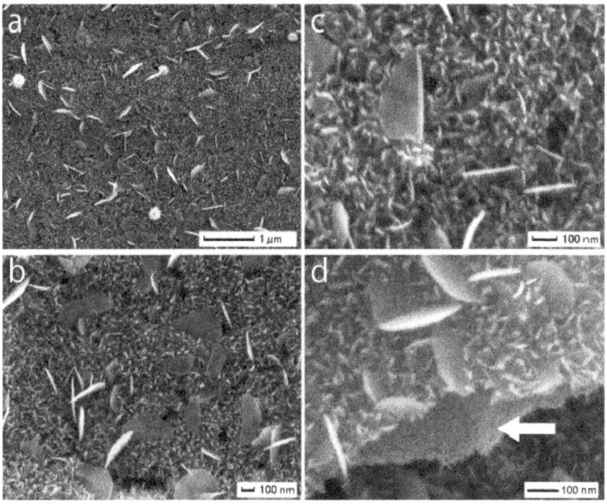


Fig. 30. SEM micrographs of (a) tin oxide nanocrystals on a PTFE film and (b, c) magnified area of (a) showing morphology of nanocrystals and (d) exfoliated tin oxide nanocrystal films showing liquid-side surface and substrate-side surface. Substrate-side surface is indicated by a white arrow. Reprinted with permission from Ref.[182], Masuda, Y. and Kato, K., 2010, *Polym. Adv. Technol.*, 21(3), 211. Copyright @Wiley

**3. Crystal phase of tin oxide nanocrystals on PTFE films**

X-ray diffraction pattern obtained from a PTFE film coated with tin oxide was similar to that from a bare PTFE film. Diffraction peaks assigned to tin oxide were not observed because of thin film thickness.

Precipitated particles were evaluated after drying. They showed broad X-ray diffraction peaks at 2θ = 27, 34, 38 and 52. They were assigned to 110, 101, 200 or 211 diffraction peaks from SnO2 (JDPDS No. 41-1445) (Fig. 3c). The peaks were wide in width due to the small crystallite size of $SnO_2$. Peaks were also observed $2\theta$ = 29.5, 32.0, 37.4, 48.0, 50.4 and 57.4. They were assigned to 101, 110, 002, 200, 112 or 211 diffraction peaks from SnO (JDPDS No. 06-0395).

XRD analysis indicated that the particles obtained from the solutions consisted of $SnO_2$ crystals and SnO crystals.

**4. XPS analysis of tin oxide nanocrystals on PTFE films**

Carbon, fluorine and oxygen were detected from bare PTFE films (Fig. 31b). Chemical ratio of them was estimated to C : F : O = 1 : 2.05 : 0.02. It was consistent with chemical composition of polytetrafluoroethylene (PTFE, C : F = 1 : 2). Small amount of oxygen was detected from surface contamination. PTFE films coated with tin oxide showed spectra of tin, oxygen, fluorine and carbon (Fig. 31a). Spectral peak corresponding to Sn $3d_{5/2}$ was observed at 487.2 eV (Fig. 31c1). The binding energy was similar to that of $SnO_2$ (486.3 eV[189],

486.5 eV[190], 486.6 eV[191], 487.3 eV) and higher than that of Sn metal (484.8 eV, 484.85 eV, 484.87 eV, 484.9 eV, 485.0 eV), which suggested tin atoms in surface coatings were positively charged by forming direct bonds with oxygen. Binding energy of O 1s centered at about 531.2 eV corresponds to that of $SnO_2$ (Fig. 31c2). Chemical ratio of the coatings was estimated to Sn : O : F : C = 1 : 1.88 : 7.25 : 3.82 and C : F = 1 : 1.90. It was indicated that the surface coatings consisted mainly of $SnO_2$ (Sn : 1 : 2). They were formed on polytetrafluoroethylene films (PTFE, C : F = 1 : 2). Difference and similarity between XPS analyses and XRD analyses suggested deposition mechanism of tin oxides. Crystal phases and chemical compositions of the surface coatings were different from those of the precipitated particles. $SnO_2$ crystallized on the films to form surface coatings, on the other hand, $SnO_2$ and SnO homogeneously crystallized to form particles in the solutions. Pure $SnO_2$ coatings were thus successfully formed on the PTFE films.

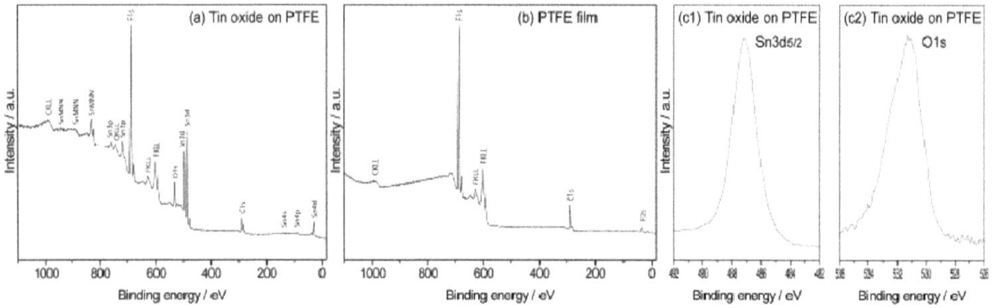


Fig. 31. XPS spectra of (a) tin oxide nanocrystals on a PTFE film and (b) a bare PTFE film. (c1) Sn $3d_{5/2}$ spectrum and (c2) O 1s spectrum of tin oxide nanocrystals on a PTFE film. Reprinted with permission from Ref.[182], Masuda, Y. and Kato, K., 2010, *Polym. Adv. Technol.*, 21(3), 211. Copyright @Wiley

## 5. Optical property of tin oxide nanocrystals on PTFE films

Quarts substrates had high transparency in the range from 200 nm to 850 nm (Fig. 32, black line). PTFE films pasted on quarts substrates were visually observed to be slight white. Their transparency was lower than that of quarts (Fig. 32, red line). Especially, it decreased as decrease of wavelength below 350 nm. PTFE films coated with tin oxide showed the same transparency as bare PTFE films in the range from 400 nm to 850 nm (Fig. 32, blue line). They decreased transparency about 25 % at 320 nm. Tin oxide particles precipitated from the solution were evaluated for comparison. They had absorption peak centered at 320 nm. This absorption was caused by tin oxide. These analyses indicated that tin oxide was deposited on PTFE films immersed in the solutions and they decrease transparency at 320 nm.

Polytetrafluoroethylene (PTFE) films were successfully coated with tin oxide nanocrystals. Tin oxide was crystallized in aqueous solutions to form nanosheet-assembled films. They were about 10 nm to 50 nm in size and about 5 nm in thickness. Large sheets of about 200 nm to 300 nm in size and about 10 nm in thickness were also crystallized on the surfaces. X-ray diffraction analysis indicated that tin oxide was a mixture of $SnO_2$ and SnO. PTFE films coated with tin oxide were transparent in the range from 400 nm to 850 nm. Tin oxide on the films had absorption centered at 320 nm. Tin oxide coated PTFE films can be pasted on desired substrates. Hybrid tin oxide-PTFE composites may be useful for increasing the potential application of tin oxide film as flexible electronics.

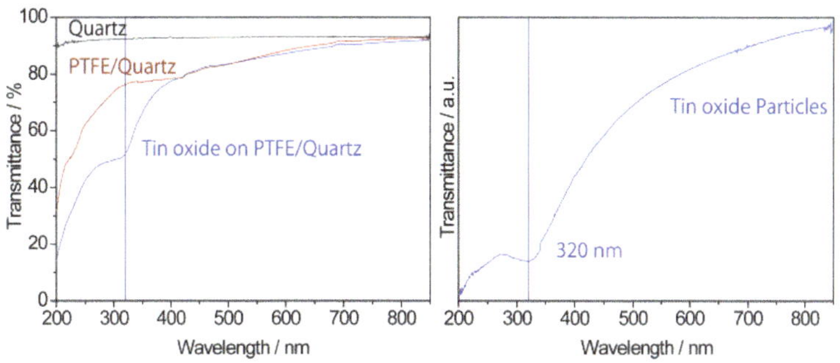


Fig. 32. UV-Vis spectra of (a) a quarts substrate (black line), (b) a PTFE film pasted on a quarts substrate (red line) and (c) tin oxide nanocrystals on a PTFE film pasted on a quarts substrate (blue line). UV-Vis spectrum of tin oxide particles. Reprinted with permission from Ref.[182], Masuda, Y. and Kato, K., 2010, *Polym. Adv. Technol.*, 21(3), 211. Copyright @Wiley

## 11. Summary

Liquid phase morphology control of metal oxide was developed by precise control of crystal growth in this study. Various nano/micro-structures of them were fabricated in solution systems. These would contribute to development of future metal oxide devices and develop new academic fields.

## 12. References


[1] Masuda, Y.; Yamada, T.; Koumoto, K. *Cryst. Growth Des.* 2008, *8*, 169-171.

[2] Tanaka, K.; Suzuki, K.; Fu, D. S.; Nishizawa, K.; Miki, T.; Kato, K. *Jpn. J. Appl. Phys.* 2004, *43*, 6525-6529.

[3] Aoyagi, S.; Kuroiwa, Y.; Sawada, A.; Kawaji, H.; Atake, T. *J. Therm. Anal. Cal.* 2005, *81*, 627-630.

[4] Luan, W. L.; Gao, L.; Guo, J. K. *Ceram. Inter.* 1999, *25*, 727-729.

[5] Wang, X. H.; Chen, R. Z.; Gui, Z. L.; Li, L. T. *Mater. Sci. Eng. B* 2003, *99*, 199-202.

[6] Sakabe, Y.; Yamashita, Y.; Yamamoto, H. *J. Euro. Ceram. Soc.* 2005, *25*, 2739-2742.

[7] Polotai, A. V.; Ragulya, A. V.; Randall, C. A. *Ferroelectrics* 2003, *288*, 93-102.

[8] Yashima, M.; Hoshina, T.; Ishimura, D.; Kobayashi, S.; Nakamura, W.; Tsurumi, T.; Wada, S. *J. Appl. Phys.* 2005, *98*.

[9] Frey, M. H.; Xu, Z.; Han, P.; Payne, D. A. *Ferroelectrics* 1998, *206*, 337-353.

[10] Frey, M. H.; Payne, D. A. *Phys. Rev. B* 1996, *54*, 3158-3168.

[11] Wada, S.; Suzuki, T.; Noma, T. *J. Ceram. Soc. Japan* 1996, *104*, 383-392.

[12] Masuda, Y.; Koumura, T.; Okawa, T.; Koumoto, K. *J. Colloid Interface Sci.* 2003, *263*, 190-195.

[13] Masuda, Y.; Koumoto, K.; Ueyama, R. Japan, 2007.

[14] Hayashi, Y.; Kimura, T.; Yamaguchi, T. *J. Mater. Sci.* 1986, *21*, 757-762.

[15] Li, X. L.; Liu, J. F.; Li, Y. D. *Mater. Chem. Phys.* 2003, *80*, 222-227.

[16] Sun, H.; Kang, S. Z.; Mu, J. *Mater. Lett.* 2007, *61*, 4121-4123.

[17] Mutin, J. C.; Dusausoy, Y.; Protas, J. *J. Solid State Chem.* 1981, *36*, 356-364.
[18] Potdar, H. S.; Deshpande, S. B.; Date, S. K. *J. Am. Ceram. Soc.* 1996, *79*, 2795-2797.
[19] Masuda, Y.; Kinoshita, N.; Koumoto, K. *J. Nanosci. Nanotechnol.*, in press.
[20] Masuda, Y.; Kinoshita, N.; Sato, F.; Koumoto, K. *Cryst. Growth Des.* 2006, *6*, 75-78.
[21] Wu, X. L.; Siu, G. G.; Fu, C. L.; Ong, H. C. *Appl. Phys. Lett.* 2001, *78*, 2285-2287.
[22] Kang, J. S.; Kang, H. S.; Pang, S. S.; Shim, E. S.; Lee, S. Y. *Thin Solid Films* 2003, *443*, 5-8.
[23] Golego, N.; Studenikin, S. A.; Cocivera, M. J. *J. Electrochem. Soc.* 2000, *147*, 1592-1594.
[24] Sberveglieri, G. *Sens. Actuators B: Chem.* 1995, *23*, 103-109.
[25] Pauporte, T.; Lincot, D. *Electrochem. Acta.* 2000, *45*, 3345-3353.
[26] Law, M.; Greene, L. E.; Johnson, J. C.; Saykally, R.; Yang, P. D. *Nature Mater.* 2005, *4*, 455-459.
[27] Masuda, Y.; Kato, K. *Cryst. Growth Des.* 2008, *8*, 2633-2637.
[28] Gong, H.; Hu, J. Q.; Wang, J. H.; Ong, C. H.; Zhu, F. R. *Sens. Actuators B: Chem.* 2006, *115*, 247-251.
[29] Moon, W. J.; Yu, J. H.; Choi, G. M. *Sens. Actuators B: Chem.* 2002, *87*, 464-470.
[30] Chang, J. F.; Kuo, H. H.; Leu, I. C.; Hon, M. H. *Sens. Actuators B: Chem.* 2002, *84*, 258-264.
[31] Tang, H. X.; Yan, M.; Zhang, H.; Li, S. H.; Ma, X. F.; Wang, M.; Yang, D. R. *Sens. Actuators B: Chem.* 2006, *114*, 910-915.
[32] Shishiyanu, S. T.; Shishiyanu, T. S.; Lupan, O. I. *Sens. Actuators B: Chem.* 2005, *107*, 379-386.
[33] Wagh, M. S.; Patil, L. A.; Seth, T.; Amalnerkar, D. P. *Mater. Chem. Phys.* 2004, *84*, 228-233.
[34] Xu, J. Q.; Pan, Q. Y.; Shun, Y. A.; Tian, Z. Z. *Sens. Actuators B: Chem.* 2000, *66*, 277-279.
[35] Paraguay, D. F.; Miki-Yoshida, M.; Morales, J.; Solis, J.; Estrada, L. W. *Thin Solid Films* 2000, *373*, 137-140.
[36] Baxter, J. B.; Aydil, E. S. *Appl. Phys. Lett.* 2005, *86*, 53114.
[37] Katoh, R.; Furube, A.; Hara, K.; Murata, S.; Sugihara, H.; Arakawa, H.; Tachiya, M. *J. Phys. Chem. B* 2002, *106*, 12957-12964.
[38] Karuppuchamy, S.; Nonomura, K.; Yoshida, T.; Sugiura, T.; Minoura, H. *Solid State Ionics* 2002, *151*, 19-27.
[39] Keis, K.; Bauer, C.; Boschloo, G.; Hagfeldt, A.; Westermark, K.; Rensmo, H.; Siegbahn, H. *J. Photochem. Photobiol. A* 2002, *148*, 57-64.
[40] Kasuga, T.; Hiramatsu, M.; Hoson, A.; Sekino, T.; Niihara, K. *Adv. Mater.* 1999, *11*, 1307-+.
[41] Oaki, Y.; Imai, H. *J. Am. Chem. Soc.* 2004, *126*, 9271-9275.
[42] Xia, Y. N.; Yang, P. D.; Sun, Y. G.; Wu, Y. Y.; Mayers, B.; Gates, B.; Yin, Y. D.; Kim, F.; Yan, Y. Q. *Adv. Mater.* 2003, *15*, 353-389.
[43] Masuda, Y.; Saito, N.; Hoffmann, R.; De Guire, M. R.; Koumoto, K. *Sci. Tech. Adv. Mater.* 2003, *4*, 461-467.
[44] Masuda, Y.; Ieda, S.; Koumoto, K. *Langmuir* 2003, *19*, 4415-4419.
[45] Masuda, Y.; Jinbo, Y.; Yonezawa, T.; Koumoto, K. *Chem. Mater.* 2002, *14*, 1236-1241.
[46] Nakanishi, T.; Masuda, Y.; Koumoto, K. *Chem. Mater.* 2004, *16*, 3484-3488.

[47] Peng, W. Q.; Qu, S. C.; Cong, G. W.; Wang, Z. G. *Cryst. Growth Des.* 2006, *6*, 1518-1522.
[48] Lin, Y. R.; Yang, S. S.; Tsai, S. Y.; Hsu, H. C.; Wu, S. T.; Chen, I. C. *Cryst. Growth Des.* 2006, *6*, 1951-1955.
[49] Zhang, H.; Yang, D. R.; Li, D. S.; Ma, X. Y.; Li, S. Z.; Que, D. L. *Cryst. Growth Des.* 2005, *5*, 547-550.
[50] Liu, K. H.; Lin, C. C.; Chen, S. Y. *Cryst. Growth Des.* 2005, *5*, 483-487.
[51] Masuda, Y.; Kato, K. *Cryst. Growth Des.* 2008, *8*, 275-279.
[52] Greene, L. E.; Law, M.; Tan, D. H.; Montano, M.; Goldberger, J.; Somorjai, G.; Yang, P. D. *Nano Lett.* 2005, *5*, 1231-1236.
[53] O'Brien, P.; Saeed, T.; Knowles, J. *J. Mater. Chem.* 1996, *6*, 1135-1139.
[54] Yamabi, S.; Imai, H. *J. Mater. Chem.* 2002, *12*, 3773-3778.
[55] Vayssieres, L.; Keis, K.; Lindquist, S. E.; Hagfeldt, A. *J. Phys. Chem. B* 2001, *105*, 3350-3352.
[56] Vayssieres, L. *Adv. Mater.* 2003, *15*, 464-466.
[57] Saito, N.; Haneda, H.; Sekiguchi, T.; Ohashi, N.; Sakaguchi, I.; Koumoto, K. *Adv. Mater.* 2002, *14*, 418-421.
[58] Saito, N.; Haneda, H.; Seo, W. S.; Koumoto, K. *Langmuir* 2001, *17*, 1461-1469.
[59] Saeed, T.; O'Brien, P. *Thin Solid Films* 1995, *271*, 35-38.
[60] Kakiuchi, K.; Hosono, E.; Kimura, T.; Imai, H.; Fujihara, S. *J. Sol-Gel Sci. Technol.* 2006, *39*, 63-72.
[61] Schwenzer, B.; Gomm, J. R.; Morse, D. E. *Langmuir* 2006, *22*, 9829-9831.
[62] Yu, H. D.; Zhang, Z. P.; Han, M. Y.; Hao, X. T.; Zhu, F. R. *J. Am. Chem. Soc.* 2005, *127*, 2378-2379.
[63] Choi, K. S.; Lichtenegger, H. C.; Stucky, G. D.; McFarland, E. W. *J. Am. Chem. Soc.* 2002, *124*, 12402-12403.
[64] Yin, M.; Gu, Y.; Kuskovsky, I. L.; Andelman, T.; Zhu, Y.; Neumark, G. F.; O'Brien, S. *J. Am. Chem. Soc.* 2004, *126*, 6206-6207.
[65] Feng, X. J.; Feng, L.; Jin, M. H.; Zhai, J.; Jiang, L.; Zhu, D. B. *J. Am. Chem. Soc.* 2004, *126*, 62-63.
[66] Yoshida, T.; Tochimoto, M.; Schlettwein, D.; Wohrle, D.; Sugiura, T.; Minoura, H. *Chem. Mater.* 1999, *11*, 2657-2667.
[67] Lee, J. Y.; Yin, D. H.; Horiuchi, S. *Chem. Mater.* 2005, *17*, 5498-5503.
[68] Kopalko, K.; Godlewski, M.; Domagala, J. Z.; Lusakowska, E.; Minikayev, R.; Paszkowicz, W.; Szczerbakow, A. *Chem. Mater.* 2004, *16*, 1447-1450.
[69] Turgeman, R.; Gershevitz, O.; Deutsch, M.; Ocko, B. M.; Gedanken, A.; Sukenik, C. N. *Chem. Mater.* 2005, *17*, 5048-5056.
[70] Turgeman, R.; Gershevitz, O.; Palchik, O.; Deutsch, M.; Ocko, B. M.; Gedanken, A.; Sukenik, C. N. *Cryst. Growth Des.* 2004, *4*, 169-175.
[71] Mirica, E.; Kowach, G.; Evans, P.; Du, H. *Cryst. Growth Des.* 2004, *4*, 147-156.
[72] Lin, Y. R.; Tseng, Y. K.; Yang, S. S.; Wu, S. T.; Hsu, C. L.; Chang, S. J. *Cryst. Growth Des.* 2005, *5*, 579-583.
[73] Gao, Y. F.; Nagai, M. *Langmuir* 2006, *22*, 3936-3940.
[74] Wu, X. D.; Zheng, L. J.; Wu, D. *Langmuir* 2005, *21*, 2665-2667.
[75] Peterson, R. B.; Fields, C. L.; Gregg, B. A. *Langmuir* 2004, *20*, 5114-5118.

[76] Liu, T. Y.; Liao, H. C.; Lin, C. C.; Hu, S. H.; Chen, S. Y. *Langmuir* 2006, *22*, 5804-5809.
[77] Masuda, Y.; Kato, K. *Cryst. Growth Des.* 2008, *8*, 3213-3218.
[78] Carlson, T.; Giffin, G. L. *J. Phys. Chem.* 1986, *90*, 5896-5900.
[79] Zhang, Z. B.; Wang, C. C.; Zakaria, R.; Ying, J. Y. *J. Phys. Chem. B* 1998, *102*, 10871-10878.
[80] Wang, R.; Hashimoto, K.; Fujishima, A. *Nature* 1997, *388*, 431-432.
[81] Choi, W. Y.; Termin, A.; Hoffmann, M. R. *J. Phys. Chem.* 1994, *98*, 13669-13679.
[82] Sung, Y. M.; Lee, J. K.; Chae, W. S. *Cryst. Growth Des.* 2006, *6*, 805-808.
[83] Kumazawa, N.; Islam, M. R.; Takeuchi, M. *J. Electroanal. Chem.* 1999, *472*, 137-141.
[84] Ferroni, M.; Carotta, M. C.; Guidi, V.; Martinelli, G.; Ronconi, F.; Sacerdoti, M.; Traversa, E. *Sens. Actuators B: Chem.* 2001, *77*, 163-166.
[85] Wagemaker, M.; Kentgens, A. P. M.; Mulder, F. M. *Nature* 2002, *418*, 397-399.
[86] Arico, A. S.; Bruce, P.; Scrosati, B.; Tarascon, J. M.; Van Schalkwijk, W. *Nature Mater.* 2005, *4*, 366-377.
[87] Guo, Y. G.; Hu, Y. S.; Maier, J. *Chem. Commun.* 2006, *26*, 2783-2785.
[88] Tokudome, H.; Yamada, Y.; Sonezaki, S.; Ishikawa, H.; Bekki, M.; Kanehira, K.; Miyauchi, M. *Appl. Phys. Lett.* 2005, *87*, 213901-213903.
[89] Nazeeruddin, M. K.; De Angelis, F.; Fantacci, S.; Selloni, A.; Viscardi, G.; Liska, P.; Ito, S.; Takeru, B.; Gratzel, M. G. *J. Am. Chem. Soc.* 2005, *127*, 16835-16847.
[90] Wang, P.; Zakeeruddin, S. M.; Moser, J. E.; Humphry-Baker, R.; Comte, P.; Aranyos, V.; Hagfeldt, A.; Nazeeruddin, M. K.; Gratzel, M. *Adv. Mater.* 2004, *16*, 1806-1811.
[91] Morrison, P. W.; Raghavan, R.; Timpone, A. J.; Artelt, C. P.; Pratsinis, S. E. *Chem. Mater.* 1997, *9*, 2702-2708.
[92] Yang, G. X.; Zhuang, H. R.; Biswas, P. *Nanostruct. Mater.* 1996, *7*, 675-689.
[93] Yu, J. C.; Yu, J. G.; Ho, W. K.; Zhang, L. Z. *Chem. Commun.* 2001, *19*, 1942-1943.
[94] Huang, W. P.; Tang, X. H.; Wang, Y. Q.; Koltypin, Y.; Gedanken, A. *Chem. Commun.* 2000, *15*, 1415-1416.
[95] Seifried, S.; Winterer, M.; Hahn, H. *Chem. Vap. Deposition* 2000, *6*, 239-244.
[96] Scolan, E.; Sanchez, C. *Chem. Mater.* 1998, *10*, 3217-3223.
[97] Wang, C. C.; Ying, J. Y. *Chem. Mater.* 1999, *11*, 3113-3120.
[98] Burnside, S. D.; Shklover, V.; Barbe, C.; Comte, P.; Arendse, F.; Brooks, K.; Gratzel, M. *Chem. Mater.* 1998, *10*, 2419-2425.
[99] Zhang, H. Z.; Finnegan, M.; Banfield, J. F. *Nano Lett.* 2001, *1*, 81-85.
[100] Sung, Y. M.; Lee, J. K. *Cryst. Growth Des.* 2004, *4*, 737-742.
[101] Perkas, N.; Pol, V.; Pol, S.; Gedanken, A. *Cryst. Growth Des.* 2006, *6*, 293-296.
[102] Deki, S.; Aoi, Y.; Hiroi, O.; Kajinami, A. *Chem. Lett.* 1996, *6*, 433-434.
[103] Deki, S.; Aoi, Y.; Yanagimoto, H.; Ishii, K.; Akamatsu, K.; Mizuhata, M.; Kajinami, A. *J. Mater. Chem.* 1996, *6*, 1879-1882.
[104] Deki, S.; Aoi, Y.; Asaoka, Y.; Kajinami, A.; Mizuhata, M. *J. Mater. Chem.* 1997, *7*, 733-736.
[105] Kishimoto, H.; Takahama, K.; Hashimoto, N.; Aoi, Y.; Deki, S. *J. Mater. Chem.* 1998, *8*, 2019-2024.

[106] Choi, S. Y.; Mamak, M.; Coombs, N.; Chopra, N.; Ozin, G. A. *Adv. Func. Mater.* 2004, *14*, 335-344.
[107] Antonelli, D. M.; Ying, J. Y. *Angew. Chem., Int. Ed.* 1995, *34*, 2014-2017.
[108] Shibata, H.; Ogura, T.; Mukai, T.; Ohkubo, T.; Sakai, H.; Abe, M. *J. Am. Chem. Soc.* 2005, *127*, 16396-16397.
[109] Shibata, H.; Mihara, H.; Mlikai, T.; Ogura, T.; Kohno, H.; Ohkubo, T.; Sakait, H.; Abe, M. *Chem. Mater.* 2006, *18*, 2256-2260.
[110] Tian, B. Z.; Liu, X. Y.; Yang, H. F.; Xie, S. H.; Yu, C. Z.; Tu, B.; Zhao, D. Y. *Adv. Mater.* 2003, *15*, 1370-+.
[111] Ryoo, R.; Joo, S. H.; Kruk, M.; Jaroniec, M. *Adv. Mater.* 2001, *13*, 677-681.
[112] Masuda, Y.; Kato, K. Japan, 2007.
[113] Masuda, Y.; Sugiyama, T.; Seo, W. S.; Koumoto, K. *Chem. Mater.* 2003, *15*, 2469-2476.
[114] Masuda, Y.; Kato, K. *Chem. Mater.* 2008, *20*, 1057-1063.
[115] Furlong, D. N.; Parfitt, G. D. *J. Colloid Interface Sci.* 1978, *65*, 548-554.
[116] Wahi, R. K.; Liu, Y. P.; Falkner, J. C.; Colvin, V. L. *J. Colloid Interface Sci.* 2006, *302*, 530-536.
[117] Kruk, M.; Jaroniec, M. *Chem. Mater.* 2001, *13*, 3169-3183.
[118] Ravikovitch, P. I.; Odomhnaill, S. C.; Neimark, A. V.; Schuth, F.; Unger, K. K. *Langmuir* 1995, *11*, 4765-4772.
[119] Lastoskie, C.; Gubbins, K. E.; Quirke, N. *J. Phys. Chem.* 1993, *97*, 4786-4796.
[120] Katagiri, K.; Ohno, K.; Masuda, Y.; Koumoto, K. *J. Ceram. Soc. Japan* 2007, *115*, 831-834.
[121] Masuda, Y.; Ohji, T.; Kato, K. *Crystal Growth & Design* 2010, *10*, 913-922.
[122] Masuda, Y.; Kato, K. *Thin Solid Films* 2008, *516*, 2547-2552.
[123] Masuda, Y.; Kato, K. *J. Cryst. Growth* 2009, *311*, 512-517.
[124] Ohsaka, T.; Izumi, F.; Fujiki, Y. *J. Raman Spectrosc.* 1978, *7*, 321-324.
[125] Gao, K. *Physica Status Solidi B-Basic Solid State Physics* 2007, *244*, 2597-2604.
[126] Cong, Y.; Zhang, J.; Chen, F.; Anpo, M. *J. Phys. Chem. C* 2007, *111*, 6976-6982.
[127] Ohsaka, T.; Yamaoka, S.; Shimomura *Solid State Communications* 1979, *30*, 345-347.
[128] Yu, J. G.; Yu, H. G.; Cheng, B.; Zhao, X. J.; Yu, J. C.; Ho, W. K. *Journal of Physical Chemistry B* 2003, *107*, 13871-13879.
[129] Ignat'eva, L. N.; Polishchuk, S. A.; Antokhina, T. F.; Buznik, V. T. *Glass Physics and Chemistry* 2004, *30*, 139-141.
[130] Nakamoto, K. *Infrared Spectra of Inorganic and Coordination Compounds*; Wiley: New York, 1970.
[131] Padmanabhan, S. C.; Pillai, S. C.; Colreavy, J.; Balakrishnan, S.; McCormack, D. E.; Perova, T. S.; Gun'ko, Y.; Hinder, S. J.; Kelly, J. M. *Chemistry of Materials* 2007, *19*, 4474-4481.
[132] Wu, J. J.; Yu, C. C. *Journal of Physical Chemistry B* 2004, *108*, 3377-3379.
[133] Perathoner, S.; Passalacqua, R.; Centi, G.; S. Su, D.; Weinberg, G. *Catalysis Today* 2007, *122*, 3-13.
[134] Sato, H.; Ono, K.; Sasaki, T.; Yamagishi, A. *Journal of Physical Chemistry B* 2003, *107*, 9824-9828.

[135] Naik, V. M.; Haddad, D.; Naik, R.; Benci, J.; Auner, G. W. In *Symposium on Solid-State Chemistry of Inorganic Materials IV held at the 2002 MRS Fall Meeting;* AlarioFranco, M. A., Greenblatt, M., Rohrer, G., Whittingham, M. S., Eds.; Materials Research Society: Boston, Ma, 2002, p 413-418.
[136] Masuda, Y.; Sugiyama, T.; Lin, H.; Seo, W. S.; Koumoto, K. *Thin Solid Films* 2001, *382*, 153-157.
[137] Huang, D.; Xiao, Z.-D.; Gu, J.-H.; Huang, N.-P.; Yuan, C.-W. *Thin Solid Films* 1997, *305*, 110-115.
[138] Zhang, F.; Mao, Y.; Zheng, Z.; Chen, Y.; Liu, X.; Jin, S. *Thin Solid Films* 1997, *310*, 29-33.
[139] Li, D.; Haneda, H.; Hishita, S.; Ohashi, N.; Labhsetwar, N. K. *Journal of Fluorine Chemistry* 2005, *126*, 69-77.
[140] Wang, C. M.; Mallouk, T. E. *Journal of Physical Chemistry* 1990, *94*, 423-428.
[141] Minero, C.; Mariella, G.; Maurino, V.; Pelizzetti, E. *Langmuir* 2000, *16*, 2632-2641.
[142] Minero, C.; Mariella, G.; Maurino, V.; Vione, D.; Pelizzetti, E. *Langmuir* 2000, *16*, 8964-8972.
[143] Vohra, M. S.; Kim, S.; Choi, W. In *International Symposium on Photochemistry at Interfaces*; Elsevier Science Sa: Sapporo, Japan, 2002, p 55-60.
[144] Fu, H. B.; Zhang, L. W.; Zhang, S. C.; Zhu, Y. F.; Zhao, J. C. *Journal of Physical Chemistry B* 2006, *110*, 3061-3065.
[145] Irie, H.; Watanabe, Y.; Hashimoto, K. *Journal of Physical Chemistry B* 2003, *107*, 5483-5486.
[146] Masuda, Y.; Kato, K. *J. Cryst. Growth* 2009, *311*, 593-596.
[147] Ginley, D. S.; Bright, C. *MRS Bull.* 2000, *25*, 15-21.
[148] Idota, Y.; Kubota, T.; Matsufuji, A.; Maekawa, Y.; Miyasaka, T. *Science* 1997, *276*, 1395 - 1397.
[149] Zhang, Y. L.; Liu, Y.; Liu, M. L. *Chem. Mater.* 2006, *18*, 4643-4646.
[150] Wang, Y.; Lee, J. Y.; Zeng, H. C. *Chem. Mater.* 2005, *17*, 3899-3903.
[151] Bose, A. C.; Kalpana, D.; Thangadurai, P.; Ramasamy, S. *Journal of Power Sources* 2002, *107*, 138-141.
[152] Pianaro, S. A.; Bueno, P. R.; Longo, E.; Varela, J., A. *J. Mat. Sci. Lett.* 1995, *14*, 692-694.
[153] Bueno, P. R.; de Cassia-Santos, M. R.; Leite, E. R.; Longo, E.; Bisquert, J.; Garcia-Belmonte, G.; Fabregat-Santiago, F. *Journal of Applied Physics* 2000, *88*, 6545-6548.
[154] Tagawa, T.; Kataoka, S.; Hattori, T.; Murakami, Y. *Applied Catalysis* 1982, *4*, 1-4.
[155] Park, P. W.; Kung, H. H.; Kim, D. W.; Kung, M. C. *Journal of Catalysis* 1999, *184*, 440-454.
[156] Chopra, K. L.; Major, S.; Pandya, D. K. *Thin Solid Films* 1983, *102*, 1-46.
[157] Sergent, N.; Gelin, P.; Perier-Camby, L.; Praliaud, H.; Thomas, G. *Sensors and Actuators B-Chemical* 2002, *84*, 176-188.
[158] Baik, N. S.; Sakai, G.; Miura, N.; Yamazoe, N. *J. Am. Ceram. Soc.* 2000, *83*, 2983-2987.
[159] Ristic, M.; Ivanda, M.; Popovic, S.; Music, S. *Journal of Non-Crystalline Solids* 2002, *303*, 270-280.

[160] Broussous, L.; Santilli, C. V.; Pulcinelli, S. H.; Craievich, A. F. *Journal of Physical Chemistry B* 2002, *106*, 2855-2860.
[161] Zhang, J. R.; Gao, L. *Journal of Solid State Chemistry* 2004, *177*, 1425-1430.
[162] Deng, Z. X.; Wang, C.; Li, Y. D. *Journal of The American Ceramic Society* 2002, *85*, 2837-2839.
[163] Leite, E. R.; Gomes, J. W.; Oliveira, M. M.; Lee, E. J. H.; Longo, E.; Varela, J. A.; Paskocimas, C. A.; Boschi, T. M.; Lanciotti, F.; Pizani, P. S.; Soares, P. C. *Journal of Nanoscience and Nanotechnology* 2002, *2*, 125-128.
[164] Leite, E. R.; Maciel, A. P.; Weber, I. T.; Lisboa, P. N.; Longo, E.; Paiva-Santos, C. O.; Andrade, A. V. C.; Pakoscimas, C. A.; Maniette, Y.; Schreiner, W. H. *Advanced Materials* 2002, *14*, 905-908.
[165] Dai, Z. R.; Gole, J. L.; Stout, J. D.; Wang, Z. L. *J. Phys. Chem. B* 2002, *106*, 1274 -1279.
[166] Liu, Y.; Liu, M. L. *Adv. Funct. Mater.* 2005, *15*, 57 - 62.
[167] Huang, J.; Matsunaga, N.; Shimanoe, K.; Yamazoe, N.; Kunitake, T. *Chem. Mater.* 2005, *17*, 3513-3518.
[168] Liu, Y. K.; Zheng, C. L.; Wang, W. Z.; Yin, C. R.; Wang, G. H. *Adv. Mater.* 2001, *13*, 1883 - 1887.
[169] Sun, J. Q.; Wang, J. S.; Wu, X. C.; Zhang, G. S.; Wei, J. Y.; Zhang, S. Q.; Li, H.; Chen, D. R. *Cryst. Growth Des.* 2006, *6*, 1584-1587.
[170] Yang, R. S.; Wang, Z. L. *J. Am. Chem. Soc.* 2006, *128*, 1466-1467.
[171] Duan, J. H.; Yang, S. G.; Liu, H. W.; Gong, J. F.; Huang, H. B.; Zhao, X. N.; Zhang, R.; Du, Y. W. *J. Am. Chem. Soc.* 2005, *127*, 6180 -6181.
[172] Ohgi, H.; Maeda, T.; Hosono, E.; Fujihara, S.; Imai, H. *Cryst. Growth Des.* 2005, *5*, 1079-1083.
[173] Uchiyama, H.; Imai, H. *Cryst. Growth Des.* 2007, *7*, 841-843.
[174] Niesz, D. E.; Bennett, R. B.; Snyder, M. J. *Am. Ceram. Soc. Bull.* 1972, *51*, 677-680.
[175] Roosen, A.; Hausener, H. *Adv. Ceram. Mater.* 1988, *3*, 131.
[176] Vasylkiv, O.; Sakka, Y. *Journal of The American Ceramic Society* 2001, *84*, 2489-2494.
[177] Niesen, T. P.; DeGuire, M. R. *J. Electroceramics* 2001, *6*, 169-207.
[178] Baes, C. F.; Mesiner, R. E. *The Hydrolysis of Cations*; John Wiley & Sons, Inc., Wiley-Interscience: New York, 1976.
[179] Ararat Ibarguena, C.; Mosqueraa, A.; Parrab, R.; Castrob, M. S.; Rodríguez-Páeza, J. E. *Mater. Chem. Phys.* 2007, *101*, 433-440.
[180] Masuda, Y.; Ohji, T.; Kato, K. *Journal of The American Ceramic Society* 2010, *93*, 2140-2143.
[181] Masuda, Y.; Kato, K. *Journal of Crystal Growth* 2009, *311*, 593-596.
[182] Masuda, Y.; Kato, K. *Polymers for Advanced Technologies* 2010, *21*, 211-215.
[183] Shukla, S.; Seal, S.; Ludwig, L.; Parish, C. *Sens. Actuators B: Chem.* 2004, *97*, 256-265.
[184] Masuda, Y.; Kato, K.; Sonezaki, S.; Ajimi, M.; Bekki, M. Japan, 2008, p P 2008-227389.
[185] Xiang, J. H.; Masuda, Y.; Koumoto, K. *Adv. Mater.* 2004, *16*, 1461-1464.
[186] Xiang, J. H.; Zhu, P. X.; Masuda, Y.; Koumoto, K. *Langmuir* 2004, *20*, 3278-3283.
[187] Masuda, Y.; Yamagishi, M.; Koumoto, K. *Chem. Mater.* 2007, *19*, 1002-1008.
[188] Masuda, Y.; Wakamatsu, S.; Koumoto, K. *J. Euro. Ceram. Soc.* 2004, *24*, 301-307.

[189] Themlin, J. M.; Chtaib, M.; Henrard, L.; Lambin, P.; Darville, J.; Gilles, J. M. *Phys. Rev. B* 1992 *46*, 2460-2466.
[190] Yan, L.; Pan, J. S.; Ong, C. K. *Mater. Sci. Eng. B* 2006, *128*, 34-36.
[191] Wagner, C. D. *Practical Surface Analysis*; 2 ed.; John Wiley, 1990; Vol. 1.

# 5

# Solution Processing of Nanoceramic $VO_2$ Thin Films for Application to Smart Windows

Yanfeng Gao[1,2,3,], Litao Kang[1], Zhang Chen[1] and Hongjie Luo[2]
*[1]State Key Laboratory of High Performance Ceramics and Superfine Microstructure, Shanghai institute of Ceramics, Chinese Academy of Sciences, Shanghai,*
*[2]Research Center for Industrial Ceramics, Shanghai institute of Ceramics, Chinese Academy of Sciences, Shanghai,*
*[3]Xinjiang Key Laboratory of Electronic Information Materials and Devices, Xinjiang Technical Institute of Physics & Chemistry, Chinese Academy of Sciences, Beijing Road, Urumqi, Xinjiang, China*

## 1. Introduction

Energy conservation has directed a global trend towards sustainable development. Due to global warming, air conditioning systems have been widely used in daily life, thus inducing a series of problems,[1] including increases in electricity consumption and carbon dioxide emissions along with the formation of other atmospheric pollutants from the electricity-generation process. Air conditioning in China accounts for 40-60% of a building's energy consumption (the exact figure depends on the area of the building), and overall, uses 28% of the total available primary energy. These figures will grow rapidly with urban development, as the case of China. One effective way to reduce the amount of electricity consumed by cooling is to apply solar-control coatings to glass windows, or so-called "energy-efficient windows" or "smart windows". Because lighting demands transparency, most of the smart windows are designed to intelligently control the amount of light and heat (mainly in the near infrared region) passing through in response to an external stimulus such as light (photochromic), heat (thermochromic) or electricity (electrochromic).[2-7] In this regard, the thermochromic smart window, typically based on a vanadium dioxide ($VO_2$) functional layer, has received particular interests due to two aspects. First, it can respond to environmental temperatures, making reversible structural changes from an infrared-transparent semiconductive crystalline phase to an infrared-blocking metallic crystalline phase. Second, the visible transparency remains almost unchangeable.

$VO_2$, which undergoes a metal-insulator transition (MIT) at a critical temperature $T_c$ (68 °C for bulk $VO_2$),[8] has attracted much attention as a thermochromic material for smart windows.[9] Owing to the MIT, $VO_2$ transforms between the monoclinic ($P2_1/c$, $M_1$) and the tetragonal ($P4_2/mnm$, R) phase, inducing a severe change in optical properties. $VO_2$ is transparent to near infrared (NIR) light at temperatures below $T_c$, but NIR-light reflective above $T_c$. The MIT is simply determined by environmental temperature and occurs fast, usually within $10^{-12}$ s (1 ps).[10-14] This character can be employed to intelligently control the NIR-light radiation which carries appropriate 45% heat energy of solar light, and makes $VO_2$ a promising material for smart windows. Though $VO_2$ single crystals cannot withstand the structural distortions associated with the MIT and break after a few cycles of the phase transition, $VO_2$ films can survive stress changes above $10^8$ cycles[15-17] and thus is suitable for applications in smart windows.

However, an expensive fabrication method, low visible transparency and insufficient energy-saving efficiency limit the application of $VO_2$ smart windows to certain architectures. The difficulties in the $VO_2$ synthesis are due partially to the existence of abundant oxide forms and polymorphous of the vanadium-oxygen system. From the vanadium-oxygen phase-diagram, there are nearly 15 to 20 other stable vanadium oxide phases besides $VO_2$, such as VO, $V_6O_{13}$, and $V_7O_{13}$. The formation of $VO_2$ occurs only at a very narrow oxygen partial pressure of 2 (± 0.2) %.[18,19]

Additionally, more than ten kinds of crystalline phases of vanadium dioxide have been reported, including tetragonal rutile-type $VO_2$ (R) (*P4/mmm*),[20] monoclinic rutile-type $VO_2$ (M) (*P21/c*),[21] triclinic $VO_2$ (*P**(2),[22] tetragonal $VO_2$ (A) (*P42/nmc*),[23-25] monoclinic $VO_2$ (B) (*C2/m*),[26] (C) $VO_2$,[27] orthorhombic $VO_2 \cdot H_2O$ (*P222*),[28] tetragonal $VO_2 \cdot 0.5H_2O$ (*I4/mmm*),[29] monoclinic $V_2O_4$ (*P21/c,*), and $V_2O_4 \cdot 2H_2O$. Only the rutile-type $VO_2$ (R/M) undergoes a fully reversible metal-semiconductor phase transition (MST) at approximately 68 °C.

The low visible transmittance originates from the strong inner-band and inter-band absorptions in the short-wavelength range for both the metallic and semiconductive states.[30,31] The transmittance values in the visible region (380-780 nm) reported for $VO_2$ thin films are quite low (~50%,[32,33] 42-45 %,[34,35] or less than 40%[19,36-39]). However, for films suitable for use in architectural windows, the visible transmittance should exceed 60%.[40,41]

Energy-saving efficiency is one of limits that drawback the application of this material. For smart window applications, the optical characteristics of $VO_2$ films in the wavelength ranging from near-infrared (NIR) to mid-infrared are usually concerned.[34,38,42-45] The change in NIR transmittance before and after the MIT is defined as NIR switching efficiency ($\Delta T$) of $VO_2$ films (typically referred to the transmission difference at a wavelength of 2000 nm), and this value is affected by several factors, for example film thickness,[45,46] doping,[36,47] microstructure[44,48,49] and stoichiometry.[50,51] Of these factors, the thickness usually affects the switching efficiency most dramatically, but increases in the thickness are usually accompanied by great losses in the visible transmittance.[46] For example, when the NIR switching efficiency reached 50%, the visible transmittance maxima were lowered to <43%,[52] <40%,[35] ~33%,[46] and ~21%,[53] respectively. Furthermore, to our knowledge, a NIR switching efficiency of 50% cannot even be achieved by increasing the film thickness.[36,51] The film becomes a static absorber if the thickness exceeds a threshold, without significant benefits on the NIR switching efficiency. In addition, the optical performance in the mid-infrared region, e.g., the contrast of both reflectance and emissivity before and after the MIT, is also highly concerned.[42,45]

The main preparation methods for $VO_2$ films are based on gas-phase reactions such as sputtering deposition,[4,5] chemical vapour deposition,[32,54] pulsed laser deposition[55] and ion implantation.[56] These methods are superior in terms of precise control of process parameters such as oxygen partial pressure (typically in the range of 0.06-0.13 Pa) and film features, including thickness and microstructure.[57,58] They are, however, restricted by expensive equipment. The low visible transmittance (integral transmittance≤40%) of $VO_2$ films originates from strong absorption and high reflectance.[36,59] Switching efficiency ($\Delta T_{sol}$) refers to a difference in the solar energy transmittance across metal-insulator transition (MIT) and is used to characterize the thermochromic properties of $VO_2$ films. This value is usually below 10% for a single layered $VO_2$ film.[4-5]

In the past several years, we have worked to develop a solution-based process for $VO_2$-based thin films with a special emphasis on their preparation, thermochromic property study and application to smart windows.[43,44,60-72] Polymer-assisted deposition (PAD) process was realized for the preparation of $VO_2$ and $VO_2$-based multilayered films. The method enables us to facially control over the film thickness, morphology and optical constants. By combining clever control of the optical parameters and/or their thickness and microstructural regulation, we obtained $VO_2$ films with high visible transmittance (40-84%), controllable Mott phase transition temperatures and high switching efficiencies (max. 15.1%). The results show that the current solution process is a powerful competitor towards practical applications of this material.

## 2. Polymer-assisted deposition of $VO_2$ films

### 2.1 The method

The process was started with vanadium oxides or inorganic salts, as shown in **Figure** 1. These raw materials were treated to make an aqueous transparent solution. To the above solution the selected, weighed soluble polymers and doping agents were added. Then the precursor $VO_2$ film was prepared by traditional solution methods, such as spinning coating or dip coating. After drying at 80 °C in air, the precursor film was annealed at 300- 600 °C in $N_2$. Readers can read references for details.[43,44]

It is also found that precursor solution without polyvinylpyrrolidone (PVP) was unstable and a large amount of precipitates formed after aging for several days. Whereas, PVP-employing solution stayed stable for several months with only a little precipitates suspending in the solution, indicating that PVP improved stability of the precursor solution, probably because the negatively charged carbonyl groups bound with aqua vanadium ions to form a relatively stable precursor solution.

To examine the interactions between PVP and aqua vanadium ions, Fourier transform infrared spectroscopy (FTIR) was employed to characterize the precursor solution with or without PVP. The results were shown in **Figure** 2. To compare, FTIR spectra of PVP (**Figure** 2a) and PVP in the presence of $H_2SO_4$ (**Figure** 2b) were also included. Generally, the strong and sharp peak at 1654 $cm^{-1}$ was assigned to the stretching vibration of -C=O, and the peak at 1292 $cm^{-1}$ was attributed to -C-N (**Figure** 2a).[73] When $H_2SO_4$ was added to PVP (**Figure** 2b), the stretching vibration of -C=O shifted to low wavenumber of 1635 $cm^{-1}$, which originates from the loosening of the -C=O double bond by coordinating between negatively charged carbonyl groups and $H^+$.[74] For precursor solution with PVP (**Figure** 2c), the frequency of -C=O stretching vibration became lower (1620 $cm^{-1}$) due to the influence of aqua vanadium ions. The coordination interactions between the carboxyl group of PVP and metal ions (Li, Ca, Co, Ag) were also reported in PVP–DMF–$MCl_n$ systems[75] as well as polymer/silver salt complex membranes.[76-78] Although the vibration at 1624 $cm^{-1}$ in the spectrum for a PVP-free precursor solution was poorly identified (**Figure** 2d), the vibration of -C-N kept almost unchanged (**Figure** 2a, b and c) for all these films, indicating that there were no interactions between amine groups and aqua vanadium ions. According to above discussion, the effects of PVP on stabilizing precursor solution were due to the interactions of the negatively charged carbonyl groups in PVP with qua vanadium ions. Furthermore, it is reported that the interactions between metal ions and the carbonyl groups from different PVP molecules increased the apparent viscosity of the PVP–DMF–$MCl_n$ solution (M = Li, Ca, Co),[75] where metal ions act as cross-linking points between different PVP molecular chains,[75] improving the film formability.

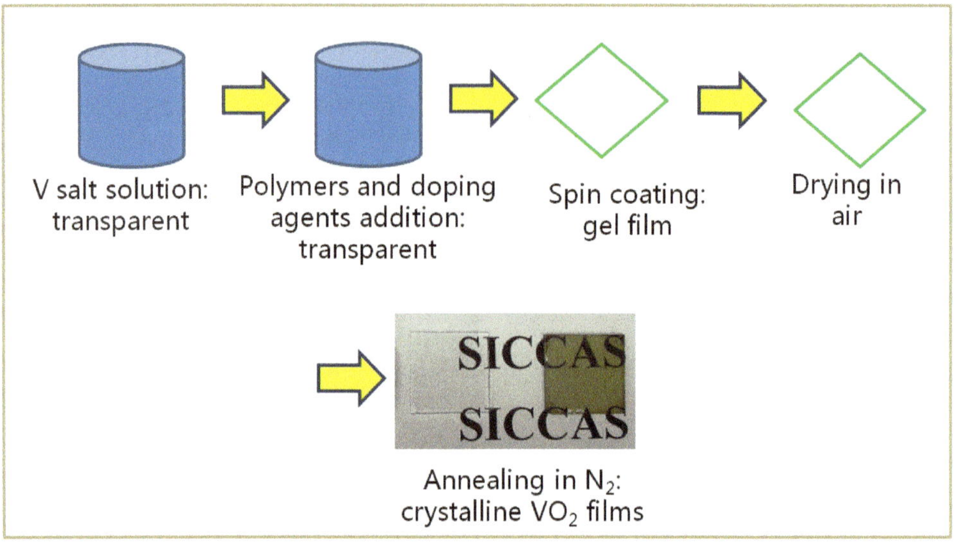


Fig. 1. Polymer-assisted deposition process for $VO_2$ films.

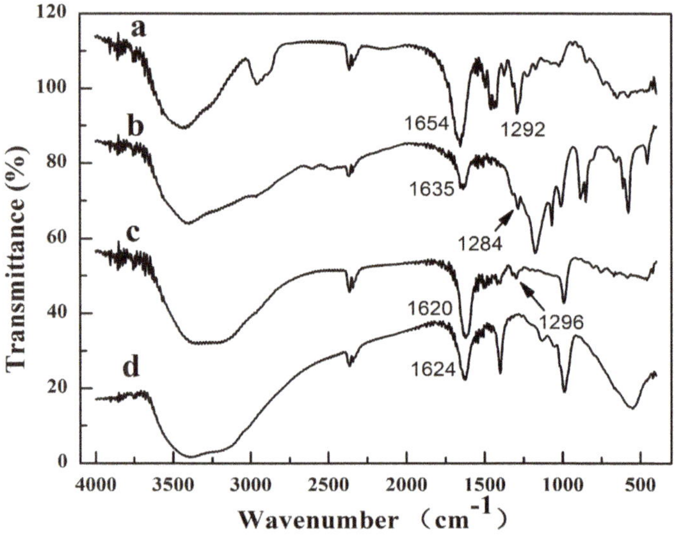


Fig. 2. FTIR spectra of PVP (a), PVP with $H_2SO_4$ (b), precursor solution with PVP (c) and PVP-free vanadium solution (d). pH values of (b, c and d) were adjusted to the same.[43a]

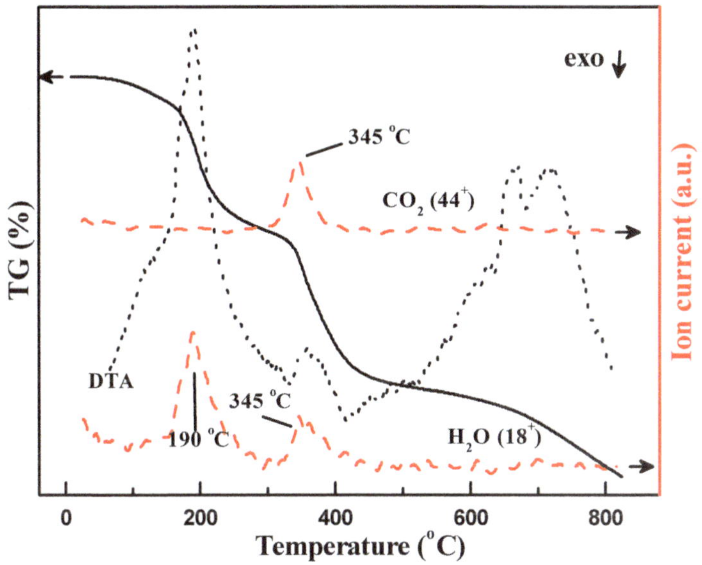


Fig. 3. Simultaneous TG, DTA and evolved gas analytical curves of 7.6 mg precursor gel containing K90 PVP with nitrogen flow of 30 mL·min$^{-1}$, heating rate of 10 °C·min$^{-1}$ in an open $Al_2O_3$ crucible.[43b]

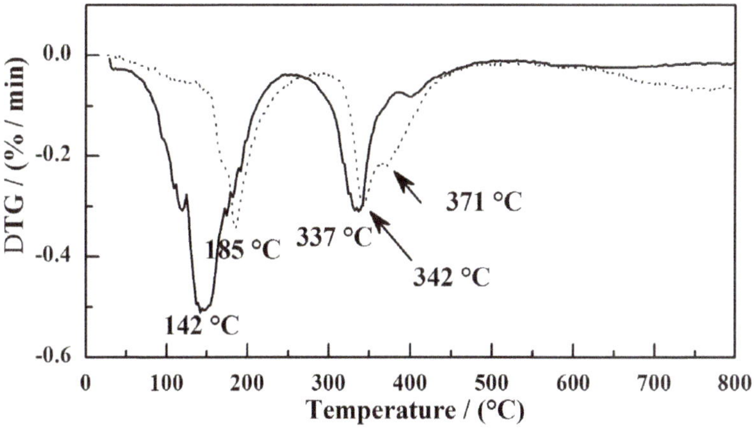


Fig. 4. DTG curves of precursor gel containing K90 (the dot line) and K30 PVP (the solid line) with nitrogen flow of 30 mL·min$^{-1}$, heating rate of 10 °C·min$^{-1}$ in open $Al_2O_3$ crucibles.[43b]

The simultaneous TG, DTA, DTG and evolved gas analytical curves of the gel contain K90 PVP in a nitrogen atmosphere are presented in **Figure** 3 and **Figure** 4 (the dot line). On the basis of MS (mass spectrum) signals, the evolution of $H_2O$ and $CO_2$ was detected (**Figure** 3).

At temperatures up to 250 °C, adsorbed and chemisorbed $H_2O$ was released, indicating by MS and DTG changes centered at 185 - 190 °C (a continuous weight loss from 100 to 250 °C in TG) along with an endothermic peak in DTA. After the evaporation of water, there is a plat stage in both MS and TG curves from 250 to 300 °C.

At 300 - 450 °C, both $H_2O$ and $CO_2$ were released, as observed distinct peaks in MS curves at 345 °C, implying the oxidization of PVP with a trace amount of oxygen in the nitrogen flow. CO is formed when oxygen is insufficient, although the MS signals for CO were failed to be detected due to the similar molecular weight of CO and $N_2$. This conclusion also can be drawn from the DTA/DTG results, in which an endothermic peak appeared at 360°C in the DTA curve (a corresponding shoulder at 371°C in the DTG curve). In the range of 450 - 650 °C, no obvious reactions occurred. Above 650 °C, a gradual weight loss in the TG curve was observed, implying that the residual carbon (which had been confirmed by Raman analysis) can further reduce $VO_2$ to $V_2O_3$ (from XRD results).

Further DTG analysis on precursor gels containing PVP of different molecular weights reveals that the decomposition sequence is related to the molecular weight of the polymer. As show in Figure 4, both DTG peaks for water desorption and polymer decomposition shifted to lower temperatures for the low molecular weight PVP (K30 vs K90). However, in both cases, heating temperatures ≥ 450 °C are needed to decompose the polymer. Therefore, the thermal analyses suggest that the appropriate annealing temperature for crystallization to thermochromic $VO_2$ should be in the range of 450 - 650 °C.

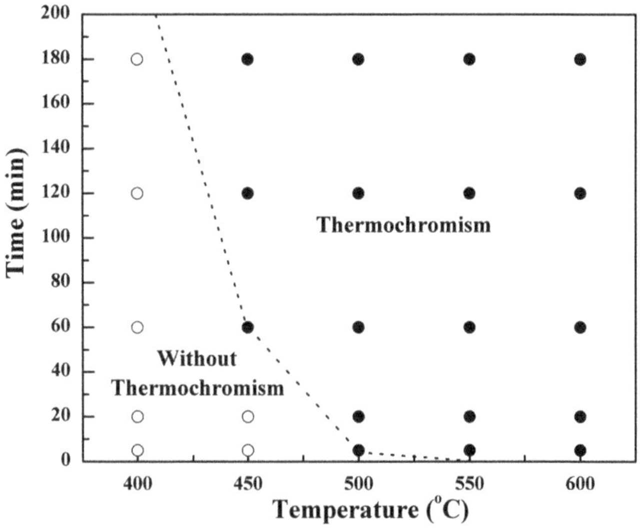


Fig. 5. Dependence of thermochromic properties on annealing parameters for $VO_2$ films from K30 PVP by annealing at a heating rate of 10 $^{o}C{\cdot}min^{-1}$ in $N_2$.[43b]

Subsequently, a series of annealing experiments were performed and the results are summarized in **Figure** 5. It was shown that at 450 °C a relatively longer annealing time was required to crystallize $VO_2$ with thermochromic properties. At 500 °C and above, the annealing time could be significantly shortened, and holding time was even unnecessary. These results are coincident with those of thermal analyses, and are very profitable for scale-up mass manufacturing.

Interestingly, when K30 PVP was replaced by K90 PVP, the annealing time at 500 °C should be prolonged to exceeding 20 min to achieve the thermochromic properties. This result shows that the degradation process of polymers effectively influences the formation of thermochromic $VO_2$, probably due to interactions between PVP and $VO^{2+}$ at the atomic scale. This conclusion supports the film-forming mechanism that we proposed in the previous work.[43a] Furthermore, these interactions ensure the formation of homogeneous hybrid precursor films after solvent evaporation.[43] Thereby, the morphologies of final films could be easily controlled via adjusting the degradation rate of polymers.

**Figure** 6 (a) shows the TEM image for precursor gel. PVP aggregates, inorganic clusters and their aggregates were clearly identified. Using this precursor solution, a homogeneous film was obtained, as show in **Figure** 6 (b). After crystallization, both PVP-free and PVP-employing films were particulate, porous, while the particle size of PVP-employing films (~100 nm in diameter) was much smaller than PVP-free ones (over ~100 nm) that showed a broad size distribution. The porosity of PVP-free films was obviously larger than PVP-employing films, typically about 63.5 % (**Figure** 7a and b) vs 17 % (**Figure** 7c and d) when it was analyzed by the software image-pro plus 5.0.

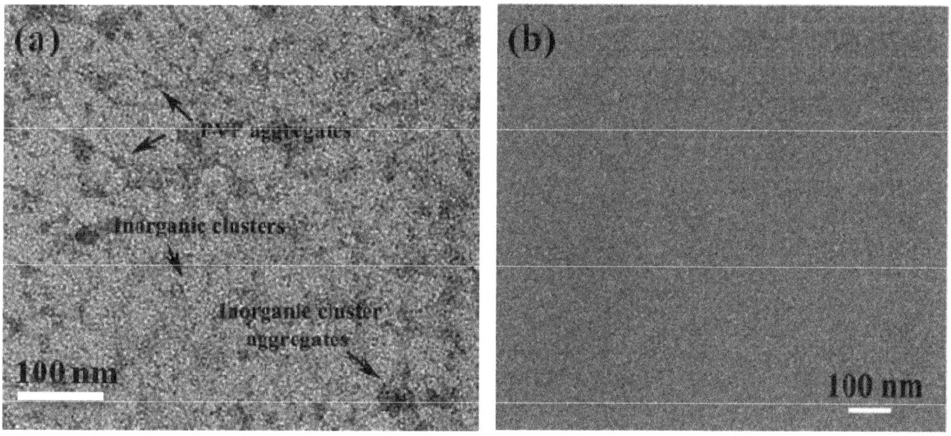


Fig. 6. (a) TEM images of precursor gel and (b) SEM image of precursor gel film.[43a]

A schematic illustration of film-forming mechanism was given in **Figure** 8. The interactions among polymer molecules via the oppositely charged groups along with that between the carbonyl groups and the metal ions ensured the formation of cross-linked high-quality gel films after the solvent evaporation. In addition, steric entangling of the polymer chains assists in the enhancement of cross linking.

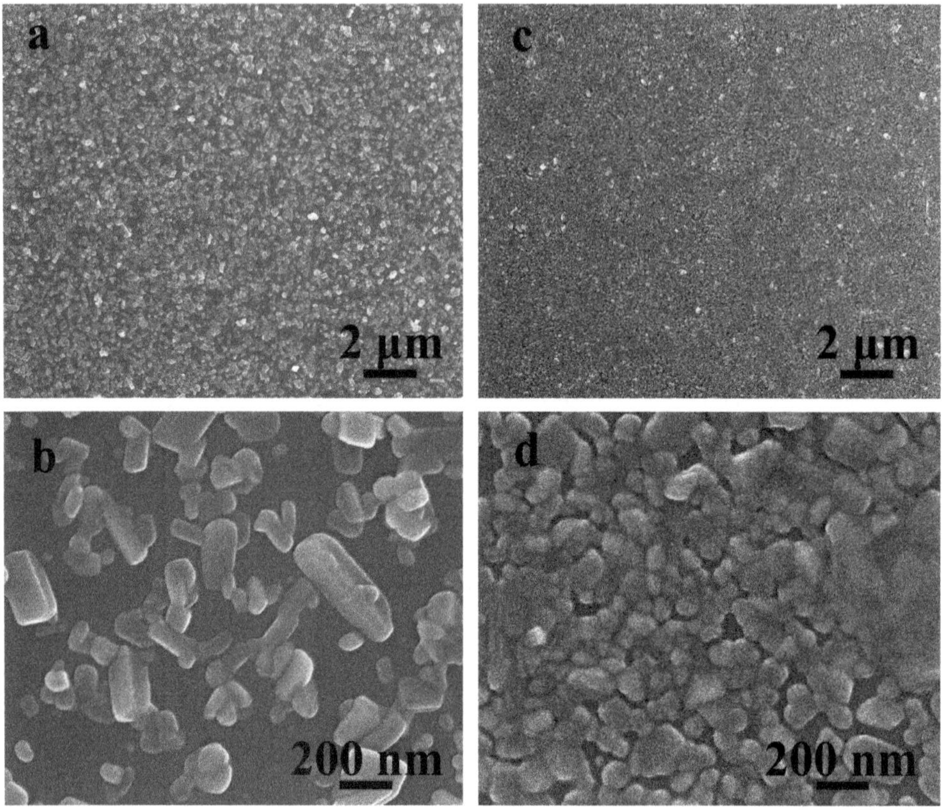


Fig. 7. SEM images of a PVP-free film (a and b) and a PVP-employing film (c and d) on fused silica substrates.[43a]

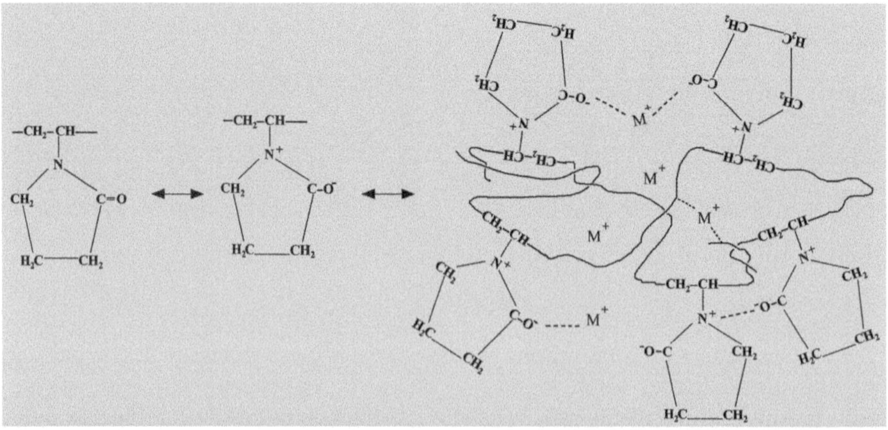


Fig. 8. A schematic illustration of the interactions between aqua vanadium ions ($M^+$) and PVP as well as the formation mechanism of precursor films.[43a]

## 2.2 The fundamental properties of single-layered $VO_2$ films

**Figure** 9 shows that the $VO_2$ film prepared by polymer-assisted deposition process had a relatively pure crystalline phase. For the film obtained by a PVP-free solution, only limited peak were detected, which were assigned to the characteristic peak of M-phase $VO_2$ (JCPDS Card No. 72-0514, P21/c, a = 0.5743 nm, b = 0.4517 nm, c = 0.5375, β = 122.61°), but two weak diffraction peaks at 9.36 ° and 12.24° were also detected, which were too skimpy to be determinately identified. For the film prepared by the PAD process, only M-phase $VO_2$ characteristic peaks (011) was detected, which suggested that the film was in a preferential orientation.

The formation of M-phase $VO_2$ was further confirmed by Raman spectra. At an output power of 1 mW (Figure 10, curve a) for a pure film, an almost complete set of Raman bands of M-phase $VO_2$ were observed. The bands agree well with references[9, 79-80] for M phase $VO_2$, with centers at 192, 222, 261, 308, 337, 391, 440, 497 and 615 and 816 $cm^{-1}$. There is at least one corresponding band for every Raman band of our film and the positions of the bands are agreed well with each other, meaning that every Raman bands we collected here can be assigned to M-phase $VO_2$. A Raman spectrum for a 1 at % W-doped $VO_2$ film (Figure 10, curve b) at 1 mW showed only weak bands, suggesting that the film was at the phase transition point due to the reduction effect of the W-doping on the transition temperature. No Raman bands appeared at output power of 20 mW for both of the films (Figure 10, curves c and d), indicating that complete phase transition from M-phase $VO_2$ to R-phase $VO_2$ occurred under this output power condition. The contribution of a fused quartz substrate has also been included (**Figure** 10, curve e) for the sake of comparison. Although an overlap of Raman bands of fused quartz and $WO_3$ at 807 $cm^{-1}$ makes it difficult to determine the presence of $WO_3$ from the enlarged Raman spectra of W-doped films, other strong Raman bands located at 716, 275 $cm^{-1}$ were not observed, excluding the absence of $WO_3$. Meanwhile, no obvious Raman bands for impure phases were found after the phase transition, indicating that there were no obvious impure vanadium oxides.

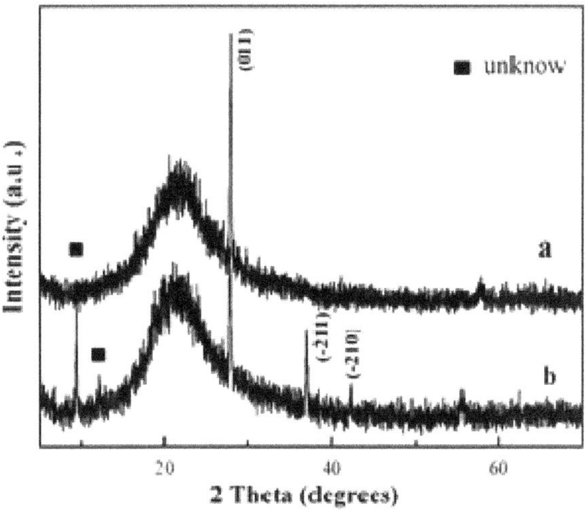


Fig. 9. XRD patterns of $VO_2$ film prepared by using a PVP-employing aqueous solution (a) and a PVP-free solution (b) on fused silica substrates.[43a]

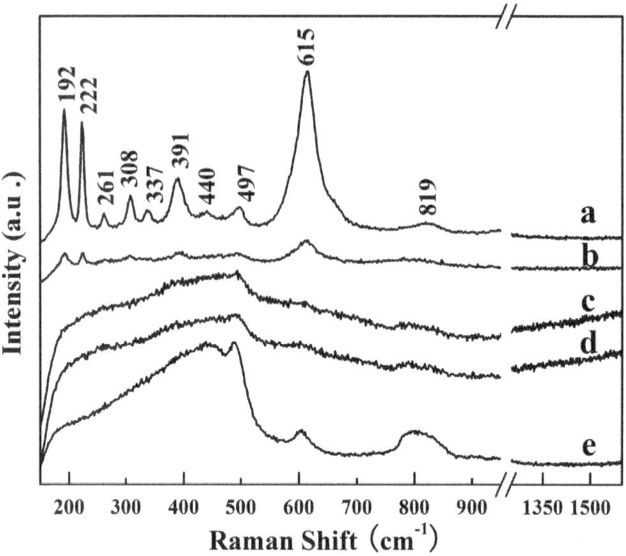


Fig. 10. Output power-dependent Raman spectra for PVP-employing $VO_2$ films on fused silica substrates (a-d) and a substrate only (e). (a) 1 mW (b) 1 mW and 1 at % W-doped film (c, e) 20 mW (d) 20 mW and 1 at % W-doped film.[43a]

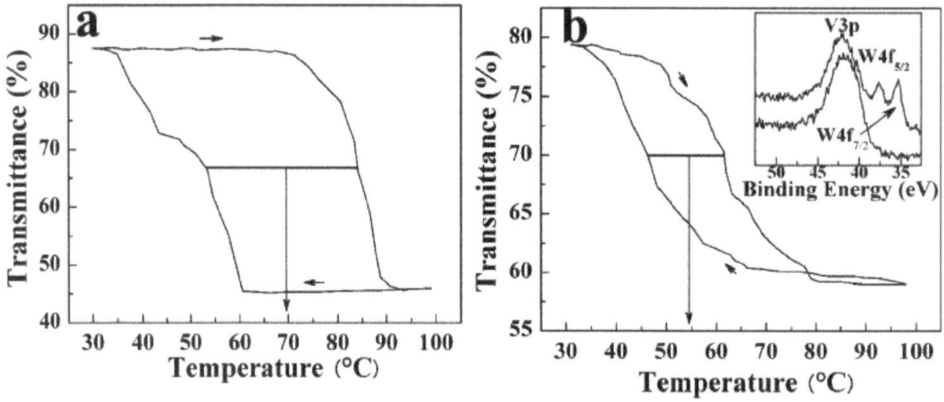


Fig. 11. Temperature dependence of optical transmittance at a fixed wavelength (2000 nm) for a pure (a) and a 1 at % W-doped (b) PVP-employing $VO_2$ film. The inset shows the $W4f_{5/2}$ and $W4f_{5/2}$ spectra for a pure (bottom) and a 1 at % W-doped (top) $VO_2$ film.[43a]

The 1 at % W-doped $VO_2$ film exhibits a hysteresis loop centered at 54 °C with a width of 16 °C (**Figure** 11b), implying a decrease in phase transition temperature of about 15 °C. This decreasing efficiency is less than other reports.[36,38] Significantly, the decreasing efficiency of phase transition temperature for 1 at % W-doped film was lower than the average efficiencies of other doping doses in references, which employed similar doping process to prepare $VO_2$ particles,[81] probably because the tungsten ion was not completely incorporated into the final $VO_2$ films. $W4f_{5/2}$ and $W4f_{7/2}$ XPS peaks with a binding energy of 37.7 and 35.3 eV, respectively, are clearly seen in Figure 11b inset for the W-doped thin films and the tungsten ion in these films is $W^{6+}$ according to the standard binding energy. The width narrowing of the hysteresis loop can be explained by the martensitic transformation model for $VO_2$.[82] W-doping increases the density of structural defects, which is a power function of the driving force, and relatively reduces the activation energy of the coordinated jumps of V cations (phase transitions take place at a certain defect density for size-fixed crystal grains[58]). Activation energy further influences the widths of hysteresis loops via degrees of supercooling or superheating. This deduction of martensitic transformation model is in agreement with various experiment results.[25,39,83]

### 2.3 Optimized thermochromic hysteresis properties of single-layered films

Thermochromic properties are important parameters for the practical application of this material. These parameters include phase transition temperature, thermochromic hysteresis and energy-saving efficiency. **Figure** 12 shows a schematic description of phase transition temperature and thermochromic hysteresis. From the transmittance ($Tr$) - temperature ($T$) data, a plot of $d(Tr)/d(T)$ - $T$ is obtained, yielding one or two peaks with well-defined maxima (see Figure 12). Each of the $d(Tr)/d(T)$ - $T$ curves has been fitted with a Gaussian function using the peak fitting module of *Originpro 7.5 software*. The temperature corresponding to the maximum $d(Tr)/d(T)$ is defined as the phase transition temperature ($T_c$) of the branch ($T_{c,h}$ and $T_{c,c}$ represent $T_c$ of heating and cooling branches, respectively). For cooling branches, the appearance of steps introduces two peaks in the $d(Tr)/d(T)$ - $T$

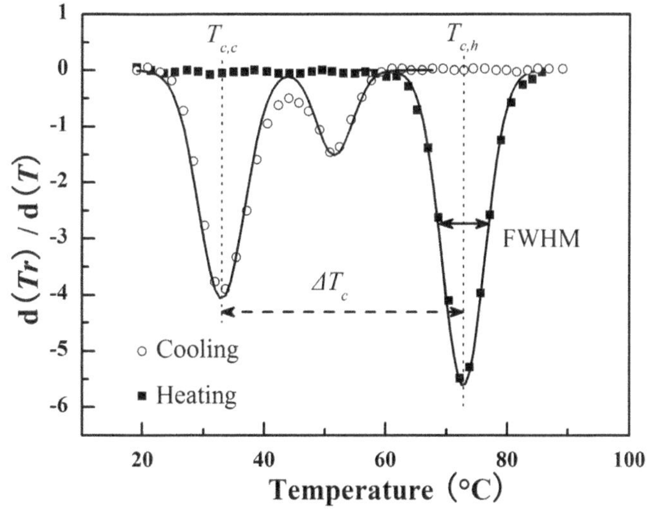


Fig. 12. A schematic description of the definition of the S-M transition parameters for heating and cooling branches using the $d(Tr)/d(T)$ - $T$ plot. $T$ and $Tr$ represent temperature and transmittance at wavelength of 2000 nm. The resulting parameters $T_{c,h}$, $T_{c,c}$, $\Delta T_c$ and FWHM of peaks are employed to express phase transition temperature of heating, cooling branches, the width of the hysteresis loop, and the slope of the transition, respectively.[44]

curves. The $T_c$ values of these branches are determined by the main peaks. The slope of the transition is expressed by the full width at half maximum (FWHM) of the peak. The width of the hysteresis loop, $\Delta T_c$, is defined as the temperature difference of $T_{c,h}$ - $T_{c,c}$.

**Figure** 13 shows typical SEM photos of $VO_2$ films heated at 600 °C for different times with a heating rate of 30 °C·min$^{-1}$. All the films are porous, consisting of irregular particles. The pore formation is attributed to the degradation of PVP and the shrinkage of the gel film during annealing. Grain boundaries change from clear (**Figure** 13a, b, c) to fuzzy (inset in **Figure** 13d) as the annealing time is prolonged. Meanwhile, the particle size varies dramatically, associated with distinct change of porosity and pore shape. To evaluate the differences of particle size among samples, the distributions of particle sizes was measured. The corresponding size distribution scatter graph as well as fitting curves with Gaussian distribution is shown in **Figure** 14. The scatter graphs were plotted by randomly measuring the dimensions of 300 grains in a given micrograph. The particle size distribution scatter graphs of **Figure** 13d failed to be obtained due to fuzzy grain boundaries. The results showed that the grain size of Sample I complied with Gaussian distribution, while the

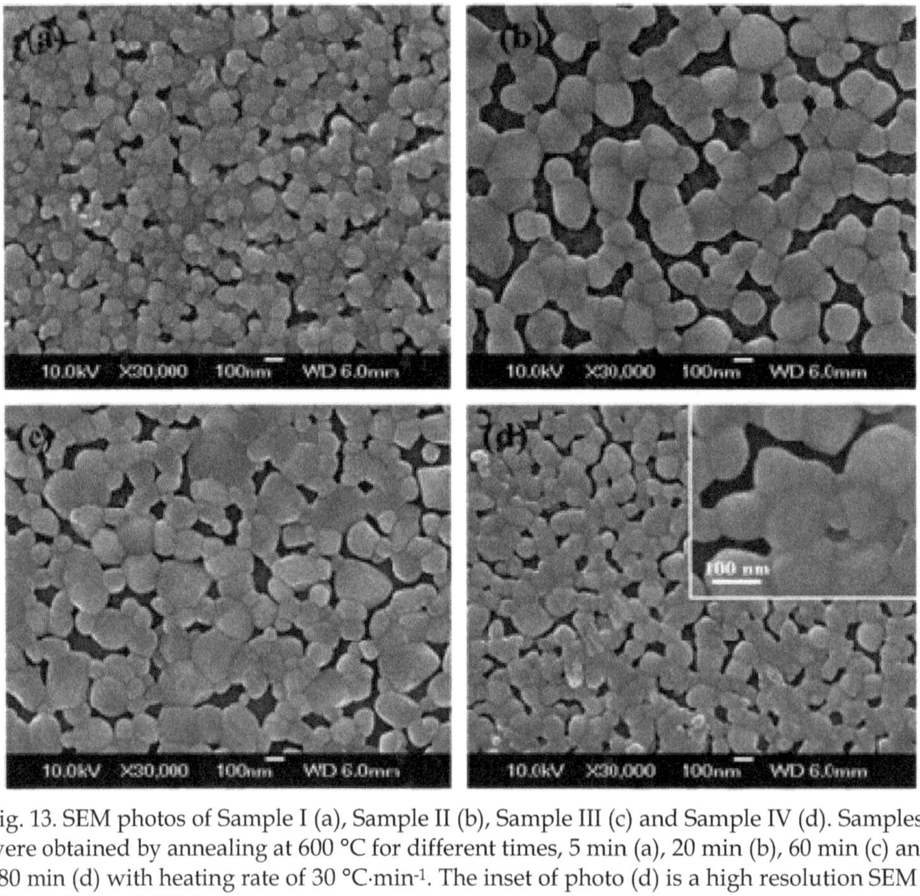


Fig. 13. SEM photos of Sample I (a), Sample II (b), Sample III (c) and Sample IV (d). Samples were obtained by annealing at 600 °C for different times, 5 min (a), 20 min (b), 60 min (c) and 180 min (d) with heating rate of 30 °C·min$^{-1}$. The inset of photo (d) is a high resolution SEM photo of sample IV.[44]

distributions for Samples II and III distinctly exhibited two maxima. The size distribution scatter graphs indicated that when annealing time increases from 5 to 20 min, the particle size increases rapidly (**Figure** 13a, b and **Figure** 14a, b). However, the $VO_2$ particles shrink noticeably for the long time annealing samples (**Figure** 13c and **Figure** 14c, d), resulting in few large particles and a broad distribution of particle sizes. The shrinkage of particle size (which is also observed in the 500 °C annealed films) and change of grain boundaries are tentatively attributed to the mass transport via surface diffusion during annealing. This mass transport has been reported as the main reason for morphology evolution of $VO_2$ films during the heat treatment of pre-deposited amorphous ones at 450 °C.[83]

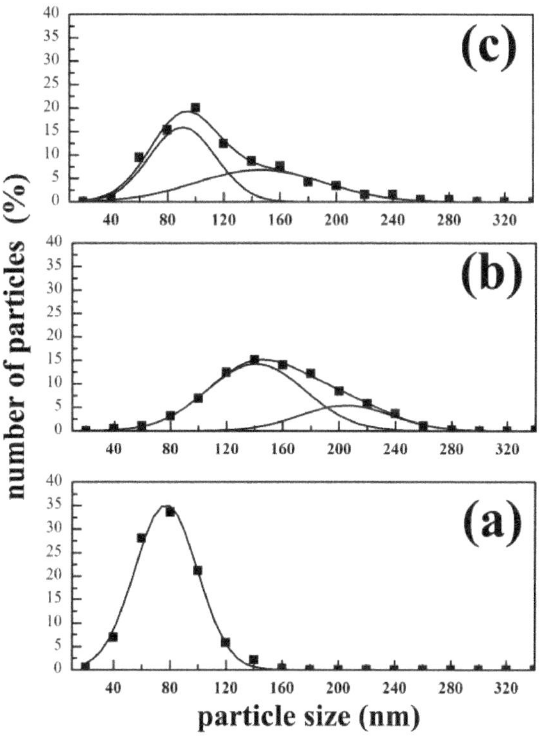


Fig. 14. Scatter graphs showing the particle size distributions in Sample I (a), Sample II (b) and Sample III (c). The solid curves are those after fitting by Gaussian distribution. Samples were obtained by annealing at 600 °C for different times, 5 min (a), 20 min (b) and 60 min (c) with heating rate of 30 °C·min$^{-1}$.[44]

Significantly, obvious steps were observed at the cooling branches for Samples I, II and III (insets of **Figure** 15a, b, c, annealing for 5, 20 and 60 min, respectively). The appearance of steps suggests another loop width. The steps are resulted from a difference in temperature of inhomogeneous occurrence of phase transition in the films due to the two-humped grain size distributions (**Figure** 14b, c)[84] or the site-selective nucleation of product phase, as observed in literature.[85] However, there are also some contradictions. First, this deduction is difficult to explain the step appearing in Sample I, which manifests a Gaussian distribution with one maximum in grain size. Second, this deduction suggests that the temperatures of the d(*Tr*)/d(*T*) - *T* peaks should vary as the size distribution changes. For Sample II and III, however, although the size distribution curves are quite different (**Figure** 14b, c), the step appears at similar temperatures on the cooling branch of the d(*Tr*)/d(*T*) - *T* curves. Therefore, it seems that these experiment results don't agree to the model suggested by Klimov, V. A et al..[84] In fact, the propagation of the phase transition through grain boundaries may possibly counteract inhomogeneous distribution of $T_c$ in the $VO_2$ films. Therefore, there is usually no step observed in relatively narrow hysteresis loops.[86-87] It is believed that any models related to the steps in the hysteresis loops should take the influence of grain boundary into account.

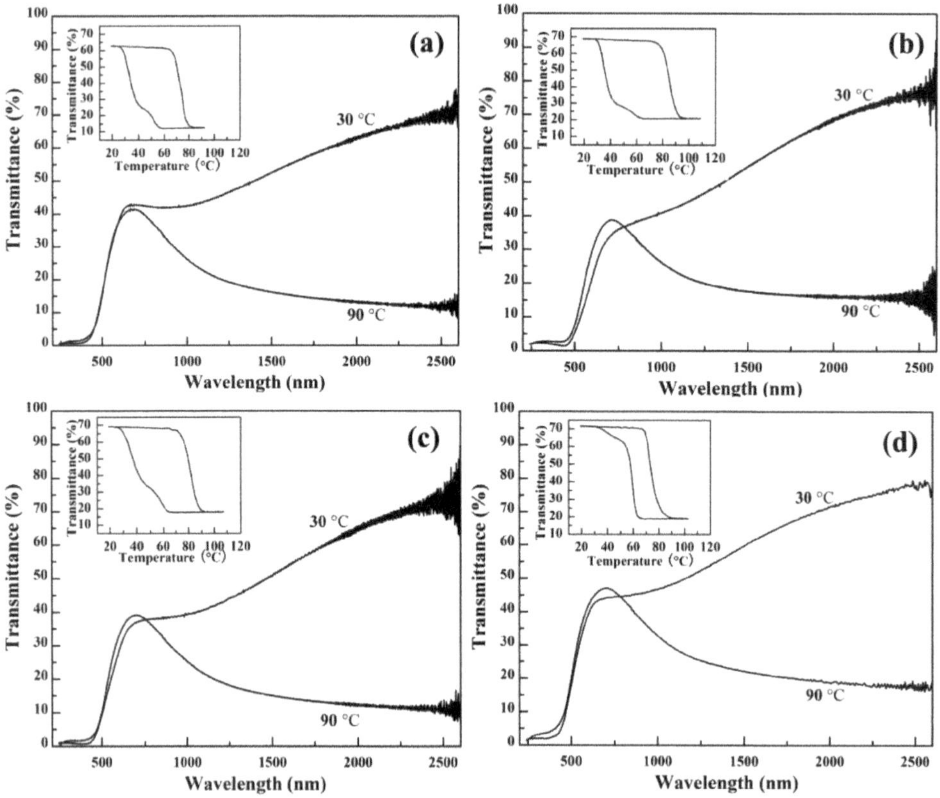


Fig. 15. Optical transmittance spectra and temperature dependence of optical transmittance (insets) at a fixed wavelength (2000 nm) for Sample I (a), Sample II (b), Sample III (c) and Sample IV (d). Samples were obtained by annealing at 600 °C for different times, 5 min (a), 20 min (b), 60 min (c) and 180 min (d) with heating rate of 30 °C·min$^{-1}$.[44]

For Sample IV (annealed for 180 min), the step seems to be depressed (inset of **Figure** 15d). This result could be explained by the coalescence of grain boundaries and the improvement of phase transition propagation through grain boundaries, which would push the step upward. The appearance of the step in only one of the two branches is associated with the asymmetry of the elementary hysteresis loops (i.e., the loops assigned to individual grains) with respect to the phase equilibrium temperature.[86] The asymmetry of the elementary hysteresis loops, in turn, is attributed to a transition temperature shift accompanying with a decrease in grain size,[88-89] or with stress at the substrate/film interface.[86, 90]

To further clarify the effects of grain size on the widths of hysteresis loops, films with similar morphologies and relatively homogeneous size distribution are needed. The current solution process enables us to achieve widely morphology control. In previous research, we have revealed that the precursor solution is in a state of solution rather than sol. After the solvent evaporation, the interactions among polymer molecules, along with those between the carbonyl groups and the metal ions, ensure the formation of cross-linked high quality gel films.[43a] In these films, the metal ions are bonded with the polymer through electrostatic interactions, forming a uniform organic-inorganic hybrid precursor film. The formation of metal oxide occurs after degradation of polymer begins.[43, 91] Therefore, the morphologies of final films could be easily controlled via adjusting the degradation rate of polymers. Meanwhile, for crystallization in a solid-state reaction, the dependence of nucleation and growth rate on temperature is usually different.[92] Thus, it is expected that the morphologies of $VO_2$ films could be tailored by variations of heat treatments. Accordingly, the films have been prepared under the same synthesis conditions as Samples I - IV, but at an annealing temperature of 500 °C. And the typical SEM photos and optical transmittance spectra, along with the hysteresis loops, are shown in Figure 16.

The sample obtained by annealing at 500 °C for 5 min showed a bluish color, but almost no thermochromic properties, and is excluded. Films produced at this annealing temperature have similar morphologies (Figure 16a, b, c). Compared to films obtained by annealing at 600 °C, in which many quasi-isolated grains appear, all of these films show grains tightly connected across boundaries. The grain sizes are smaller than those treated at 600 °C. The film obtained by annealing for 20 min (Sample V) consists of connected, irregular particles (Figure 16a). Typically, the largest dimension of each particle is around 100 nm. These particles possess flat surfaces, indicating that the mass transport during annealing is feeble for a short annealing time. The film produced by annealing for 60 min (Sample VI) shows similar granular morphologies with a relatively large roughness, and the particle size is reduced notably (Figure 16b). Prolonging the annealing time to 180 min (Sample VII) results in increases in porosity and surface roughness (Figure 16c), indicating the enhancement in mass transport that changes the surface morphology of these particles.[83, 92]

To further verify our discussion on grain boundaries, two films (Sample I and Sample VIII) were scraped off from their substrates and observed by TEM (Figure 17). **Figure** 17a, b and c present TEM photos of Sample I. It is shown that the sample is assembled from irregular particles. Distinct grain boundaries are observed in high resolution TEM photos (**Figure** 17a, b and c), indicating that the grain boundaries are very loose in this film. The Fourier transform (FFT) patterns of corresponding crystal grains (**Figure** 17 insets) reveal diffraction spots, verifying the crystallinity. Significantly, the diffraction spots in the FFT patterns become hazy or disordered in the vicinity of grain boundaries. This experimental phenomenon can be ascribed to the poor crystallization of $VO_2$ under short annealing time

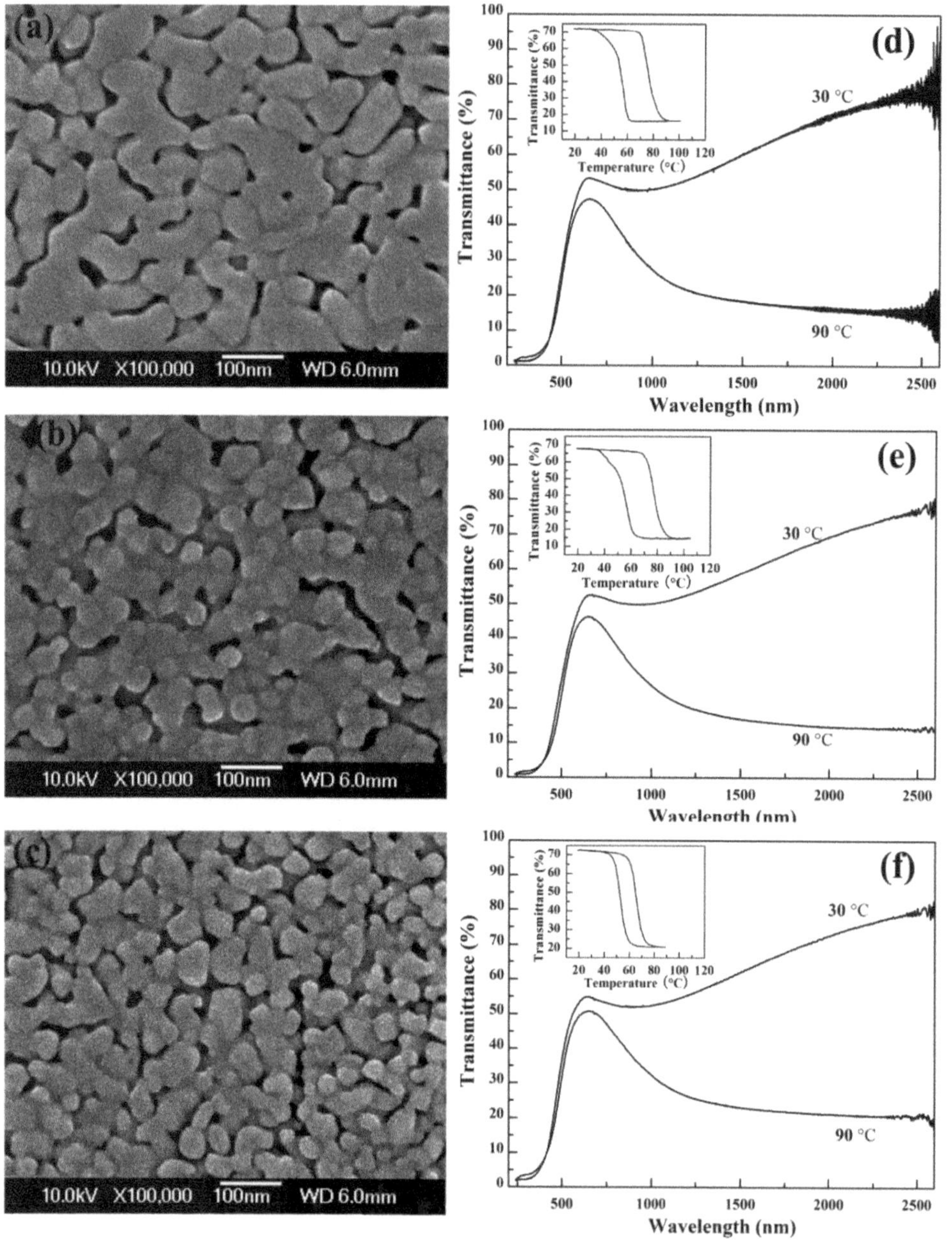


Fig. 16. SEM photos and optical transmittance spectra of Samples V (a and d), VI (b and e) and VII (c and f). Insets show corresponding hysteresis loops at 2000 nm. Samples were obtained by annealing at 500 °C for different times, 20 min (a and d), 60 min (b and e) and 180 min (c and f). Heating rate is the same, 30 °C·min$^{-1}$.[44]

(600 °C, 5 min). Generally, spaces about 2-5 nm in width were observed between grains, and they were attributed to poorly crystallized phases and/or voids. It is widely reported that $VO_2$ films with second or amorphous phases at grain boundaries demonstrate a broadened hysteresis,[82, 93, 94] which is definitely consistent with our experiments.

Sample VIII, on the other hand, consisted of smooth connected particles and interconnected pores among particles (**Figure** 17d, e). This morphology should be a result of mass transport and aggregation of the primary $VO_2$ crystals during annealing.[83] The high magnification TEM photo clearly demonstrates that the grain boundaries are very sharp (**Figure** 17f). In fact, it seems that the lattice fringes pass through grain boundaries. The FFT patterns (**Figure** 17f insets) of the corresponding crystal grains are composed of two groups of bright diffraction spots, confirming the existence of boundaries and the high crystallinity of these crystals. The TEM photos are fully consistent with the SEM results (**Figure** 13a and **Figure** 16a) and the deductions from optical measurements (**Figure** 15a inset and **Figure** 16b inset).

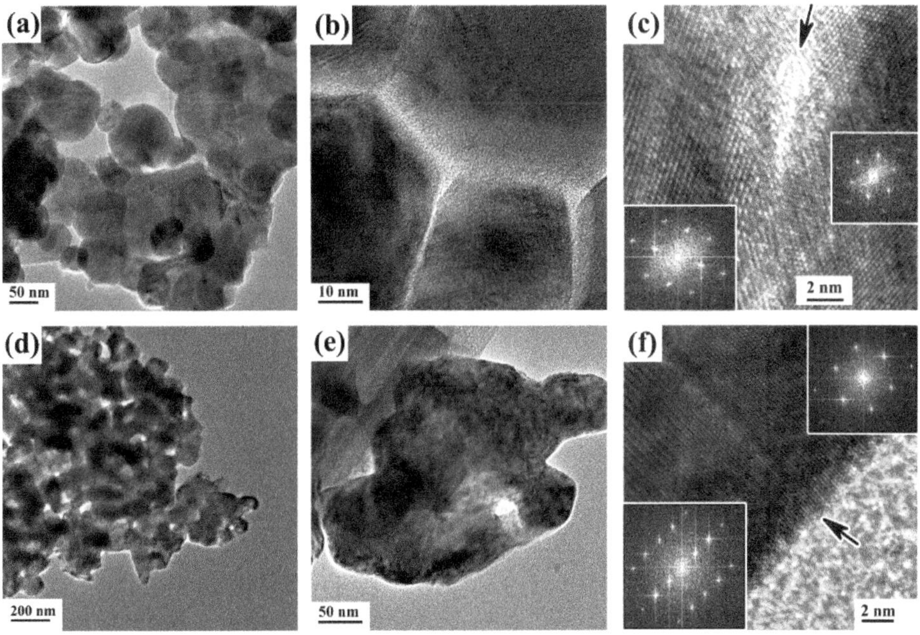


Fig. 17. TEM photos of Sample I (a, b and c) and Sample VIII (d, e and f). The insets of photos c and f are FFT patterns of corresponding crystal grains. Samples were obtained by annealing at 600 °C for 5 min (a, b and c) with a heating rate of 30 °C·min$^{-1}$, and 500 °C for 180 min (d, e and f) with a heating rate of 3 °C·min$^{-1}$, respectively.[44]

To evaluate the optical and thermochromic properties of the samples, the integral transmittances of films annealed with heating rate of 30 °C·min$^{-1}$ were calculated in the visible region (380 - 780 nm), so were the integral infrared transmittance reductions before and after the S-M phase transition in the wavelength of 1500 - 2500 nm. The results are shown in **Figure** 18. According to the theory suggested by J. Narayan and V. M. Bhosle,[57] the amplitude of the transition should deteriorate as the defect content increases. It is expected that the 500 °C annealed samples should be less crystallized and thus contain high defect content. Accordingly, they should exhibit deteriorated infrared transmittance reduction compared with the 600 °C annealed samples. This phenomenon was not obviously detected in our films, even though the hysteresis widths of the films vary considerably. In fact, it is clearly revealed that the 500 °C annealed samples show better visible transmittance than those annealed at 600 °C, while the infrared transmittance reduction remains comparable. For instance, the infrared transmittance reduction of sample VI (53.8%, 500 °C, 60 min annealed sample) is a little larger than that of sample II (51.2%, 600 °C, 20 min annealed sample). Meanwhile, the integral visible transmittance of sample VI (34.5%) is much higher than that of sample II (20.2%).

The difference in the microstructure and crystallinity had effects on the optical properties of these films. Hemispherical reflectance spectra of Sample II and VI were collected from 240 to 2600 nm (**Figure** 19). It was found that the integral hemispherical reflectance of Sample VI (18.7%) was 3.5% larger than that of Sample II (15.2%) in the visible region. Because both of the integral transmittance and integral reflectance for Sample II are smaller than those of Sample VI in the visible region, one can conclude that the absorption and/or scattering of Sample II are much larger. The hemispherical transmittance of the sample was recorded, in order to eliminating the influence of scattering. It was still shown that both the hemispherical transmittance and the hemispherical reflectance of Sample II are less than those of Sample VI in the visible region, indicating that the absorption of Sample II is larger than that of Sample VI.

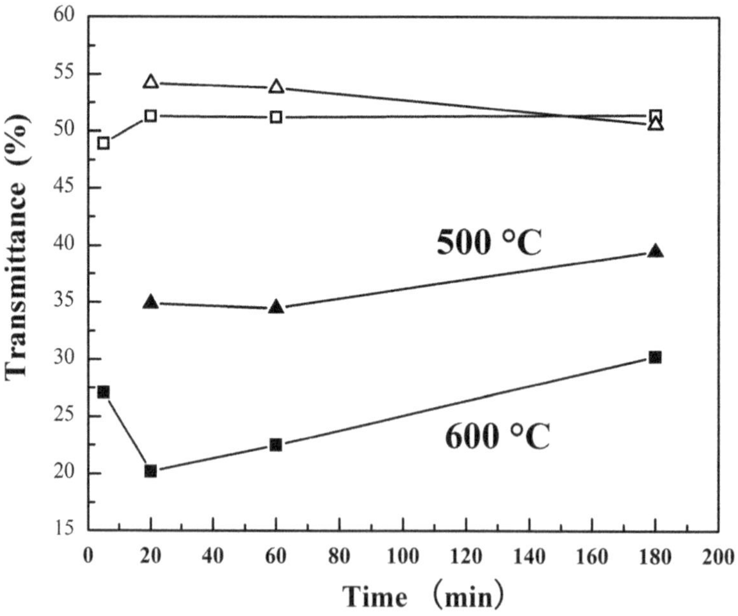


Fig. 18. Integral visible transmittance (in a region of 380 - 780 nm, filled polygons) and integral infrared transmittance reduction (in a region of 1500 - 2500 nm, open polygons) of films annealed at 600 °C (squares, Samples I - IV) and 500 °C (triangles, Samples V - VII) with heating rate of 30 °C·min$^{-1}$. [44]

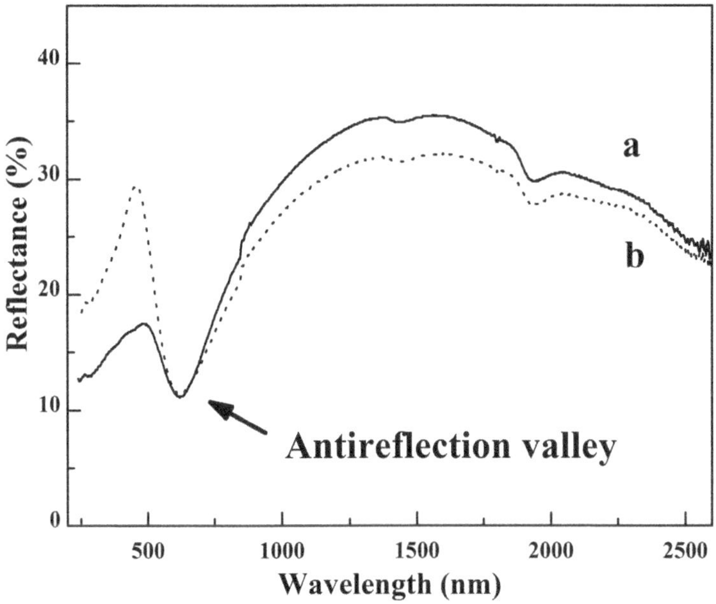


Fig. 19. Reflectance spectra of Sample II (a) and Sample VI (b). Samples were obtained by annealing at 600 °C for 20 min (a), 500 °C for 60 min (b) with heating rate of 30 °C·min$^{-1}$. [44]

## 2.4 Optimized visible transmittance of single-layered films

For the practical application of $VO_2$-based smart windows, a low luminous transmittance ($T_{lum}$) and solar modulating ability ($\Delta T_{sol}$) are two major drawbacks. The $T_{lum}$ and $T_{sol}$ values were obtained from

$$T_{lum,sol}(\tau) = \int \varphi_{lum,sol}(\lambda) T(\lambda,\tau) d\lambda \, / \int \varphi_{lum,sol}(\lambda) d\lambda \,,$$

where $\varphi_{lum}$ is the spectral sensitivity of the light-adapted eye and $\varphi_{sol}$ is the solar irradiance spectrum for an air mass of 1.5 (corresponding to the sun standing 37° above the horizon). $\Delta T_{sol}$ is obtained from

$$\Delta Tsol = T_{sol,l} - T_{sol,h} \,,$$

where $l$ and $h$ denote low- and high-temperature, respectively.

Strategies to improve $T_{lum}$ and $\Delta T_{sol}$ have been investigated, including Mg- or F-doping,[33, 95] multilayer-stack design,[35,96] and composite film construction.[97] Besides a depression in transition temperature, Mg- or F-doping causes a blueshift in the absorption edge of $VO_2$ films (from around 445 to 415 nm).[95] This change results in a significant increase of the luminous transmittance at the expense of infrared modulating ability (wavelength ≥ 1000 nm).[33] Similar trade-offs between the luminous transmittance and thermochromic properties have also been observed in $VO_2$-based multilayer films.[35] A $VO_2$-$SiO_2$ composite film shows a high visible transmittance but weak infrared regulation ability.[97] Five-layered $TiO_2/VO_2/TiO_2/VO_2/TiO_2$ films with optically optimized structures have a relatively higher luminous transmittance and solar modulating ability.[98] However, incorporating dielectric layers with a certain refractive index and thickness into complicated stack structures is a difficult technological challenge. Optical calculations suggest that $VO_2$ nanoparticles distributed in a dielectric matrix have higher $T_{lum}$ and $\Delta T_{sol}$ than pure $VO_2$ films.[99] This study further predicts that a limit for noticeable solar energy modulation is $T_{lum}$ = 40%, and $\Delta T_{sol} \leq 10\%$.[99] We have recently confirmed experimentally the effects of composite matrix.[68] $VO_2$-$ZrV_2O_7$ composite films with similar thickness of about 95 nm showed decreased particle sizes and significantly enhanced luminous transmittances (from 32.3% at Zr/V=0 to 53.4% at Zr/V=0.12) with increasing Zr/V rations.

However, the influence of porosity on optical properties of single-layered $VO_2$ films, which should have priority over the stack structure of $VO_2$-based multilayers, is seldom reported. Pores with air in $VO_2$ films can be considered as a secondary component that should have similar effects on improving $T_{lum}$ and $\Delta T_{sol}$. In this study, calculations confirm that optical constants and film thickness have important effects on the thermochromic properties of these films. An optical-admittance recursive method was used to simulate the spectral transmittance using the optical constants of $VO_2$ and a fused-silica-glass substrate. The optical constant is assumed to be linearly dependent on the volume fraction.[100] **Figure** 20 shows the computed luminous transmittance and solar modulation of $VO_2$ films with different thicknesses and porosities. The results indicate that $\Delta T_{sol}$ could be increased without decreasing $T_{lum}$ by increasing the porosity, which is derived from the depression of the reflection.

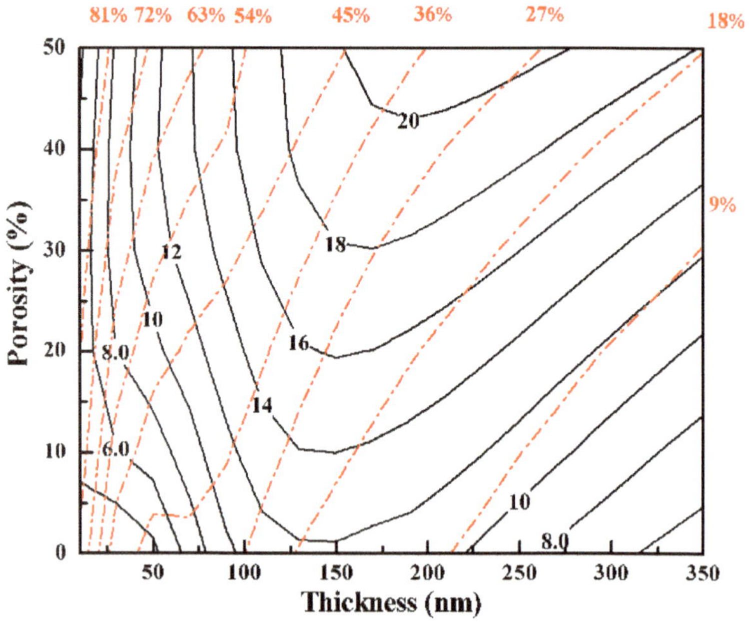


Fig. 20. Computed luminous transmittance in an insulator state ($T_{lum,l}$, red dot lines) and solar modulation between MIT ($\Delta T_{sol}$, black solid lines) as a function of the thickness and porosity of the $VO_2$ films.[69]

To validate the above prediction, we prepared $VO_2$ films by PAD. X-ray diffraction and Raman spectra confirmed the formation of a monoclinic (M) phase with a trace amount of $V_2O_5$. **Figure** 21a shows a top-down SEM image of $VO_2$ films, which reveals that the sample consisted of interconnected $VO_2$ particles and irregular nano-pores. The size of particle and pore ranges from 20-70 and 15-80 nm with a mean value of 38 and 28 nm, respectively. The feature size of the film is well below the wavelength of visible and infrared light, favoring the improvement of optical quality. The nano-porous feature of the films is observable (inset of **Figure** 21a), which is also supported by the low *n* and *k* values of ellipsometry results compared with those in the literature (**Figure** 21b).[46] For example, the value of *n* is 2.2 for our $VO_2$ film, which is around 3 in the literature.[46] The result also shows that $T_{lum}$ in the current research is much higher (by ≥12%) than that in the literature with comparable $\Delta T_{sol}$ due to the different optical constants (**Figure** 21b).[46] These results are attributed to the degradation of PVP and the shrinkage of the gel film during annealing.[43a]

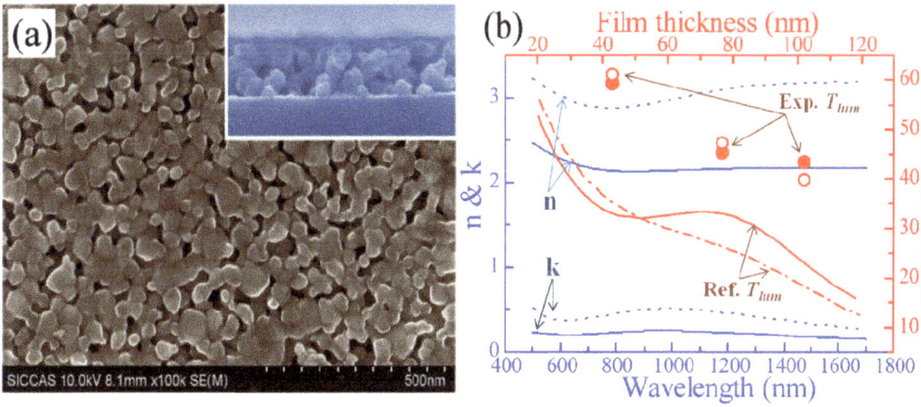


Fig. 21. Part (a) shows a SEM image of a 147-nm-thick $VO_2$ films on quartz glass. The inset of (a) is a cross-sectional SEM image. Part (b) shows the experimental (solid lines) and reference (dotted lines) optical constants at 20 °C as well as experimental (Exp. $T_{lum}$) and reference (Ref. $T_{lum}$) integral luminous transmittance of the film at 20 (solid symbols) and 90 °C (open symbols), respectively. The reference data are redrawn from Reference 46.[69]

**Figure** 22 shows the transmittance and reflectance spectra of typical $VO_2$ films. The MIT transition is clearly observed as a dramatic infrared-transmittance change with temperature (**Figure** 22a). $T_{lum}$ reduces steadily with increasing film thickness, which is ascribed to the strong absorption of $VO_2$ in this region.[44, 99] The change in the infrared transmittance of $VO_2$ films at 90 °C with different film thicknesses shows a similar behavior. Nevertheless, the infrared transmittance at 20 °C for a 428-nm-thick film is obviously higher than that of a 215-nm-thick film after 1700 nm (**Figure** 22a). These changes in the transmittance spectra correspond to reflectance valleys in **Figure** 22b, suggesting the existence of a self-antireflection effect in these thicknesses due to the destructive interference of light reflected from the film-substrate and the air-film interfaces.

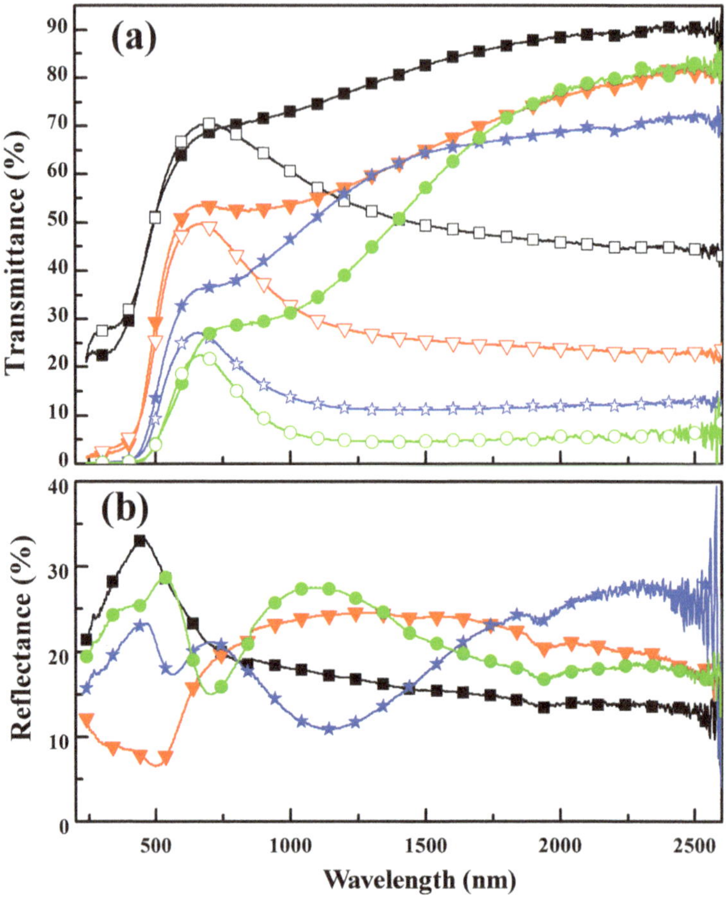


Fig. 22. Thickness dependence of transmittance (a) and reflectance (b) spectra for typical samples with different thicknesses (43, 102, 215 and 428 nm corresponding to the black squares, the red triangles, the blue pentagrams and the green circles, respectively). The transmittance was measured at both 20 °C (lines with solid symbols) and 90 °C (lines with open symbols) and the reflectance was measured at only 20 °C.[69]

The change of the optical constants of $VO_2$ across the MIT can effectively modulate the infrared transmittance and shift the position of the reflectance valley at 20 °C, leading to a significant enhancement of the IR modulating ability at a certain wavelength. This phenomenon could be harnessed to boost the thermochromic properties of a single-layer film in selected wavelength ranges. For instance, a 215-nm film shows a transmittance change of 50% (from 61.1% to 11.1%) at 1350 nm across the MIT, the highest value at this wavelength to our knowledge. Furthermore, the enhancement of the IR modulating ability could be adjusted to longer wavelengths by a simple regulation of the film thickness. A 428-nm-thick $VO_2$ film exhibits a IR transmittance change of 76.5% (from 83% to 6.5%) at 2500 nm, even prior to the optimized result of sputtering films (74%).[19]

Another interesting phenomenon is that the changes in luminous transmittance ($\Delta T_{lum}$) across the MIT are thickness dependent. For thin films, the visible transmittance at 20 °C is generally lower than that at 90 °C (**Figure** 22 and Table 1) and, vice versa. The visible transmittance at 20 °C for the above 100-nm-thick films exceeds that at 90 °C (**Figure** 22b and Table 1). This reversion in $\Delta T_{lum}$ is ascribed to interference effects and was also reported by Xu et al.[46]

In view of the fact that solar energy is mainly in the visible region with a peak at 550 nm, the $\Delta T_{lum}$ reversion across the MIT effectively influences $\Delta T_{sol}$. For instance, $\Delta T_{sol}$ increased by only 0.6% (from 6.4% to 7.0%) as the film thickness increased from 43 nm to 77 nm. Nevertheless, compared with the 77-nm-thick film, $\Delta T_{sol}$ largely increased from 7.0% to 14.1% for the 102-nm-thick film. Meanwhile, the $T_{lum,l}$ remained almost unchangeable (45.3% and 43.3% for the 77- and 102-nm-thick films, respectively). Further increasing the film thickness to 147 and 215 nm increased $\Delta T_{sol}$ gently to 16.6% and 18.6%, accompanied by an evident depression in $T_{lum}$. Both $T_{lum}$ and $\Delta T_{sol}$ were reduced for the 428-nm-thick film. Therefore, the optimized thickness for films prepared by this system to balance $T_{lum}$ and $\Delta T_{sol}$ is 100 nm. The single-layer film of this thickness shows comparable $T_{lum}$ and $\Delta T_{sol}$ values to those of five-layered $TiO_2/VO_2/TiO_2/VO_2/TiO_2$ films with optically optimized structures ($T_{lum,l}$ = 45%, $T_{lum,h}$ = 42% and $\Delta T_{sol}$ = 12%, from $T_{sol,l}$ = 52% to $T_{sol,h}$ = 40%).[100]

| Sample | Thickness (nm) | $T_{sol,l}$ (%) | $T_{sol,h}$ (%) | $\Delta T_{sol}$ (%) | $T_{lum,l}$ (%) | $T_{lum,h}$ (%) | $\Delta T_{lum}$ (%) |
|---|---|---|---|---|---|---|---|
| I | 43 | 62.1 | 55.7 | 6.4 | 59.2 | 61.1 | -1.9 |
| II | 77 | 48.1 | 41.1 | 7.0 | 45.3 | 47.2 | -1.9 |
| III | 102 | 42.9 | 28.8 | 14.1 | 43.3 | 39.9 | 3.4 |
| IV | 147 | 39.4 | 22.8 | 16.6 | 33.7 | 29.4 | 4.3 |
| V | 215 | 32.9 | 14.3 | 18.6 | 28.7 | 19.2 | 9.5 |
| VI | 428 | 23.0 | 9.2 | 13.8 | 11.5 | 12.6 | -1.1 |

Table 1. Optical properties of typical samples with different thicknesses; *l* and *h* mean low-temperature (20 °C) and high-temperature (90 °C), respectively.[69]

In summary, nanoporous thermochromic $VO_2$ films with low optical constants and tunable thicknesses have been prepared by polymer-assisted deposition. The film porosity and thickness change the interference relationship of light reflected from the film-substrate and the air-film interfaces, strongly influencing the optical properties of these $VO_2$ films. Our optimized single-layered $VO_2$ films exhibit high integrated luminous transmittance ($T_{lum,l}$ = 43.3%, $T_{lum,h}$ = 39.9%) and solar modulation ($\Delta T_{sol}$ = 14.1%, from $T_{sol,l}$ = 42.9% to $T_{sol,h}$ = 28.8%), which are comparable to those of five-layered $TiO_2/VO_2/TiO_2/VO_2/TiO_2$ films ($T_{lum,l}$ = 45%, $T_{lum,h}$ = 42% and $\Delta T_{sol}$ = 12%, from $T_{sol,l}$ = 52% to $T_{sol,h}$ = 40%, from *Phys. Status Solidi A-Appl. Mat.* **2009,** *206,* 2155-2160.). Optical calculations suggest that the performance could be further improved by increasing the porosity.

### 2.5 Optimized optical properties of $VO_2$-based double or multi-layered films

The current techniques used to improve visible transparency mainly include Mg doping,[95] formation of mixtures ($VO_2/SiO_2$),[97] regulation of the thickness of $VO_2$ films[63] and deposition of antireflective layers.[35,52,96] Among these techniques, $VO_2$-based multi-layered structures containing antireflection layers show better optical performance, especially a balance between luminous transmittance ($T_{lum}$) and switching efficiency ($\Delta T_{sol}$). Moreover, the antireflective layer can protect $VO_2$ from oxidation and add new functions such as photocatalysis.[101,102] $\Delta T_{lum}$ represents the improvement of $T_{lum}$ after antireflecion. An integrated improvement of 23 % (from 32 to 55 %) in $T_{lum}$ can be achieved for $VO_2/ZrO_2$ double layers using $ZrO_2$ as an antireflective layer.[52] A $TiO_2/VO_2/TiO_2$ three layer shows $\Delta T_{sol}$ =2.9 % and $\Delta T_{lum}$ = 27% (increased from 31 to 58%).[35] A $TiO_2/VO_2/TiO_2/VO_2/TiO_2$ five layers can improve $\Delta T_{sol}$ to 12.1% (6.7 % for the single $VO_2$ film).[98] All of these films were prepared by gas-phase deposition. And, it seems difficult to improve $\Delta T_{lum}$ and $\Delta T_{sol}$ simultaneously at a higher $\Delta T_{lum}$ level for $VO_2$ films. The $\Delta T_{sol}$ of the $TiO_2/VO_2/TiO_2$ three layer decreased from 3.9 to 2.9 %, while the $T_{lum}$ of the $TiO_2/VO_2/TiO_2/VO_2/TiO_2$ five layer just increased by 4 % (from 41 to 45 %).

The present work was conducted to develop an all-solution preparation process for $VO_2$-based double-layered films. The study began with the optical optimization of the structure of the double-layered films by combining basic interference principles with spectral

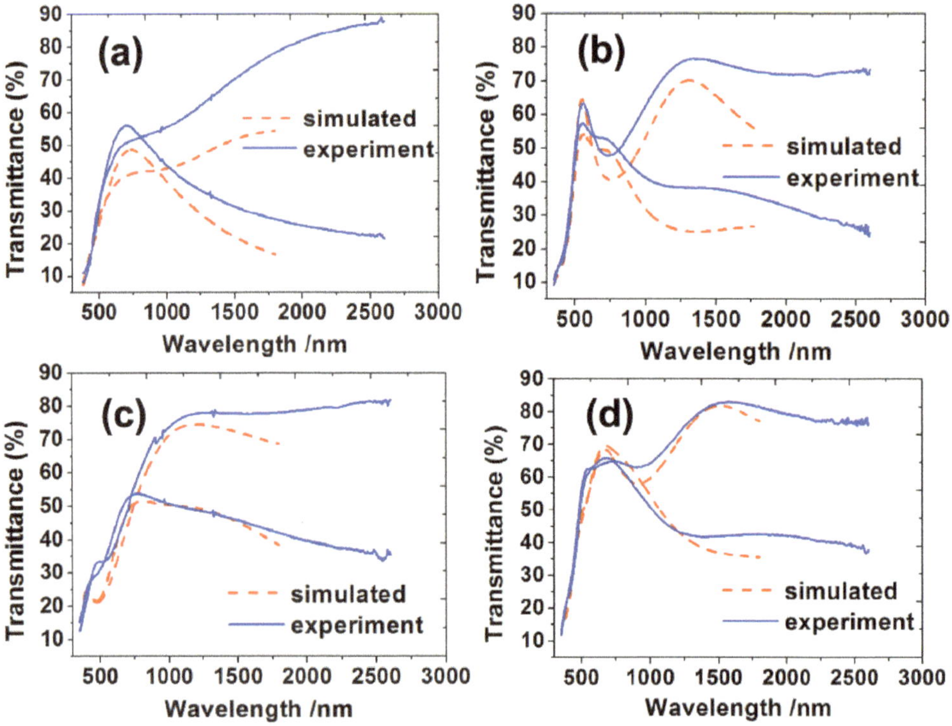


Fig. 23. Simulation and experimental results for various structured $VO_2$ films. (a) single $VO_2$(55 nm); (b) $VO_2$(55 nm)/$TiO_2$(210 nm) quarter-3-quarter-waved double-layer; (c) $VO_2$(55 nm)/$SiO_2$(100 nm)/$TiO_2$(70 nm) quarter-quarter-quarter triple-layer; (d) $VO_2$(55 nm)/$TiO_2$(210 nm)/$SiO_2$(100 nm) quarter-3-quarter-quarter-waved triple-layer. Data used for simulation: RI of $TiO_2$ and $SiO_2$ were 2.5 and 1.45; the thickness for $VO_2$, quarter-waved $TiO_2$, 3-quarter-waved $TiO_2$, and $SiO_2$ was 45, 56, 168, and 98 nm.[71]

simulation, followed by all-solution preparation and investigation of the improvement in $T_{lum}$ and $\Delta T_{sol}$ to confirm the computational predictions. By adding a quarter-waved optical thickness $TiO_2$ film on $VO_2$, the integrated luminous reflectance ($R_{lum}$) of $VO_2$ was reduced dramatically from 31.2 % to 3.0 %, and $T_{lum}$ is close to that of the $TiO_2/VO_2/TiO_2$ layer films.[102] These results are comparable to the films prepared by gas-phase deposition,[4, 102] but $\Delta T_{sol}$ is slightly higher (6.9 % vs. 6.0 %). In addition, methods to improve $\Delta T_{sol}$ while maintaining a high $\Delta T_{lum}$ were explored; the highest $\Delta T_{sol}$ was 15.1 % for optimised double-layered films, which still showed $T_{lum}$ = 49.5 % at 20 °C and 44.8 % at 90 °C.

The simulation results in **Figure** 23 show that the changing trend of wavelength-dependent transmittance agrees well with the experimental results, and are also similar to the results in the reference.[98] These results suggest that although veracity is not high enough by using optical constants of gas-phase-derived $VO_2$ films, the simulation still can give a trend prediction on the optical properties.

To improve $T_{lum}$, the value and position of reflection minima are two key factors. The simulation results show that $VO_2$ can be simplified as a RI-fixed dielectric film. For a double-

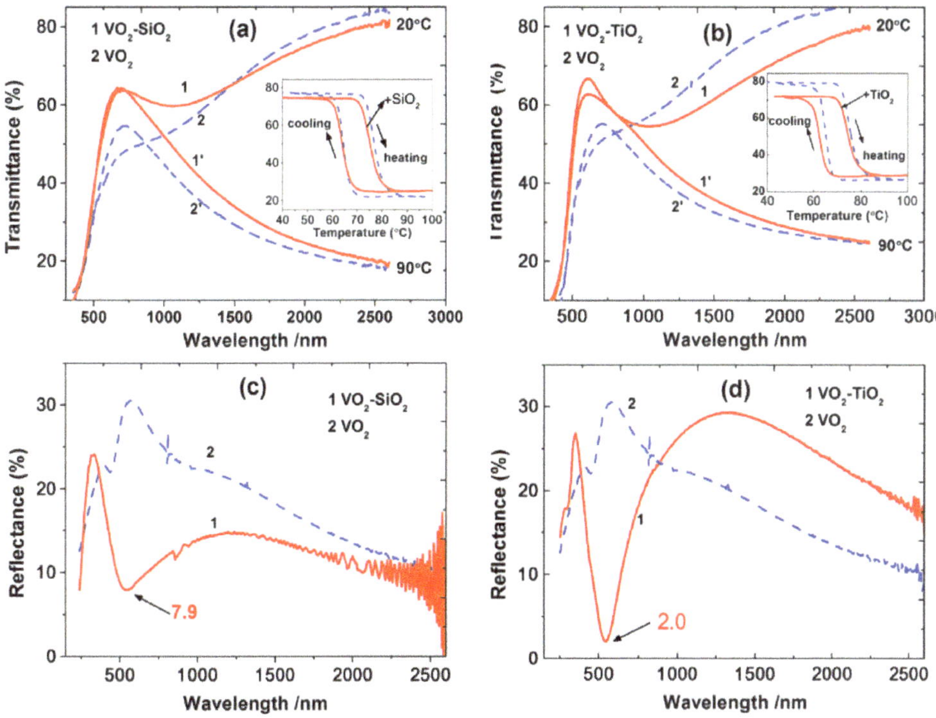


Fig. 24. Transmittance (at 20 and 90 °C) and reflectance (20 °C) spectra before (blue, dotted lines) and after (red, solid lines) adding $SiO_2$ films (a, c: transmittance and reflectance spectra for sample B2, using $SiO_2$ as antireflection layer) or $TiO_2$ films (b, d: transmittance and reflectance spectra for sample B1, using $TiO_2$ as antireflection layer) on $VO_2$ films. Insets (a) and (b) are hysteresis loops.[71]

layered film containing two dielectric layers, reflectance minima can be achieved by adjusting the thickness and RI.[103] **Figure** 24 shows the transmittance and reflectance spectra of $VO_2/SiO_2$ (B2) and $VO_2/TiO_2$ (B1) double-layered films. Hysteresis loops at 2000 nm before and after coating show that the antireflective coatings have little effects on the thermochromism of $VO_2$ films. $TiO_2$ films had a better antireflection efficiency than $SiO_2$ because the RI of $TiO_2$ is closer to an ideal value for $VO_2$.[104] For the simplified $VO_2$-based double-layered structure, the ideal $n_t$ should be 2.14 to achieve a reflection minimum, $n_u$ (2.6) is the RI of $VO_2$ in the middle of the visible wavelength range measured by ellipsometry, and $n_s$ (1.45) is the RI of quartz glass. Theoretical simulation results by optical admittance recursive method showed the minimum reflection at 20 °C appearing at $n_t$=2.10 due to the influence of the extinction coefficient, whereas the minimum $R_{lum}$ (2.8 %, representing luminous reflection) appears at $n_t$=2.04. For $TiO_2$, $n_t$=1.94, the minimum reflectance is 0.5 %, and $R_{lum}$ is 3.0 %; for $SiO_2$, $n_t$=1.43, the minimum reflectance is 8.4 %, and $R_{lum}$ is 13.6 %. Experiments results show a reflectance minimum of sample B2 is 7.9 %, and antireflection peaks appeared at 610 nm (20 °C) and 560 nm (90 °C). The minimum reflectance of sample B1 was about 2 % and 4.2 % with antireflection peaks appearing at 560 nm (20 °C) and 520 nm (90 °C). Note that the reflectance is a sum of the values for the film side and the backside of the substrate, actually, the minimum reflectance of the sample on the film side is about 1 % and $R_{lum}$ is about 3.0 %, close to the prediction. $R_{lum}$ of B1 is even lower than that of $TiO_2/VO_2/TiO_2$ three-layer structure sample. A $VO_2/SiO_2$ double-layered film (sample D, $VO_2$ thickness is 30 nm) showed enhanced performance ($T_{lum}$ of 78.1 and 76.1 % at 20 and 90 °C, respectively). The highest $T_{lum}$ was shown for a $VO_2/TiO_2$ double-layered sample, which had $T_{lum}$ of 84.8 %, but $\Delta T_{sol}$ (0.9 %) is too small to practically apply. The switching efficiency ($\Delta T_{sol}$) is not only a function of transmittance but also solar energy. Most of the solar energy is distributed in the visible (380-780 nm) and short-wave near-

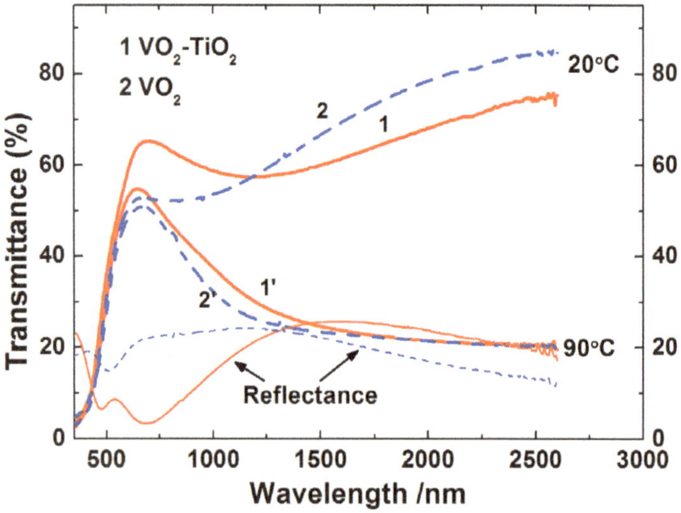


Fig. 25. Transmittance at 20 (1and 2) and 90 °C (1′ and 2′), and reflectance spectra (20 °C) of sample A, (95 nm in thickness) without (blue, dotted lines) and with $TiO_2$ antireflection layer (red, solid lines).[71]

infrared (780-1100 nm) regions. Visible transmittance is almost constant across MIT of $VO_2$ films, so the transmittance difference in the short-wave near-infrared region makes a major contribution to $\Delta T_{sol}$.

One attempt to improve the performance difference at short-wave near-infrared wavelengths is to shift the reflectance minima toward longer wavelengths by regulation of the thickness of antireflective layers. $\Delta T_{lum}$ is caused by the different RI of $VO_2$ films at 20 and 90 °C, which influences the value and position of reflectance minima. The half-quarter-waved structure has two reflection minima with one minimum at longer wavelengths, which can improve $\Delta T_{sol}$. The $VO_2$ film (sample A, 95nm) with a 73nm $TiO_2$ antireflective layer gave two antireflection peaks at 20 °C at about 475 nm and 685 nm (**Figure** 25) and led to an increase in $\Delta T_{sol}$ of 2.0 %. $\Delta T_{sol}$ could be further improved by shifting antireflection peaks to longer wavelengths, but this treatment sacrifices $\Delta T_{lum}$ (reflective minimum departure from the middle of visible wavelength range).

To simultaneously increase $T_{lum}$ and $\Delta T_{sol}$, one way is to design an antireflective layer that can form two antireflection peaks both in the visible wavelength range and the short-wave near-infrared wavelength range. For this purpose, a 3-times quarter-waved thick $TiO_2$ (210 nm) film was prepared. Based on the simplified model (using the optical constants of $VO_2$ at 560 nm at 20 °C) two antireflection peaks (correspondingly, two transmittance maxima) should appear at 1680 nm and 560 nm at 20 °C. Experimentally, because the optical constants of $VO_2$ films are wavelength-dependent and existence of extinction coefficient, the first antireflection peak shifted to 1250 nm (20 °C) and 2400 nm (90 °C) (**Figure** 26). However, the second antireflection peak fell in the middle of the visible range (570 and 540 nm for 20 and 90 °C, respectively), and the transmittance difference in the short-wave near-infrared is enlarged due to different $\Delta T_{lum}$ at 20 and 90 °C. High $\Delta T_{lum}$ values close to those of the quarter-waved samples (17.3 and 11.8 % at 20 and 90 °C) and an improvement of 3.7 % (7.2 to 10.9 %) for $\Delta T_{sol}$ were achieved.

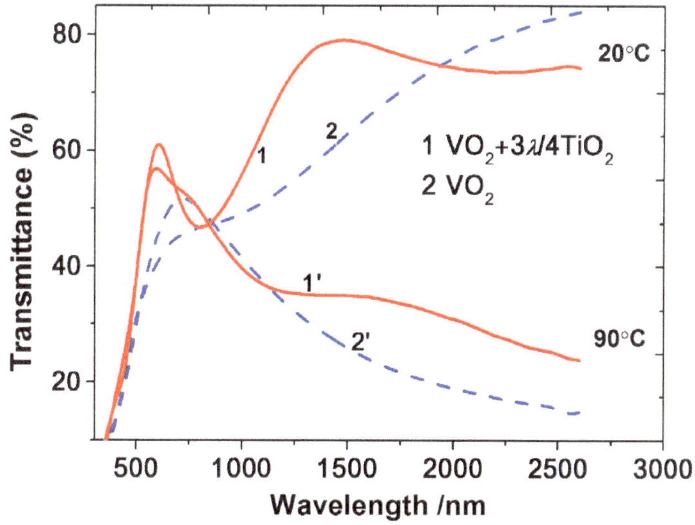


Fig. 26. Transmittance spectra at 20 °C (1, 2) and 90 °C (1′, 2′) of $VO_2$ film (sample B3) after adding a 3 times quarter-wave thick $TiO_2$ ($3\lambda/4$) film (1 and 1′).[71]

This work reveals an impressive improvement on the visible transmittance and switching efficiency, and is an important technical breakthrough toward the practical application of $VO_2$-based smart windows. For a double-layered system, luminous transmittance and switch efficiency could be greatly improved by regulation of the RI and thickness of the films. Single quarter-waved $TiO_2$ films on $VO_2$ could reduce $R_{lum}$ from 31.2 to 4.2 %, improving $T_{lum}$ up to 21.2 % at 20 °C (from 40.3 to 61.5 %). Sample with $T_{lum}$ of 78.1 % was further obtained by optimising the thickness of a $VO_2$ film and adding an antireflective coating, which still had $\Delta T_{sol}$ of 7.5 %. The highest $\Delta T_{sol}$ achieved was 15.1 %. Films with balanced luminous transmittance ($T_{lum}$=58.0 and 53.9 % at 20 and 90 °C, respectively) and switching efficiency (10.9 %) were prepared.

## 3. Summary remarks and outlook

This chapter introduces a solution method, polymer assisted deposition process, for the preparation of $VO_2$ and $VO_2$-based multilayered films. The method enables us to facially control over the film thickness, morphology and optical constants. By combining the optical design in respect to the materials selection and/or their thickness and microstructural control, we obtained $VO_2$ films with high visible transmittance (40-84%), controllable Mott phase transition temperatures and high switching efficiencies (max. 15.1%). The results show that the current solution process is a powerful competitor towards practical applications of this material.

## 4. Acknowledgements

This study was supported in part by the Century Program (One-Hundred-Talent Program) of the Chinese Academy of Sciences, National Key Basic Research Project (NKBRP, 2009CB939904), the National Natural Science Foundation of China (NSFC, Contract No: 50772126, 51172265, 51032008), Shanghai Key Basic Research Project (09DJ1400200), Shanghai Basic Research Project (08JC1420300), The key project of the Chinese Academy Sciences (Grant. No. 4912009YC006), the Opening Project of Xinjiang Key Laboratory of Electronic Information Materials and Devices, and Shanghai Talent Project of Science and Technology (Pujiang Talent Program, 09PJ1410700).

## 5. References

[1] Parkin, I. P.; Manning, T. D. *J. Chem. Educ.* 2006, *83*, 393.
[2] Irie, M. *Chem. Rev.* 2000, *100*, 1685.
[3] Maaza, M.; Bouziane, K.; Maritz, J.; McLachlan, D. S.; Swanepool, R.; Frigerio, J. M.; Every, M. *Opt. Mate.* 2000, *15*, 41.
[4] Jin, P.; Xu, G.; Tazawa, M.; Yoshimura, K. *Jpn. J. Appl. Phys.* 2002, *41*, L278.
[5] Mlyuka, N. R.; Niklasson, G. A.; Granqvist, C. G. *Sol. Energ. Mat. Sol. Cells.* 2009, *93*, 1685.
[6] Lee, M. J.; Park, Y.; Suh, D. S.; Lee, E. H.; Seo, S.; Kim, D. C.; Jung, R.; Kang, B. S.; Ahn, S. E.; Lee, C. B.; Seo, D. H.; Cha, Y. K.; Yoo, I. K.; Kim, J. S.; Park, B. H. Adv. Mater. 2007, *19*, 3919.
[7] Baek, I. G.; Kim, D. C.; Lee, M. J.; Kim, H. J.; Yim, E. K.; Lee, M. S.; Lee, J. E.; Ahn, S. E.; Seo, S.; Lee, J. H.; Park, J. C.; Cha, Y. K.; Park, S. O.; Kim, H. S.; Yoo, I. K.; Chung, U.

I.; Moon, J. T.; Ryu, B. I.; Ieee. In *Ieee International Electron Devices Meeting 2005, Technical Digest*, 2005; 769.
[8] Morin, F. J.; *Phys. Rev. Lett.* 3 (1959) 34-36.
[9] Manning, T. D.; Parkin, I. P.; Clark, R. J. H.; Sheel, D.; Pemble, M. E.; Vernadou, D.; *J. Mater. Chem.* 12 (2002) 2936-2939.
[10] Cavalleri, A.; Tóth, C.; Siders, C. W.; Squier, J. A.; Ráksi, F.; Forget, P.; Kieffer, J. C.; *Phys. Rev. Lett.* 87 (2001) 237401,1-4.
[11] Baum, P.; Yang, D. S.; Zewail, A. H. *Science* 2007, *318*, 788.
[12] Lysenko, S.; Rua, A.; Vikhnin, V.; Fernandez, F.; Liu, H. *Phys. Rev. B* 2007, *76*.
[13] Cavalleri, A.; Rini, M.; Chong, H. H. W.; Fourmaux, S.; Glover, T. E.; Heimann, P. A.; Kieffer, J. C.; Schoenlein, R. W. *Phys. Rev. Lett.* 2005, *95*, 067405.
[14] Becker, M. F.; Buckman, A. B.; Walser, R. M.; Lepine, T.; Georges, P.; Brun, A. *J. App. Phys.* 1996, *79*, 2404.
[15] Beteille, F.; Livage, J. *J. Sol-Gel Sci. Technol.* 1998, *13*, 915.
[16] Guzman, G.; Beteille, F.; Morineau, R.; Livage, J. *J. Mater. Chem.* 1996, *6*, 505.
[17] Livage, J. *Solid State Ionics* 1996, *86 (8)*, 935-942.
[18] Nag, J.; Haglund, R. F. *Journal of Physics-Condensed Matter* 2008, *20*, 264016.
[19] Balu, R.; Ashrit, P. V. *Applied Physics Letters* 2008, *92*, 021904.
[20] Andersson.G. *Acta Chem. Scand.* 1954, *8*, 1599.
[21] Andersson, G. *Acta Chem. Scand.* 1956, *10*, 653.
[22] Matsuishi.T. *Jpn. J. Appl. Phys* 1967, *6*, 1060.
[23] Oka, Y.; Yao, T.; Sato, S.; Yamamoto, N. *J. Solid State Chem* 1998, *140*, 219.
[24] Théobald, F. *J. Less-CommonMet* 1977, *53*, 55.
[25] Galy, J. *J.Solid State Chem* 1999, *148*, 224.
[26] Oka, Y. *J. Solid State Chem.* 1990, *86*, 116.
[27] Hagrman, D.; Zubieta, J.; Warren, C. J.; Meyer, L. M.; Treacy, M. M. J.; Haushalter, R. C. *J.Solid State Chem* 1998, *138*, 178.
[28] Gui, Z.; Fan, R.; Chen, X. H.; Wu, Y. C. *J. Solid State Chem* 2001, *157*, 250.
[29] Liu, C. H.; Zapien, J. A.; Yao, Y.; Meng, X. M.; Lee, C. S.; Fan, S. S.; Lifshitz, Y.; Lee, S. T. *Adv Mater* 2003, *15*, 838.
[30] Qazilbash, M. M.; Schafgans, A. A.; Burch, K. S.; Yun, S. J.; Chae, B. G.; Kim, B. J.; Kim, H. T.; Basov, D. N. *Phys. Rev. B* 2008, *77*, 115121.
[31] Verleur, H. W.; Barker, A. S.; Berglund, C. N. *Phys. Rev. Lett* 1968, *172*, 788.
[32] Manning, T. D.; Parkin, I. P.; Pemble, M. E.; Sheel, D.; Vernardou, D. *Chem. Mater* 2004, *16*, 744-749.
[33] Burkhardt, W.; Christmann, T.; Franke, S.; Kriegseis, W.; Meister, D.; Meyer, B. K.; Niessner, W.; Schalch, D.; Scharmann, A. *Thin Solid Films* 2002, *402*, 226-231.
[34] Jin, P.; Tanemura, S. *Jpn. J. Appl. Phys., Part 1* 1994, *33*, 1478-1483.
[35] Jin, P.; Xu, G.; Tazawa, M.; Yoshimura, K. *Appl. Phys. A* 2003, *77*, 455-459.
[36] Binions, R.; Hyett, G.; Piccirillo, C.; Parkin, I. P. *J. Mater. Chem* 2007, *17*, 4652-4660.
[37] Takahashi, I.; Hibino, M.; Kudo, T. *Jpn. J. Appl. Phys., Part 1* 2001, *40*, 1391-1395.
[38] Manning, T. D.; Parkin, I. P. *J. Mater. Chem* 2004, *14*, 2554-2559.
[39] Vernardou, D.; Pemble, M. E.; Sheel, D. W. *Chem. Vap. Deposition* 2007, *13*, 158-162.
[40] Sobhan, M. A.; Kivaisi, R. T.; Stjerna, B.; Granqvist, C. G. *Sol. Energy Mater. Sol. Cells* 1996, *44*, 451-455.

[41] Blackman, C. S.; Piccirillo, C.; Binions, R.; Parkin, I. P. *Thin Solid Films* 2009, *517*, 4565-4570.

[42] Guinneton, F.; Sauques, L.; Valmalette, J. C.; Cros, F.; Gavarri, J. R. *J. Phys. Chem. Solids* 2001, *62*, 1229-1238.

[43] (a) Kang, L. T.; Gao, Y. F.; Luo, H. J. *ACS Appl. Mater. Interfaces* 2009, *1*, 2211-2218. (b) Kang, L. T.; Gao, Y. F.; Luo, H. *Materials Science Forum*, 2011, 687, 791

[44] Kang, L. T.; Gao, Y. F.; Zhang Z. T.; Du J.; Cao, C. X.; Chen, Z.; Luo, H. J. *J. Phys Chem C.* 2010, *114 (4)*, 1901–1911.

[45] Guinneton, F.; Sauques, L.; Valmalette, J. C.; Cros, F.; Gavarri, J. R. *Thin Solid Films* 2004, *446*, 287-295.

[46] Xu, G.; Jin, P.; Tazawa, M.; Yoshimura, K. *J. Appl. Phys., Part 1* 2004, *43*, 186-187.

[47] Chen, B.; Yang, D. F.; Charpentier, P. A.; Zeman, M. *Sol. Energy Mater. Sol. Cells* 2009, *93*, 1550-1554.

[48] Guinneton, F.; Sauques, L.; Valmalette, J. C.; Cros, F.; Gavarri, J. R. *J. Phys. Chem. Solids* 2005, *66*, 63-73.

[49] Brassard, D.; Fourmaux, S.; Jean-Jacques, M.; Kieffer, J. C.; El Khakani, M. A. *Appl. Phys. Lett* 2005, *87*, 051910

[50] Griffiths, C.H.; Eastwood, H. K. *J. Appl. Phys* 1974, 2201-2206.

[51] Kusano, E.; Theil, J. A. *J. Vac. Sci. Technol., A* 1989, *7*, 1314-1317.

[52] Xu, G.; Jin, P.; Tazawa, M.; Yoshimura, K. *Sol. Energy Mater. Sol. Cells* 2004, *83*, 29-37.

[53] Lappalainen, J.; Heinilehto, S.; Saukko, S.; Lantto, W.; Jantunen, H. *Sens. Actuators, A* 2008, *142*, 250-255.

[54] Chain, E.E.; *Appl. Opt.*, 1991, 30, 2782-2787.

[55] Choi, H.S.; Ahn, J.S.; Jung, J.H.; Noh, T.W.; Kim, D.H.; *Phys. Rev. B: Condens. Matter Mater. Phys*, 1996, 54, 4621-4628.

[56] Lopez, R.; Boatner, L.A.; Haynes, T.E.; Feldman, L.C.; Haglund, R.F.; *J. Appl. Phys.*, 2002, 92, 4031-4036.

[57] Narayan, J.; Bhosle, V.M.; *J. Appl. Phys.*, 2006, 100, 103524.

[58] Petit, C.; Frigerio, J.M.; Goldmann, M.; *J. Phys.: Condens. Matter*, 1999, 11, 3259-3264.

[59] Kakiuchida, H.; Jin, P.; Tazawa, M.; *Sol. Energy Mater. Sol. Cells*, 2008, 92, 1279-1284.

[60] Cao, C.; Gao, Y.-F.; Luo, H.J.; *J. Phys. Chem. C*, 2008, 112 (48), 18810–18814.

[61] Bai,L. Y.; Gao, Y. F.; Li , W. D.; Luo, H. J.; Jin, P.; *J. Ceram. Soc. Japan*, 2008, 116(3), 395-399 .

[62] Zhang, Z.; Gao, Y. F.; Chen, Z.; Du, J.; Cao, C.; Kang L. T.; Luo, H.J.; *Langmuir* , 2010, 26 (13), 10738–10744.

[63] Kang, L. T.; Gao, Y. F.; Chen, Z.; Du, J.; Zhang Z.; Luo, H.J.; *Sol. Energy Mater. Sol. Cells*, 2010, 94, 2078–2084.

[64] Du, J.; Gao, Y. F.; Luo, H. J.; Kang, L. T.; Chen, Z.; Zhang Z.; Cao, C.; *Sol. Energy Mater. Sol. Cells*, 2011, 95 (2), 469-475.

[65] Cao, C.; Gao, Y. F.; Kang, L. T.; Luo, H. J.; *CrystEngComm*, 2010, 12 (12), 4048 – 4051.

[66] Dai, L.; C.; Gao, Y. F.; Luo, H. J.; *Sol. Energy Mater. Sol. Cells*, 2011, 95 (2), 712-715.

[67] Zhang, Z.; Gao, Y. F.; Kang L. T.; Du, J.; Luo, H.J.; *J. Phys. Chem. C*, 2010, *114*, 22214–22220.

[68] Du, J.; Gao, Y. F.; Luo, H. J.; Zhang Z.; Kang, L. T.; Chen, Z.; *Sol. Energy Mater. Sol. Cells*, 2011, 95, 1604-1609.

[69] Kang, L. T.; Gao, Y. F.; Chen, Z.; Du, J.; Zhang Z.; Luo, H.J.; *ACS Applied Materials & Interfaces*, 2011, *3(2)*, 135-138.
[70] Ni, J.; Jiang, W.; Yu, K.; Gao, Y. F.; Zhu, Z.; *Electrochimica Acta*, 2011, 56, 2122-2126.
[71] Chen, Z.; Gao, Y. F.; Kang, L. T.; Du, J.; Zhang Z.; Luo, H. J.; Miao, H.; Tan, G.; *Sol. Energy Mater. Sol. Cells*, 2011, 95(9), 2677-2684.
[72] Kang, L. T.; Gao, Y. F.; Luo, H.J.; Wang, J.; Zhu, B.; Zhang Z.; Du, J.; Kanehira, M.; Zhang, Y.; Thermochromic Properties and Low Emissivity of ZnO:Al/$VO_2$ Double-Layered Films with a Lowered Phase Transition Temperature, *Sol. Energy Mater. Sol. Cells*, in press.
[73] Yuan, X. P.; Li, C. C.; Guan, G. H.; Xiao, Y. N.; Zhang, D. *J. Appl. Polym. Sci.* 2009, 111, 566–575.
[74] Wang, Z. D.; Xu, Z. L.; Yang, Y. G.; Li, S. X. *Mater. Rev. (Chongqing, P. R. China) Special Issue VII*, 2006, 20, 53–57 (in Chinese).
[75] Hao, C. W.; Zhao, Y.; Zhou, Y.; Zhou, L. J.; Xu, Y. Z.; Wang, D. J.; Xu, D. F. *J. Polym. Sci., Part B: Polym. Phys.* 2007, 45, 1589–1598.
[76] Kim, J. H.; Kim, C. K.; Won, J.; Kang, Y. S. *J. Membr. Sci.* 2005, 250, 207–214.
[77] Kim, J. H.; Min, B. R.; Won, J.; Kang, Y. S. *Chem.sEur. J.* 2002, 8, 650–654.
[78] Hong, S. U.; Jin, J. H.; Won, J.; Kang, Y. S. *Adv. Mater.* 2000, 12, 968–971.
[79] Kim, H. T.; Chae, B. G.; Youn, D. H.; Kim, G.; Kang, K. Y.; Lee, S. J.; Kim, K.; Lim, Y. S. *Appl. Phys. Lett.* 2005, 86, 242101.
[80] Petrov, G. I.; Yakovlev, V. V.; Squier, J. *Appl. Phys. Lett.* 2002, 81, 1023–1025.
[81] Shi, J. Q.; Zhou, S. X.; You, B.; Wu, L. M. *Sol. Energy Mater. Sol. Cells* 2007, 91, 1856–1862.
[82] Lopez, R.; Haynes, T. E.; Boatner, L. A.; Feldman, L. C.; Haglund, R. F. *Phys. Rev. B* 2002, 65, 224113.
[83] Suh, J. Y.; Lopez, R.; Feldman, L. C.; Haglund, R. F. *J. Appl. Phys.* 2004, 96, 1209–1213.
[84] Klimov, V. A.; Timofeeva, I. O.; Khanin, S. D.; Shadrin, E. B.; Ilinskii, A. V.; Silva-Andrade, F. *Tech. Phys.* 2002, 47 (9), 1134–1139.
[85] Frenzel, A.; Qazilbash, M. M.; Brehm, M.; Chae, B. G.; Kim, B. J.; Kim, H. T.; Balatsky, A. V.; Keilmann, F.; Basov, D. N. *Phys. ReV. B* 2009, 80 (11), 115115.
[86] Aliev, R. A.; Klimov, V. A. *Phys. Solid State* 2004, 46 (3), 532- 536.
[87] Aliev, R.; Andreev, V.; Kapralova, V.; Klimov, V.; Sobolev, A.; Shadrin, E. *Phys. Solid State* 2006, 48 (5), 929–934.
[88] Whittaker, L.; Jaye, C.; Fu, Z. G.; Fischer, D. A.; Banerjee, S. *J. Am. Chem. Soc.* 2009, 131 (25), 8884–8894.
[89] Donev, E. U.; Lopez, R.; Feldmant, L. C.; Haglund, R. F. *Nano Lett.* 2009, 9 (2), 702–706.
[90] Muraoka, Y.; Hiroi, Z. *Appl. Phys. Lett.* 2002, 80 (4), 583–585.
[91] Jia, Q. X.; McCleskey, T. M.; Burrell, A. K.; Lin, Y.; Collis, G. E.; Wang, H.; Li, A. D. Q.; Foltyn, S. R. *Nat. Mater.* 2004, 3 (8), 529–532.
[92] Kingery, W. D. *Introduction To Ceramics;* John Wiley and Sons, Inc.: New York, 1960.
[93] Lopez, R.; Boatner, L. A.; Haynes, T. E.; Haglund, R. F.; Feldman, L. C. *Appl. Phys. Lett.* 2001, 79 (19), 3161–3163.
[94] Sidorov, A. I.; Vinogradova, O. P.; Obyknovennaya, I. E.; Khrushchova, T. A. *Tech. Phys. Lett.* 2007, 33 (7), 581–582.
[95] Mlyuka, N.R.; Niklasson, G.A.; Granqvist, C.G. *Appl. Phys. Lett.*, 2009, 95, 171909.
[96] Lee, M.H. Cho, J.S. *Thin Solid Films*, 2000, 365, 5-6.

[97] H.-K. Chen, H.-C. Hung, T.C.K. Yang, S.-F. Wang, *J. Non-Cryst. Solids*, 2004, 347, 138-143.

[98] N.R. Mlyuka, G.A. Niklasson, C.G. Granqvist, *Physica Status Solidi a-Applications and Materials Science*, 2009, 206, 2155-2160.

[99] Li, S.-Y.; Niklasson, G. A.; Granqvist, C. G. *J. Appl. Phys.* 2010, 108, 063525.

[100] Lu, C. L.; Cheng, Y. R.; Liu, Y. F.; Liu, F.; Yang, B. *Adv. Mater.* 2006, 18, 1188–1192.

[101] P. Evans, M.E. Pemble, D.W. Sheel, H.M. Yates, *J. Photochem. Photobiol. A-Chem.*, 2007, 189, 387-397.

[102] Saitzek, S.; Guinneton, F.; Sauques, L.; Aguir, K.; Gavarri, J.R.; *Opt. Mater.*, 2007, 30, 407-415

[103] Heavens, O.S. *Optical Properties of Thin Solid Films*, Courier Dover Publications, 1991.

[104] Babulanam, S.M.; Eriksson, T.S.; Niklasson, G.A.; Granqvist, C.G. *Solar Energy Materials*, 1987, 16, 347-363.

# 6

# Chemical Nanomanipulation of Two-Dimensional Nanosheets and Its Applications

Minoru Osada and Takayoshi Sasaki
*International Center for Materials Nanoarchitectonics (MANA), National Institute for Materials Science (NIMS), Tsukuba, Ibaraki, Japan*

## 1. Introduction

Two-dimensional (2D) nanosheets obtained *via* exfoliation of layered compounds have attracted intensive research in recent years, opening up new fields in the science and technology of 2D nanomaterials.[1-6] These 2D nanosheets, which possess atomic or molecular thickness and infinite planar dimensions, are emerging as important new materials due to their unique properties. Research in such exotic 2D systems recently intensified as a result of emerging progress in graphene (carbon nanosheet)[1, 2] and novel functionalities in oxide nanosheets.[3-5] In particular, oxide nanosheets are exceptionally rich in both structural diversity and electronic properties, with potential application in areas ranging from catalysis to electronics. Now, by using the exfoliation approach, it is possible to investigate dozens of different 2D oxide nanosheets in search of new phenomena and applications.

One of the important and attractive aspects of the exfoliated nanosheets is that various nanostructures can be fabricated using them as 2D building blocks.[7-18] It is even possible to tailor superlattice-like assemblies, incorporating into the nanosheet galleries a wide range of materials such as organic molecules, polymers, and inorganic and metal nanoparticles. Sophisticated functionalities or nanodevices may be designed through the selection of nanosheets and combining materials, and precise control over their arrangement at the molecular scale.

In this chapter, we review the current research on oxide nanosheets. Our particular focus is placed on recent progress that has been made in the synthesis and properties of oxide nanosheets, highlighting emerging functionalities in electronic applications.

## 2. Synthesis of oxide nanosheets

Various nanosheets based on transition-metal oxides have been synthesized by delaminating the precursor crystals of layered oxide into their elemental layers (Table 1).

Chemical exfoliation is the most facile route for making isolation of single layers (oxide nanosheets) separately from thicker layered compounds (Fig. 1). These procedures have attracted much attention as an efficient method for preparing single layers with lateral sizes of up to several micrometers. Pioneering works in this line appeared in the 1990s by Sasaki

| | | |
|---|---|---|
| Ti oxide | $Ti_{0.91}O_2$, $Ti_{0.87}O_2$, $Ti_{0.8}Co_{0.2}O_2$, $Ti_{0.6}Fe_{0.4}O_2$, $Ti_{(5.2-2x)/6}Mn_{x/2}O_2$ ($0 \le x \le 0.4$) $Ti_{0.8-x/4}Fe_{x/2}Co_{0.2-x/4}O_2$ ($0 \le x \le 0.8$) | Semiconducting, Photocatalytic, Dielectric, Ferromagnetic |
| Mn oixde | $MnO_2$ | Redoxable |
| Nb/Ti and Ta oxide | $Nb_6O_{17}$, $Nb_3O_8$, $TiNbO_5$, $Ti_2NbO_7$, $Ti_5NbO_{14}$, $TaO_3$ | Photocatalytic, Dielectric |
| Perovksite | $Gd_{1.4}Eu_{0.6}Ti_3O_{10}$, $Bi_4Ti_3O_{12}$, $LaNb_2O_7$, $La_{0.9}Eu_{0.05}Nb_2O_7$, $SrTa_2O_7$, $La_{0.7}Tb_{0.3}Ta_2O_7$, $Eu_{0.56}Ta_2O_7$, $Ca_2Nb_3O_{10}$, $Sr_2Nb_3O_{10}$, $Ca_2Ta_3O_{10}$, $Sr_2Ta_3O_{10}$ | Photocatalytic, Dielectric, Photoluminescence |
| Mo oxide | $MoO_2$ | Conducting |
| Ru oxide | $RuO_{2.1}$, $RuO_2$ | Redoxable, Conducting |
| W oxide | $W_2O_7$, $Cs_4W_{11}O_{36}$ | Redoxable Photochromic |

Table 1. Library of oxide nanosheets

*et al.*,[19, 20] reporting the successful delamination of layered titanates into single titanate nanosheets. Prompted by the findings related to functional oxide nanosheets, several strategies on functional oxides can be found in the literature.

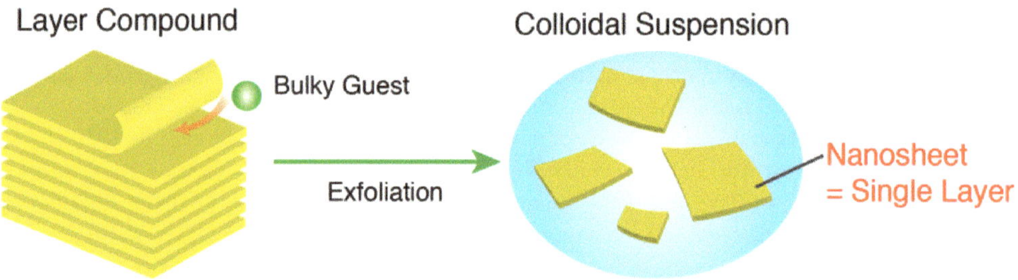


Fig. 1. Schematic illustration for the exfoliation of a layer compound into nanosheets.

In the case of metal oxides, protonation usually resulted in electrostatic repulsions that facilitated exfoliation. By this procedure, single layers of Ti oxides,[19-22] Mn oxides,[23] Nb/Ta oxides,[24-26] Mo oxides,[27] Ru oxides,[28] and W oxides,[29] as well as sheets of several perovskites[7, 9, 30-35] have been separated from bulk samples (Fig. 2).

In these cases, a chemical intercalator that assists the separation of layers and hampers the reassembly of the bulk lamellar material is always required. Tetrabutylammonium (TBA) is the most commonly used intercalator, but also tetrametylammonium and ethylammonium have been used successfully for these purposes. Layered transition-metal oxides such as $Cs_{0.7}Ti_{1.825}\Upsilon_{0.175}O_4$ ($\Upsilon$: vacancy), $K_{0.45}MnO_2$, and $KCa_2Nb_3O_{10}$ can be used as the starting material for the nanosheet.[19, 20, 23, 31] A common feature of these host compounds is cation-exchange properties involving interlayer alkali metal ions, which are a key to facilitating exfoliation. As the first step to delamination, these layered materials are acid-exchanged into protonated forms such as $H_{0.7}Ti_{1.825}\Upsilon_{0.175}O_4 \bullet H_2O$, $H_{0.13}MnO_2 \bullet 0.7H_2O$, and

$HCa_2Nb_3O_{10}\cdot 1.5H_2O$, in which the interlayer alkali metal ions can be completely removed under suitable conditions while maintaining the layered structure. The resulting protonic oxides are subsequently delaminated through reaction with a solution containing TBA ions, producing turbid colloidal suspensions of $Ti_{0.91}O_2$, $MnO_2$, and $Ca_2Nb_3O_{10}$ nanosheets. Such an exfoliation process is quite general: exfoliation of the other layered host compounds proceeds in a similar fashion.

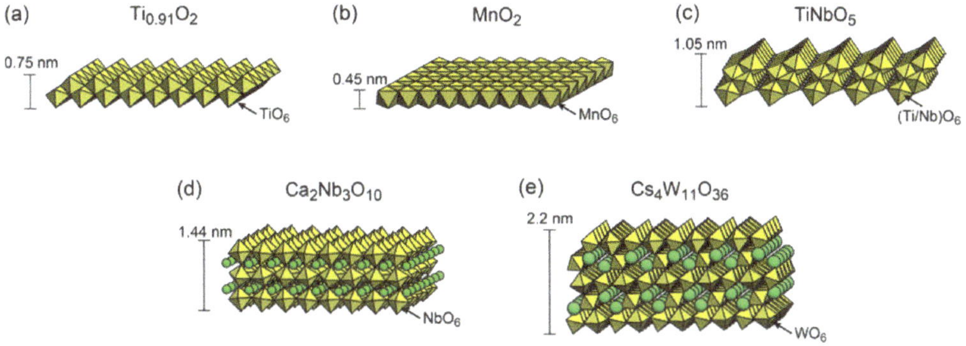


Fig. 2. Strucutres of selected oxide nanosheets. (a) $Ti_{0.91}O_2$, (b) $MnO_2$, (c) $TiNbO_5$, (d) $Ca_2Nb_3O_{10}$, (e) $Cs_4W_{11}O_{36}$.

These materials have prompted many efforts to elucidate their structural properties. The formation of unilamellar nanosheets was confirmed by direct observation with atomic force microscopy (AFM), x-ray diffraction (XRD), and transmission electron microscopy (TEM).[36-40] Fig. 3 depicts an AFM image for $Ti_{0.87}O_2$ nanosheets. The AFM data clearly reveals a sheet-like morphology, which is inherent to the host layer in the parent compounds. The average thickness was 0.93 ± 0.1 nm. This value is nearly comparable to the crystallographic thickness of the host layer in the corresponding parent compound, supporting the formation of unilamellar nanosheets. On the other hand, the lateral size depends on the choice of starting materials. For nanosheets derived from polycrystalline powder samples, the lateral size ranges from submicrometers to several tens of micrometers. After tuning the exfoliation conditions by using flux-grown single crystals, the technique provides high-quality nanosheet crystallites up to ~100 μm in size, which is suitable for electronic applications.[21]

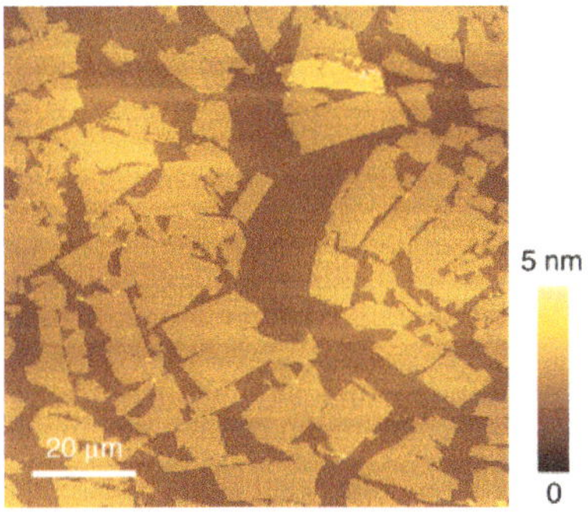


Fig. 3. AFM image of $Ti_{0.87}O_2$ nanosheets dispersed on a Si substrate.

## 3. Chemical nanomanipulation of oxide nanosheets

Oxide nanosheets are an important and promising component for creating new materials. Oxide nanosheets have a high 2D anisotropy of the crystallites: thickness is ~1 nm wheras lateral size ranges from submicrometers to ~100 μm. In addition, these nanosheets are obtained as negatively charged crystallites that are dispersed in a colloidal suspension. These aspects make the nanosheets a suitable building block for designing nanostructured films. In practice, colloidal nanosheets can be organized into various nanostructures or combined with a range of foreign materials at the nanometer scale by applying wet-process synthetic techniques involving flocculation and layer-by-layer (LbL) self-assembly. Through these processes, oxide nanosheets can be combined with a wide range of polyions such as organic polyelectrolytes, metal complexes, clusters and even oppositely charged nanosheets, which is a major advantage of this approach. Furthermore, control of particulate shape as thin flakes and hollow spheres has been achieved through freeze- or spray-drying techniques.

One of the highlights is the fabrication of nanocomposite films of organic polymer/nanosheet materials that exhibit useful properties. Several groups have demonstrated that the electrostatic LbL self-assembly *via* sequential adsorption and Langmuir-Blodgett (LB) procedure are effective for this purpose (Fig. 4).

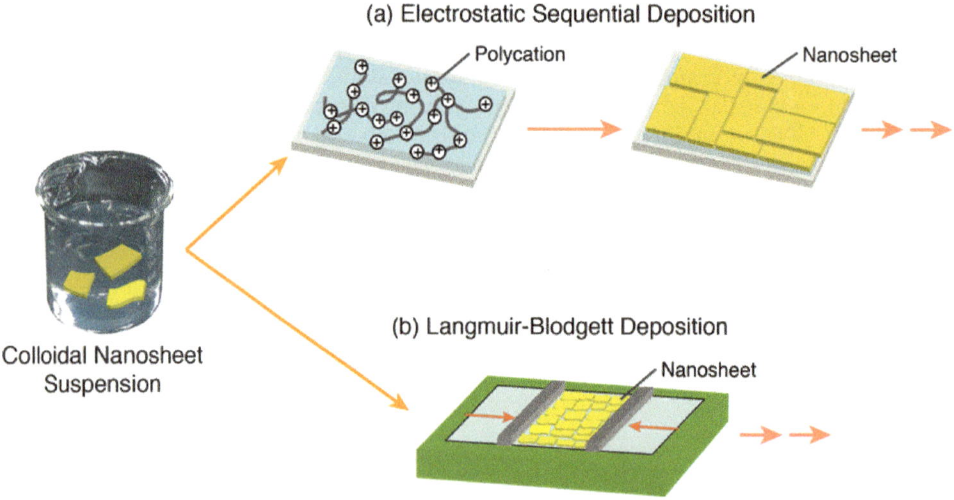


Fig. 4. Schematic illustration for chemical nanomanipulation of oxide nanosheets. (a) Electrostatic sequential deposition and (b) Langmuir-Blodgett deposition.

Electrostatic sequential deposition is one of the most powerful methods of fabricating nanostructured multilayer films with precisely controlled composition, thickness and architecture on a nanometer scale [Fig. 4(a)]. This technique, often called "molecular beaker epitaxy", has been first developed by Decher[41] and applied to various charged materials. In this LbL process, a multilayer assembly can be built up by alternately dipping the substrate in a colloidal suspension of nanosheets and an aqueous solution of suitable polyelectroytes. Polycations such as poly (diallyldimethylammonium chloride) (PDDA) and poly (ethylenimine) are usually used as a counterpart of the oxide nanosheets. Fig. 5(a) depicts an example of the multilayer film of $(PDDA/Ti_{0.91}O_2)_{10}$ on a quartz glass substrate, showing UV-visible absorption spectra in the fabrication process. The absorption peak at 265 nm, attributable to the $Ti_{0.91}O_2$ nanosheets, was progressively enhanced as the number of deposition cycles increased, clearly indicating the repeated adsorption of nearly equal amounts of nanosheets. XRD data provides further evidence for the formation of multilayer films by the evolution of Bragg peaks and their progressive enhancement. Other characterizations by ellipsometry, FT-IR and AFM all support the growth of multilayer nanocomposite films. Multilayer films of other nanosheets were fabricated by similar procedures.

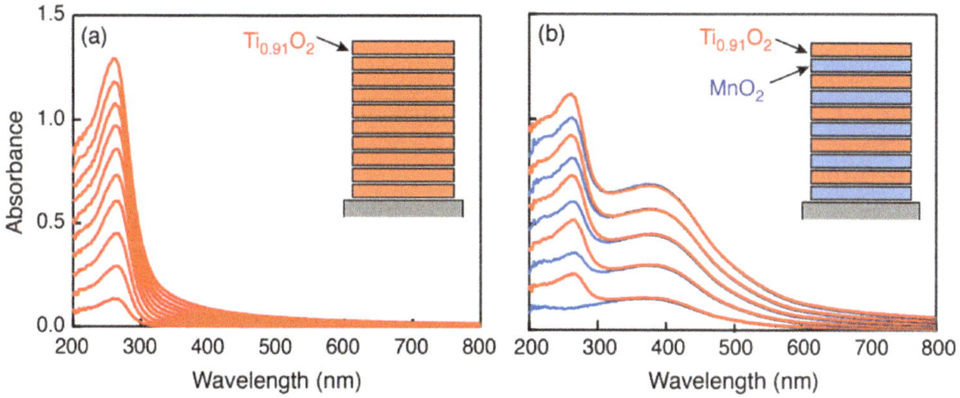


Fig. 5. UV-visible absorption spectra in the multilayer buildup processes for $(Ti_{0.91}O_2/PDDA)_{10}$ (a) and $(PDDA/Ti_{0.91}O_2/PDDA/MnO_2)_{10}$ (b). The insets indicate the designed stacked structures of the nanosheets.

Such LbL assembly of various nanosheets also allows us to tailor superlattices or heterostructures by tuning the number of nanosheets and their stacking sequences. Fig. 5(b) shows UV-visible absorption spectra for the superlattice assembly composed of $MnO_2$ and $Ti_{0.91}O_2$ nanosheets. The observed spectral changes clearly indicate that the films grew as designed. The superlattice approach makes it possible to design complex functions that cannot be achieved using a single material.

LB deposition has been proved much simple and effective as another approach for organizing 2D nanosheets [Fig. 4(b)].[42-45] LB film deposition, the formation of a floating monolayer on water surface in a Langmuir trough followed by an appropriate level of compression, is preferable for achieving dense packing or neat tiling. Through vertical-dipping/lifting, the monolayer is deposited onto a flat substrate in LbL fashion. Pioneering work[42] has demonstrated that exfoliated nanosheets could float by adhering to amphiphilic ammonium cations at the air/water interface through electrostatic interaction, and thus the ordinary LB procedure is applicable for fabricating nanosheet films. Although LB technique has been used for decades, its application for nanoparticles and nanorods is often frustrated by defects ranging from pinholes to larger reorganization of the layers. In the case of nanosheets, the LB technique provides nearly perfect mono- and multilayer films with atomically flat surfaces. The LB-based LbL approach with the use of an atomically flat substrate is effective for fabricating atomically uniform and highly dense nanofilms of oxide nanosheets. Fig. 6 shows a cross-sectional high-resolution TEM image of a 5-layer (7.5 nm thick) $Ca_2Nb_3O_{10}$ film on a $SrRuO_3$ substrate.[46] The image clearly reveals a stacking structure corresponding to the LbL assembly of nanosheets. Such LB-deposited nanofilms are very suitable for a number of applications in electronic devices.

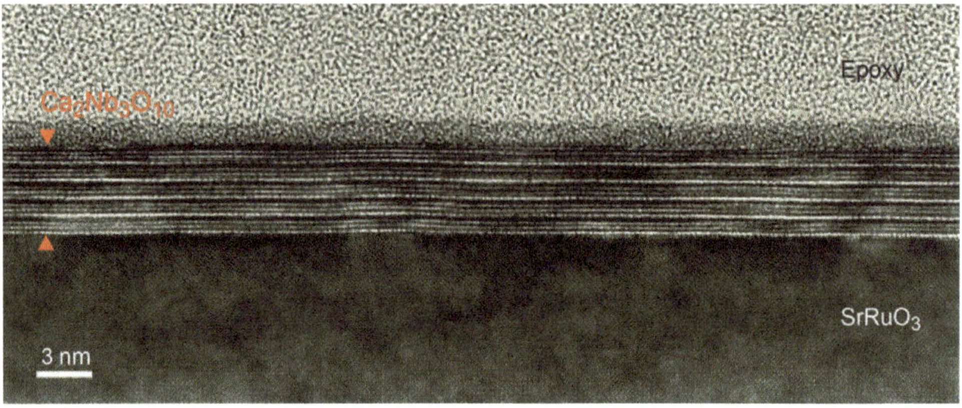


Fig. 6. Cross-sectional high-resolution TEM image of a 5-layer (7.5 nm thick) $Ca_2Nb_3O_{10}$ film on a $SrRuO_3$ substrate. Note that the film/substrate interface is atomically flat without an interfacial layer between $Ca_2Nb_3O_{10}$ and $SrRuO_3$ substrate. The nanofilms of this quality show an excellent dielectric property as will be discussed in section 4.1.

A clear benefit of these LbL approaches is the engineering of the clean interface, which appears to be a key step in the design of film properties. Currently, physical depositions such as vapor deposition and laser ablation are the main methods of fabricating oxide films. These techniques, however, usually require a complex and difficult deposition process involving high-temperature postannealing (> 600°C), which can cause degradation in the film-substrate interface arising from both nonstoichiometry and thermal stress. In that scence, the solution-based bottom-up fabrication using oxide nanosheets provides new opportunities for room-temperature fabrication of oxide nanoelectronics, while eliminating integration problems encountered in current film-growth techniques.

## 4. Electronic applications

The development of a wide range of nanosheets with various properties is very important in the design of nanodevices with sophisticated functionality. Currently, extensive effort is being made to develop oxide nanosheets with new physical and chemical properties. The range of applications of nanoassemblies could therefore be widened significantly. Here, we describe the current status of researches on oxide nanosheets, highlighting emerging functionalities in electronic applications.

### 4.1 Electronic devices

In nanosheets, 2D structures created by lateral confinement can potentially lead to not only the modification of electronic structures but also the modulation of electron-transport phenomena that arise from the quantum confinement effect. Research in such exotic 2D systems recently intensified as a result of emerging progress in graphene and its novel functionalities.[1, 2, 47] In graphene, a number of unique conducting phenomena have already been found, such as anomalous quantum Hall effect, bipolar supercurrent, etc.

Despite the similar 2D structural nature, oxide nanosheets are quite different electronically (Table 1). Most oxide nanosheets synthesized to date are $d^0$ transition metal oxides (with $Ti^{4+}$, $Nb^{5+}$, $Ta^{5+}$, $W^{6+}$), where the empty *d* orbitals of metal mix with the filled *p* orbitals of the ligands.[48] Such $d^0$ materials are not electronically interesting, but semiconducting or insulating materials. In current research on oxide nanosheets, experimental efforts have thus focused on their use as a semiconducting host or dielectric.

$Ti_{0.91}O_2$ nanosheets possess semiconducting properties similar to those of bulk $TiO_2$, such as rutile and anatase except for some modifications due to size quantization.[49] $Ti_{0.91}O_2$ nanosheets generate anodic photocurrent by band gap excitation under light irradiation with wavelengths shorter than 320 nm, corresponding to wider band gap energy of 3.8 eV.[50] In contrast, $MnO_2$ nanosheets have a broad absorption peak centered at 372 nm, which results from *d-d* transitions in the $MnO_2$ nanosheets.[13] Various interesting and useful properties have also been developed by organizing or assembling these oxide nanosheets into composite materials or multilayer films. $Ti_{0.91}O_2$ nanosheets flocculated with lanthanide cations emitted intense photoluminescence at room temperature through effective energy transfer from the semiconducting nanosheet host.[15, 51] Highly stable photoinduced charge separation was attained in a composite film of restacked $Ti_{0.91}O_2$ nanosheets and mesoporous silica or clay minerals, in which electron donors and acceptors are spatially separated at a distance of micrometers.

Another enticing possibility is the use of oxide nanosheets in high-$\kappa$ dielectrics, a key material for future semiconducting technology. For example, $Ti_{0.87}O_2$ and $Ca_2Nb_3O_{10}$ nanosheets act as high-$\kappa$ nanoblocks, and their multilayer assemblies exhibit low leakage current density (<$10^{-7}$ $Acm^{-2}$) with a high dielectric constant of >100 even for thicknesses as low as 10 nm.[43, 46, 52-55] Fig. 7 summarizes the $\varepsilon_r$ values for oxide nanosheets and various high-$\kappa$ oxides. In the ultrathin region (< 20 nm), the $\varepsilon_r$ values of $Ti_{0.87}O_2$ and $Ca_2Nb_3O_{10}$ nanosheets are larger than the values reported for any other high-$\kappa$ materials. It should be noted that the high $\varepsilon_r$ values of $Ti_{0.87}O_2$ and $Ca_2Nb_3O_{10}$ nanosheets persist even in the < 10 nm region, which is in sharp contrast to a size-induced dielectric collapse in $(Ba_{1-x}Sr_x)TiO_3$.[56, 57] These results suggest that $Ti_{0.87}O_2$ and $Ca_2Nb_3O_{10}$ are a very promising candidate for high-$\kappa$ applications such as high-density capacitors and gate dielectrics.

Oxide nanosheets are reported to be an excellent material for electric batteries. In particular, owing to their unique 2D morphology, it is expected that laterally confined 2D nanosheet crystals can significantly enhance the host capabilities of active electrode materials. $RuO_{2.1}$ nanosheets showed high performance as electrochemical supercapacitors.[28] Reassembled $Ti_{0.91}O_2$ or $MnO_2$ nanosheets, either with or without carbon, are reported to have as large a capacity as Li-ion batteries.[14, 58-60] Multilayer films of $MnO_2$ nanosheets prepared on ITO substrate also exhibited electrochromic behavior associated with the electrochemical redox process between $Mn^{3+}$ and $Mn^{4+}$.[61] This electrochromic efficiency is estimated to be 64.2 $cm^2$ $C^{-1}$ at 385 nm, which is a relatively high value among manganese oxides.

An alternative route to nanosheet-based electronics is to consider oxide nanosheet not as an active component for nanodevices but as a seed layer, in which 1-nm-thick monolayer films can be used for design and orientation control of crystal films.[62-64] The idea is to exploit the advantage of oxide nanosheets having high thermal stability even in ultrathin form. Such a technique is expected to have great potential for advances in thin-film technology.

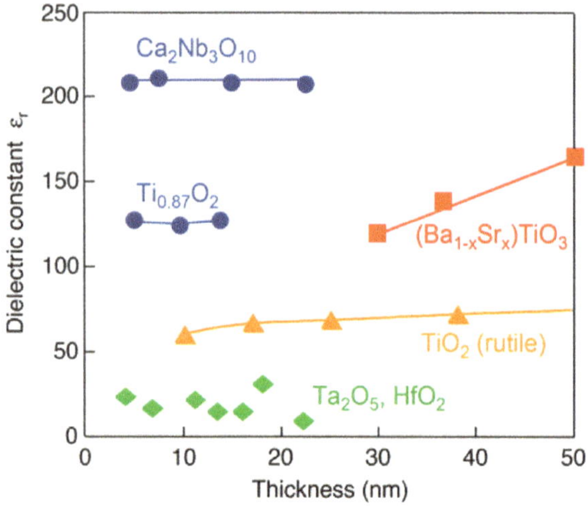


Fig. 7. The $\varepsilon_r$ values for oxide nanosheets and various high-$\kappa$ oxides.

### 4.2 Spin-electronic devices

Recent interest in room-temperature (RT) ferromagnetic semiconductors and low-dimensional magnetic nanostructures (motivated by possible application in spin-electronic devices) has stimulated research in the synthesis and characterization of $TiO_2$ nanosheet based materials. Titania nanosheets substituted with magnetic elements (Co, Fe, Mn) are ferromagnetic at room temperature.[65] The magnetization of $Ti_{0.8}Co_{0.2}O_2$ nanosheets is anisotropic due to the 2D nature, and a maximum magnetic moment of 1.4 $\mu_B$/Co for $H$ // film is obtained, which is greater than the spin moment of 1$\mu_B$/Co theoretically expected for low-spin $Co^{2+}$ as well as that in Co-doped anatase with semiconducting (0.3 $\mu_B$/Co) and insulating (1.1 $\mu_B$/Co) grounds.[65, 66] Similar ferromagnetic properties have also been reported in a series of substituted and co-substituted titania nanosheets, including $Ti_{0.8}Co_{0.2}O_2$, $Ti_{0.6}Fe_{0.4}O_2$,[67] $Ti_{0.8-x/4}Fe_{x/2}Co_{0.2-x/4}O_2$ ($0 \le x \le 0.8$)[68], and $Ti_{(5.2-2x)/6}Mn_{x/2}O_2$ ($0 \le x \le 0.4$)[69], and $Co_{1/3}Al_{2/3}(OH)_2$. Spin-glass behavior was recently observed in the dried aggregate of tetramethylammonium (TMA)/$MnO_2$ nanosheets, in which the geometrical frustration was caused by the triangular arrangement of the mixed-valence $Mn^{4+}$/$Mn^{3+}$ ions in the $MnO_2$ layer.[70]

Concerning applications, ferromagnetic nanosheets have become a pivotal architectural element in magneto-optical (MO) and magneto-electronic devices, because low-dimensional nanostructures make use of the advantage offered by spin-polarized electrons and realize the integration of ferromagnetic materials into nanoelectronics. Indeed, the 2D nature of the electronic state of ferromagnetic nanosheets leads to a gigantic MO response, superior to that of bulk systems (Fig. 8). Multilayer films of $Ti_{0.8}Co_{0.2}O_2$ and $Ti_{0.6}Fe_{0.4}O_4$ nanosheets exhibited a robust MO effect (~$10^4$ deg $cm^{-1}$) near the absorption edge at 280 nm, the shortest operating wavelength attained so far.[67] More interestingly, alternating stacking $(Ti_{0.8}Co_{0.2}O_2/Ti_{0.6}Fe_{0.4}O_2)_5$ caused a strong enhancement in MO response (~$3 \times 10^5$ deg $cm^{-1}$) at 400-550 nm, which stems from the interlayer $d$-$d$ transitions ($Co^{2+}$–$Fe^{3+}$) between adjacent nanosheets.[67] A similar MO response (~$2\times10^5$ deg $cm^{-1}$) at 400–750 nm was also observed in (Co/Fe)-cosubstituted titania nanosheets, $Ti_{0.8-x/4}Fe_{x/2}Co_{0.2-x/4}O_2$ ($x$ = 0.2, 0.4, 0.6).[68] These

MO materials are also important from a practical viewpoint as a key component for optical isolators in optical communication and data storage devices. In particular, such a large MO response including the blue light region offers potential for short-wavelength MO applications.

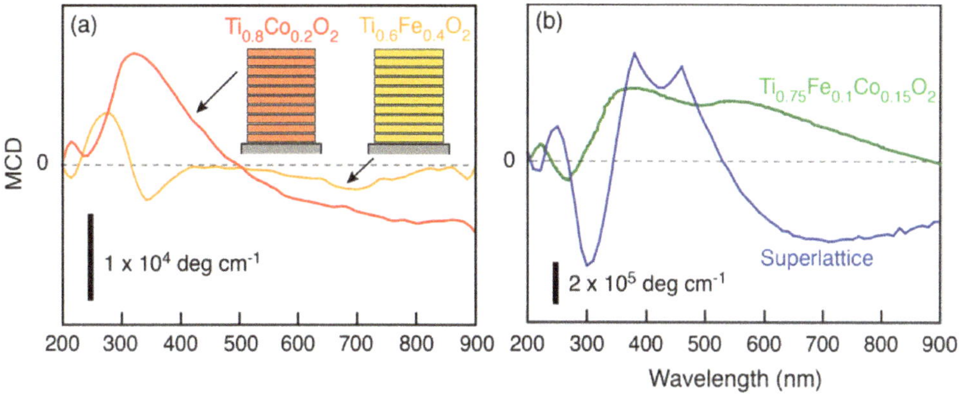


Fig. 8. (a) Magneto-optical spectra for multilayer assemblies of $(Ti_{0.8}Co_{0.2}O_2)_{10}$ and $(Ti_{0.6}Fe_{0.4}O_2)_{10}$. (b) Magneto-optical spectra for $(Ti_{0.8}Co_{0.2}O_2/Ti_{0.6}Fe_{0.4}O_2)_5$ superlattice and $(Ti_{0.75}Fe_{0.1}Co_{0.15}O_2)_{10}$. We used magnetic circular dichroism (MCD) spectroscopy for the characterization of nanosheets. The MCD spectra were measured at RT on the basis of the difference in the absorption of right and left circularly polarized light. 1° of MCD corresponds to a 7% difference of optical absorption.

These ferromagnetic nanosheets are also a model experimental system for future spintronics studies, and their assembly has great potential for the rational design and construction of complex nanodevices, even combined with transparent electronics and molecular devices. Although we focus here only on MO devices, the assembled structure is naturally viewed as a tunnel junction, which could obviously be used in novel devices such as spin-tunneling switches, spin valves and optical interconnectors.

## 5. Conclusion

The current status of research on oxide nanosheets was reviewed. A variety of physical and chemical properties of oxide nanosheets have been developed to functionalize nanosheets for electronic and spin-electronic applications, and further studies will yield new information on their physics.

2D nanosheets also teach us how to handle and process 2D nanomaterials and develop nanotechnology in general. Although we have focused here only on high-$\kappa$ properties in oxide nanosheets, 2D nanosheets exist in a whole class of functional materials, including metals, semiconductors, ferromagnetic, redox-active, photoluminescence, photochromic etc. 2D nanosheets with having regulated 2D would create the unconventional interactions of electrons as well as the confinements of electrons and ions inside the 2D nanospace or quantum well. Through new chemical design of 2D nanosheets, we can expect new or unprecedented functionalities in the 2D confined system. Furthermore, we can utilize nanosheet-based LbL technology as a new tool to develop advanced fusion functions by promoting the cooperative interaction between organized components, which are difficult to attain with the current synthetic techniques and thin-film technologies.

Oxide nanosheets provide an ideal model to study phenomena in 2D systems. Previously restricted to theoretical study, 2D nanomaterials with their exotic properties are now open to experimentation using the individual 2D system. Graphene has already been found to exhibit a number of unique phenomena such as anomalous quantum Hall effect, bipolar supercurrent, half-metallic, etc. Although current experimental and theoretical efforts mainly focus on graphene, similar properties may be available with oxide nanosheets. We hope that all aspects described here demonstrate the great potential of oxide nanosheets, introducing more exciting physics and wide-ranging applications.

## 6. Acknowledgements

The authors are grateful to Dr. K. Takada, Dr. Y. Ebina, Dr. R. Ma, Dr. T. C. Ozawa, Dr. T. Shibata, Dr. K. Fukuda, Dr. K. Akatsuka, Dr. B-W. Li, and Genki Takanashi (NIMS), Prof. H. Funakubo (TIT), Prof. T. Kiguchi (Tohoku U.), and Prof. K. Ono (KEK) for their experimental contributions and stimulating discussions over the years of the authors' researches of oxide nanosheets. This work was in part supported by World Premier International Research Center Initiative (WPI Initiative on Materials Nanoarchitronics), MEXT, CREST, JST, the Industrial Technology Research Grant Program (06A22702d), NEDO, the Grant-in-Aid for Scientific research (Fusion Materials 2106), MEXT, and Murata Foundation, Japan.

## 7. References

[1] Novoselov, K. S.; Geim, A. K.; Morozov, S. V.; Jiang, D.; Zhang, Y.; Dubonos, S. V.; Grigorieva, I. V.; Firsov, A. A., Electric field effect in atomically thin carbon films. *Science* 2004, 306, 666-669.

[2] Geim, A. K.; Novoselov, K. S., The rise of graphene. *Nature Materials* 2007, 6, 183-191.

[3] Sasaki, T., Fabrication of nanostructured functional materials using exfoliated nanosheets as a building block. *J. Ceram. Soc. Jpn.* 2007, 115, 9-16.

[4] Osada, M.; Sasaki, T., Exfoliated oxide nanosheets: new solution to nanoelectronics. *J. Mate. Chem.* 2009, 19, 2503.

[5] Ma, R. Z.; Sasaki, T., Nanosheets of oxides and hydroxides: Ultimate 2D charge-bearing functional crystallites. *Adv. Mater.* 2010, 22, 5082-5104.

[6] Golberg, D., NANOMATERIALS Exfoliating the inorganics. *Nature Nanotech.* 2011, 6, 200-201.

[7] Schaak, R. E.; Mallouk, T. E., Self-assembly of tiled perovskite monolayer and multilayer thin films. *Chem. Mater.* 2000, 12, 2513-2516.

[8] Sasaki, T.; Ebina, Y.; Tanaka, T.; Harada, M.; Watanabe, M.; Decher, G., Layer-by-layer assembly of titania nanosheet/polycation composite films. *Chem. Mater.* 2001, 13, 4661-4667.

[9] Schaak, R. E.; Mallouk, T. E., Perovskites by design: A toolbox of solid-state reactions. *Chem. Mater.* 2002, 14, 1455-1471.

[10] Ebina, Y.; Sasaki, T.; Harada, M.; Watanabe, M., Restacked perovskite nanosheets and their Pt-loaded materials as photocatalysts. *Chem. Mater.* 2002, 14, 4390-4395.

[11] Wang, Z.-S.; Sasaki, T.; Muramatsu, M.; Ebina, Y.; Tanaka, T.; Wang; Watanabe, M., Self-assembled multilayers of titania nanoparticles and nanosheets with polyelectrolytes. *Chem. Mater.* 2003, 15, 807-812.

[12] Wang, Z.-S.; Ebina, Y.; Takada, K.; Watanabe, M.; Sasaki, T., Inorganic multilayer assembly of titania semiconductor nanosheets and Ru complexes. *Langmuir* 2003, 19, 9534-9537.

[13] Wang; Omomo, Y.; Sakai, N.; Fukuda, K.; Nakai, I.; Ebina, Y.; Takada, K.; Watanabe, M.; Sasaki, T., Fabrication and characterization of multilayer ultrathin films of exfoliated $MnO_2$ nanosheets and polycations. *Chem. Mater.* 2003, 15, 2873-2878.

[14] Wang; Takada, K.; Kajiyama, A.; Onoda, M.; Michiue, Y.; Zhang; Watanabe, M.; Sasaki, T., Synthesis of a Li,Mn-oxide with disordered layer stacking through flocculation of exfoliated $MnO_2$ nanosheets, and its electrochemical properties. *Chem. Mater.* 2003, 15, 4508-4514.

[15] Xin, H.; Ma, R.; Wang, L. Z.; Ebina, Y.; Takada, K.; Sasaki, T., *Appl. Phys. Lett.* 2004, 85, 4187.

[16] Wang, L.; Ebina, Y.; Takada, K.; Sasaki, T., Ultrathin hollow nanoshells of manganese oxide. *Chem. Commun.* 2004, 1074-1075.

[17] Wang, L. Z.; Ebina, Y.; Takada, K.; Kurashima, K.; Sasaki, T., A new mesoporous manganese oxide pillared with double layers of alumina. *Adv. Mater.* 2004, 16, 1412-1416.

[18] Wang; Ebina, Y.; Takada, K.; Sasaki, T., Ultrathin films and hollow shells with pillared architectures fabricated *via* layer-by-layer self-assembly of titania nanosheets and aluminum keggin ions. *J. Phys. Chem. B* 2004, 108, 4283-4288.

[19] Sasaki, T.; Watanabe, M.; Hashizume, H.; Yamada, H.; Nakazawa, H., Macromolecule-like aspects for a colloidal suspension of an exfoliated titanate. Pairwise association of nanosheets and dynamic reassembling process initiated from it. *J. Am. Chem. Soc.* 1996, 118, 8329-8335.

[20] Sasaki, T.; Watanabe, M., Osmotic swelling to exfoliation. Exceptionally high degrees of hydration of a layered titanate. *J. Am. Chem. Soc.* 1998, 120, 4682-4689.

[21] Tanaka, T.; Ebina, Y.; Takada, K.; Kurashima, K.; Sasaki, T., Oversized titania nanosheet crystallites derived from flux-grown layered titanate single crystals. *Chem. Mater.* 2003, 15, 3564-3568.

[22] Sugimoto, W.; Terabayashi, O.; Murakami, Y.; Takasu, Y., Electrophoretic deposition of negatively charged tetratitanate nanosheets and transformation into preferentially oriented $TiO_2$(B) film. *J. Mater. Chem.* 2002, 12, 3814-3818.

[23] Omomo, Y.; Sasaki, T.; Wang, L. Z.; Watanabe, M., Redoxable nanosheet crystallites of $MnO_2$ derived via delamination of a layered manganese oxide. *J. Am. Chem. Soc.* 2003, 125, 3568-3575.

[24] Takagaki, A.; Sugisawa, M.; Lu, D. L.; Kondo, J. N.; Hara, M.; Domen, K.; Hayashi, S., Exfoliated nanosheets as a new strong solid acid catalyst. *J. Am. Chem. Soc.* 2003, 125, 5479-5485.

[25] Takagaki, A.; Yoshida, T.; Lu, D.; Kondo, J. N.; Hara, M.; Domen, K.; Hayashi, S., Titanium niobate and titanium tantalate nanosheets as strong solid acid catalysts. *J. Phys. Chem. B* 2004, 108, 11549-11555.

[26] Fukuda, K.; Nakai, I.; Ebina, Y.; Ma, R. Z.; Sasaki, T., Colloidal unilamellar layers of tantalum oxide with open channels. *Inorg. Chem.* 2007, 46, 4787-4789.

[27] Kim, D. S.; Ozawa, T. C.; Fukuda, K.; Ohshima, S.; Nakai, I.; Sasaki, T., Soft-Chemical Exfoliation of $Na_{0.9}Mo_2O_4$: Preparation and Electrical Conductivity Characterization of a Molybdenum Oxide Nanosheet. *Chem. Mater.* 2011, 23, 2700-2702.

[28] Sugimoto, W.; Iwata, H.; Yasunaga, Y.; Murakami, Y.; Takasu, Y., Preparation of ruthenic acid nanosheets and utilization of its interlayer surface for electrochemical energy storage. *Angew. Chem. Int. Ed.* 2003, 42, 4092-4096.

[29] Fukuda, K.; Akatsuka, K.; Ebina, Y.; Ma, R.; Takada, K.; Nakai, I.; Sasaki, T., Exfoliated nanosheet crystallite of cesium tungstate with 2D pyrochlore structure: Synthesis, characterization, and photochromic properties. *ACS Nano* 2008, 2, 1689-1695.

[30] Treacy, M. M. J.; Rice, S. B.; Jacobson, A. J.; Lewandowski, J. T., Electron microscopy study of delamination in dispersions of the perovskite-related layered phases $K[Ca_2Na_{n-3}Nb_nO_{3n-1}]$: evidence for single-layer formation. *Chem. Mater.* 1990, 2, 279-286.

[31] Ebina, Y.; Sasaki, I.; Watanabe, A., Study on exfoliation of layered perovskite-type niobates. *Solid State Ionics* 2002, 151, 177-182.

[32] Ozawa, T. C.; Fukuda, K.; Akatsuka, K.; Ebina, Y.; Sasaki, T., Preparation and Characterization of the $Eu^{3+}$ Doped Perovskite Nanosheet Phosphor: $La_{0.90}Eu_{0.05}Nb_2O_7$. *Chem. Mater.* 2007, 19, 6575-6580.

[33] Ida, S.; Ogata, C.; Unal, U.; Izawa, K.; Inoue, T.; Altuntasoglu, O.; Matsumoto, Y., Preparation of a blue luminescent nanosheet derived from layered perovskite $Bi_2SrTa_2O_9$. *J. Am. Chem. Soc.* 2007, 129, 8956-8959.

[34] Ozawa, T. C.; Fukuda, K.; Akatsuka, K.; Ebina, Y.; Sasaki, T.; Kurashima, K.; Kosuda, K., $(K_{1.5}Eu_{0.5})Ta_3O_{10}$: A far-red luminescent nanosheet phosphor with the double perovskite structure. *J. Phys. Chem. C* 2008, 112, 17115-17120.

[35] Ida, S.; Ogata, C.; Eguchi, M.; Youngblood, W. J.; Mallouk, T. E.; Matsumoto, Y., Photoluminescence of perovskite nanosheets prepared by exfoliation of layered oxides, $K_2Ln_2Ti_3O_{10}$, $KLnNb_2O_7$, and $RbLnTa_2O_7$ (Ln : lanthanide ion). *J. Am. Chem. Soc.* 2008, 130, 7052-7059.

[36] Fukuda, K.; Nakai, I.; Oishi, C.; Nomura, M.; Harada, M.; Ebina, Y.; Sasaki, T., Nanoarchitecture of semiconductor titania nanosheets revealed by polarization-dependent total reflection fluorescence X-ray absorption fine structure. *J. Phys. Chem. B* 2004, 108, 13088-13092.

[37] Xu, F. F.; Bando, Y.; Ebina, Y.; Sasaki, T., Modification of crystal structures in perovskite-type niobate nanosheets. *Philosophical Magazine a-Physics of Condensed Matter Structure Defects and Mechanical Properties* 2002, 82, 2655-2663.

[38] Xu, F. F.; Bando, Y.; Ebina, Y.; Sasaki, T., Stacking-fault pyramids formed in perovskite-type niobate nanosheet aggregates under electron irradiation. *Philosophical Magazine Letters* 2003, 83, 367-373.

[39] Xu, F. F.; Ebina, Y.; Bando, Y.; Sasaki, T., Structural characterization of $(TBA,H)Ca_2Nb_3O_{10}$ nanosheets formed by delamination of a precursor-layered perovskite. *J. Phys. Chem. B* 2003, 107, 9638-9645.

[40] Xu, F. F.; Ebina, Y.; Bando, Y.; Sasaki, T., *In-situ* transmission electron microscopic study of perovskite-type niobate nanosheets under electron-irradiation and heating. *J. Phys. Chem. B* 2003, 107, 6698-6703.

[41] Decher, G., Fuzzy nanoassemblies: Toward layered polymeric multicomposites. *Science* 1997, 277, 1232-1237.

[42] Muramatsu, M.; Akatsuka, K.; Ebina, Y.; Wang, K. Z.; Sasaki, T.; Ishida, T.; Miyake, K.; Haga, M., Fabrication of densely packed titania nanosheet films on solid surface by use of Langmuir-Blodgett deposition method without amphiphilic additives. *Langmuir* 2005, 21, 6590-6595.

[43] Osada, M.; Akatsuka, K.; Ebina, Y.; Kotani, Y.; Ono, K.; Funakubo, H.; Ueda, S.; Kobayashi, K.; Takada, K.; Sasaki, T., Langmuir-Blodgett fabrication of nanosheet-based dielectric films without an interfacial dead layer. *Jpn. J. Appl. Phys.* 2008, 47, 7556-7560.

[44] Akatsuka, K.; Haga, M.; Ebina, Y.; Osada, M.; Fukuda, K.; Sasaki, T., Construction of highly ordered lamellar nanostructures through Langmuir-Blodgett deposition of molecularly thin titania nanosheets tens of micrometers wide and their excellent dielectric properties. *ACS Nano* 2009, 3, 1097-1106.

[45] Li, B. W.; Osada, M.; Ozawa, T. C.; Ebina, Y.; Akatsuka, K.; Ma, R. Z.; Funakubo, H.; Sasaki, T., Engineered interfaces of artificial perovskite oxide superlattices *via* nanosheet deposition process. *ACS Nano* 2010, 4, 6673-6680.

[46] Osada, M.; Akatsuka, K.; Ebina, Y.; Funakubo, H.; Ono, K.; Takada, K.; Sasaki, T., Robust High-k response in molecularly thin perovskite nanosheets. *ACS Nano* 2010, 4, 5225-5232.

[47] Geim, A. K.; MacDonald, A. H., Graphene: Exploring carbon flatland. *Physics Today* 2007, 60, 35-41.

[48] Sato, H.; Ono, K.; Sasaki, T.; Yamagishi, A., First-principles study of two-dimensional titanium dioxides. *J. Phys. Chem. B* 2003, 107, 9824-9828.

[49] Sakai, N.; Ebina, Y.; Takada, K.; Sasaki, T., Electronic band structure of titania semiconductor nanosheets revealed by electrochemical and photoelectrochemical studies. *J. Am. Chem. Soc.* 2004, 126, 5851-8.

[50] Sasaki, T.; Watanabe, M., Semiconductor nanosheet crystallites of quasi-$TiO_2$ and their optical properties. *J. Phys. Chem. B* 1997, 101, 10159-10161.

[51] Xin, H.; Ebina, Y.; Ma, R.; Takada, K.; Sasaki, T., Thermally stable luminescent composites fabricated by confining rare earth complexes in the two-dimensional gallery of titania nanosheets and their photophysical properties. *J. Phys. Chem. B* 2006, 110, 9863-9868.

[52] Osada, M.; Ebina, Y.; Funakubo, H.; Yokoyama, S.; Kiguchi, T.; Takada, K.; Sasaki, T., High-κ dielectric nanofilms fabricated from titania nanosheets. *Adv. Mater.* 2006, 18, 1023-1027.

[53] Osada, M.; Akatsuka, K.; Ebina, Y.; Funakubo, H.; Kiguchi, T.; Takada, K.; Sasaki, T., Solution-based fabrication of high-kappa dielectric nanofilms using titania nanosheets as a building block. *Jpn. J. Appl. Phys.* 2007, 46, 6979-6983.

[54] Li, B.-W.; Osada, M.; Ozawa, T. C.; Ma, R.; Akatsuka, K.; Ebina, Y.; Funakubo, H.; Ueda, S.; Kobayashi, K.; Sasaki, T., Solution-based fabrication of perovskite nanosheet films and their dielectric properties. *Jpn. J. Appl. Phys.* 2009, 48, 09KA15.

[55] Li, B. W.; Osada, M.; Ozawa, T. C.; Akatsuka, K.; Ebina, Y.; Ma, R.; Ono, K.; Funakubo, H.; Sasaki, T., A-site-modified perovskite nanosheets and their integration into high-k dielectric thin films with a clean interface. *Jpn. J. Appl. Phys.* 2010, 49.

[56] Werner, M. C.; Banerjee, I.; McIntyre, P. C.; Tani, N.; Tanimura, M., Microstructure of (Ba,Sr)$TiO_3$ thin films deposited by physical vapor deposition at 480 degrees C and its influence on the dielectric properties. *Appl. Phys. Lett.* 2000, 77, 1209-1211.
[57] Hwang, C. S., Thickness-dependent dielectric constants of (Ba,Sr)$TiO_3$ thin films with Pt or conducting oxide electrodes. *J. Appl. Phys.* 2002, 92, 432-437.
[58] Suzuki, S.; Miyayama, M., Lithium intercalation properties of octatitanate synthesized through exfoliation/reassembly. *J. Phys. Chem. B* 2006, 110, 4731-4734.
[59] Suzuki, S.; Miyayama, M., Lithium intercalation properties of reassembled titanate/carbon composites. *J. Electrochem. Soc.* 2007, 154, A438-A443.
[60] Suzuki, S.; Miyayama, M., Microstructural controls of titanate nanosheet composites using carbon fibers and high-rate electrode properties for lithium ion secondary batteries. *J. Power Sources* 2011, 196, 2269-2273.
[61] Sakai, N.; Fukuda, K.; Omomo, Y.; Ebina, Y.; Takada, K.; Sasaki, T., Hetero-nanostructured films of titanium and manganese oxide nanosheets: Photoinduced charge transfer and electrochemical properties. *J. Phys. Chem. C* 2008, 112, 5197-5202.
[62] Shibata, T.; Ebina, Y.; Ohnishi, T.; Takada, K.; Kogure, T.; Sasaki, T., Fabrication of anatase thin film with perfect c-axis orientation on glass substrate promoted by a two-dimensional perovskite nanosheet seed layer. *Crystal Growth & Design* 2010, 10, 3787-3793.
[63] Shibata, T.; Fukuda, K.; Ebina, Y.; Kogure, T.; Sasaki, T., One-nanometer-thick seed layer of unilamellar nanosheets promotes oriented growth of oxide crystal films. *Adv. Mater.* 2008, 20, 231-+.
[64] Shibata, T.; Ohnishi, T.; Sakaguchi, I.; Osada, M.; Takada, K.; Kogure, T.; Sasaki, T., Well-controlled crystal growth of zinc oxide films on plastics at room temperature using 2D nanosheet seed layer. *J. Phys. Chem. C* 2009, 113, 19096-19101.
[65] Osada, M.; Ebina, Y.; Fukuda, K.; Ono, K.; Takada, K.; Yamaura, K.; Takayama-Muromachi, E.; Sasaki, T., Ferromagnetism in two-dimensional $Ti_{0.8}Co_{0.2}O_2$ nanosheets. *Phys. Rev. B* 2006, 73.
[66] Kotani, Y.; Taniuchi, T.; Osada, M.; Sasaki, T.; Kotsugi, M.; Guo, F. Z.; Watanabe, Y.; Kubota, M.; Ono, K., X-ray nanospectroscopic characterization of a molecularly thin ferromagnetic Ti1-xCoxO2 nanosheet. *Appl. Phys. Lett.* 2008, 93.
[67] Osada, M.; Ebina, Y.; Takada, K.; Sasaki, T., Gigantic magneto-optical effects in multilayer assemblies of two-dimensional titania nanosheets. *Adv. Mater.* 2006, 18, 295-+.
[68] Osada, M.; Itose, M.; Ebina, Y.; Ono, K.; Ueda, S.; Kobayashi, K.; Sasaki, T., Gigantic magneto-optical effects induced by (Fe/Co)-cosubstitution in titania nanosheets. *Appl. Phys. Lett.* 2008, 92.
[69] Dong, X. P.; Osada, M.; Ueda, H.; Ebina, Y.; Kotani, Y.; Ono, K.; Ueda, S.; Kobayashi, K.; Takada, K.; Sasaki, T., Synthesis of Mn-substituted titania nanosheets and ferromagnetic thin films with controlled doping. *Chem. Mater.* 2009, 21, 4366-4373.
[70] Kai, K.; Yoshida, Y.; Kageyama, H.; Saito, G.; Ishigaki, T.; Furukawa, Y.; Kawamata, J., Room-temperature synthesis of manganese oxide monosheets. *J. Am. Chem. Soc.* 2008, 130, 15938-15943.

7

# Self-Organization of Mesoscopically-Ordered Parallel Rare-Earth Silicide Nanowire Arrays on Si(110)-16×2 Surface

Ie-Hong Hong
*Department of Electrophysics, National Chiayi University, Chiayi, Taiwan*

## 1. Introduction

One-dimensional nanowires have attracted significant attention due to their exotic physical properties and potential applications in nanoelectronic, nanophotonic, molecular-electronic, and magnetoelectronic devices (Agarawal, 2008; Gambardella *et al.*, 2002; Melosh *et al.*, 2003; Segovia *et al.*, 1999; Snijders, & Weitering, 2010; Yeom *et al.*, 2005; Zeng *et al.*, 2008). In particular, rare-earth metal silicide nanowires are excellent candidates as low-resistance interconnections or as fully-silicided nanoelectrodes for attaching electrically active nanostructures within a nanodevice because of their high conductivity and extremely low Schottky barrier height on *n*-type silicon and their compatibility with current silicon-based integrated circuit technology. As the continuing miniaturization encounters the lithographical limits, self-assembly (Shchukin & Bimberg, 1999) is being regarded as an effective bottom-up approach for growing one-dimensional nanowires because of its potential advantages of low cost, simplicity and high throughput. Moreover, it is believed that the basic building blocks for future complementary metal-oxide-semiconductor (CMOS) architectures will feature the integration of nanostructures based on self-assembly (Lu & Lieber, 2007; Zhirnov & Herr, 2001). In this regard, the self-assembly of epitaxial rare-earth silicide nanowires on silicon surfaces is the subject of numerous studies. It has been shown that Gd, Dy, Er, Y, and Ho can self-assemble into epitaxial silicide nanowires on a planar Si(001)-2×1 surface with large aspect ratios (length > 1 μm, width < 10 nm) by utilizing the anisotropic lattice mismatch between the rare-earth metal silicides and the Si substrate; these rare-earth silicide nanowires grow to arbitrary length in the direction of small (<1%) mismatch and are limited in the lateral direction of large (~8%) mismatch (Chen *et al.*, 2000, 2002; Iancu *et al.*, 2009; Lee & Kim, 2003; Liu & Nogami, 2003a; Nogami *et al.*, 2001; Ohbuchi & Nogami, 2006; Ye *et al.*, 2009; Zeng *et al.*, 2008; Zhang *et al.*, 2009; Zhou *et al.*, 2006).

However, due to the alternating domains of 2×1 and 1×2 reconstruction on the neighboring single-height atomic terraces of the nominally flat Si(001) surface, rare-earth silicide nanowire always grow in two orthogonal directions on the planar Si(001) substrate. These silicide nanowires formed with two perpendicular orientations will intercept each another to terminate further growth of these orthogonal nanowires. Consequently, it is difficult to obtain parallel-aligned and long nanowires for the application in nanoelectronic devices. Therefore, several studies have attempted to use a single-domain vicinal Si(001) surface as a template to grow a large array of unidirectional rare-earth silicide nanowires (Lee *et al.*, 2006; Liu & Nogami, 2003b; Preinesberger, 2005 ; Yeom *et al.*, 2005). Although a massively parallel array of straightly-aligned nanowires can be produced by such an approach, the resulting nanowires are randomly-distributed, and different in size. Even an attempt to use a highly-anisotropic vicinal Si(557) surface as a nanotemplate fails to self-organize a highly-regular parallel rare-earth silicide nanowire array over mesoscopic areas of Si(557) surface (Wanke *et al.*, 2008). For practical applications of rare-earth silicide nanowires in nanoelectronic architectures, it is of paramount importance to self-organize a massively periodic array of parallel-aligned silicide nanowires with uniform distribution, single orientation and identical size over a macroscopic area on silicon surfaces.

We have recently demonstrated that the most straightforward bottom-up approach for growing a well-ordered, highly-integrated two-dimensional networks and one-dimensional parallel arrays is to use a Si(110)–16×2 surface as a nanopatterned template (Hong *et al.*, 2009a, 2009b, 2011a, 2011b, 2011c). Because the 16×2 reconstruction consists of a unique periodic up-and-down sequence of terraces with a perfect periodicity of 5 nm and an equal width of 2.5 nm (An *et al.*, 2000), this naturally-patterned nanotemplate is thus very promising for the self-ordered growth of extremely-regular parallel nanowire arrays with high nanowire density and good uniformity over mesoscopic areas. Moreover, Si(110) wafer has been recognized as an ideal substrate for next-generation CMOS nanodevices (Yang *et al.*, 2006; Mizuno *et al.*, 2005), because CMOS transistors fabricated on Si(110) wafers exhibit both higher hole mobility and ultra-dense spatial integration than ones fabricated on Si(100) wafers (Cheng *et al.*, 2006). Thus, a larger-area self-organization of highly-integrated, well-regular, parallel epitaxial nanowire arrays on Si(110) surface can provide an efficient wafer-scale integration for the construction of future nanowire-based CMOS nanodevices.

In this chapter, we present that a large-scale self-organization of highly-regular, parallel arrays of epitaxial rare-earth metal silicide nanowires, covering a mesoscopic area exceeding 1×1 μm$^2$, can be realized through the heteroepitaxial growth of rare-earth metal silicides along the periodic terraces of a Si(110)-16×2 surface. Here, we chose Gd, Ce and Er metals to self-form massively highly-ordered rare-earth silicide nanowire arrays on Si(110)-16×2 surface, based on the following reasons: (1) The half-filled *4f* electronic state of Gd atom contains seven spin-up electrons, Gd atoms can act as a source of spin-polarized electrons injected into silicon and create spin chains via the magnetic coupling between the large magnetic moments of *4f* electrons in the adjacent atomic chaisn (McChesney *et al.*,; Žutić, 2006). Also, $GdSi_{1.7}$ epilayers exhibit magnetic ordering below 44 K (Hogg *et al.*, 2002) and $Gd_5(Si_{1-x}Ge_x)_4$ has a giant magnetoresistance within the range from 270 K to 290 K (Pecharsky & Gschneidner, 2001). (2) $CeSi_x$ (where $1.0 \leq x \leq 2.0$) compounds have been reported to have many intriguing physical properties, such as intermediate valence, Kondo lattice, heavy-fermion superconductivity and giant magnetoresistance (Münzenberg *et al.*, 2007; Shaheen & Mendoza, 1999; Smith *et al.*, 2005; Yokota *et al.*, 2002); however, so far no Ce-silicide nanowires can be formed by using the anisotropic lattice-mismatched growth because of the poor lattice mismatches between silicides of the light rare-earth metal Ce and Si(001) substrate along two orthogonal crystalline directions. (3) Er silicides not only possess the lowest resistance of all rare earth metal silicides, but also exhibit anomalous Hall effect below 60 K (Pierre *et al.*, 1994) and present strong magnetic dichroism due to the large magnetocrystalline anisotropy at room temperature (Castrucci *et al.*, 1995).

Our results demonstrate the fact that our developed template-directed self-organization process based on the nanotemplate of Si(110)-16×2 superstructure can be applicable for the spontaneous formation of massively well-regular parallel arrays of other epitaxial rare-earth metal silicide nanowires, and open up the possibility for self-ordered, large-area, parallel arrays of magnetic atomically-precise rare-earth metal silicide nanowires on Si(110)–16×2 surface, which could be integrated into high-density magnetoelectronic or spintronic nanodevices. Moreover, the excellent regularity of our fabricated well-ordered parallel rare-earth silicide nanowire arrays is comparable to that fabricated lithographically (Melosh *et al.*, 2003), and is much better than that produced by the anisotropic lattice mismatch growth. This advantage allows us to further carry out detailed investigations of electronic, chemical and magnetic properties of such massive self-ordered parallel rare-earth silicide nanowire arrays by means of photoemission spectroscopy (Segovia *et al.*, 1999; Yeom *et al.*, 2005) and magnetic circular dichroism (Gambardella *et al.*, 2002).

## 2. Experimental methods

The experiments were performed in a commercial, ultrahigh vacuum, variable-temperature STM system with a base pressure of ~$5\times10^{-11}$ mbar. The device-quality flat Si(110) substrates (n-type, ~10 Ωcm) were used as one-dimensional nanotemplates for the self-organization of parallel rare-earth silicide nanowire arrays. Atomically clean Si(110)-16×2 surface was prepared by well-established annealing procedures (Hong *et al.*, 2009a, 2009b), and was confirmed by STM observation of the long-range 16×2 superstructure, as shown in Fig. 1. Parallel Gd-silicide nanowire arrays were produced by depositing ~6 Å of high purity (99.95%) Gd onto Si(110) surface at 750 °C at a deposition rate of ~0.1 Å/min and subsequently quenching to room temperature. Parallel Ce-silicide nanowire arrays were produced by depositing ~8 Å of high-purity (99.95%) Ce metals onto Si(110) surface at 400 °C at a deposition rate of ~0.2 Å/min and subsequently annealing at 700 °C for 20 minutes, then slowly cooling down to room temperature. Parallel Er-silicide nanowire arrays were produced by depositing ~10 Å of high-purity (99.95%) Er metals onto Si(110) surface at 650 °C at a deposition rate of ~0.2 Å/min and subsequently annealing at 750 °C for 30 minutes. Gd, Ce and Er metals were evaporated individually from a water-cooled electron-beam evaporator with an internal flux meter; their deposition coverages were determined *in situ* using a quartz-crystal thickness monitor. During the deposition of these rare-earth metals, the system pressure was maintained below $1 \times 10^{-9}$ mbar. The STM measurements were acquired in the constant current mode with the atomic-resolution electrochemically etched nickel tips. The I-V curve was performed by recording the tunneling current ($I_t$) while ramping the sample bias ($V_s$) at a specific location on the nanowire; and was then numerically differentiated to obtain the dI/dV curve (i.e., STS spectrum).

## 3. Results and discussion

### 3.1 One-dimensional naturally-patterned nanotemplate of Si(110)-16×2 surface

#### 3.1.1 Atomically clean Si(110)-16×2 surface

Figure 1 represents typical STM topographic images of the atomically clean Si(110)–16×2 surface at different magnifications. As visibly seen in Figs. 1(a) and 1(b), the parallel-aligned up-and-down terraces of the 16×2 superstructure extending over a wide area exhibit a mesoscopic ordering and clearly reveal a unique highly-anisotropic character of this reconstruction. Such uniform grating-like terraces are very suitable to be used as a novel nanotemplate for the large-scale nanofabrication of well-ordered, high-density, parallel-aligned nanowires covering macroscopic areas. A magnified image (50 × 50 nm²) of a part of Fig. 1(b) is shown in Fig. 1(c). It obviously shows that zigzag chains appear in the grating-like up-and-down terraces. An atomic-resolution image (12 × 12 nm²) of Fig. 1(d), magnified from Fig. 1(c), shows a pair of Si pentagons forming zigzag chains on the upper terraces of 16×2 reconstruction are clearly resolved, consistent with the pervious result (An *et al.*, 2001; Stekolnikov *et al.*, 2004). Figure 2 displays the cross-sectional profile across the up-and-down terraces of 16×2 reconstruction along the line scans **C** in Fig. 1(c). The typical width and average height of these periodic upper terraces are 2.2 ± 0.1 nm and 0.29 ± 0.01 nm, respectively, and the regular periodicity is 5.0 ± 0.1 nm. These measurements are consistent with the previous results (An *et al.*, 2001; Yamamoto *et al.*, 2000).

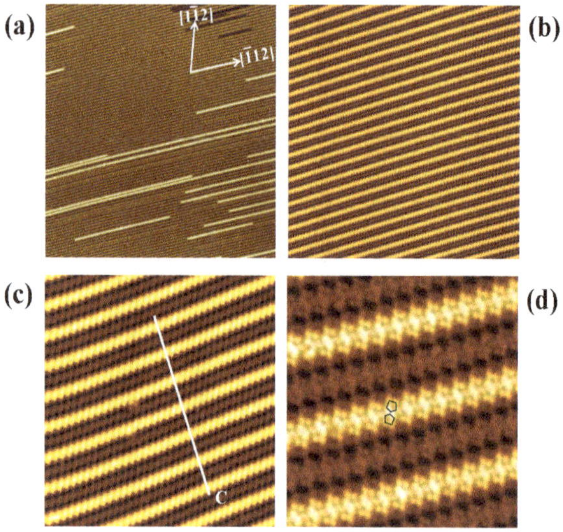


Fig. 1. (a)−(c) STM topographic images of the atomically clean Si(110)-16×2 surface taken at different magnifications: (a) 500 × 500 nm² ($V_s$ = +3 V, $I_t$ = 30 pA), (b) 125 × 125 nm², (c) 50 × 50 nm², and (c) 15 × 15 nm² ($V_s$ = +1.5 V and $I_t$ = 10 pA).

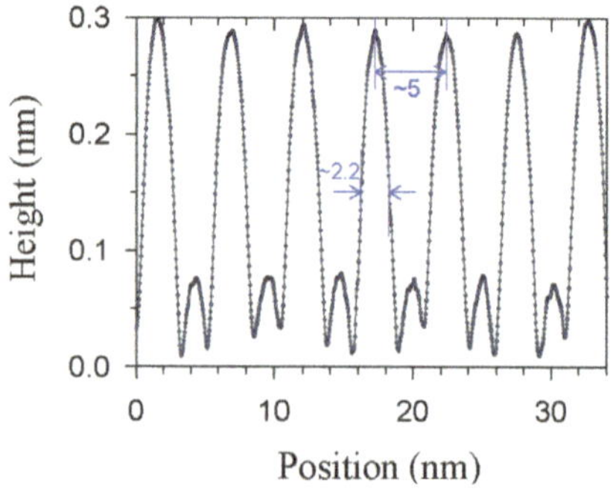


Fig. 2. Cross-sectional profile across the up-and-down terraces of 16×2 reconstruction along the white line indicated in Fig. 1(c).

### 3.1.2 Parallel silicon nanowire array grown naturally on Si(110) surface

Since both the width and height of these upper terraces of Si(110)-16×2 surface are nearly close to those of silicon nanowires formed on Ag(110) surface (Sahaf *et al.*, 2007; Padova *et al.*, 2008), the long-range, parallel-aligned upper terraces of 16×2 reconstruction can be recognized as massively-parallel silicon nanowires grown naturally on Si(110) surface. The identification of the upper Si(110) terraces as silicon nanowires formed on Si(110) surface is similar to the self-formation of massive Si atomic lines on the Si-terminated $\beta$-SiC(100)-3×2 surface (Soukiassian *et al.*, 1997). Silicon nanowires has been regarded as the most promising basic building blocks for the bottom-up assembly of integrated electronic and photonic nanodevices because they have the advantage of easy integration into the existing silicon-based semiconductor industry (Agarawal, 2008; Lu & Lieber, 2007).

## 3.2 Parallel Gd-silicide nanowire array self-organized on Si(110)-16×2 surface

Figures 3(a)−3(e) show the representative STM topographic images of parallel Gd-silicide nanowire array at different magnifications. As clearly observed in Figs. 3(a)−3(e), these parallel-aligned, straight and nearly defect-free nanowires cover a mesoscopic area with a typical lateral separation beyond 1.0 μm and are most elongated along the $[\bar{1}\ 1\ 2]$ direction. In Fig. 3(d), each Gd-silicide nanowire consists of double bead chains separated by a bean chain. The double bead chains are higher than the bean chain by a difference of ~25 pm, as shown in the cross-sectional profile of Fig. 4. Each bead chain is composed of rounded protrusions marked by the red circles of ~1.5 nm in diameter, which are too large to be interpreted as individual atoms. Additionally, the adjacent substrate between the neighboring nanowires still shows a zigzag chain structure. All individual nanowires are atomically precise nanostructures, essentially identical to one another over the entire macroscopic area of Si(110) surface. Their width remains atomically precise along the nanowire and is typically 4.0 ± 0.1 nm [see Fig. 4]. Therefore, these nanowires possess an extraordinarily high aspect ratio of greater than 250. Moreover, this massively parallel array shows a regular periodicity of 7.2 ± 0.2 nm [see Fig. 4]. Such a perfect one-dimensional self-organized parallel nanowire array with good uniformity and alignment over a mesoscopic area represents a well-established long-range spatial ordering. This large-area Gd-silicide nanowire array also exhibits a high nanowire density and thus can be applied for the wafer-scale integration of high-density nanoelectronic devices. The growth mechanism of such a highly-regular parallel array of uniformly-spaced Gd-silicide nanowires on Si(110)-16×2 surface is driven by the heteroepitaxial growth of Gd silicides on these periodic grating-like upper terraces of Si(110)-16×2 superstructure, as explained in detail in our previous publication (Hong *et al.*, 2009a). This perfect self-organization of a large-area parallel array consisting of atomically-precise Gd-silicide nanowires with regular periodicity is observed for the first time.

Figures 3(e) and 3(f) show an enlarged part of the STM image in Fig. 3(d), recorded at the sample bias $V_s$ = +2.0 V and −2.0 V, respectively. We note that both empty-state and filled-state STM images [i.e., Figs. 3(e) and 3(f)] represent the same area of the sample. In dual-bias images, the parallel, registry-aligned nanowires are well resolved. The empty-state image [Fig. 3(e)] clearly shows that the rounded protrusions of the double bead chains in Fig. 3(d) converts to pentagons; however, the pentagons in the left and right bead chains are oppositely oriented, which is the same as the pair of Si pentagons of Si(110)-16×2 superstructure [see Fig. 1(d)]. Moreover, the bean chain disappears and instead becomes to a trench. This result strongly suggests that the Si pentagon pair on the upper terraces of 16×2 reconstruction was split into two individual Si pentagons due to a strong reaction of Si and Gd upon the adsorption of Gd on the upper terraces. Consequently, a trench was formed in the middle of Gd-silicide nanowires and Gd atoms were embedded in the trench (i.e., Gd atoms were adsorbed in the bean chain of Fig. 3(d)), similar to the Si-deficient structural model (Lee & Kim, 2003). The width of Gd-silicide nanowire thus becomes to be ~4 nm from the Si terrace width of ~2.2 nm. In the filled-state image [Fig. 3(f)], each Gd-silicide nanowire appears to consist of two sawtooth rows of distinct atomic arrangements on both sides and a pair of double linear chains with an up-down buckling configuration in the middle of nanowire; moreover, the contrast of these two sawtooth rows is much lower than that of double linear chains. Notice that the dark trench between the double bead chains in Fig. 3(e) inverts to the bright double linear chains in Fig. 3(f) when the bias polarity is reversed. Considering the charge transfer from Si to Gd atoms by electronegativity, the bright double linear chains in Fig. 3(f) are primarily contributed from the electronic states localized on Gd atoms, suggesting that Gd atoms are located on the bean chain in Fig. 3(d). That is, the two sawtooth rows and the double linear atomic chains consist mainly of Si and Gd atoms, respectively. The appearance of the well-ordered internal structures in Gd-silicide nanowires strongly verifies that these straightly-aligned, long Gd-silicide nanowires grow epitaxially on Si(110) surface.

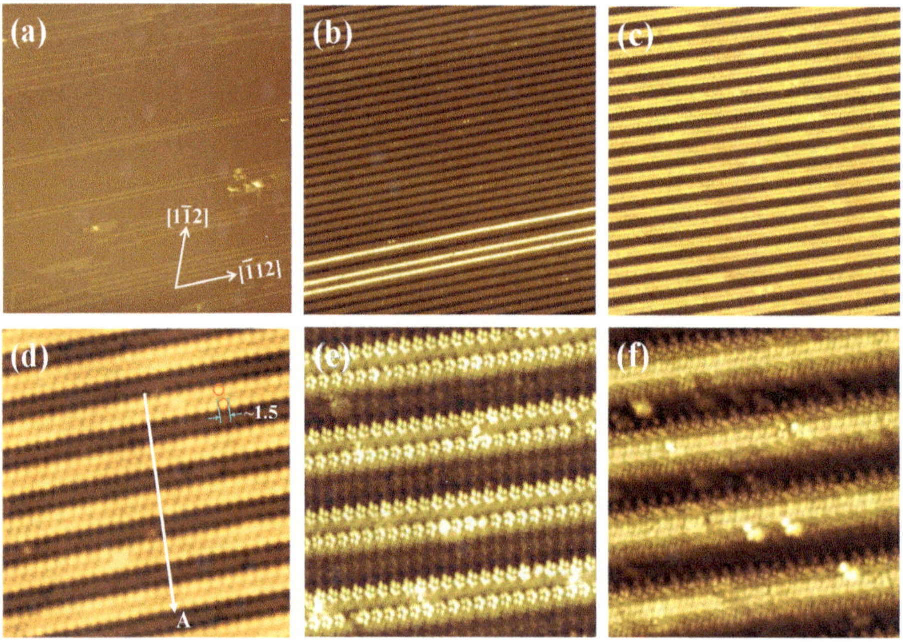


Fig. 3. (a)−(d) STM topographic images of parallel Gd-silicide nanowire array on Si(110)-16×2 surface taken at different magnifications: (a) 0.8 × 0.8 μm² ($V_s$ = +3 V, $I_t$ = 30 pA), (b) 250 × 250 nm², (c) 110 × 110 nm², and (d) 50 × 50 nm². (e), (f) Dual-polarity images (25 × 25 nm²) of parallel Gd-silicide nanowire array acquired at +2 V and −2 V, respectively, and at $I_t$ = 10 pA.

As seen in Figs. 3(e) and 3(f), the two bead chains of empty orbitals on both sides of nanowire are in antiphase with the double linear chains of filled orbitals in the middle of nanowire. This polarity dependence of STM images clearly reveals that each Gd-silicide nanowire really consists of a bundle of chain-like structures with a peculiar charge arrangement of alternating filled and empty orbitals. These dual-polarity STM images clearly show that these parallel-aligned Gd-silicide nanowires are well-defined and periodically arranged in both filled-state and empty-state images. The periodic intensity modulation exhibited by this parallel array in dual-bias images clearly reveals a perfect spatial-organization of the charge distribution. Such a well-regular charge modulation of this parallel Gd-silicide nanowire array could be exploited in nanoelectronic devices.

Figure 4 displays the cross-sectional profiles of the line scan **A** across the parallel Gd-silicide nanowire array in Fig. 3(d). This cross-sectional profiles show clearly that the shape distribution of the parallel Gd-silicide nanowire array is well-defined and uniformly-spaced. The unprecedented uniformity exhibited by this parallel array reflects a perfect spatial-organization characteristic of our performed one-dimensional self-assembling process. As shown in the topographic profile in Fig. 4, the nanowires are typically 4.0 ± 0.1 nm in width and have an average periodicity of 7.2 ± 0.2 nm; the heights observed in the empty-state image and the filled-state image are, respectively, 270 ± 10 pm and 250 ± 10 pm, similar to the height obtained by Lee *et al* (Lee & Kim, 2003). It is noted that the nanowire height is nearly equal to that of Si upper terraces on Si(110)-16×2 surface, which suggests that Gd atoms may incorporate into Si upper terraces to form one-dimensional Gd-silicide nanowires, similar to the proposed structure of Si-deficient rare-earth metal silicide nanowires (Chen *et al.*, 2002; Eames *et al.*, 2010).

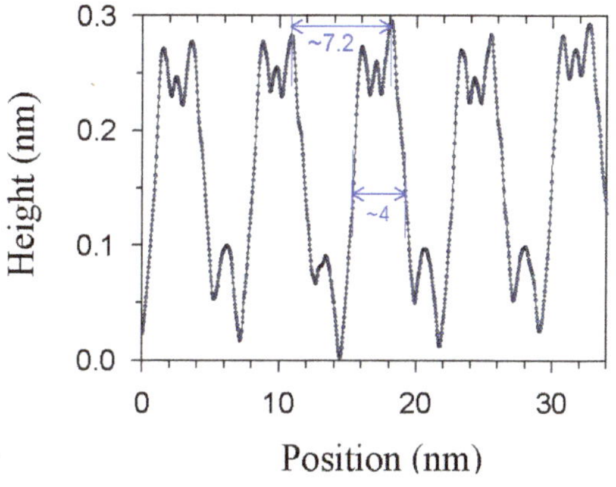


Fig. 4. Cross-sectional profile **A** across the parallel Gd-silicide nanowire array along the white line indicated in Fig. 3(d)

To gain insight into the local electronic structure of Gd-silicide nanowires, STS spectra (i.e., dI/dV curves) were measured on top of the individual Gd-silicide nanowires and the adjacent substrate between neighboring nanowires, as displayed in Fig. 5. A third dI/dV curve taken on clean Si(110) surface is also shown for reference. Each dI/dV curve is laterally averaged over 20 individual spectra taken within the corresponding region. These STS spectra allow us to compare the surface conductivity and local density of states (LDOS) in the areas selected on the surface. Curve **I** measured on the self-organized parallel Gd-silicide nanowires show non-zero conductance near the zero bias, indicating that Gd-silicide nanowires have one-dimensional metallic character along their length. As clearly revealed in Curve **I**, the Gd-silicide nanowires are characterized by a set of the discrete, distinct electronic states.

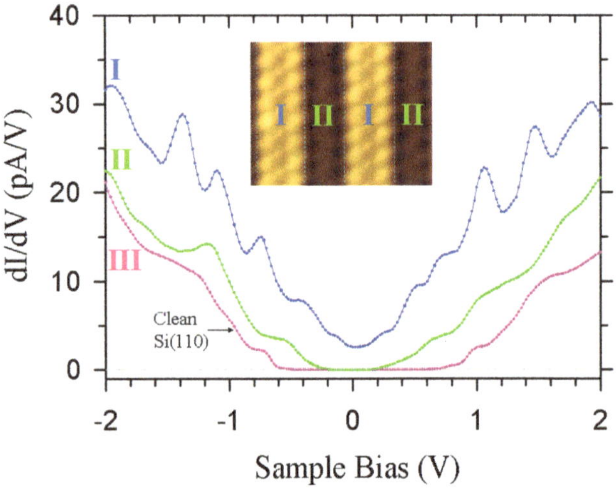


Fig. 5. Averaged dI/dV curves of the parallel Gd-silicide nanowires (Curve **I**), the adjacent substrate (Curve **II**), and the clean Si(110) surface (Curve **III**). Each dI/dV curve was laterally averaged over their corresponding areas in the inserted STM image ($15 \times 10$ nm$^2$).

These sharp peaks in the STS spectrum of individual Gd-silicide nanowire are characteristic of the Van Hope singularities expected for the LDOS of one-dimensional materials (Iancu *et al.*, 2009; Zeng *et al.*, 2008). Curve **II** taken over the adjacent substrate shows semiconducting behavior with a band gap of ~0.5 eV, significantly smaller than ~1.2 eV for clean Si(110) surface [see Curves **III**]. The reduced gap opening of the substrate together with the zigzag chain structure suggests that Si(110)-16×2 surface is reconstructed with Gd.

### 3.3 Parallel Ce-silicide nanowire arrays self-organized on Si(110)-16×2 surface

Figure 6 shows the STM topographic images of the parallel Ce-silicide nanowire array at different magnifications. As visibly observed in Figs. 6(a)−6(e), these parallel-aligned, straight and nearly defect-free nanowires cover a mesoscopic area with a typical lateral distance exceeding 1.3 μm and are elongated along the $[\bar{1}\,1\,2]$ direction. These nanowires thus possess an extraordinarily high aspect ratio. Also, this massively parallel nanowire array shows a regular periodicity and a high integration density. Moreover, these parallel nanowires are essentially identical to one another over the entire macroscopic area of the Si(110) surface. This result clearly reveals that these parallel nanowires grew along the periodic terraces of Si(110)-16×2 superstructure. Such large-area parallel Ce-silicide nanowire array could be applied in the wafer-scale integration of high-density nanoelectronic architectures on Si(110) surface. As clearly seen in Fig. 6(e), each Ce-silicide nanowire consists of double zigzag chains with distinct morphologies. In the atomic-resolution image of Fig. 6(f), we can visibly find that the elemental structures of the double zigzag chains in the Ce-silicide nanowire are completely different from those of up-and-down terraces on Si(110)-16×2 surface; moreover, the width of all individual nanowires remains atomically precise along their lengths. This perfect one-dimensional self-organized parallel array with good uniformity and alignment over a mesoscopic area of 1.3×1.3 $\mu m^2$ reveals a well-established long-range spatial ordering. The spontaneous formation of such a large-area, highly-ordered parallel array consisting of atomically-precise Ce-silicide nanowires with a regular periodicity has never been reported before.

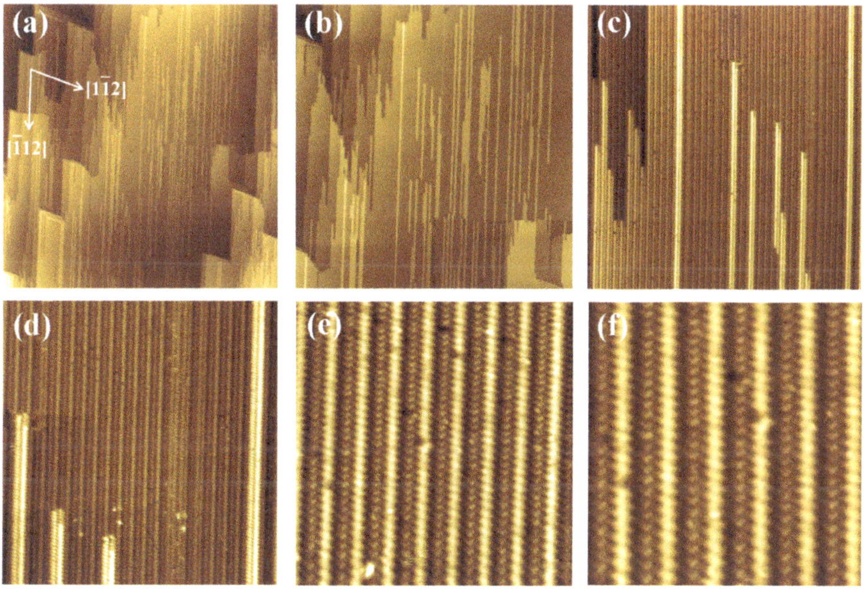


Fig. 6. A series of different magnified STM topographic images of the parallel Ce-silicide nanowire array: (a) 1.3 × 1.3 $\mu m^2$ ($V_s$ = 2.0 V, $I_t$ = 0.1 nA); (b) 850 × 850 $nm^2$; (c) 250 × 250 $nm^2$; (d) 125 × 125 $nm^2$; (e) 50 × 50 $nm^2$ ($V_s$ = 2.0 V, $I_t$ = 20 pA); and (f) 35 × 35 $nm^2$.

Figures 7(a) and 7(b) show dual-polarity STM images of an enlarged area of the parallel Ce-silicide nanowire array in Fig. 6(e), recorded at the sample bias $V_s$ = +1.5 V and −1.5 V. The empty-state image [Fig. 7(a)] at +1.5 V clearly shows a set of double zigzag chains on each Ce-silicide nanowire and the right zigzag chain appears much brighter than the left chain. In the filled-state image [Fig. 7(b)] at −1.5 V, the Ce-silicide nanowires consist of two linear rows of distinct atomic arrangements and the right linear row is also brighter than the left row. These dual-bias STM images evidently show that Ce-silicide nanowires are registry-aligned and really consist of a bundle of double chain-like structures with different morphologies. Figure 7(c) plots the superposition of the cross-sectional profiles of both line scans **E** and **F** across the empty-state and filled-state images of parallel Ce-silicide nanowires in Figs. 7(a) and 7(b), respectively. The section profiles of both lines **E** and **F** clearly show that these parallel Ce-silicide nanowires are well defined and periodically positioned. The unprecedented regularity exhibited by this parallel array in dual-bias images also reflects a perfect spatial-ordering of the charge distribution.

As clearly revealed in Fig. 7(c), these parallel-aligned nanowires are identical to 5.0 ± 0.2 nm in width and have an equal periodicity of 6.0 ± 0.2 nm. Moreover, the topographic maxima of the double zigzag chains in the empty-state image and the double linear rows in the filled-state image are spatially coincident; indicating that the height difference between the right chains and left chains is due to the topographic effect, not the electronic effect. The left–right asymmetry observed in the dual-polarity images of each Ce-silicide nanowire indicates that the building blocks for Ce-silicide nanowires are asymmetrically composed. Additionally, the widths of the left and right zigzag chains are 2.0 ± 0.2 and 3.0 ± 0.2 nm, respectively; while the double linear rows are equal to 2.5 ± 0.1 nm in width, similar to the width (2.2 ± 0.1 nm) of Si terraces of the Si(110)-16×2 reconstruction [see Fig. 2].

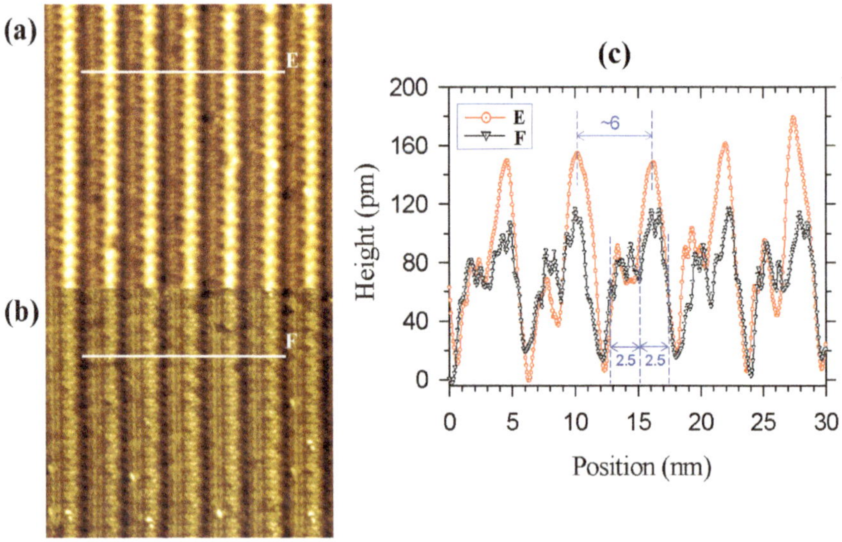


Fig. 7. Dual polarity STM images (45 × 35 nm$^2$) of the parallel Ce-silicide nanowires acquired at +1.5 V (a) and −1.5 V (b), respectively, and at 20 pA. (c) Cross-sectional profiles **E** and **F** across parallel-aligned Ce-silicide nanowires of the empty- and filled-state images along the white lines indicated in (a), respectively.

These results strongly suggest that Ce atoms nucleated concurrently along the upper and lower terraces of Si(110)-16×2 surface to form the one-dimensional Ce-silicide nanowires consisting of double chain row structures with different heights. The Ce-silicide nanowires are characterized by a well-defined shape indicating epitaxial growth, which is promising for the basic building block of nanoelectronic devices.

It is very interesting to explore the local electronic structures of the parallel Ce-silicide nanowire array. STS spectra (i.e., dI/dV curves) were measured on the individual Ce-silicide nanowires and also on the adjacent substrate between the neighboring nanowires, as displayed in Fig. 8. A third dI/dV curve taken on clean Si(110) surface is also shown for comparison. Each dI/dV curve was laterally averaged over 30 individual spectra taken within the corresponding areas indicated in the inserted STM image (20 × 10 nm$^2$). Curve **I** for the left chain shows an energy gap of ~0.68 eV, suggesting a semiconducting property along its length. These results clearly reveal that self-organized parallel Ce-silicide nanowires are semiconducting. Curve **II** for the adjacent substrate shows a lager energy gap of ~0.9 eV, smaller than the gap of ~1.2 eV obtained for the clean Si(110)-16×2 substrate [see Curve **III**].

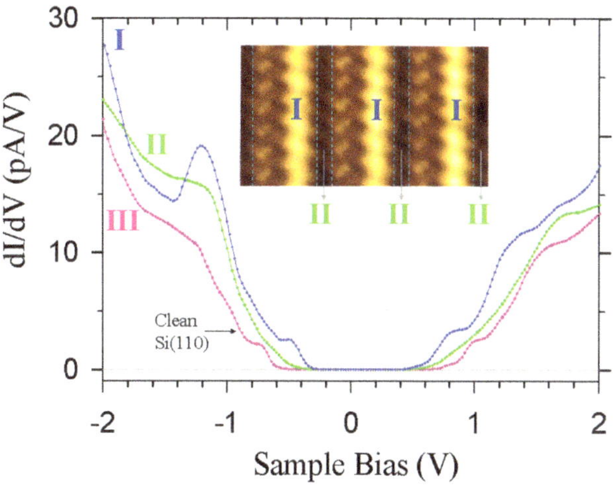


Fig. 8. Averaged STS spectra of the right zigzag chain taken on the individual Ce-silicide nanowires (Curve **I**), the adjacent substrate between neighboring nanowires (Curve **II**), and the clean Si(110) surface (Curve **III**).

The reduced gap opening of the substrate and the featureless LDOS means that the Si(110) surface was covered with a small amount of Ce. The semiconducting behavior of Ce-silicide nanowires is in contrast to the metallic character of other rare-earth silicide nanowires (Hong *et al.*, 2011a; Lee & Kim, 2003; Yeom *et al.*, 2005; Zeng *et al.*, 2008) and $CeSi_x$ Kondo compounds (Malterre *et al.*, 1993; Mimura *et al.*, 2005). The suppression of LDOS near the Fermi energy (i.e., zero bias) can be ascribed to the formation of a Mott-Hubbard insulating states at these Ce-silicide nanowires (Hong *et al.*, 2011c).

### 3.4 Parallel Er-silicide nanowire arrays self-organized on Si(110)-16×2 surface

Figure 9 shows a series of different magnified STM topographic images of parallel Er-silicide nanowire arrays. As clearly observed in Fig. 9, these parallel-aligned, straight andnearly defect-free nanowires cover a mesoscopic area with a typical length beyond 1.0 μm and are most elongated along the $[\bar{1}\,1\,2]$ direction. In Fig. 9(e), each Er-silicide nanowire consists of double bead chains, but unlike to the morphology of Gd-silicide nanowire with the bean chain in the middle of nanowire. Each bead chain is composed of round protrusions with a typical diameter of ~2.0 nm. Additionally, the adjacent substrate between the neighboring nanowires still shows a zigzag chain structure. All individual nanowires are atomically precise nanostructures, essentially identical to one another over the entire macroscopic area of Si(110) surface.Their width remains atomically precise along the nanowire and is typically 4.2 ± 0.1 nm [see Fig. 10].Therefore, these nanowires possess an extraordinarily high aspect ratio of greater than 250. Moreover, this massively parallel array shows a regular periodicity of 8.0 ± 0.2 nm [see Fig. 10]. Such a perfect one-dimensional self-organized parallel nanowire array with good uniformity and alignment over a mesoscopic area represents a well-established long-range spatial ordering. This large-area Er-silicide nanowire array also exhibits a high nanowire density and thus can be applied for the wafer-scale integration of high-density nanoelectronic devices.

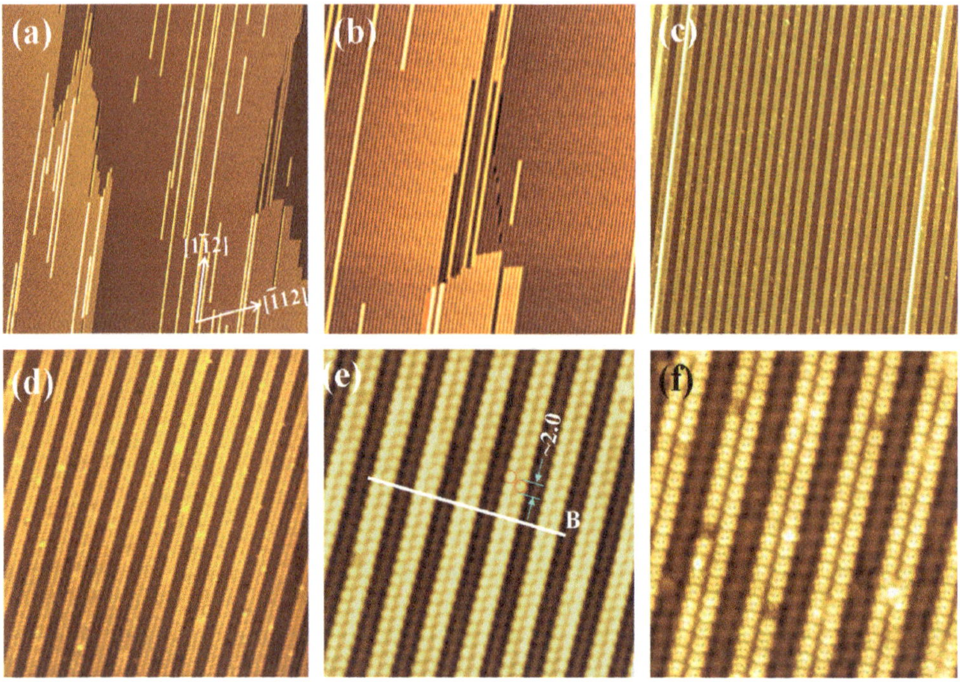


Fig. 9. (a)−(d) STM topographic images of parallel Er-silicide nanowire array on Si(110)-16×2 surface taken at different magnifications: (a) 1 × 1 μm² ($V_s$ = +3 V, $I_t$ = 30 pA), (b) 500 × 500 nm², (c) 200 × 200 nm², (d) 110 × 110 nm², (e) 60 × 60 nm² ($V_s$ = +3 V, $I_t$ = 10 pA), and (f) 36 × 36 nm² ($V_s$ =+2 V, $I_t$ = 10 pA).

Figure 10 displays the cross-sectional profiles of the line scan B across the parallel Er-silicide nanowire array in Fig. 9(e). This cross-sectional profiles show clearly that the shape distribution of the parallel Er-silicide nanowire array is well-defined and uniformly-spaced. The unprecedented uniformity exhibited by this parallel array reflects a perfect spatial-organization characteristic of our performed one-dimensional self-assembling process. As shown in the topographic profile in Fig. 10, the nanowires are typically 4.2 ± 0.1 nm in width and have a representative periodicity of 8.0 ± 0.2 nm; the average heights is 230 ± 20 pm, similar to the height obtained by Chen *et al.* (Chen *et al.*, 2002) The growth mechanism of such a highly-regular parallel array of uniformly-spaced Er-silicide nanowires on Si(110)-16×2 surface is also driven by the heteroepitaxial growth of Er silicides along the perfect grating-like upper terraces of Si(110)-16×2 superstructure, similar to the self-organized growth of Gd-silicide nanowires on Si(110)-16×2 surface.

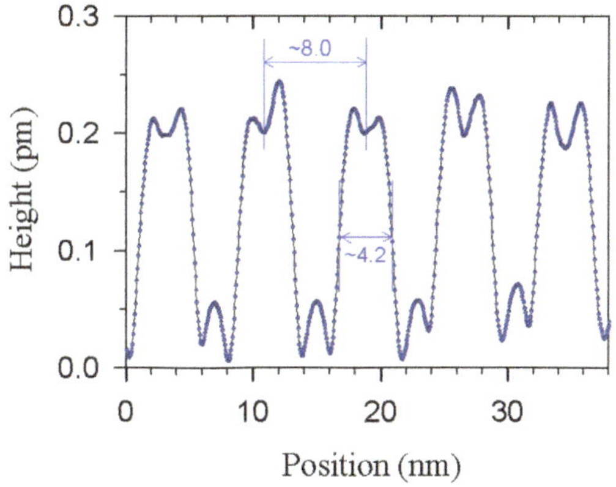


Fig. 10. Cross-sectional profile **B** across the parallel Er-silicide nanowire array along the white line indicated in Fig. 9(e)

Figure 11 displays the representative STS spectra of the self-organized Er-silicide nanowires (Curve **I**) and the adjacent substrate between neighboring nanowires (Curve **II**). Each dI/dV curve is laterally averaged over 30 individual spectra taken within the corresponding region of the inserted STM image. Curves **I** shows non-zero conductance near the zero bias, indicating that Er-silicide nanowires exhibit metallic nature along their length. Curve **II** depicts metallic-like behavior with a narrower band gap of ~0.12 eV, significantly smaller than ~1.2 eV for the clean Si(110) surface [see Curves **III**]. The metallic-like character of the substrate together with the zigzag chain structure suggests that Si(110) surface is reconstructed with Er.

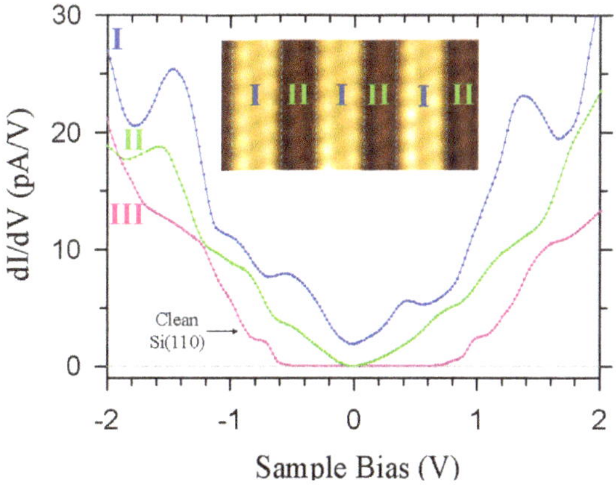


Fig. 11. Averaged dI/dV curves of the parallel Er-silicide nanowires (Curve **I**), the adjacent substrate (Curve **II**), and the clean Si(110) surface (Curve **III**). Each dI/dV curve was laterally averaged over their corresponding areas in the inserted STM image (25× 12 nm$^2$).

### 3.5 Parallel transition metal silicide nanowires self-assembled on Si(110)-16×2 surface

As clearly demonstrated in Figs. 3-11, the massively parallel arrays of various epitaxial rare-earth metal silicide nanowires on Si(110)-16×2 surface always grow along a unidirectional orientation of [$\bar{1}$ 1 2 ]with uniform distribution and identical size. However, the growth mode of rare-earth metal silicide nanowires on Si(110) surface is completely different from that of the transition-metal silicide nanowires on Si(110) surface. As shown in Fig. 12, the parallel-aligned Ni-silicide and Fe-silicide nanowires can not form well-ordered parallel nanowire arrays with a regular periodicity and identical dimensions on Si(110) surface, as reported in the previous results (He *et al.*, 2004; Liang *et al.*, 2006). The difference can be attributed to the fact that transition-metal silicide nanowires prefer to grow into the Si(110) substrate along the [$\bar{1}$ 1 0 ] direction via an "endotaxial" growth mechanism (He *et al.*, 2004) and do not favor to nucleate along the [$\bar{1}$ 1 2 ]-oriented terraces of Si(110)-16×2 reconstruction.

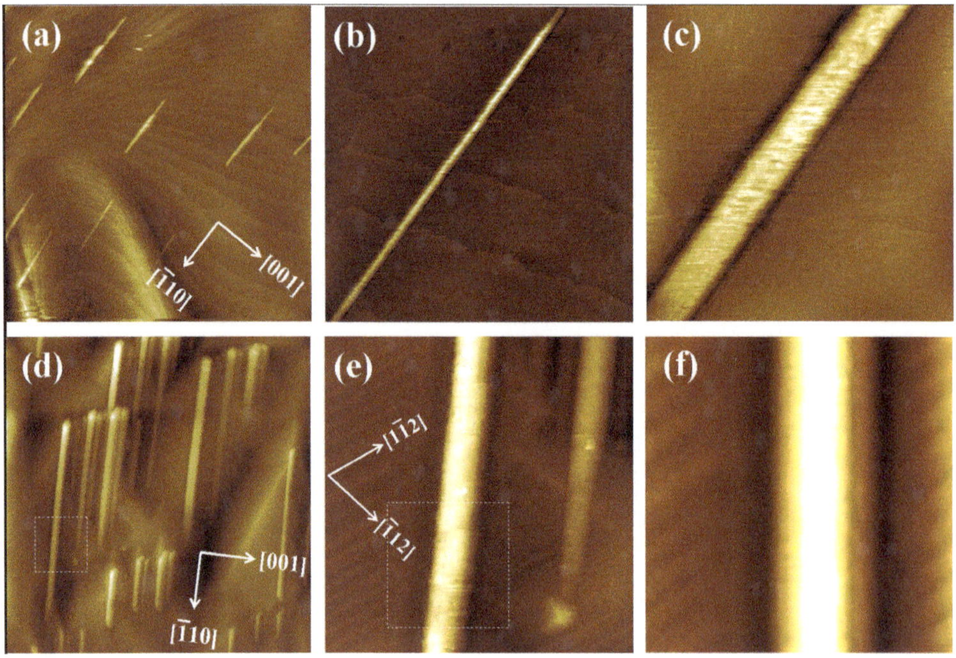


Fig. 12. STM topographic images of parallel Ni-silicide nanowires on Si(110) surface [(a)−(c)] and parallel Fe-silicide nanowires on Si(110) surface [(d)−(f)] at different magnifications: (a) 7 × 7 μm², (b) 1× 1 μm², (c) 500 × 500 nm², (d) 800 × 800 nm², (e) 125 × 125 nm², and (f) 45 × 45 nm².

## 4. Conclusion

We have successfully developed a simple and efficient bottom-up nanofabrication for the self-organization of mesoscopically-ordered parallel arrays, consisting of uniformly-spaced identical-size epitaxial rare-earth metal silicide nanowires over a mesoscopic area on Si(110)-16×2 surface. Three perfectly-ordered parallel arrays of atomically-precise Gd-silicide nanowires, Ce-silicide nanowires, and Er-silicide nanowires are presented to demonstrate the versatility of this novel approach. These massively parallel rare-earth silicide nanowire arrays were self-organized with widths and pitches (center-to-center distances) as small as 4~5 nm and 6~8 nm, respectively, and lengths exceeding 1 μm, which were achieved through the heteroepitaxial growth of rare-earth metal silicides along the periodic terraces of the Si(110)-16×2 surface. The ability to form such large-area, highly-ordered parallel arrays of epitaxial rare-earth metal silicide nanowires on the Si(110) nanotemplate represents a simple step towards the bottom-up nanofabrication of high-density, well-defined interconnections and functional nanowire-based CMOS nanodevices in a straightforward, cost-effective and high-throughput process. As they have the advantage of easy integration into the existing silicon-based integrated-circuit technology, we believe that our developed large-scale self-organization of epitaxial rare-earth metal silicide nanowire arrays on Si(110)-16×2 surface will be an important Si-compatible nanofabrication process for highly-integrated, well-defined parallel active nanoarchitectures with diverse functions over mesoscopic areas, which could be used to realize the future innovations in Si-based nanoelectronics.

## 5. Acknowledgements

This work is financially supported by the National Science Council of Taiwan under Grant Nos. 97-2738-M-415-001 and 97-2112-M-415-003-MY3.

## 6. References

Agarawal, R. (2008). Heterointerfaces in Semiconductor Nanowires. *Small*, Vol. 4, No. 11, (April 2008), pp. 1872-1893, ISSN 1613-6829.

An, T.; Yoshimura, M.; Ono, I. & Ueda, K. (2000). Elemental Structure in Si(110)-"16x2" Revealed by Scanning Tunneling Microscopy. *Physical Review B*, Vol. 61, No. 4 (January 2000), pp. 3006-3011, ISSN 1098-0121.

Castrucci, P.; Yubero, F. ; Vicentin, F. C.; Vogel, J. & Sacchi, M. (1995). Surface Crystal Field at The Er/Si(111) Interface Studied by Soft-X-Ray Linear Dichroism. *Physical Review B*, Vol. 52, No. 19, (November 1995), pp. 14035–14039, ISSN 1098-0121.

Chen, Y.; Ohlberg, D. A. A.; Medeiros-Ribeiro, G.; Chang, Y. A. & Williams, R. S. (2000). Self-Assembled Growth of Epitaxial Erbium Disilicide Nanowires on Silicon(001). *Applied Physics Letters*, Vol. 76, No. 26, (June 2000), pp. 4004-4006, ISSN 0003-6951.

Chen, Y.; Ohlberg, D. A. A. &Williams, R. S. (2002). Nanowires of Four Epitaxial Hexagonal Silicides Grown on Si(001). *Journal of Applied Physics*, Vol. 91, No. 5, (March 2002), pp. 3213-3218, ISSN 0021-8979, ISSN 0021-8979.

Cheng, W.; Teramoto, A.; Hirayama, M.; Sugawa, S. & Ohmi, T. (2006). Impact of Improved High-Performance Si(110)-Oriented Metal–Oxide–Semiconductor Field-Effect Transistors using Accumulation-Mode Fully Depleted Silicon-on-Insulator Devices. *Japanese Journal of Applied Physics*, Vol. 45, No. 48, (April 2006), pp. 3110-3116, ISSN 0021-4922.

Eames, C.; Probert, M. I. J. & Tear, S. P. (2010). The Structure and Growth Direction of Rare Earth Silicide Nanowires on Si(100). *Applied Physics Letters*, Vol. 96, No. 24, (June 2010), pp. 241903, ISSN 0003-6951.

Gambardella, P.; Dallmeyer, A.; Maiti, K.; Malagoli, M. C., Eberhardt, W.; Kern, K. & Carbone, C. (2002). Ferromagnetism in One-Dimensional Monatomicmetal Chains. *Nature*, Vol. 416, (March 2002), pp. 301-303, ISSN 0028-0836.

Hogg, S. M.; Vantomme, A. & Wu, M. F. (2002). Growth and Electrical Characterization of $GdSi_{1.7}$ Epilayers Formed by Channeled Ion Beam Synthesis. *Journal of Applied Physics*, Vol. 91, No. 6, (March 2002), pp. 3664-3668, ISSN 0021-8979.

He, Z.; Smith, D. J. & Bennett, P. A. (2004). Endotaxial Silicide Nanowires. *Physical Review Letters*, Vol. 93, No. 25, (December 2004), pp. 256102, ISSN 0031-9007.

Hong, I.-H.; Yen, S.-C. & Lin, F.-S. (2009). Two-Dimensional Self-Organization of an Ordered Au Silicide Nanowire Network on a Si(110)-16×2 Surface. *Small*, Vol. 5, pp. 1855–1861, ISSN 1613-6829.

Hong, I.-H.; Liao, Y.-C. & Yen, S.-C. (2009). Self-Organization of a Highly Integrated Silicon Nanowire Network on a Si(110)-16×2 Surface by Controlling Domain Growth. *Advanced Functional Materials*, Vol. 19, (November 2009), pp. 3389-3395, ISSN 161-3028.

Hong, I.-H.; Tsai, Y.-F. & Chen, T.-M. (2011). Self-Organization of Mesoscopically Ordered Parallel Gd-Silicide Nanowire Arrays on a Si(110)-16×2 Surface: A Massively Parallel Active Architecture. *Applied Physics Letters*, Vol. 98, No. 19, (May 2011), pp. 193118, ISSN 0003-6951.

Hong, I.-H. (2011). Self-Organization of Two-dimensional Highly-Regular Nanowire Meshes on Si(110)-16×2 Surface, Nova Science, *Mesh Networks*, to be published. ISBN 978-1-62100-150-8

Hong, I.-H.; Y.-C. Liao & Tsai, Y.-F. Electronic Structure of Strongly-Correlated Ce-Silicide Nanowires Self-Organized on Si(110)-16×2 Surface: One-Dimensional Mott-Hubbard Insulating State. Submitted to *Physical Review B*, ISSN 1098-0121.

Iancu, V.; Kent, P. R. C.; Zeng, C. G.; Weitering, H. H. (2009). Structure of $YSi_2$ Nanowires from Scanning Tunneling Spectroscopy and First Principles. *Applied Physics Letters*, Vol. 95, No. 12, (September 2009), pp. 123107, ISSN 0003-6951.

Lee, D. & Kim, S. (2003). Formation of Hexagonal Gd Disilicide Nanowires on Si(100). *Applied Physics Letters*, Vol. 82, No. 16, (April 2003), pp. 2619-2621, ISSN 0003-6951.

Lee, D.; Lim, D. K.; Bae, S. S.; Kim, S.; Ragan, R.; Ohlberg, D. A. A.; Chen, Y. & Williams, R. S. (2006). Unidirectional Hexagonal Rare-Earth Disilicide Nanowires on Vicinal Si(100)-2×1. *Applied Physics A: Materials Science & Processing*, Vol. 80, No. 6, (March 2005), pp. 1311–1313, ISSN 0947-8396.

Liang, S.; Islam, R.; Smith, D. J.; Bennett, P. A.; O'Brien, J. R. & Taylor, B. (2006). Magnetic iron silicide nanowires on Si(110). *Applied Physics Letters*, Vol. 88, No. 11, (March 2006), pp. 113111, ISSN 0003-6951.

Liu, B. Z. & Nogami, J. (2003). Growth of Parallel Rare-Earth Silicide Nanowire Arrays on Vicinal Si(001). *Nanotechnology*, Vol. 14, (June 2003), pp. 873–877, ISSN 0957-4484.

Liu, B. Z. & Nogami, J. (2003). A Scanning Tunneling Microscopy Study of Dysprosium Silicide Nanowire Growth on Si(001). *Journal of Applied Physics*, Vol. 93, No. 1, (January 2003), pp. 593-599, ISSN 0021-8979.

Lu, W. & Lieber, C. M. (2007). Nanoelectronics from The Bottom Up. *Nature Materials*, Vol. 6, (November 2007), pp. 841-850, ISSN 1476-1122.

McChesney, J. L.; Kirakosian, A.; Bennewitz, R.; Crain, J. N.; Lin, J.-L. & Himpsel, F. J. (2002). Gd Disilicide Nanowires Attached to Si(111) Steps. *Nanotechnology*, Vol. 13, (July 2002), pp. 545–547, ISSN 0957-4484.

Malterre, D.; Grioni, M.; Weibel, P.; Dardel, B. & Baer, Y. (1993). Correlation between The Kondo Temperature and The Photoemission Spectral Function in The $CeSi_x$ ($1.6 \leq x \leq 2$) System. *Physical Review B*, Vol. 48, No. 14, (October 1993), pp. 10599, ISSN 1098-0121.

Melosh, N. A. Boukai, A.; Diana, F.; Gerardot, B.; Badolato, A.; Petroff, P. M.; & Heath, J. R. (2003). Ultrahigh-Density Nanowire Lattices and Circuits. *Science*, Vol. 300, (April 2003), pp. 112-115, ISSN 0036-8075.

Mimura, K.; Noguchi, S.; Suzuki, M.; Higashiguchi, M.; Shimada, K.; Ichikawa, K.; Taguchi, Y.; Namatame, H.; Taniguchi, M. & Aita, O. (2005). Temperature Dependence of High-Resolution Resonant Photoemission Spectra of CeSi. *Journal of Electron Spectroscopy and Related Phenomena*. Vol. 144-147, (June 2005), pp. 715, ISSN 0368-2048.

Mizuno, T.; Sugiyama, N.; Tezuka, T.; Moriyama, Y.; Nakaharai, S. & Takagi, S.-I. (2005). [110]-surface strained-SOI CMOS devices. *IEEE Transaction on Electron Devices*, Vol. 52, No. 3, (March 2005), pp. 367-374, ISSN 0018-9383.

Münzenberg, M.; Felsch, W. & Schaaf, P. (2007). Tuning the *4f* State Occupancy of Ce in Highly Correlated CeSi/Fe Multilayers: An X-Ray Absorption Spectroscopy Study. *Physical Review B*, Vol. 76, No. 1, (July 2007) pp. 014427, ISSN 1098-0121.

Nogami, J.; Liu, B. Z.; Katkov, M. V.; Ohbuch, C. & Birge, N. O. (2001). Self-Assembled Rare-Earth Silicide Nanowires on Si(001). *Physical Review B*, Vol. 63, No. 23, (May 2001), pp. 233305, ISSN 1098-0121.

Ohbuchi, C. & Nogami, J. (2002). Holmium Growth on Si(001): Surface Reconstructions and Nanowire Formation. *Physical Review B*, Vol. 66, No. 16, (October 2002), pp. 165323, ISSN 1098-0121.

Padova, P.D.; Quaresima, C.; Perfetti, P.; Olivieri, B.; Leandri, C.; Aufray, B.; Vizzini, S. & Lay, G. L. (2008). Growth of Straight, Atomically Perfect, Highly Metallic Silicon Nanowires with Chiral Asymmetry. *Nano Letters*, Vol. 8, No. 1, (November 2007), pp. 271-275, ISSN 1530-6984.

Pierre, J.; Auffrettll, S.; Chroboczek, J. A. & Nguyen, T. T. A. (1994). Magnetotransport in the Rare Earth Silicides $RSi_{2-x}$, *Journal of Physics: Condensed Matter*, Vol. 6, No. 1, (January 1994), pp. 79-92, ISSN 0953-8984.

Pecharsky, V. K. & Gschneidner Jr., K. A. (2001). $Gd_5(Si_{1-x}Ge_x)_4$: An Extremum Material. *Advanced Materials*, Vol. 13, No. 9, (May 2001), pp. 683-686, ISSN 1521-4095.

Preinesberger, C.; Pruskil, G.; Becker, S. K.; Dähnea, M.; Vyalikh, D. V.; Molodtsov, S. L.; Laubschat, C. & Schiller, F. (2005). Structure and Electronic Properties of Dysprosium-Silicide Nanowires on Vicinal Si(001). *Applied Physics Letters*, Vol. 87, No. 8, (August 2005), pp. 083107, ISSN 0003-6951.

Qiu, D.; Zhang, M.-X. & Kelly & P. M. (2009). Crystallography of Self-Assembled $DySi_2$ Nanowires on Si Substrate. *Applied Physics Letters*, Vol.94, No. 8, (February 2009), pp. 083105, ISSN 0003-6951.

Sahaf, H.; Masson, L.; Léandri, C.; Aufray, B.; Lay, G. L. & Ronci F. (2007). Formation of A One-Dimensional Grating at the Molecular Scale by Self-Assembly of Straight Silicon Nanowires. *Applied Physics Letters*, Vol. 90, No. 26, (June 2007), pp. 263110, ISSN 0003-6951.

Segovia, P.; Purdie, D.; Hengsberger, M. & Baer, Y. (1999). Observation of Spin and Charge Collective Modes in One-Dimensional Metallic Chains. *Nature*, Vol. 402, (December 1999), pp. 504-507, ISSN 0028-0836.

Shaheen, S. A. & Mendoza, W. A. (1999). Origin of Multiple Magnetic Transitions in $CeSi_x$ Systems. *Physical Review B*, Vol. 60, No. 13, (October 1999), pp. 9501-9505, ISSN 1098-0121.

Shchukin, V. A. & Bimberg, D. (1999). Spontaneous Ordering of Nanostructures on Crystal Surfaces. *Reviews of Modern Physics*, Vol.*71*, No. 4, (July 1999), pp. 1125-1171, ISSN 0034-6861.

Smith, J. S.; Zan, J. A.; Lin, C. L. & Li, J. (2005). Electric, Thermal, and Magnetic Properties of $CeSi_x$ with $1.57<x\leq 2.0$. *Journal of Applied Physics*, Vol. 97, No. 10, (May 2005), pp. 10A905, ISSN 0021-8979.

Snijders, P. C. & Weitering, H. H. (2010). *Colloquium*: Electronic instabilities in self-assembled atom wires. *Reviews of Modern Physics*, Vol. 82, No. 1, (January 2010), pp. 307-329, ISSN 0034-6861.

Soukiassian, P.; Semond, F.; Mayne, A. & Dujardin, G. (1997). Highly Stable Si Atomic Line Formation on the β-SiC(100) Surface. *Physical Review Letters*, Vol. 79, No. 13, (September 1997), pp. 2498-2501, ISSN 0031-9007.

Stekolnikov, A. A.; Furthmüller, J. & Bechstedt, F. (2004). Long-Range Surface Reconstruction: Si(110)-16×2. *Physical Review Letters*, Vol. 93, No. 13, pp. 136104, ISSN 0031-9007.

Yamamoto, Y.; Sueyoshi, Sato, T. T. & Iwatsuki, M. (2000). High-Temperature Scanning Tunneling Microscopy Study of the '16×2'⇔(1×1) Phase Transition on an Si(110) Surface. *Surface Science*, Vol. 466, (August 2000), pp. 183-188, ISSN 0036-8075.

Yang, M.; Chan, W. C.; Shi, L.; Fried, D. M.; Stathis, J. H.; Chou, A. I.; Gusev, E.; Ott, J. A.; Burns, L. E.; Fischetti, M. V. & Ieong, M. (2006). Hybrid-Orientation Technology (HOT): Opportunities and Challenges. *IEEE Transaction on Electron Devices*, Vol. 53, No. 5, (May 2006), pp. 965-978, ISSN 0018-9383.

Ye, G.; Crimp, M. A. & Nogami, J. (2009). Self-Assembled Gd Silicide Nanostructures Grown on Si(001). *Journal of Applied Physics*, Vol. 105, No. 10, (May 2009), pp. 104304, ISSN 0021-8979.

Yeom, H. W.; Kim, Y. K.; Lee, E. Y.; Ryang, K.-D.; Kang, & P. G. (2005), Robust One-Dimensional Metallic Band Structure of Silicide Nanowires. *Physical Review Letters*, Vol. 95, No. 20, (November 2005), pp. 205504, ISSN 0031-9007.

Yokota, T.; Fujimura, N. & Ito, T. (2002). Effect of Substitutionally Dissolved Ce in Si on the Magnetic and Electric properties of Magnetic Semiconductor $Si_{1-x}Ce_x$ Films. *Applied Physics Letters*, Vol. 81, No. 21, (Noveber 2002), pp. 4023-4025, ISSN 0003-6951.

Zeng, C. G.; Kent, P. R .C.; Kim, T.-H.; Li, A.-P.; Weitering, H. H. (2008). Charge-Order Fluctuations in One-Dimensional Silicides. *Nature Materials*, Vol. 7, (July 2008), pp. 539-542, ISSN 1476-1122.

Zhirnov, V. V. & Herr, D. J. C. (2001). New Frontiers: Self-Assembly and Nanoelectronics. *Computer*, Vol. 34, No. 1, (January 2001), pp. 34-43, ISSN 0018-9162.

Zhou, W.; Zhu, Y. Ji T.; Hou, X. & Cai, Q. (2006). Formation and Evolution of Erbium Silicide Nanowires on Vicinal and Flat Si(001). *Nanotechnology*, Vol. 17, (January 2006), pp. 852–858, ISSN 0957-4484.

Žutić, I. (2006). Gadolinium Makes Good Spin Contacts. *Nature Materials*, Vol. 5, (October 2006), pp. 771-772, ISSN 1476-1122.

Wanke, M.; Löser, K.; Pruskil, G. & Dähne, M. (2009). Structural and Electronic Properties of Rare Earth Silicide Nanowires on Si(557). *Physical Review B*, Vol. 79, No. 15, (April 2009), pp. 155428, ISSN 1098-0121.

# 8

# Nanofabrication and Characterization of Plasmonic Structures

Yongqi Fu[1], Fengzhou Fang[2] and Zongwei Xu[2]
[1]*School of Physical Electronics, University of Electronic Science and Technology of China,*
[2]*Key Laboratory of Precision Measuring Technology & Instruments Centre of MicroNano Manufacturing Technology, Tianjin University, China*

## 1. Introduction

The nanofabrication processes can be divided into two well defined approaches: 1) 'top-down' and 2) 'bottom-up'. The 'top-down' approach uses traditional methods to guide the synthesis of nanoscale materials. The paradigm proper of its definition generally dictates that in the 'top-down' approach it all begins from a bulk piece of material, which is then gradually or step-by-step removed to form objects in the regime of nanometer-size scale. Well known techniques such as photo lithography, electron beam lithography, anodization, and ion- and plasma-etching, that will be later described, all belong to this type of approach. The top-down approach for nanofabrication is the method firstly suggested by Feynman in his famous American Physical Society lecture in 1959.

Top down fabrication can be likened to sculpting from a block of stone. A piece of the base material is gradually eroded until the desired shape is achieved. That is, you start at the top of the blank piece and work your way down removing material from where it is not required. Nanotechnology techniques for top down fabrication vary but can be split into mechanical and chemical fabrication techniques. The most top down fabrication technique is nanolithography. In this process, required material is protected by a mask and the exposed material is etched away. Depending upon the level of resolution required for features in the final product, etching of the base material can be done chemically using acids or mechanically using ultraviolet light, and x-rays or electron beams. This is the technique applied to the manufacture of computer chips.

Bottom up fabrication can be described as building a brick house. Instead of placing bricks one-by-one at a time to produce a house from bottom, bottom up fabrication technology places atoms or molecules one-by-on at a time to build the desired nanostructure. Such processes are time consuming and so self assembly techniques appeared where the atoms arrange themselves as required. Self assembling nanomachines are regularly mentioned by science fiction writers but significant obstacles including the laws of physics will need to be overcome or circumvented before this becomes a reality. Other areas involving bottom up fabrication are already quite successful. Manufacturing quantum dots by self-assembly quantum dots has rendered the top down lithographic approach to semiconductor quantum dot fabrication virtually obsolete. One of the basic 'bottom-up' techniques is chemical precipitation by which nanoparticles of metals, alloys, and oxides etc. are prepared in aqueous or organic solutions. There are several ways to obtain nanoscale precipitates. They can be derived by

1. A controlled phase transformation (*i.e.*, liquid state diffusion) guided by the free energy diagrams or
2. By controlling the solid state diffusion: following a composite route approach mixing, *e.g.*, two different materials and stirring them mechanically.
3. Other approaches can be found in exploiting internal oxidation of materials or thin film deposition of coatings or sputtering.

For fabrication of plasmonic nanostructures, the most commonly used methods are self-assembly of nanoparticles or monomer/polymer molecules and electrochemical deposition (includes electroplating and electro-deposition). This chapter involves both top-down and bottom-up technologies which are being used for fabrication of plasmonic nanostructures and devices.

## 2. Nanofabrication of plasmonic structures

In this chapter, we targeted two commonly used top-down approaches for fabrication of subwavelength metallic structures: focused ion beam (FIB) technology and laser interference photolithography. The former can realized fine nanofabrication over a local tiny area in one-step only. But it is a technique with high expenditure and small localized fabrication area. The latter can realize large area fabrication and cost effective. But it needs pattern transformation from photoresist into substrate. As a typical bottom-up technique, self-assembly monolayer is introduced for nanofabrication of metallic particles array.

Some fabrication examples were presented. Problems existing in the fabrication processes were addressed as well.

### 2.1 Focused ion beam technology

A. Metallic nanoparticles array

As well known, LSPR-enhanced transmission plays a crucial role for the spectrum during detection. The LSPR effect modifies electromagnetic (EM) fields and indirectly affects the spectrum width which determines detection resolution for biosensing. Therefore, investigating the EM fields is necessary while design the metallic nanoparticles arrays. A few reports were found for design of the metallic nanoaprticles arrays in EM fields analysis points of view. In this paper, we proposed an approach for design of the metallic nanostructure arrays with E-field enhancements through the EM fields calculation. To calculate the EM fields of the metallic nanoaprticles array, we need to consider resonance wavelength $\lambda_{SPP}$ firstly because the LSPR can only be excited in monochromatic incident light with linear polarization (TM mode for asymmetric structures). However, currently, besides Mie theory for sphere and ellipse particles, there is no theoretical model can be used to directly calculate the resonance wavelength for the metallic nanoparticles array with complicated shapes such as triangular, rhombic, star, and prolate etc. A computational numerical analysis must be employed to calculate the resonant wavelength $\lambda_{SPP}$.

Considering this, our approach is that firstly calculate the spectrum (extinction efficiency) of the array using FDTD algorithm. Then peak wavelength of the spectrum is regards as the resonant wavelength $\lambda_{SPP}$. After that, we use the resonant wavelength $\lambda_{SPP}$ as the incident light to calculate the corresponding EM fields of the particles array. In order to fully understand the propagation properties and EM fields distribution, three-dimensional (3D) computational numerical calculation is carried out on the basis of the FDTD algorithm.

Focused ion beam (FIB) direct milling technique has advantages of one-step fabrication process, no material selectivity, and locally tiny area milling. Two-dimensional patterns with arbitrary shapes can be fabricated. Therefore, FIB technique provides large design freedom for the nanoparticles designing that is incomparable for the chemistry-based techniques. The milling experiments were carried out by our FIB machine (FEI Nova 200) with a liquid gallium ion source. This FIB machine is integrated with a scanning electron microscope (SEM) and a gas-assistant etching (GAE) functions. This machine used a focused $Ga^+$ ion beam with energy of 30 keV, a probe current of 1pA~20 nA. For the smallest beam currents, the beam was focused down to 7 nm in diameter at full width and half maximum (FWHM). Using a computer program the milling process is carried out by means of varying the ion dose for different relief depths. The defined area for the FIB with bitmap function and zero-overlap scanning was 1.5×1.5 $\mu m^2$. Processing parameters in our FIB experiment is 30 keV ion energy, 10 pA beam current, and 5 μs dwell time. Considering the inherent ion beam broadening effect, firstly, a bow-tie structure was fabricated using the FIB direct milling for optimization of the process parameters. Figure 1 is the FIB fabricated rhombic particles array. It can be seen that the bow-tie-shaped nanoparticles array was well etched with sharp corners and apexes.

Figure 2 (a) shows the X-Y cross-section and three dimensional (3D) plot of a single cross-shaped gold nanoparticle. Along the $x$-axis direction, the distance along the external vertices is denoted as long axis $a$, and the distance along the internal vertices is denoted as short axis $b$, $h$=60 nm. Our previous theoretical study revealed that the nanoparticles with such shape can generate significant SPPs enhanced E-field intensity with coupled "hot-spots" which is a key factor for SERS.

Using a computer program the milling process is carried out by means of varying the ion dose for different relief depths. The defined area for the FIB with bitmap function and zero-overlap scanning was 4×4 $\mu m^2$. Processing parameters in our FIB experiment is 30 keV ion energy, 10 pA beam current, and 5 μs dwell time. Figure 2 (b) is SEM micrograph of the FIB fabricated particles array. It can be seen from the inset image that the adjacent spatial gap is 29.2 nm. The cross-shaped nanoparticles array was well etched with sharp corners and apexes.

A hybrid spider-web-like rhombic nanoparticles array was designed, as shown in Figs. 3 (a) and (b). We adopted this specific pattern based on the following considerations:

1. Ensure interconnections of metallic nanowires for each rhombic particle.
2. Leave extra space for FIB fabrication due to its inherent line broadening effect in horizontal direction, and the web lines themselves having tip features from which "hot spots" can be formed.
3. The lines with two different sizes were designed so as to form irregular web lines in which some lines are broken and the others are connected with the rhombic particles because of the FIB broadening effect.

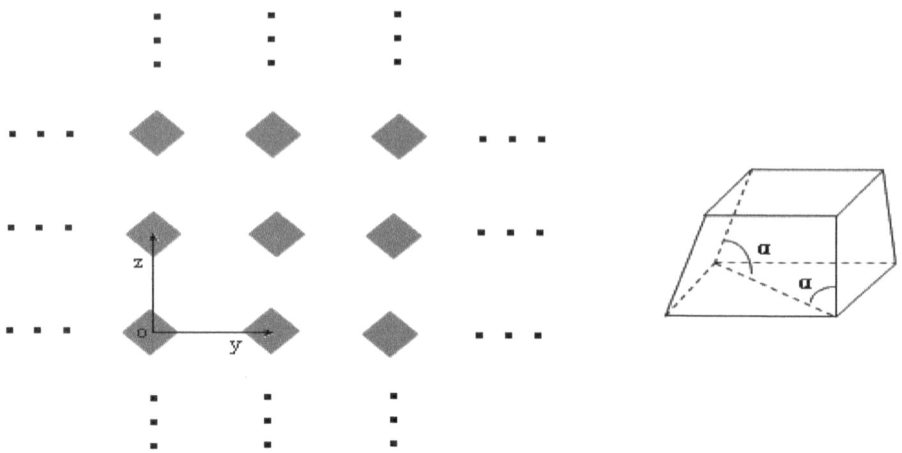


(a)

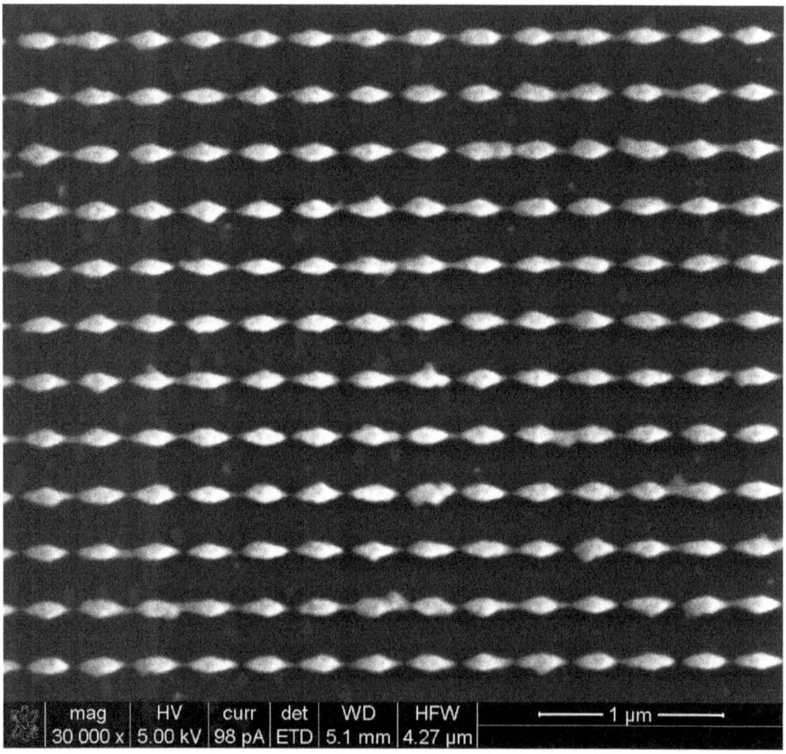


(b)

Fig. 1. (a) Symmetrical two-dimensional infinite rhombic silver nanostructures. (b) SEM micrograph of the FIB fabricated rhombic nanoparticles array. Inset image is zoom-in of the array

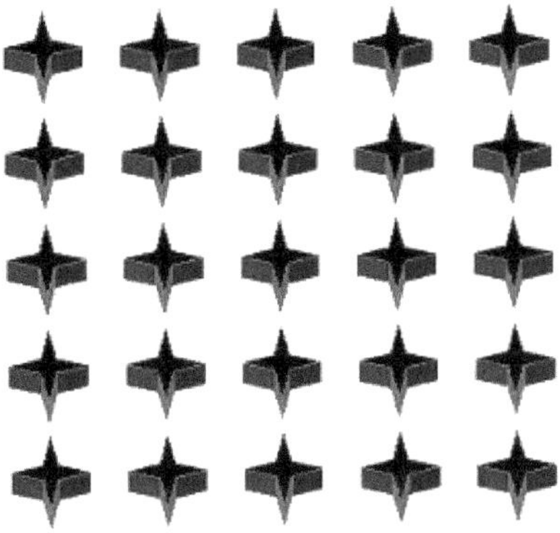

(a)

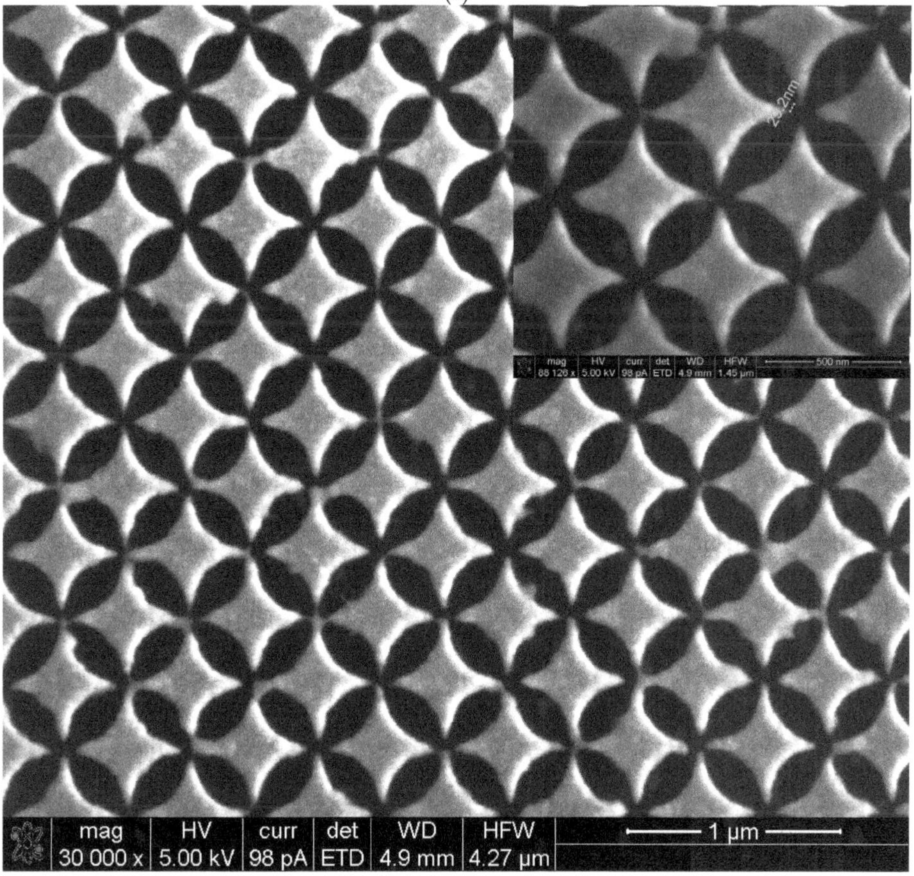


(b)

Fig. 2. (a) Schematic diagram of three-dimensional distribution of the cross-shaped array. (b) Figure 3 SEM micrograph of the FIB fabricated cross-star nanoparticles array. Inset image is zoom in of the array with measured adjacent apex-to-apex space size of 29.2 nm

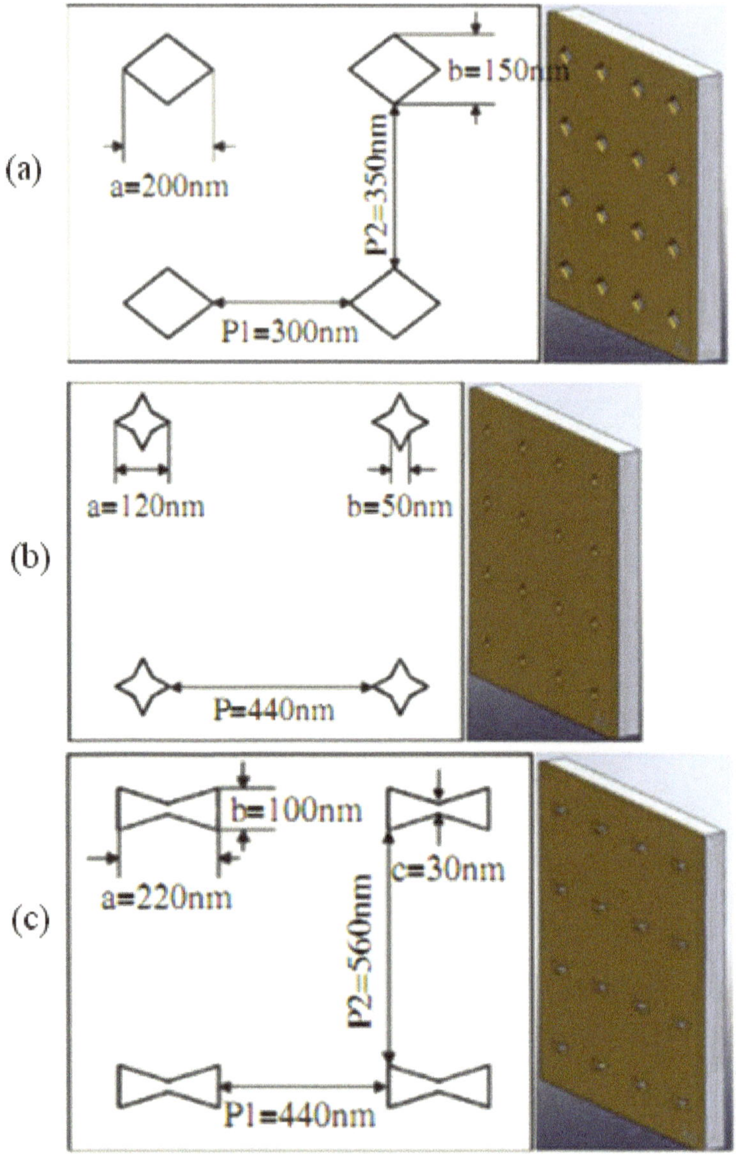


Fig. 3. Schematic drawing of the nanohole arrays with (a) rhombus, (b) cross star, and (c) butterfly nanostructures.

Height of the Au rhombic particles is 60 nm supported on a glass substrate. The FIB fabricated array area is 20×20 μm². Using a computer program the milling process is carried out by means of varying the ion dose for different relief depths. The defined area for the FIB with bitmap function and zero-overlap scanning was 20×20 μm². Processing parameters in our FIB experiment is 30 keV ion energy, 10 pA beam current, and 5 μs dwell time.

B. Metallic nanoholes array

Since the extraordinary optical transmission phenomenon was investigated in 1998 [1], nanoholes array have attracted great interest for its unique properties, as well as its potential practical significance for photonic devices and the applications including sensor, optical filtering and lithography [2-10]. The extraordinary optical transmission phenomenon is generally attributed to the excitation of localized Surface Plasmon Polaritons (SPPs) at the edge of the noble metal, and is strongly influenced by the edge shapes of the metal [6]. There are two main surface plasmon modes excited on the metal film. One is the extended surface mode depending on the periodicity; the other is the localized waveguide mode depending on the structure of the hole. The enhanced transmission intensity through the hole is due to the coupling of the two kinds of surface plasmon modes [7, 11]. In comparison to the metallic nanoparticles that behave as radiation, the metallic nanoholes act as surface wave propagation. The extraordinary transmission properties of the cylindrical, rectangular and square hole nanostructure based on simulation and experiment was reported before [8, 12].
In this chapter, we experimentally investigated the near field optical properties of three different shaped gold nanoholes arrays which were corrugated through an Au film supported on quartz substrate. The near optical transmission properties is obtained using a near-field scanning optical microscope (NSOM), and it is observed that the experimental results are in agreement with the Finite Difference Time Domain (FDTD) calculation results. Using a computer program the milling process is carried out by means of varying the ion dose for different relief depths. The defined area for the FIB with bitmap function and zero-

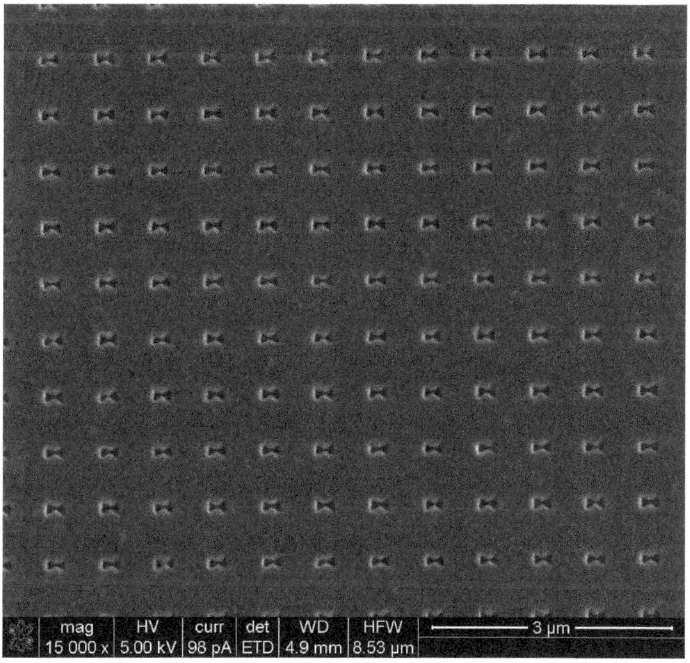


Fig. 4. Scanning electron microscope image of the butterfly nanohole arrays.

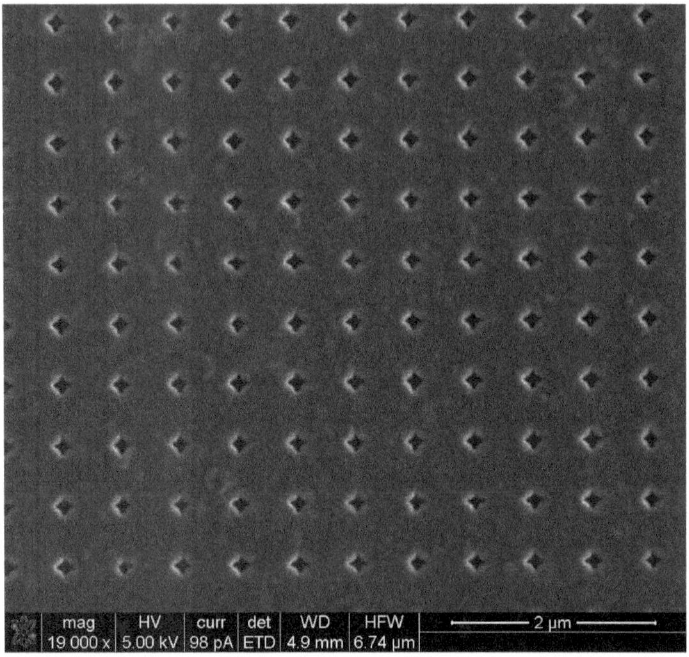


Fig. 5. Scanning electron microscope image of the cross star nanohole arrays.

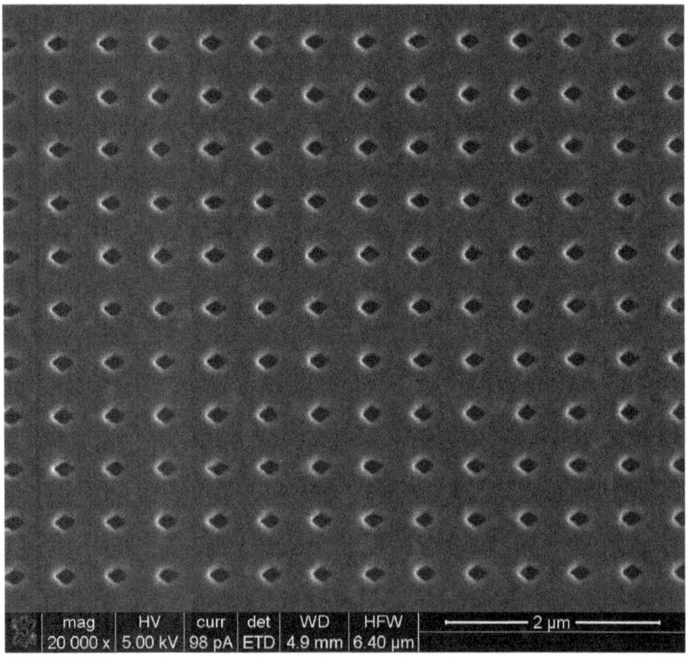


Fig. 6. Scanning electron microscope image of the rhombus nanohole arrays.

overlap scanning was 25×25 μm$^2$. Processing parameters in our FIB experiment is 30 keV ion energy, 10 pA beam current, and 5 μs dwell time. Figure 3 is a schematic drawing of the nanoholes arrays with (a) rhombus, (b) cross star, and (c) butterfly nanostructures. Figure 4, 5 and 6 is SEM micrograph of the FIB fabricated nanoholes array with different shapes of bow-tie, cross-star, and rhombic, respectively.[13]

C. Plasmonic lenses

A novel structure called plasmonic micro-zone plate-like (PMZP) or plasmonic lens with chirped slits is put forth to realize superfocusing. It was proposed by our group [14, 15]. Unlike conventional Fresnel zone plate (CFZP), a plasmonic structure was used and combined with a CFZP. Configuration of the PMZP is an asymmetric structure with variant periods in which a thin film of Ag is sandwiched between air and glass. The PMZP is a device that a quartz substrate coated with Ag thin film which is embedded with a zone plate structure with the zone number N < 10. Figure 7 (a) is an example of schematic diagram of the structure. Fabrication and characterization of the plasmonic lens was done, as shown in Figs. 7 (a)-(d) [16]. The Au thin film of 200 nm in thickness was coated on quart substrate using e-beam evaporation technique. The lens was fabricated using focused ion beam (FEI Quanta 200 3D dual beam system) direct milling technique, as shown in Fig. 7 (b). Geometrical characterization was performed using an atomic force microscope (Nanoscope 2000 from DI company). Figure 7 (c) shows topography of the FIB fabricated plasmonic lens. The optical measurement was performed with a near-field optical microscope (MultiView 2000$^{TS}$ from Nanonics Inc. in Israel) where a tapered single mode fiber probe, with an aperture diameter of 100 nm, was used working in collection mode. The fiber tip was raster scanned at a discrete constant height of 500 nm, 1.0 μm, 1.5 μm, 2.0 μm, 2.5 μm, 3.0 μm, 3.2 μm, 3.5 μm, 3.7 μm, 4 μm, 4.5 μm, and 5 μm, respectively, above the sample surface, and allowing us to map the optical intensity distribution over a grid of 256×256 points spanning an area of 20×20 μm$^2$. Working wavelength of the light source is 532 nm (Nd: YAG laser with power of 20 mW).

Additionally, a typical lock-in amplifier and optical chopper were utilized to maximize the signal-to-noise ratio. Figure 7 (d) shows the measured three-dimensional (3D) electric field intensity distribution of the lens at propagation distance of 2.5 μm.

To further improve focusing quality of the circular holes-based plasmonic lens, an elliptical nanoholes-based plasmonic lens was put forth, as shown in Fig. 8 (a) [17]. The plasmonic lens is composed of elliptical pinholes with different sizes distributed in different rings with variant periods. Long-axis of the ellipse is defined as $a_n=3w_n$, whereas $w_n$ is width of the corresponding ring width, and $n$ is the number of rings. A thin film of Ag coated on the glass substrate is perforated by the pinholes. The numbers of $w_n$ and radius for different rings are listed in Table1.

| Ring No. | 1 | 3 | 3 | 4 | 5 | 6 | 7 | 8 |
|---|---|---|---|---|---|---|---|---|
| Ring radius(μm) | 1.41 | 2.11 | 2.78 | 3.43 | 4.08 | 4.73 | 5.37 | 6.01 |
| $w_n$ (nm) | 245 | 155 | 116 | 93 | 78 | 67 | 59 | 53 |

Table 1. The numbers of $w_n$ for different rings with orders from inner to outer (designed $f$=1 μm)

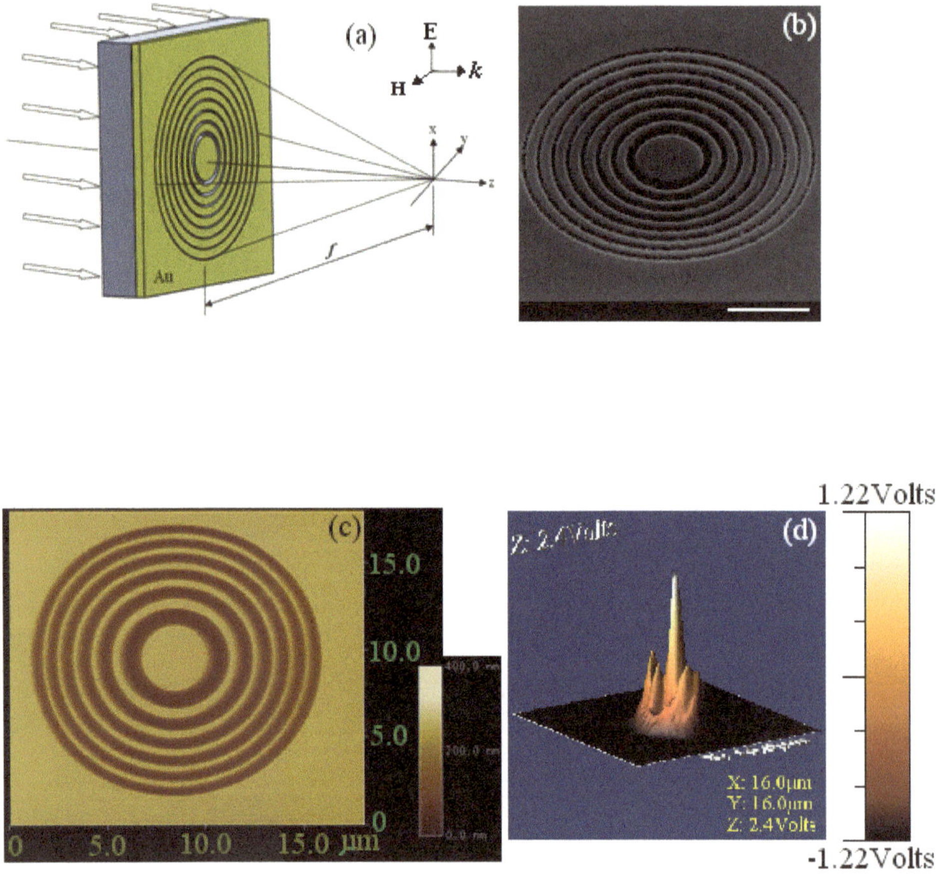


Fig. 7. (a) Schematic diagram of the sandwiched plasmonic lens with chirped circular slits corrugated on Au film. Width of the outmost circular slit is 95 nm. Lens dimension (outer diameter) is 12 μm. (b) Scanning electron microscope image of the lens fabricated using focused ion beam milling technique. The scale bar is 4 μm. (c) AFM measurement result: topography of the fabricated lens. (d) NSOM characterization result of the lens: 2D E-field intensity distribution at propagation distance of 2.5 μm

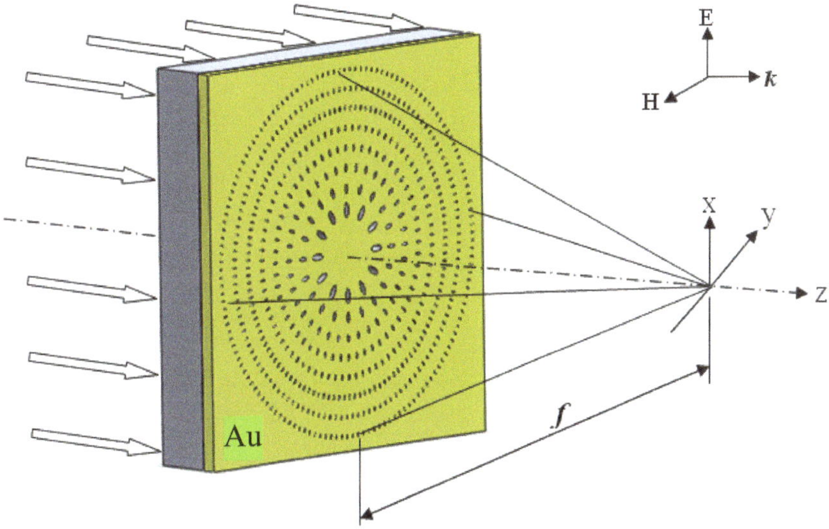


(a)

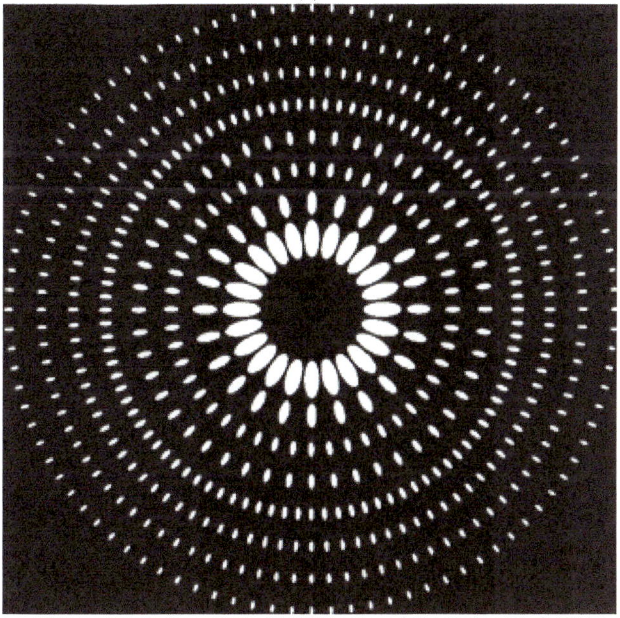

(b)

Fig. 8. (a) Schematization of the pinhole array with focal length *f*. Lateral central distance *L* determines of wave coupling between the neighbored holes. The pinholes are uniformly distributed along the zones. It is illuminated by various waves with 633 nm incident wavelength. And we have different polarization states such as TM, EP and RP. (b) layout of elliptical pinholes with total 8 rings $\delta = 0.6$ being used in our computational numerical calculation.

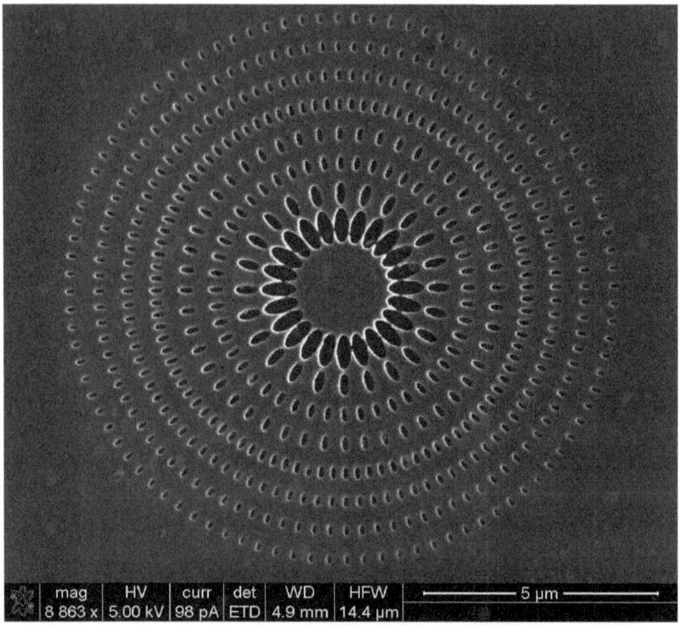


(a)

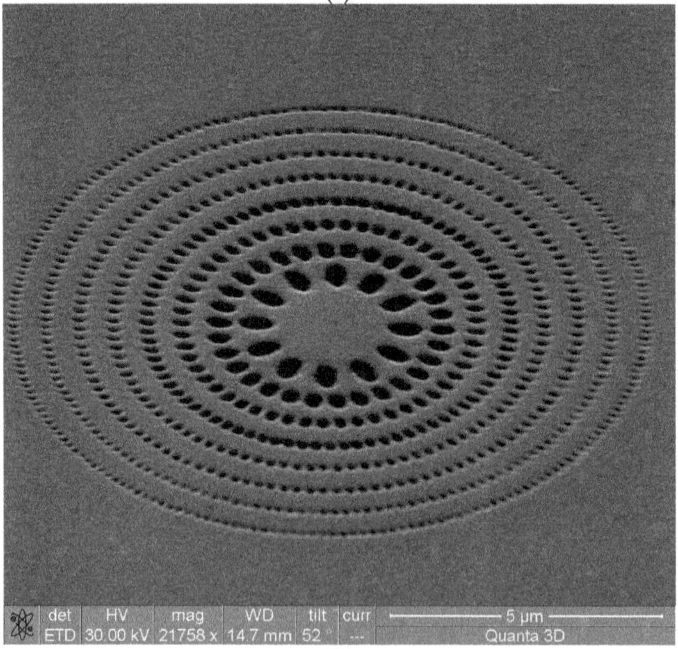


(b)

Fig. 9. (a) Top view of SEM micrograph of the elliptical nanopinholes-based plasmonic lens. (b) Bird view of SEM image of the elliptical nanopinholes.

As an example, the authors studied the case of 200 nm thickness Au film coated on quartz substrate and designed a nanostructure with 8 rings on the metal film (see Fig. 8 (b)). The pinholes are completely penetrated through the Au film. The number of pinholes from inner to outer rings is 8, 20, 36, 55, 70, 96, 107, and 140, respectively. Outer diameter of the ring is 12.05 μm. Radius of the rings can be calculated by the formula $r_n^2 = 2nf\lambda + n^2\lambda^2$, where $f$ is the focal length for working wavelength of $\lambda$ =633 nm which we used in our works. For simplicity, we define a ratio of short-axis to long-axis $\delta = b/a$ (where $a$ is the length of long-axis of the elliptical pinholes, and $b$ the short-axis). The used metal here is Ag with dielectric constant of $\varepsilon_m = -17.24 + i0.498$ at $\lambda$ =633 nm, and $\varepsilon_d$=1.243 for glass. The incident angle $\theta$ is 0° (normal incidence). In our analysis, we simulated the cases of the ratios: $\delta$ = 0.1, 0.2, and 0.4, respectively. The ultra-enhanced lasing effect disappears when the ratio $\delta \rightarrow 1$ (circular pinholes). Orientation of the pinholes is along radial direction. The pinholes symmetrically distribute in different rings with variant periods. It can generate ultra-enhanced lasing effect and realize a long focal length in free space accordingly with extraordinarily elongated depth of focus (*DOF*) of as long as 13 μm under illumination of plane wave in linear y-polarization.

The lens was fabricated using focused ion beam directly milling technique using beam current of 10 pA and ion energy of 30 keV, as shown in Figs. 9 (a) and (b).[18]

## 2.2 Laser interference photolithography

A Lloyd's mirror interferometer system was built, as shown in Fig. 10. [19] Light source is He-Cd laser with 442 nm working wavelength. The laser beam is filtered and expanded by a spatial filter which is composed of Lens 1 and a pinhole. The expanded beam is collimated by Lens 2. A part of the incident beam is reflected back by the mirror which is positioned in normal to the substrate and interferes with the other non-reflection beam to form the interference patterns. Since the beam is only split with a short path length near the substrate, this setup is very insensitive to the mechanical vibration caused instabilities. Hence no extra feedback control system is required to stabilize the interference fringe patterns [20]. Exposure time is controlled by a shutter. If the light intensity of the collimated beam is $I_0$, the radiation on the surface is given by

$$I_{2-beam} = 2I_0[1 + \cos(2kx\sin\theta)] \quad (1)$$

where $k = 2\pi/\lambda$, here $\theta$ is the half angle between the two beams. The grating period $d$ is given by

$$d = \lambda/(2\sin\theta) \quad (2)$$

After the first time exposure, the substrate is rotated over 90° and exposed again with the same exposure time. The distributed intensity on the surface is written as

$$I_{double-2b} = 2I_0[2 + \cos(2kx\sin\theta) + \cos(2ky\sin\theta)] \quad (3)$$

The period of the interference pattern generated by this method can be easily tuned by means of changing the incidence angle on the mirror as well as wavelength of the light source.

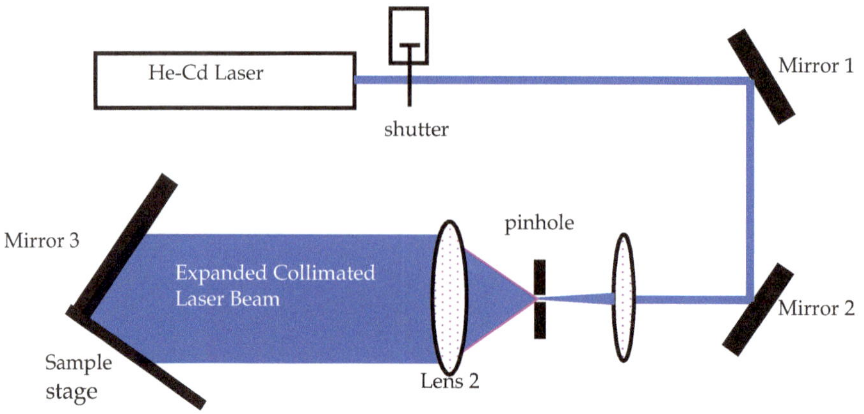


Fig. 10. Experimental setup of the Lloyd's mirror laser interference lithography

The photoresist was exposed by the collimated beam from the Lloyd's mirror interferometer system. A 1D grating pattern of the photoresist layer was formed after the first time exposure and development. Exposure dose was measured to be 1.5 $mW/cm^2$ in normal incidence. The developer adopted in our experiments is AR 300-35 (Allresist Co.). AFM measurement results after the exposure can be seen in Fig. 11 (a). In order to obtain a 2D grid pattern with the dots array in different sizes on the resist, second exposure with varied exposure time can be performed. The diameter of the holes in the photoresist depends on the exposure time. The dependence gives a possible tool to tune the microhole size at a constant space. Corresponding AFM micrograph is shown in Fig. 11 (b) and (c) respectively. In order to analyze the pattern period, an AFM measurement was carried out, which shows that the period is ~400 nm.

### 2.3 Self-assembly monolayer

An extended nanosphere lithography (NSL) technique was used to create the surface-confined rhombic Ag nanoparticles supported on a glass substrate, as shown in Fig. 12 (a).[21] Figure 13 shows the fabrication process of the nanoparticles. Firstly, the glass substrate was cleaned. Then the self-assembly of size-monodisperse (the sphere size of the chemical solutions which is spin-coated as a monolayer in which the sphere size determines the generated rhombic particle size.) polystyrene nanospheres (500 nm, 2%), glass nanospheres (200 nm, 1%), and distilled water were mixed in 5:1:50 and applied to form a monolayer (see Fig. 12 (b)), and followed by hydrofluoric acid to etch off the glass nanospheres, and PS nanospheres are remained on the substrate (see Fig. 13 (c)). After that, the Ag particles were deposited through the nanosphere masks using thermal or electron beam evaporation technique. After removal of the polystyrene nanospheres by sonication in absolute ethanol for 3 min., well ordered two-dimensional rhombic nanoparticle arrays were finally obtained on the substrates. By changing the nanosphere diameter and the deposited metal thickness, the nanoparticles with different in-plane width, out-of-plane height and interparticle space can be derived.

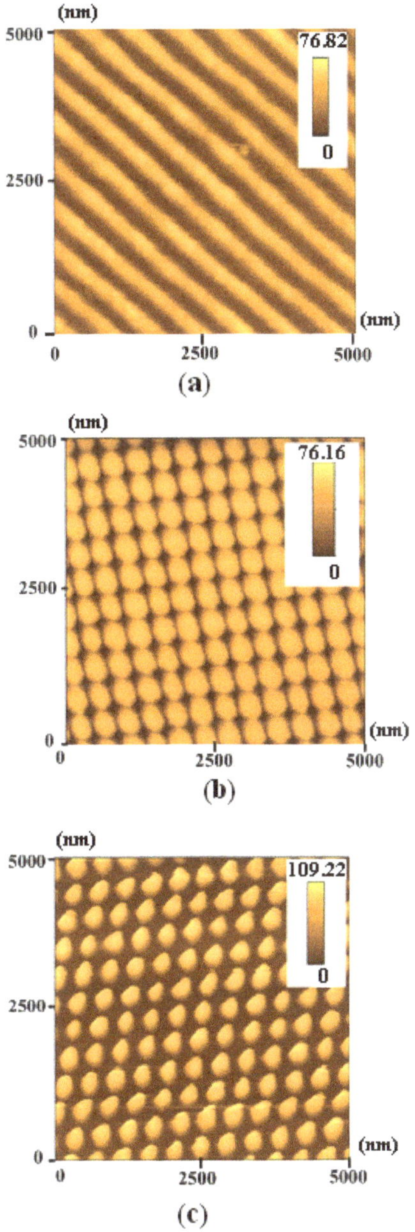


Fig. 11. AFM micrograph of the photoresist layer that remains after exposure and development in the laser interference setup (a) 1D grating structure after first exposure, (b) 2D grid pattern after second exposure with short exposure time of 8 s., (c) 2D dots array after second exposure with longer exposure time of 13 s.

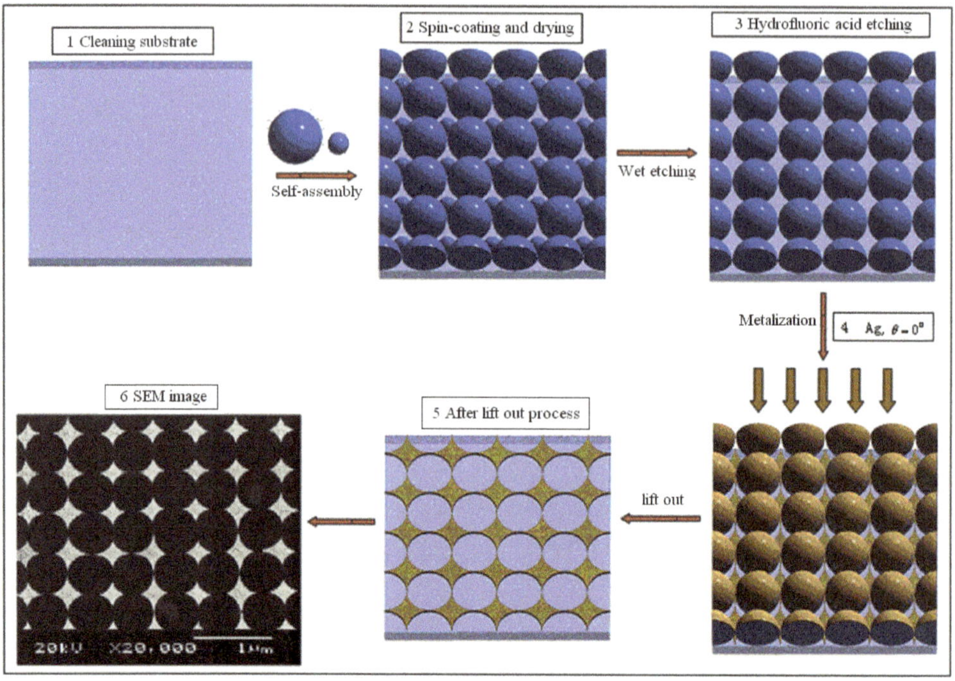


(a)

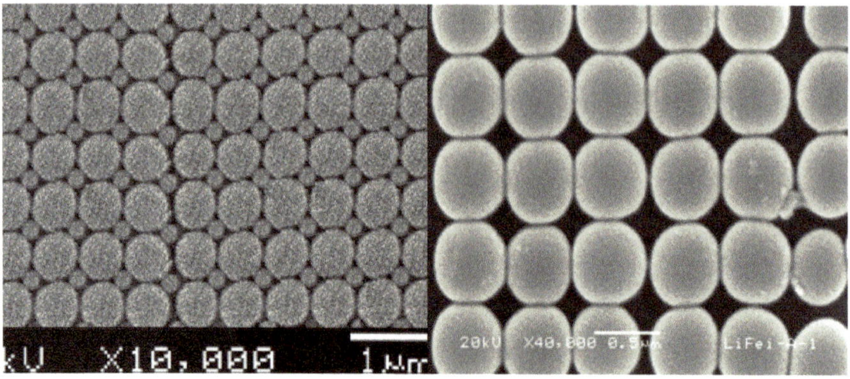


(b) (c)

Fig. 12. (a) The progress of the nanoparticle preparation. (1) glass substrates cleaned, (2) monodisperse polystyrene nanospheres and glass nanospheres were drop coated and dried, (3) hydrofluoric acid was used to etch off the glass nanospheres, (4) Ag metal was vapor deposited onto the sample, (5) lift out the polystyrene nanospheres, (6) representative scanning electron microscope image of substrate. Micrographs of SEM for (b) monolayer consisted of PS nanospheres and glass nanospheres; (c) polystyrene nanospheres being used as a template for next step: Ag metallization.

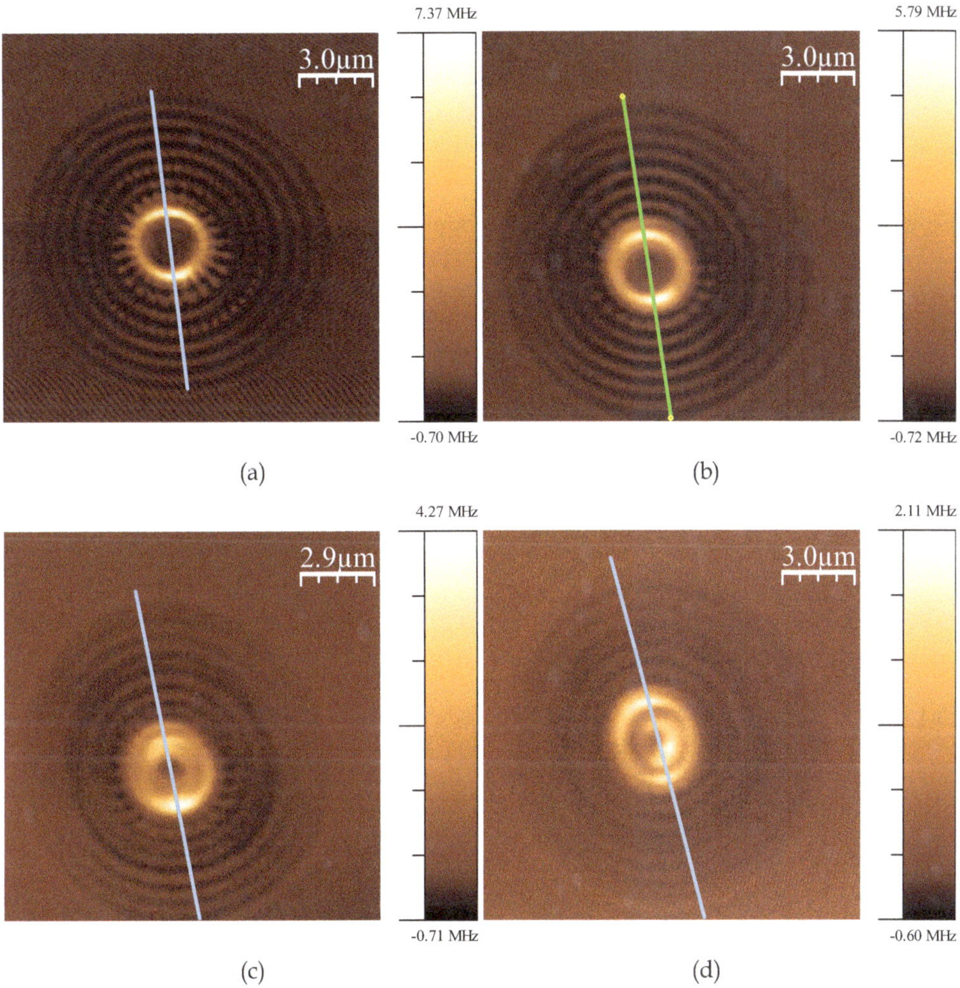


Fig. 13. NSOM probed E-field intensity distribution along different propagation distance Z (shown as the lines in the images), working in sample scan mode. (a) Z=10 nm; (b) Z=0.5 μm; (c) Z=1.5 μm; (d) Z=2.0 μm. The straight lines here indicate positions of cross-section profiles.

We measured the surface modality of the samples by using JSM-5900LV Scanning Electron Microscope (SEM). The Ag nanorhombus have in-plane widths of ~140 nm as measured by the SEM and out-of-plane heights of ~40 nm as measured by sidestep apparatus. We can see that the size of the Ag rhombic nanoparticles fabricated is not identical as the limit of the SAM method, which accounts for the errors in our following experiment.

The fabricated Ag rhombic nanoparticles have larger surface area than that of the Ag triangular nanoparticles [16] when we used the same diameter polystyrene spheres to fabricate them. The surface area of Ag rhombic nanoparticles is 2.5 times larger than the triangular structure. As mentioned above, Ag rhombic nanoparticles lead to a stronger enhancement of the localized surface plasmon resonance, and thus result in the improvement of the detection sensitivity and efficiency of the binding reaction between the antigen and antibody.

## 3. Characterization of plasmonic structures

Normally, characterization of plasmonic nanostructures and nanophotonic devices includes geometrical and optical characterizations. Currently, scanning probe microscope (SPM) is a key approach for the geometrical characterization. SPM is a general name for the probe scanning technology-based point-to-point measurement method. It involves atomic force microscope (AFM), scanning tunneling microscope (STM), and near-field scanning optical microscope (NSOM) or scanning near-field optical microscope (SNOM). Amongst, AFM is a most commonly used tool for the geometrical characterization. Optical characterization technology includes NSOM, surface enhanced Raman spectroscopy (SERS), confocal microscopy, and multiphoton microscopy. NSOM and SERS are the most popular tools for the optical characterization. The following sections will describe them one-by-one in detail combining with corresponding experimental results of the relevant plasmonic structures and devices.

### 3.1 Near-field scanning optical microscopy

A near-field scanning optical microscope (NSOM) has been adopted for the purpose of evaluating optical performance of the lens. The theoretical calculation-based analyses were completed for the plasmonic lens consisting of the Au film in Ref. [16]. To obtain the field distribution across planes of different heights slightly above the structure, we used NSOM (MutiView2000$^{TS}$, from Nanonics Inc.) working in collection mode, namely, the lens is excited by laser beam from far-field and the transmitted light is detected by the fiber probe in near-field region. Light source is a pumped YAG laser with 532 nm working wavelength. The light is focused to a spot at the surface of the lens through an optical microsope. Thus it is strong enough to be detected at surface of the exit side of the lens. We manually set height of the probe so as to keep the probe scans at a constant working distance. The NSOM probe used in this experiment is 200 nm in diameter. As a consequence, image resolution of the AFM scanning is limited especially for probing the topography of the FIB fabricated structure. But the image is still clear, as shown in Fig. 7 (b). Considering the radius of the lens and resolution limits, scanning region was defined as 15×15 μm, and 256 points per scan line. Both tip scan mode and sample scan (moving stage) modes were used to probe the lens respectively.

In our experimental study, we found that the two scan modes, fiber tip/probe scan and sample scan of the NSOM system have different characteristics for probing our plasmonic lenses. Our NSOM system illuminates the sample with Gaussian beam from laser source through an objective lens (10×, and $NA$=0.25). For the tip scan mode, the focused beam region is formed in free space already once the incident Gaussian beam illuminates the pinholes structure. Meanwhile, the fiber probe collects the optical signal originating from the whole structure of the lens while scans after exit plane of the lens in free space. Thus in this case, it functions like a dark-field optical microscope to pick-up the optical signal from dark background. But for the sample scan mode, the sample moves together with the stage for scanning while the fiber probe is fixed to collect optical signal transmitted from the lens.

It means that the whole structured lens area can be fully illuminated by the Gaussian beam. The sample moves relevant to the probe step-by-step or point-by-point. Therefore, central spot of the Gaussian beam at each scan position will be modulated by the pinholes structure while the sample and stage driven by the step-motors. The final scan image will be like an image from a bright-field optical microscope with uniform illumination. Thus the signal intensity collected by the probe is higher for the sample scan mode in comparison to that of the tip scan mode. Considering these features, we can use sample scan for probing transmission property of the lens and tip scan for characterization of focusing performance.

The experiment was carried out for the purpose of revealing focusing phenomenon of the novel lens. Firstly, we verified the focusing effect using NSOM with sample scan mode to collect focusing signal of the incident light from the lens. Figures 13 (a)-(d) are NSOM probed E-field intensity distributions along different propagation distances at z=10 nm, 0.5 μm, 1 μm, and 2 μm, respectively. As can be seen, a clear near-field image was obtained at z=10 nm (see Fig. 13 (a)). A focusing area is gradually formed at z=2 μm (see Fig. 13 (d)).

It is worthy to pint that for the near-field characterization of the plasmonic structures using NSOM, coupling between optical fiber probe and metallic surface of the plasmonic structures will play a negative role. The coupling directly influences accuracy of the NSOM probed near-field intensity distribution. A possible approach to solve the problem is that use near-field phase measurement by means of interference. Then we use the measured near-field phase distribution to restore the corresponding intensity distribution which is the real result without influence of the coupling on the real intensity distribution because the coupling does not influence the phase distribution.

### 3.2 Spectrometry of metallic nanoparticles and nanoholes

#### 3.2.1 Optical spectroscopy

In order to observe the sensor characteristic of this kind of biochips, the extinction spectrum was measured by a Sciencetech 9055 spectrophotometer. To obtain Visible-IR transmittance spectrum, a white light source (400~700 nm) was employed in the experimental system. Incident light was transmitted through a multimode optical fiber and reached collimating lens firstly, and then illuminated onto the biochip. The transmitted light beam through the sample was collected with an identical focus lens which is attached to the multimode fiber. Then a monochromator was used to separate the light from the multimode fiber to be monochromatic light beam. After that, the signal was sent to a personal computer which is integrated with an analog photomultiplier. The transmittance spectra were directly displayed on the screen of the computer. The measuring process was divided into three steps:

1. Measuring of the background light without any input sources and samples;
2. Measuring the light source intensity;
3. Collecting the transmitted light with the presence of the sample which is placed perpendicular to the incident light.

The transmittance $T$ can be written as $T=(s-b)/(r-b)$, where $s$, $r$, and $b$ denote the intensity of the sample, reference, and background, respectively. The extinction spectra $E$ for each step were achieved by equation $E=-\log T$. In addition, the transmittance spectra can be directly plotted and displayed in computer without further data processing.

The LSPR spectra of the Ag nanochips in each processing step is presented in Fig. 14 with incidence wavelength ranging from 400 nm to 700 nm, the resulting extinction spectrum of the bare Ag nanoparticles is depicted in Fig. 14 case a, where the LSPR $\lambda_{max}$ was measured to be 558.5 nm. Similarly, Fig. 14 case b shows the extinction spectrum after modification of the Ag nanoparticles with 1 mM 1:3 11-MUA/1-OT. To ensure a well-ordered SAM on the Ag nanoparticles, the sample was incubated in the thiol solution for 24 h. After carefully rinsing and thoroughly drying with $N_2$ gas, the corresponding peak transmission with LSPR wavelength $\lambda_{max}$ was measured to be 572 nm. Compared to the bare Ag nanoparticles, $\lambda_{max}$ in this surface functionalization step is red-shifted by approximately 13.5 nm. Next, 1 mM biotin was covalently attached via amide bond formation with a two-unit poly (ethylene glycol) linker to the carboxylated surface sites. Obtained LSPR spectrum is indicated in Fig. 14 case c, which shows that the peak transmission occurs at site of 594.5 nm. It corresponds to an additional 22.5 nm redshift from the second peak (red line). We finally plotted the extinction spectrum in Figs. 14 after the reaction between 100 nM streptavidin and 1 mM biotin. It is found that the maximum wavelength have a 16.5 nm redshift from 594.5 nm up to 611 nm. In contrast, using the same diameter PS spheres, the uniform height of the Ag nanoparticles, under the same metrical condition and processing time, we fabricated the

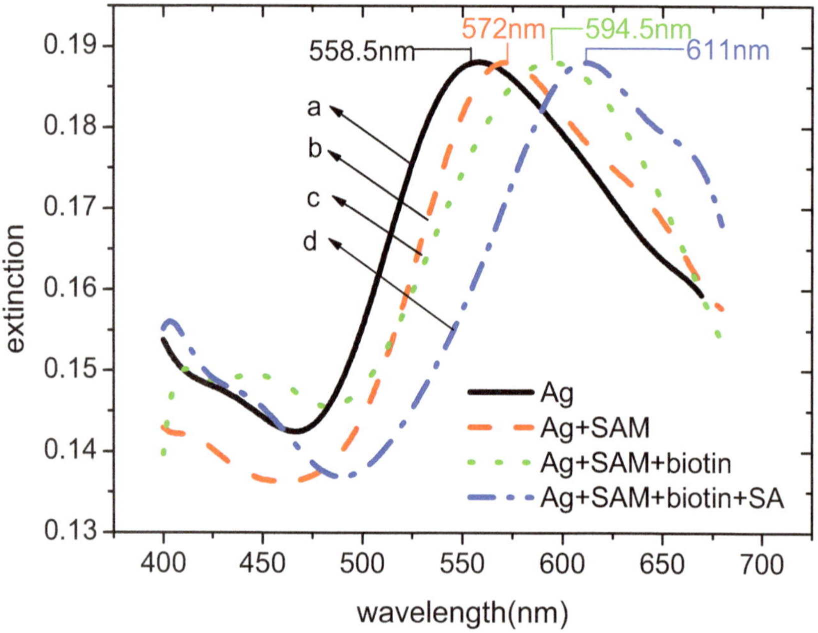


Fig. 14. LSPR spectra of each step in the surface modification of NSL-derived Ag nanoparticles to form a biotinylated Ag nanobiosensor and the specific binding of SA. (a) Ag nanoparticles before chemical modification, $\lambda_{max}$ = 558.5 nm. (b) Ag nanoparticles after modification with 1 mM 1:3 11-MUA/1-OT, $\lambda_{max}$ = 572 nm. (c) Ag nanoparticles after modification with 1 mM biotin, $\lambda_{max}$ =594.5 nm. (d) Ag nanoparticles after modification with 100 nM SA, $\lambda_{max}$ = 611 nm. All extinction measurements were collected in air environment.

traditional Ag triangular nanosensor, and measured the extinction spectrum. However, in our experiments, for the traditional Ag triangular nanosensor, a 7.5 nm redshift in the reaction between 100 nM SA and 1 mM biotin is observed. The shift of 16.5 nm mainly attributed to the "hot spot" of the rhombic particles mentioned before which plays a dominant role for improving detection sensitivity here.

By analyzing the obtained results, we apparently found that the extinction spectrum varies with each step of the surface functionalization for the Ag nanoparticles. The spectral shifting in those cases can be explained by the change of the local effective refractive index originating from the surface modification in each step.

### 3.2.2 Surface enhanced Raman spectroscopy

After nanofabrication, the sample was cleaned using deionized water. Then two samples were dipped with time of 24 h and 36 h respectively in the chemical solution of Rhodamine 6G with concentration of 10 μM/L. A Raman spectroscopy with model of Renishaw Invia 2000 from UK was employed for Raman spectrum measurement. Working wavelength of laser source ($Ar^+$ laser with power of 10 mW) is 800 nm.

Figure 15 (a) is measured Raman spectra of the rhombic nanoparticles array with SERS enhancement for the samples were dipped in 10 μMl/L Rhodamine 6G. In order to verify influence of the dipping time on Raman intensity and SERS enhancement, we prepared two samples which were dipped in the Rhodamine 6G with different time of 24h and 36h, respectively. Then they were cleaned using deionized water. The following conclusions regarding rhombic particles can be drawn in terms of the Raman spectra:

1. The rhombic particles array can distinguish the peaks of Raman spectrum with apparent positions at 611cm$^{-1}$, 1184 cm$^{-1}$, 1310 cm$^{-1}$, 1360 cm$^{-1}$, 1509 cm$^{-1}$, 1597 cm$^{-1}$, and 1650 cm$^{-1}$, respectively.
2. Background noise exists in the measured SERS signal. It may attribute to the purity of chemical solution and defects from ambient while tests the samples. But the drawbacks caused influence on SERS enhancement is a little.
3. Considering that the samples were cleaned using deionized water, remained Rhodamine 6G molecule are far lower than the previous concentration of 10 μMl/L. It is reasonable to believe that sensitivity of the rhombic particles is high for the SERS enhancement-based detection.
4. Dipping time is relevant to the Raman spectrum. It can affect both Raman intensity and peak position/Raman shift.

Figures 15 (a) and (b) are measured Raman spectra for Au rhombic nanoparticles array, and cross-star-like nanoparticles array, respectively. It can be seen that the pure spider-web-like nanostructure and the hybrid array have extraordinary Raman enhancement in comparison to the other two Au nanoparticles arrays. The Raman shifts of 611cm$^{-1}$, 1184 cm$^{-1}$, 1310 cm$^{-1}$, 1364 cm$^{-1}$ and 1509 cm$^{-1}$ can be clearly observed from Fig. 15 (b). It may attribute to the significant contribution from the spider-web-like nanostructure with numerous Au nanowires and "nano-probes" (the broken wires with sharp tips) whereas the SPPs enhanced E-field and numerous "hot-spots" can be formed. In addition, the rhombic Au particles array with designed "bow-tie"-like apex-to-apex structures in horizontal direction whereas numerous hot-spots can be formed is another reason for the extreme enhanced Raman intensity.

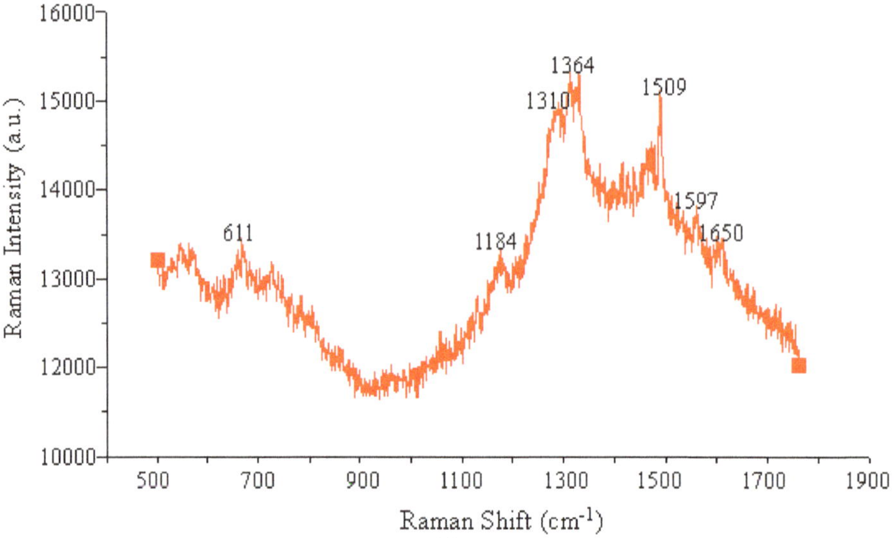


(a)

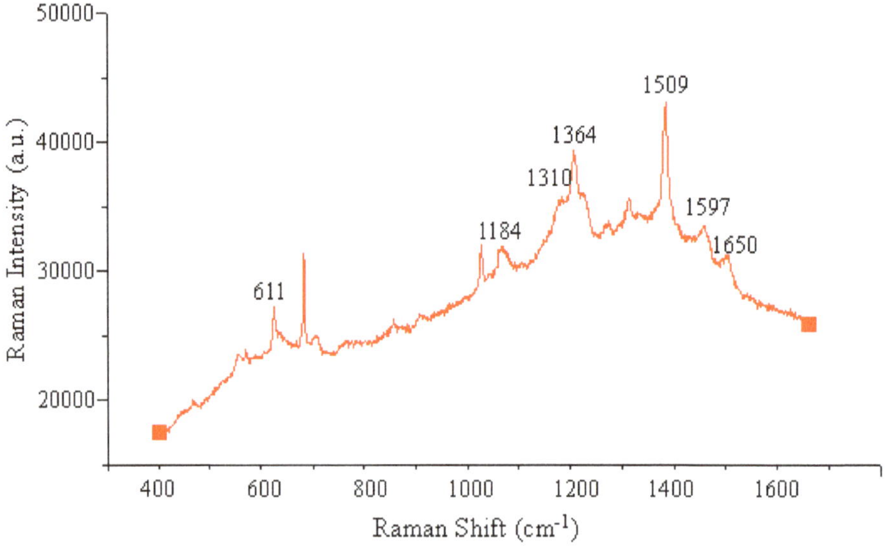


(b)

Fig. 15. Measured Raman spectra for (a) rhombic nanoparticles array (see Fig. 1(b))); (b) hybrid spider-web-like rhombic Au nanoparticles array.

## 4. References

[1] T. W. Ebbesen, H. J. Lezee, H. F. Ghaemi, T. Thio, and P. A. Wolff, Extraordinary optical transmission through subwavelength hole arrays. Nature, 391, 667-669 (1998).

[2] C. Genet and T. W. Ebbesen, Light in tiny holes. Nature, 445, 39-46 (2007).

[3] E. Laux, C.Genet, and T. W. Ebbesen, Enhanced optical trasmission at the cutoff transition. Opt Express 17, 6920-6930 (2009).

[4] Zhichao Ruan, and Min Qiu, Enhanced Transmission through Periodic Arrays of Subwavelength Holes: The Role of Localized Waveguide Resonances. Phys. Rev. Lett. 96. 233901 (2006).

[5] A. S. Vengurlekar, Extraordinary optical transmission through metal film with subwavelength holes and slits. Current Science, 98, 167-172 (2010).

[6] K. J. Klein Koerkamp, S. Enoch, F. B. Segerink, N. F. Van Hulst, and L. Kuipers, Strong Influence of Hole Shape on Extraordinary Transmission through Periodic Arrays of Subwavelength Holes. Phys. Rev. Lett. 92. 183901 (2004).

[7] Li Hai-jun, Zhang Xiao-dong, Wang Min-rui, Lin Wen-kui, Shi Wen-hua, Zhong Fei, and Zhang Bao-shun, Effects of the periodicity of the subwavelength hole arrays and hole shape of a thin gold film on the optical transmission characteristics. Optoelectron. Lett. 6(3), 78-80 (2010).

[8] A. Degiron and T. W. Ebbesen, The role of localized surface plasmon modes in the enhanced transmission of periodic subwavelength apertures. J. Opt. A: Pure Appl. Opt. 7, S90-S96 (2005).

[9] Andrey K. Sarychev, Viktor A. Podolskiy, A. M. Dykhne, and Vladimir M. Shalaev, Resonance transmittance through a metal film with subwavelength holes. IEEE Journal Of Quantum Electronics, Vol. 38, No. 7, July 2002.

[10] Brian Leathem (BSc. Simon Fraser University, 2001) Thesis-optical transmission through nanohole arrays.Chap.4, p30.

[11] L. Martín-Moreno, F. J. García-Vidal, H. J. Lezec, K. M. Pellerin, T. Thio, J. B. Pendry, and T.W. Ebbesen, Theory of Extraordinary Optical Transmssion through Subwavelength Hole Arrays. Phy. Rev. Lett. 86, 1114 (2001).

[12] Yong-Jun Bao, Ru-Wen Peng, Da-Jun Shu, Mu Wang, Xiang Lu, Jun Shao, Wei Lu, and Nai-Ben Ming, Role of Interference between Localized and Propagating Surface Waves on the Extraordinary Optical Transmission Through a Subwavelength-Aperture Array. Phys. Rev. Lett. 101, 087401 (2008).

[13] Rongjing He, Xiuli Zhou, Yongqi Fu, Near Field Optical Experimental Investigation of Gold Nanohole Array, Plasmonics, 6, (2011). (in press) DOI: 10.1007/s11468-010-9183-2.

[14] Yongqi Fu, W. Zhou, L.E.N.Lim, C.L. Du, X.G.Luo, Plasmonic microzone plate: Superfocusing at visible regime. Appl. Phys. Lett. 91, 061124 (2007).

[15] Yongqi Fu, Wei Zhou, Lim Enk Ng Lennie, Propagation properties of a plasmonic micro-zone plate in near-field region. J. Opt. Soc. Amer. A, 25, 238-249 (2008).

[16] Yongqi Fu, Yu Liu, Xiuli Zhou, Shaoli Zhu, Experimental demonstration of focusing and lasing of plasmonic lens with chirped circular slits, Opt. Express 18 (4), 3438–3443 (2010).

[17] Yongqi Fu, Xiuli Zhou, Yu Liu, Ultra-enhanced lasing effect of plasmonic lens structured with elliptical nano-pinholes distributed in variant period. Plasmonics 5(2), 111-116 (2010).

[18] Yiwei Zhang, Yongqi Fu, Yu Liu, Xiuli Zhou, Experimental study of metallic elliptical nanopinhole structure-based plasmonic lenses. Plasmonics, 5(4), (2010). (in press) DOI: 10.1007/s11468-010-9191-2.

[19] Haiying Li, Xiangang Luo, Chunlei Du, Xunan Chen, and Yongqi Fu, Ag particles array fabricated using laser interference technique for biosensing, Sensors and Actuators B Chemical 134, 940–944 (2008).
[20] Cees J. M. van Rijn, Laser interference as a lithographic nanopatterning tool, J. Microlith., Microfab., Microsyst. 5(1), 121-126( 2006).
[21] Shaoli Zhu, Fei Li, Chunlei Du, Yongqi Fu, A novel bio-nanochip based on localized surface plasmon resonance spectroscopy of rhombic nanoparticles, Nanomedicine Vol. 3, No. 5, 669-677 (October 2008).

9

# Fabrication of Photonic Crystal Cavities for Terahertz Wave Resonations

Soshu Kirihara
*Osaka University,*
*Japan*

## 1. Introduction

Photonic crystals with periodic arrangement of dielectric media can exhibit forbidden gaps in electromagnetic wave bands through Bragg diffraction (Ohtaka; 1979, Yablonovitch; 1987, John; 1987, Temelkuran; 2000). The prohibited frequency ranges are especially called photonic band gaps. If the periodicity is changed locally by introducing a defect, localized modes appear in the band gap (Ho; 1990, Soukoulis; 1999, Noda; 1999, Kirihara; 2002a, Kanehira; 2005). Such localization function of electromagnetic waves can be applied to various devices, for example resonators, waveguides, and antennas. Three dimensional photonic crystals with a diamond structure are regarded as the ideal photonic crystal since they can prohibit the propagation of electromagnetic waves for any directions in the band gap (Ho; 1990, Kirihara; 2002b). However, due to the complex structure, they are difficult to fabricate. In our previous investigations, we have succeeded in fabricating micrometer order diamond structures by using stereolithography method of a computer aided design and manufacturing (CAD/CAM) processes (Chen; 2007a, 2007b, 2007c, 2008, Kanaoka; 2008, Miyamoto; 2008). Subsequently, structural modifications of the diamond lattice structures to control the terahertz wave propagations were investigated by using the CAD/CAM process practically (Takano; 2005, Kirihara; 2009, 2008a, 2008b). In near future industries, electromagnetic waves in a terahertz frequency range with micrometer order wavelength will be expected to apply for various types of novel sensors which can detect gun powders, drugs, bacteria in foods, micro cracks in electric devices, cancer cells in human skin and other physical, chemical and living events (Kirihara; 2009b, Exter; 1989, Clery; 2002, Kawase; 2003, Woodward; 2003, Wallace; 2004, Oyama; 2008). In this chapter, the novel stereolithography process to fabricate the micro diamond photonic crystals by using the ceramic slurry with the nanoparticles will be introduced. And, the resonation and localization properties of the terahertz waves into various types of the structural defects introduced according to theoretical electromagnetic simulations will be demonstrated.

## 2. Photonic crystals

### 2.1 Band gap formation

Photonic crystals composed of dielectric lattices form band gaps for electromagnetic waves. These artificial crystals can totally reflect light or microwave at a wavelength comparable to the lattice spacings by Bragg deflection as shown in Fig. 1. Two different standing waves oscillating in the air and dielectric matrix form higher and lower frequency bands in the first and second Brillouin zones, respectively. The band gap width can be controlled by varying structure, filling ratio, and dielectric constant of the lattice. Structural modifications by introducing defects or varying the lattice spacing can control the propagation of light or microwaves. The band diagram of the photonic crystal along symmetry lines in the Brillouin zone is drawn theoretically. The Maxwell's equations (1) and (2) can be solved by means of plane wave propagation (PWE) method (Haus; 1994), where ω and c denote frequency and light velocity, respectively. Electronic and magnetic field Eω (r) and Hω(r) are described with the following plane wave equations (3) and (4), respectively. The periodic arrangement of dielectric constant ε(r) can be obtained as equation (5) from the crystal structure. G and k are reciprocal vector and wave vector, respectively.

$$\left[\nabla\times\left(\frac{1}{\varepsilon(r)}\right)\nabla\times\right]\mathbf{H}_{\omega}(r)=\left(\frac{\omega}{c}\right)^2\mathbf{H}_{\omega}(r) \tag{1}$$

$$\left[\frac{1}{\varepsilon(r)}\nabla\times\nabla\times\right]\mathbf{E}_{\omega}(r)=\left(\frac{\omega}{c}\right)^2\mathbf{E}_{\omega}(r) \tag{2}$$

$$\mathbf{H}_{k,n}(r)=\sum_{\mathbf{G}}\mathbf{H}_{k,n}(\mathbf{G})e^{i(k+G)\cdot r} \tag{3}$$

$$\mathbf{E}_{k,n}(r)=\sum_{\mathbf{G}}\mathbf{E}_{k,n}(\mathbf{G})e^{i(k+G)\cdot r} \tag{4}$$

$$\frac{1}{\varepsilon(r)}=\sum_{\mathbf{G}}\frac{1}{\varepsilon(\mathbf{G})}e^{iG\cdot r} \tag{5}$$

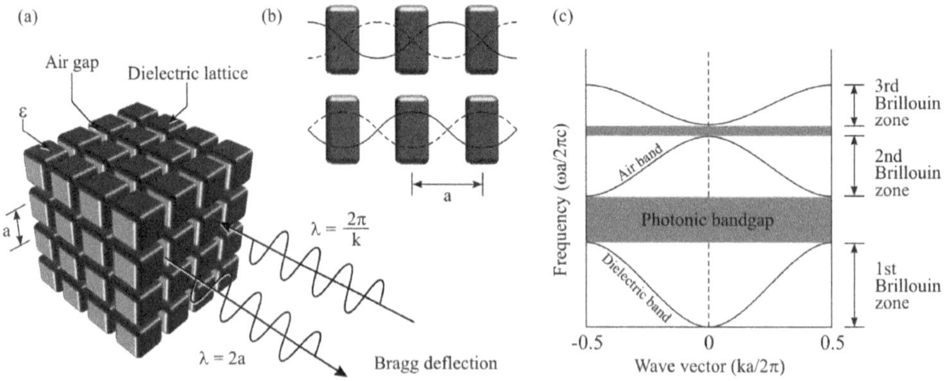


Fig. 1. Principles of photnic band gap formations in periodic arrangements of dielectric materials. Two different standing waves with higher and lower frequencies are formed in a photonic crystal, and a forbidden gap is formed between these frequencies.

## 2.2 Artificial crystal geometries

Typical photonic crystal structures were shown in Fig. 2. A woodpile structure (a) with simple structure of stacked rods can form the perfect photonic band gap. Photonic crystals composed of GaAs or InP were fabricated by using semiconductor process techniques (Noda; 2000). A light wave circuit (b) in the periodic structure of arranged AlGaAs pins is processed by using electron beam lithography and etching techniques (Baba; 2001). A layered structure (c) composed of Si and $SiO_2$ with the different dielectric constants realize light wave polarization and super prism effects (Kosaka; 1999). These layers are stacked by using self-organized growing in alternate spattering and etching. An inverse opal structure (d) is composed of air spheres with FCC structure in $TiO_2$, Si, Ge or CdS matrix (Cregan; 1999). At first, polystyrene spheres are arranged by using self-organization in colloidal solutions. Then, the slurry of these dielectric media is infiltrated into the periodic structure and sintered. The optical fiber (e) with photonic crystal structure can guide light efficiently along the central core (Vos; 1996). Silica fibers and glass capillaries were bundled by wire drawing at high temperature. Diamond type photonic crystals (f) composed of $TiO_2$, $SiO_2$ or $Al_2O_3$ can be fabricated by using stereolithography and successive sintering process. The wider perfect band gap is obtained in microwave and terahertz wave frequency ranges.

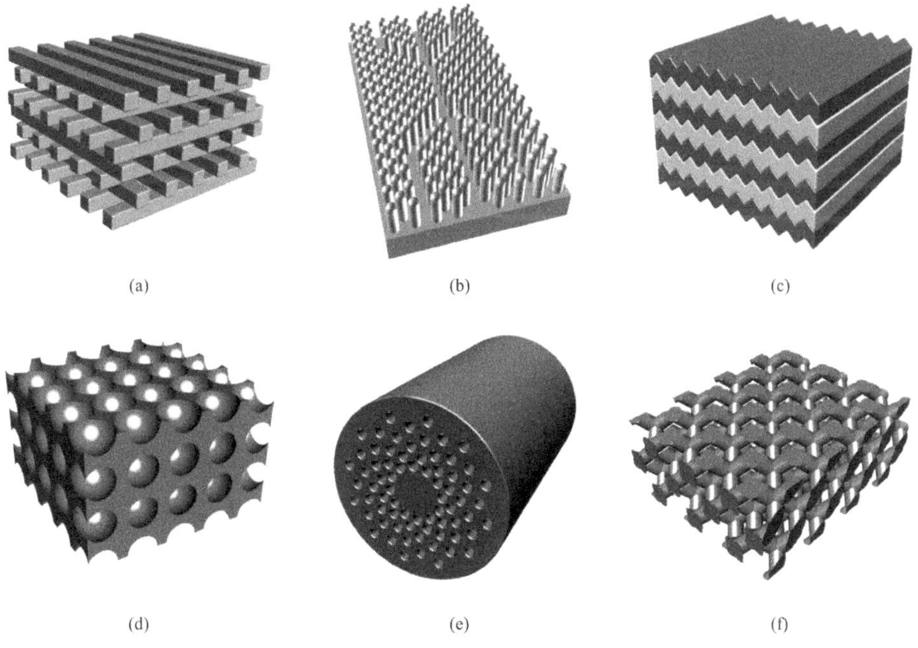


Fig. 2. Typical periodic structures of the photonic crystals with woodpile structure (a), patterned substrate (b), stacked layer (c), inverse opal structure (d), bundled fiber (e), and diamond structure (f).

## 2.3 Utility applications

Figure 3 shows expected applications of photonic crystal for light and electromagnetic wave control in various wavelength ranges (Kawakami; 2002). Air guides formed in a photonic crystal with nanometer order size will be used as the light wave circuit in the perfect reflective structure. When a light emitting diode is placed in an air cavity formed in a photonic crystal, an efficient laser emission can be enhanced due to the high coherent resonance in the micro cavity. While, millimeter order periodic structures can control microwaves effectively. Directional antennas and filters composed of photonic crystals can be applied to millimeter wave radar devices for intelligent traffic system (ITS) and wireless communication system. The perfect reflection of millimeter wave by photonic crystal will be useful for barriers to prevent wave interference. Terahertz waves with micrometer order wavelength are expected to apply for various types of sensors to detect gun powders, drugs, bacteria in foods, micro cracks in electric devices, cancer cells in human skin and other physical, chemical and living events. The micrometer order photonic crystals can be applied for the terahertz wave cavities, filters and antennas.

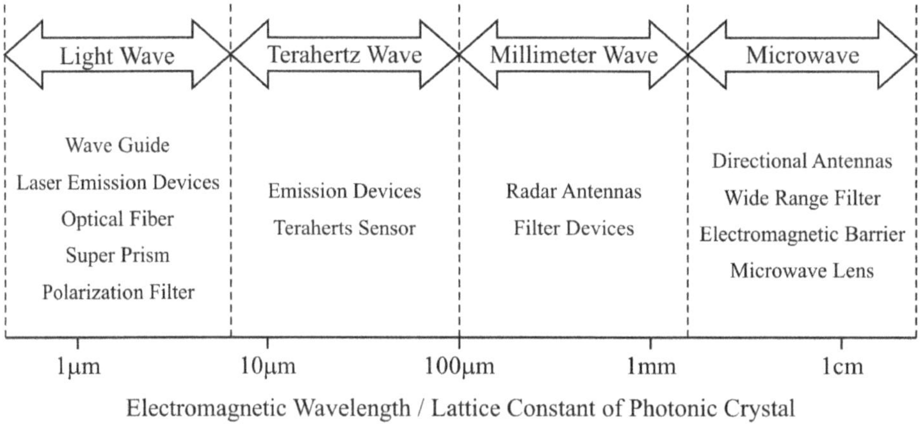


Fig. 3. Expected applications of the photonic crystal in various electromagnetic wavelengths.

## 3. Terahertz wave

### 3.1 Novel usefulness

The terahertz waves have received extensive attentions and investigations since they have a lot of interesting and applicable features in various fields such as materials, communication, medicine, and biology (Yamaguchi; 2005, Fischer; 2002, Oyama; 2009, Wallace; 2004, Hineno; 1974). It is possible to detect high explosives and ceramic blades hidden in bags, clothes, and envelopes by using terahertz waves since they can penetrate plastic, paper, and clothes without radiation damage to living bodies. It is also possible to identify toxic drugs because they have spectral fingerprints or absorption spectra. Moreover, they can distinguish cancerous areas from healthy areas due to the different absorption rates. In recent years, the terahertz waves have attracted considerable attentions as novel analytical light sources. Because the electromagnetic wave frequencies from 0.1 to 10 THz can be synchronized with collective vibration modes of saccharide or protein molecules, the terahertz wave spectroscopy are expected to be applied to various types of sensors for detecting harmful substances in human bloods, cancer cells in the skins and micro bacteria in vegetables. Moreover, the terahertz sensing technologies for aqueous phase environments in nature fields to detecting dissolved matters directly are extremely interesting topics. However, the terahertz waves are difficult to transmit into the water solvents with millimeter order thickness corresponding to the several wavelengths due to electromagnetic absorptions (Kutteruf; 2003).

### 3.2 Effective control

In the previous investigations of other research groups, Fabry-perot cavities have been applied for terahertz wave resonators. However, it is theoretically unfeasible to achieve the effective single mode operation in the micro Fabry-perot resonators since electromagnetic losses of the increasing in inverse proportion to the cavity size (Akahane; 2003). Compare with this, photonic crystals can realize the single mode resonation without the electromagnetic losses through the appropriate structural design. Especially, the three dimensional diamond lattices are regarded as the ideal photonic crystal structures. Figure 4-(a) and (b) show a unit cell of the diamond photonic crystal and a electromagnetic band diagram calculated by using plane wave expansion method simulator (RSOFT Design Group Co. Ltd., USA, BandSOLVE), respectively. The artificial crystal can totally reflect the terahertz wave with the corresponding wavelength to the lattice spacing through the Bragg diffraction. The complete photonic band gap is formed to prohibit electromagnetic wave expansions for all crystal directions. In the recent investigations, the diamond structures introduced a point or plane defect were formed successfully, and localized modes with several wavelengths were observed in the cavity regions. From these obtained results, the diamond photonic crystals can be applied to low loss terahertz wave resonator.

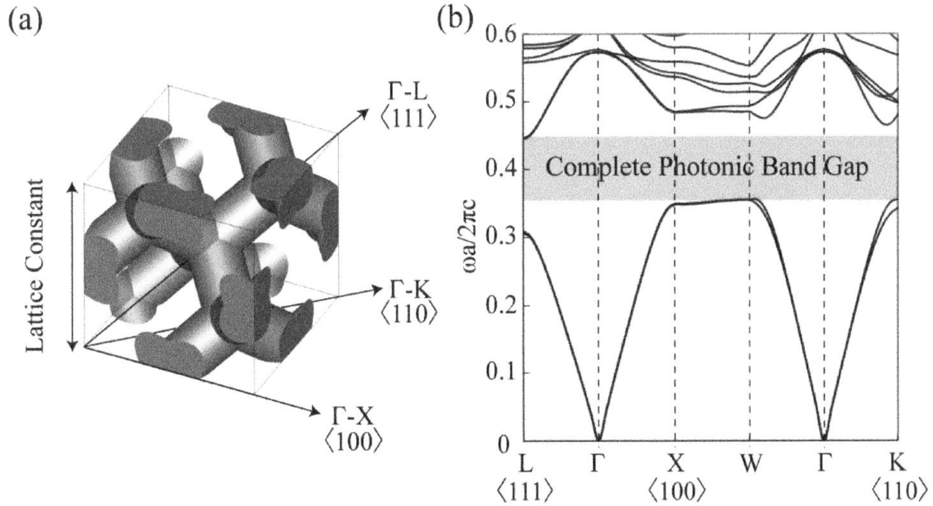


Fig. 4. A graphic image of a unit cell of the diamond structure (a), and a photonic band diagram of diamond lattice structure calculated by plane wave expansion (PWE) method (b).

### 3.3 Device concepts

In our investigation group, we designed and fabricated the terahertz wave resonators composed of the defects introduced photonic crystals. The micrometer order alumina lattices with the diamond structures can resonate and localized the terahertz waves strongly into the water solvents in order to realize the higher analytical precisions. In the photonic crystal resonator, a micro glass cell including the water solvent was put between two diamond lattice structures as a plane defect. The micro structures of the dielectric lattice and resonation cavity were modelled theoretically. Subsequently, the real resonator module was fabricated successfully by using the micro stereolithographic procedure. The incident terahertz wave was reflected multiply to achieve longer optical distance in the liquid region and localized to amplify the sensing signal in the transmission spectrum. The terahertz wave properties of the resonation profiles in the transmission spectra and the localized modes in the plane defects through fabricated resonators were observed and visualized successfully by using a time domain spectroscopy and a finite difference time domain simulator, respectively.

## 4. Geometric design

### 4.1 Diffraction lattices

The electromagnetic band properties of the diamond photonic crystals were calculated theoretically to determine the geometric parameters by using the plane wave expansion method [16]. Figure 5 shows the variations of complete band gap widths as the function of aspect ratio in the dielectric lattice. In the calculation, the alumina ceramics of 9.8 in dielectric constant was assumed as the lattice material. The aspect ratio was optimized as 1.5 to create the wider band gap. Subsequently, the gap frequency can be shifted for the lower range according in inverse proportion to the lattice spacing as shown in Fig. 6. To open the band gapes in the terahertz frequencies, micro scale structural periods need to be created. The lattice constant was designed as 375 μm corresponding to the band gap frequencies from 0.3 to 0.6 THz.

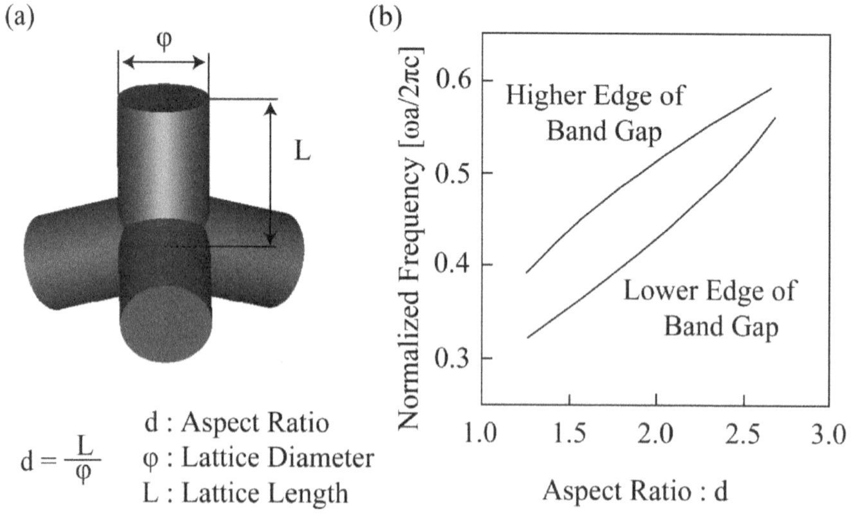


Fig. 5. Definition of aspect ratio of diamond structure (a), and the band gap width as a function of aspect of aspect ratio (b).

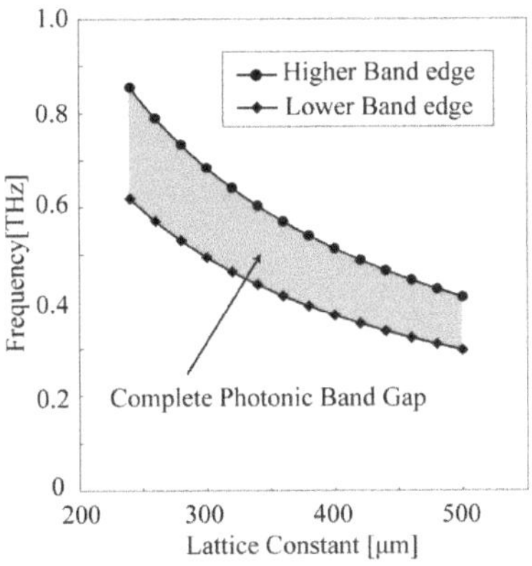


Fig. 6. Variations of electromagnetic bands as functions of lattice constants in the photonic crystals with the diamond structures.

### 4.2 Resonation cavity

Resonance efficiencies of the terahertz wave resonator with the diamond lattices were optimized by using transmission line modelling simulator (Flomerics Co. Ltd., UK, Micro-Stripes Ver. 7.5) of a finite difference time domain method. Figure 7 shows a computer graphic model of the terahertz wave resonator. The micro glass cell including the water was sandwiched as the plane defect between two alumina photonic crystals with the diamond lattice. The defect thickness of the resonator and the period numbers of the diamond units were selected as the principle parameters to control the resonance characteristics. The defect thickness enables to tune the resonance frequencies in the band gap as shown in Fig. 8. The plane defect was designed as the water cell of 470 μm in thickness composed of two quartz plates of 160 μm and an aqueous cavity of 150 μm. The period numbers of the diamond lattices enables to adjust the resonance qualities as shown in Fig. 9 The resonance qualities can be enhanced by increasing the period numbers, however, the localized mode of the transmission peak becomes lower through the perfect confinement of the electromagnetic wave in the defect domain. The diamond lattices composed of two units in period number were optimized and designed in order to detect the sharp localized mode peak in the transmission spectrum.

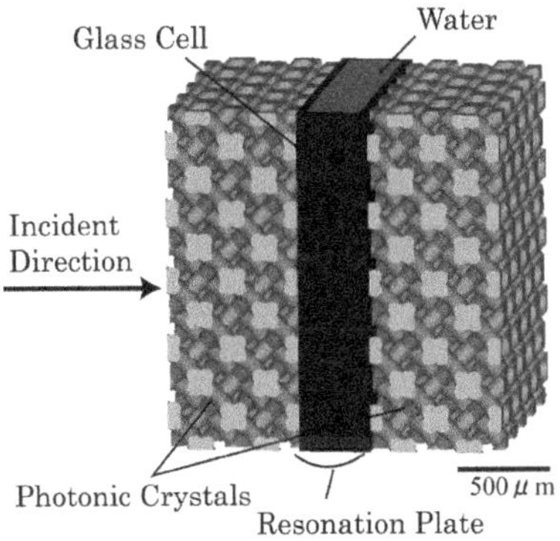


Fig. 7. A computer graphic model of a terahertz wave micro reactor.

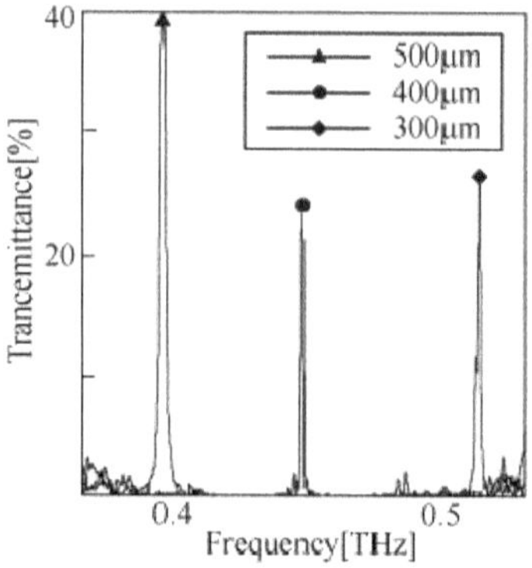


Fig. 8. Calculated localized mode formations through electromagnetic waves resonations in micro cells with various thicknesses.

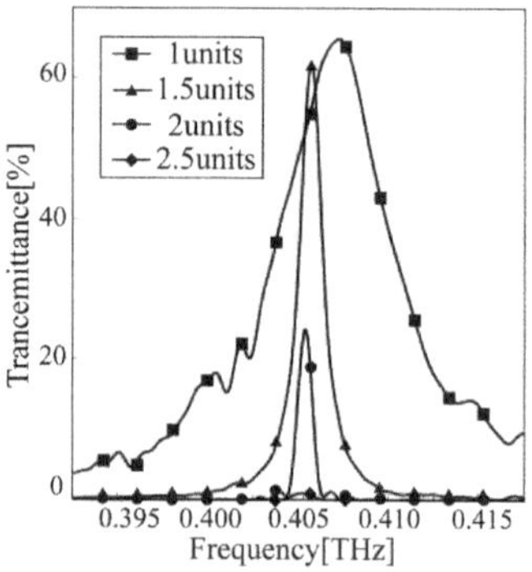


Fig. 9. Calculated peak profiles of multiple resonation modes in water cells of plane defects put between photonic crystals with various layer numbers.

## 5. Smart processing

### 5.1 Dielectric components

The photonic crystal resonators of the terahertz wave were fabricated by using computer aided design and manufacturing of smart processing. The diamond lattices were designed by using a computer graphic software (Toyota Caelum Co. Ltd., Japan, Think-Design Ver. 9.0). The lattice constant of the diamond structure and the aspect ratio of the dielectric lattice were 500 μm and 1.5, respectively. The whole size of the crystal component was 5×5×1 mm consisting of 10×10×2 unit cells. The designed model was converted into stereolithographic files of a rapid prototyping format and sliced into a series of two dimensional cross sectional data of 15 μm in layer thickness. These data were transferred into a micro stereolithographic equipment (D-MEC Co. Ltd., Japan, SI-C 1000). Figure 10 shows a schematic illustration of the micro stereolithography system. Photosensitive acrylic resins including alumina particles of 170 nm in average diameter at 40 % in volume content were supplied on a glass substrate from a dispenser nozzle by the air pressure. This paste was spread uniformly by using a mechanically controlled knife edge. The thickness of each layer was set at 15 μm.

Two dimensional solid patterns are obtained by a light induced photo polymerization. High resolution image has been achieved by using a digital micro mirror device. In this optical device, micro aluminium mirrors of 14 μm in edge length were assembled with 1024×768 in numbers. Each mirror can be tilted independently by piezoelectric actuating. Through the layer by layer stacking under the computer control, the acrylic resin component with the alumina particles dispersion was obtained. The composite precursor was dewaxed at 600 °C for 2 hs and sintered at 1500 °C for 2 hs in the air atmosphere. Subsequently, in order to obtain a plane defect between the two diamond structures, a micro glass cell was also fabricated by using the micro stereolithography.

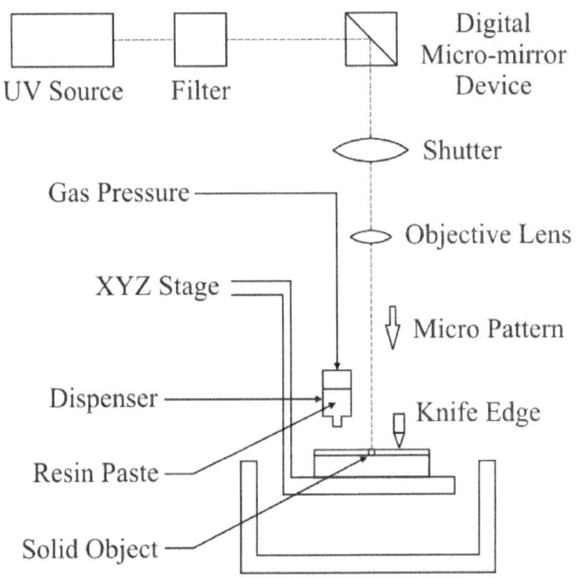


Fig. 10. A schematically illustrated free forming system of a micro stereolithography machine by using computer aided design and manufacturing (CAD/CAM) processes.

### 5.2 Device assembling

Figure 11-(a) shows the schematically illustrated components of the resonance cells. The quartz plates of 160 μm in thickness were inserted into the photosensitive acrylic resins in the stacking and exposing process. Finally, the micro resonators cell was put between the diamond photonic crystals, and the terahertz wave resonator was integrated successfully by using acrylic resin flames as shown in Figure 11-(b). These flames were glued together by using the photo sensitive liquid resin and the ultraviolet exposure solidification. Water solutions were infused through catheters connected on the top side of the resonance cell. The transmission properties of incident terahertz waves were analyzed by using the terahertz spectroscopy. The distributions of electric field intensities in the resonator were simulated and visualized at the localized frequency by using the transmission line modelling.

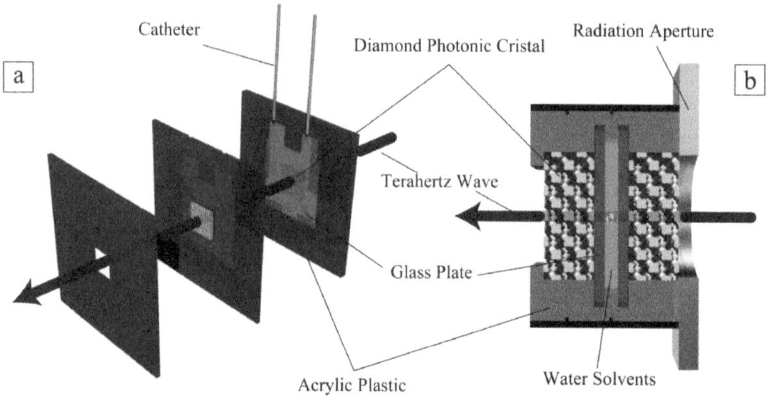


Fig. 11. A schematic illustration of the parts of the resonance cells (a), and fabricated THz wave resonator (b). The resonator was successfully created by integrating the diamond structures into the cells to realize multiple resonations of the THz wave in aqueous phase.

### 5.3 Spectroscopic evaluations

A terahertz wave attenuation of transmission amplitudes through the diamond photonic crystals were measured by using a terahertz time domain spectrometer (TDS) apparatus (Advanced Infrared Spectroscopy Co. Ltd., Japan, Pulse-IRS 1000). Figure 12 shows the schematic illustration of the measurement system. Femto second laser beams were irradiated into a micro emission antenna formed on a semiconductor substrate to generate the terahertz wave pulses. The terahertz waves were transmitted through the micro patterned samples perpendicularly. The dielectric constant of the bulk samples were measured through a phase shift counting. Diffraction and resonation behaviours in the dielectric pattern were calculated theoretically by using a transmission line modelling (TLM) simulator (Flomerics, UK, Microstripes Ver. 7.5) of a finite difference time domain (FDTD) method.

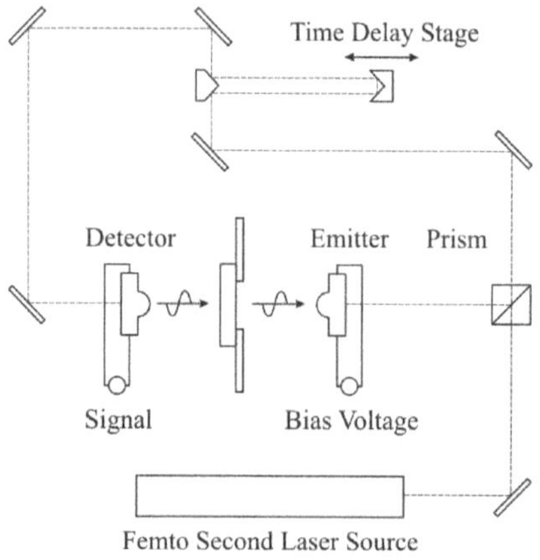


Fig. 12. The schematically illustrated measuring system of a terahertz wave analyzer by using a time domain spectroscopic (TDS) detect method.

## 6. Electromagnetic behaviours

### 6.1 Diamond structures

An alumina dispersed resin precursor fabricated by the micro stereolithography is shown in Fig. 13. The lattice constant of the formed diamond structure was 500 μm. The spatial resolution was approximately 0.5 %. The weight and color changes as a function of

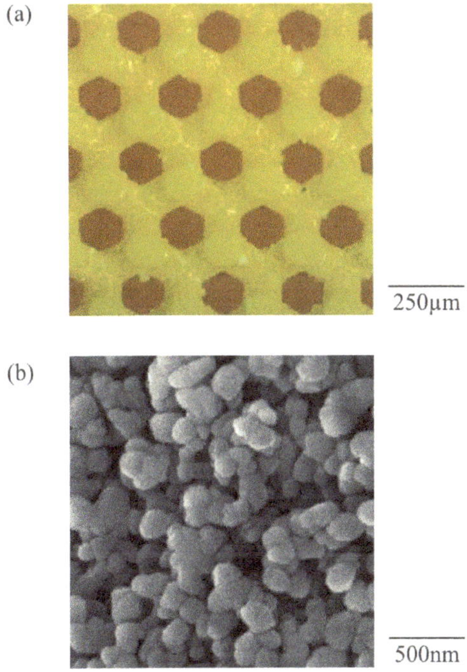


Fig. 13. The photonic crystal with the diamond structure composed of acrylic lattices with nano alumina particles dispersion formed by the micro stereolithography.

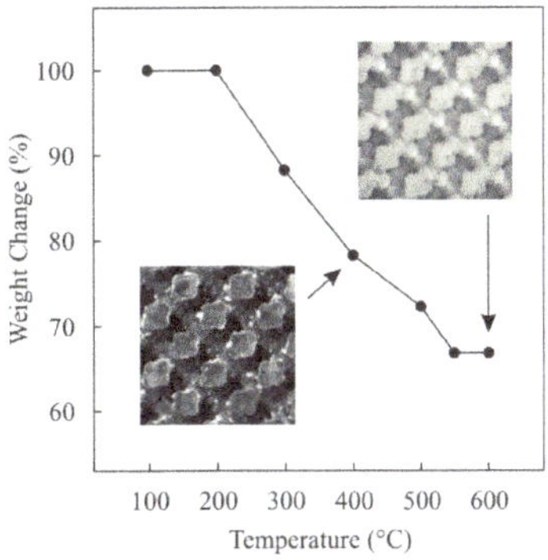


Fig. 14. The weight change as a function of temperature and the lattice color after dewaxing at 400 and 600°C in heating temperatures.

temperature are shown in Fig. 14. The sample color changed into black at 400°C due to carbonizing of resin. It became white at 600°C suggesting burning out of resin. Thus, the dewaxing process is considered to start at 200°C and complete at 600°C. The dewaxing temperature was optimized to be 600°C. Through the dewaxing and sintering processes, ceramic diamond structures were successfully obtained. Figure 15 shows (111), (100) and (110) planes of the sintered diamond structure composed of the micrometer order alumina lattice. The lattice constant was measured as 375 μm. The deformation and cracking were not observed. The linear shrinkage on the horizontal axis was 23.8 % and that on the vertical axis was 24.6 %. It is possible to obtain the uniform shrinkage by designing an appropriate elongated structure in the vertical direction for compensation to the gravity effect. The relative density reached 97.5 %. Dense alumina microstructure was formed, and the average grain size was approximately 2 μm. The measured dielectric constant of the lattice was about 10.

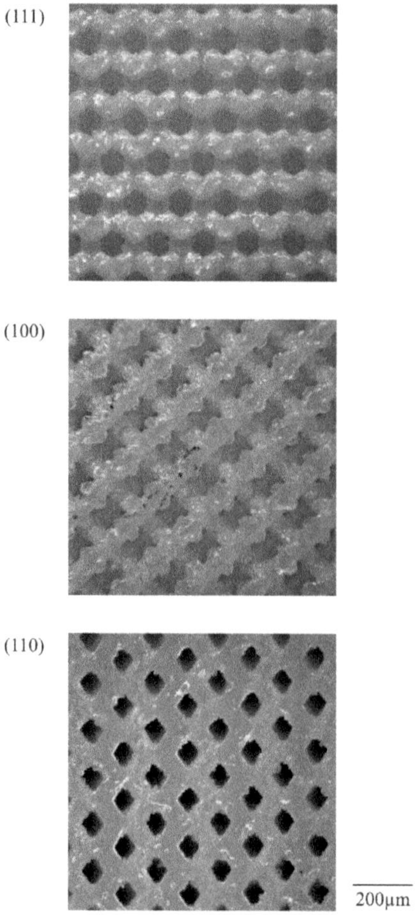


Fig. 15. Crystal planes of (111), (100) and (110) in alumina photonic crystals with the diamond structure fabricated by using the micro stereolithography. Sintering density of alumina lattices is 98.5 %. Dielectric constant of the lattice is 10.

## 6.2 Band structures

The terahertz wave attenuations of the transmission amplitudes through the alumina diamond structures for Γ-L <111>, Γ-X <100> and Γ-K <110> crystal directions are shown in Fig. 16. The forbidden gaps are formed at the frequency rages of 0.32 - 0.49, 0.35 - 0.53 and 0.35 - 0.52 THz in transmission spectra for Γ-L <111>, Γ-X <100> and Γ-K <110> directions, respectively. A common band gap was observed in every direction at the frequency range from 0.35 to 0.50 THz, where the electromagnetic wave cannot transmit through the crystal and is totally

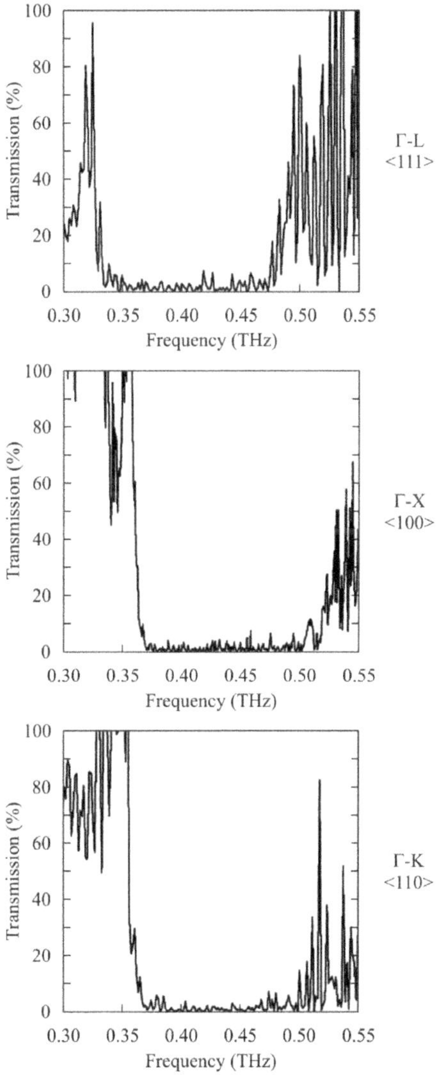


Fig. 16. Terahertz wave attenuations of transmission amplitude for Γ-L <111>, Γ-X <100> and Γ-K <110> directions in the alumina photonic crystal with the diamond structure by using terahertz time domain spectroscope.

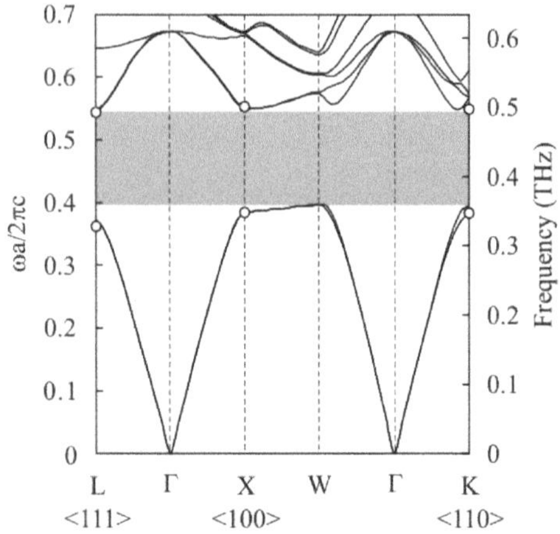


Fig. 17. A photonic band diagram calculated by plane wave expansion (PWE) method. Opened circles are measured edge frequencies of photonic band gaps. The dielectric constant of the lattice was 10. The frequency range with gray color indicates the perfect band gap in common for all directions.

reflected in all directions. The measured band gap frequencies were compared with calculation results by the plane wave expansion method as shown in Fig. 17. The band diagram of the photonic crystal along symmetry lines in the Brillouin zone is drawn theoretically by the PWE method. The opened circles mean the higher and lower edges of the measured band gaps. These frequency ranges of opaque regions corresponded to the calculation. According to the photonic band diagram, it was demonstrated that a complete photonic band gap opened between 0.35 and 0.49 THz. When a gap is formed, there are two types of the standing wave modes with the wavelength corresponding to periodicity of the dielectric lattices at the frequencies of the each band edges. The lower frequency mode concentrates the wave energy in the dielectric region, whereas the higher frequency mode concentrates in the air region.

### 6.3 Localized modes

The integrated terahertz wave resonator is shown in Fig. 18. The two diamond lattice components were attached on the quartz glasses, and these two glass plates were arranged with 150 μm in parallel interval. The tolerance for the transmission direction of the electromagnetic wave was converged within 5 μm. The cell capacity was 0.02 ml. Figure 19-(a) shows the measured transmission spectra for the resonators. Distilled water or ethanol was infused into the micro cells. In the case of distilled water, two localized modes of transmission peaks were observed at 0.410 and 0.491 THz in frequencies in the photonic band gap. In the case of ethanol, an amplification peak was observed at 0.430 THz. The measured band gap ranges and the localized mode frequencies have good agreement with the simulated results by the transmission line modelling as shown in Fig. 19-(b). In the transmission spectrum through the photonic crystal resonator including the water, the localized modes of the higher and lower peak frequencies are defined mode A and B, respectively. And, the localized mode peak in the transmission spectrum through the ethanol is defined mode C. The cross sectional profiles of the electric field intensity

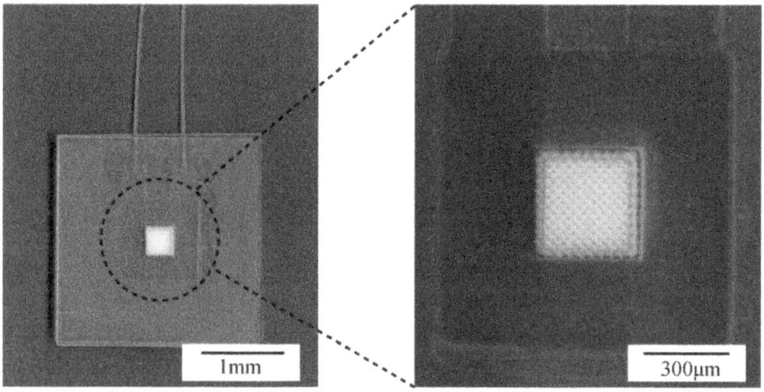


Fig. 18. A terahertz wave resonator with micro liquid cell sandwiched between diamond photonic crystals. Distilled water and ethanol were infused through the catheters implanting in the top side of the cells respectively.

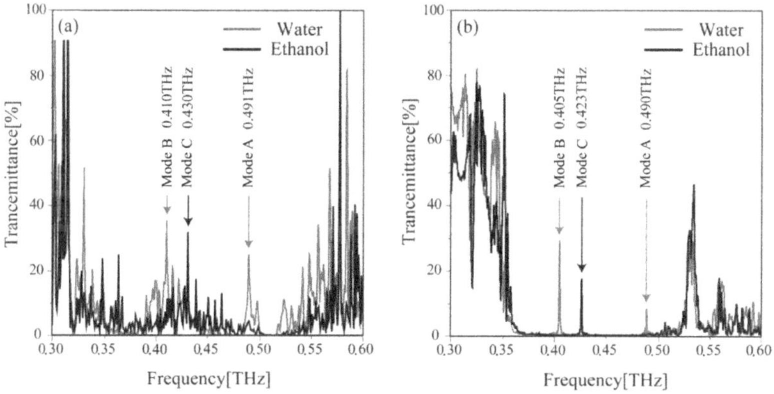


Fig. 19. Transmission spectra for the resonators infusing distilled water and ethanol. The spectra (a) and (b) are measured and calculated properties by using the terahertz wave time domain spectroscopy and a transmission line modeling methods, respectively.

corresponding to the localized mode A, B and C were simulated and visualized theoretically as shown in Fig. 20-(a), (b) and (c), respectively. The terahertz wave was propagated from the left to the right side. The white and black areas show that the electric field intensity is high and low, respectively. The incident terahertz waves were resonated and localized strongly through the multiple reflections in the liquid cell between two diffraction lattices with the diamond structures. Concerning with the localized mode A as shown in Fig. 20-(a), the standing wave with five nodes was observed in the vicinity area of the glass cell. And, the localized mode B with the lower frequency and the longer wavelength is composed of the standing wave with four nodes between the diffraction lattices as shown in Fig. 20-(b). Concerning with the localized mode C in the ethanol as shown in Fig. 20-(c), the standing wave with four nodes has the similar resonance profiles between the diffraction lattices as the localized mode B in the water. Figure 20-(d) shows the dielectric constants and

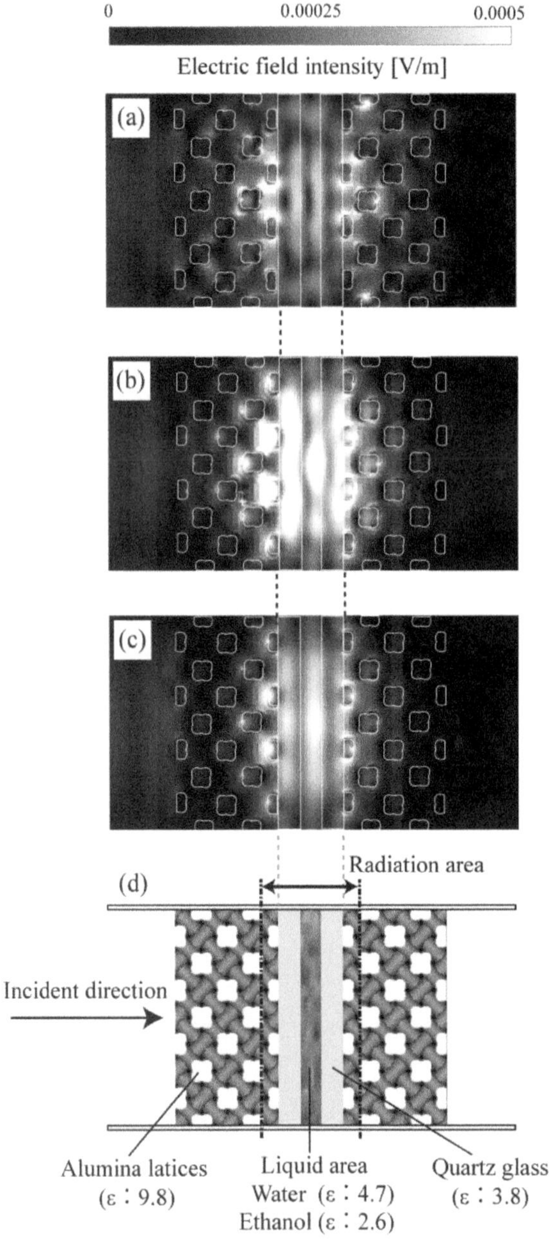


Fig. 20. Electric field distributions of localized modes simulated by transmission line modeling (TLM) method of a finite difference time domain (FDTD). The localized mode A, B and C, are shown in (a) , (b) and (c), respectively. Dielectric constants and composite dimensions of the photonic crystal resonator are shown in (d).

composite dimensions of the photonic crystal resonator. The dielectric constants of water and ethanol are 4.7 and 2.6 in the terahertz wave frequency range [17,18]. By using the terahertz wave time domain spectroscopy, the dielectric constants of alumina and quartz grass were measured as 9.8 and 3.8. In comparison with the resonance areas of the localized mode B and C, the effective dielectric constants decrease from 3.02 to 2.59, and the optical lengths decrease from 1.13 to 1.05 mm, respectively. Therefore, the resonance peak frequencies of the localized mode B and C with the similar standing wave profiles are shifted clearly from 0.410 to 0.430 THz by replacing the water with ethanol in the resonator. Form these results, the fabricated phonic crystal resonator is considered to be a promising candidate as the novel analytical device to determine the dissolved components in the aqueous solution by using the terahertz spectroscopy.

## 7. Conclusions

Terahertz wave resonators of a micro glass cell put between two photonic crystals composed of alumina lattices with a diamond structure were fabricated successfully by using micro stereolithography. Transmission spectra were measured through the photonic crystal resonators including pure water or ethanol. Localized modes of sharp transmission peaks were observed in the photonic band gaps. In a distribution profile of electric field intensity simulated by using transmission line modelling, the strong localized modes were formed through multiple reflections in the liquid regions between the diffraction lattices. Moreover, the localized mode peak was shifted clearly from higher to lower frequencies through replacing the pure water with the ethanol. The fabricated photonic crystal resonator is considered to be a promising candidate for novel analytical devices to detect the compositional variations in natural aqueous phase environments.

## 8. References

Akahane, Y., Asano, T., Song, B. & Noda, S. (2003). High-Q Photonic Nanocavity in a Two-Dimensional Photonic Crystal. Nature, Vol.425, (September 2003), pp.944-947.

Baba, T. & Fukuya, N. (2001). Light Propagation Characteristics of Defects Waveguides in a Photonic Crystal Slab. Photonic Crystals and Light Localization (edited M. Soukoulis), Kluwer Academic Publishers, Netherlands.

Chen, W,; Kirihara, S. & Miyamoto, Y. (2007a). Fabrication and Measurement of Micro Three-Dimensional Photonic Crystals of $SiO_2$ Ceramic for Terahertz Wave Applications. Joiurnal of the American Ceramic Society, Vol.90, No.7, (July, 2007), pp.2078-2081.

Chen, W.; Kirihara, S. & Miyamoto, Y. (2007b). Three-dimensional Microphotonic Crystals of ZrO2 Toughened Al2O3 for Terahertz wave applications. Applied Physics Letter, Vol.91, No.15, (October 2007b), pp.153507-1-3.

Chen, W.; Kirihara, S. & Miyamoto, Y. (2007c). Fabrication of Three-Dimensional Micro Photonic Crystals of Resin-Incorporating $TiO_2$ Particles and their Terahertz Wave Properties. Journal of the American Ceramic Society, Vol.90, No.1, (January 2007c), pp.92-96.

Chen, W.; Kirihara, S. & Miyamoto, Y. (2008). Static Tuning Band Gaps of Three-dimensional Photonic Crystals in Subterahertz Frequencies. Applied Physics Letters, Vol.92, (December 1990), pp.183504-1-3.

Clery, D. (2002). Brainstorming Their Way to an Imaging Revolution. Science, Vol.297, (August 2002), pp. 761- 763.

Cregan, R.; Mangan, B., Night, J., Birks, T., P. Russell, S., Roberts, P. & Allan, D. (1999). Singlemode Photonic Bandgap Guidance of Light in Air. Science, Vol.285, No.5433, ( September 1999), pp.1537-1539.

Fischer B M, Walther M and Jepsen P Uhd 2002 Far-infrared vibrational modes of DNA components studied by terahertz time-domein spectroscopy Phys. Med. Biol. 47 3807-14.

Haus, J. (1994). A Brief Review of Theoretical Results for Photonic Band Structures. Journal of Modern Optics, Vol.41, No.2, (February 1994), pp.195-207.

Hineno, M. & Yoshinaga, H. (1974). Far-Infrared Spectra of Mono-, Di- and Tri-saccharides in 50-16

Ho, K.; Chan, C. and Soukoulis, C. (1990). Existence of a Photonic Gap in Periodic Dielectric Structures. Physical Review Letter, Vol.65, No.25, (December 1990), pp. 3152-3165.

John, S. (1987). Strong Localization of Photons in Certain Disordered Dielectric Superlattices. Physical Review Letter, Vol.58, No.23, (June 1987), pp.2486-2489.

Kanaoka, H.; Kirihara, S. and Miyamoto, Y. (2008). Terahertz Wave Properties of Alumina Microphotonic Crystals with a Diamond Structure. Journal of Materials Research, Vol.23, No.4, (April 2008), pp.1036-1041.

Kanehira, S.; Kirihara, S. & Miyamoto, Y. (2005). Fabrication of $TiO_2$-$SiO_2$ Photonic Crystals with Diamond Structure. Journal of the American Ceramic Society, Vol.88, No.6, (June 2005), pp.1461-1464.

Kawakami, S. (2000). Photonic Crystals, CMC, Tokyo.

Kawase K.; Ogawa, Y., Watanabe, Y. & Inoue, H. (2003). Non-destructive Terahertz Imaging of Illicit Drugs Using Spectral Fingerprints. Optics Express, Vol.11, Iss.20, (October 2003), pp.2549-2554.

Kirihara, S. & Miyamoto, Y. (2009) Terahertz Wave Control Using Ceramic Photonic Crystals with Diamond Structure Including Plane Defects Fabricated by Micro-stereolithography. The International Journal of Applied Ceramic Technology, Vol.6, No.1, (January, 2009), pp.41-44.

Kirihara, S.; Miyamoto, Y., Takenaga, K., Takeda, M. and Kajiyama, K. (2002). Fabrication of Electromagnetic Crystals with a Complete Diamond Structure by Stereolithography. Solid State Communications, Vol.121, No.8, (March, 2002), pp.435-439.

Kirihara, S.; Niki, T. & Kaneko, M. (2009a). Terahertz Wave Behaviors in Ceramic and Metal Structures Fabricated by Spatial Joining of Micro-stereolithography. Journal of Physics, Vol.165, No.1(April 2009), pp.12082-1-12082-6

Kirihara, S.; Niki, T. & Kaneko, M. (2009b). Three-dimensional Material Tectonics for Electromagnetic Wave Control by Using Micoro-stereolithography. Ferroelectrics, Vol. 387, (February 2009), pp.102-111.

Kirihara, S.; Takeda, M., Sakoda, K. and Miyamoto, Y. (2002a). Control of Microwave Emission from Electromagnetic Crystals by Lattice Modifications. Solid State Communications, Vol.124, No.4, (October 2002) pp.135-139,.

Kirihara, S.; Tsutsumi, K. & Miyamoto, Y. (2009c), Localization Behavior of Microwaves in Three-dimensional Menger Sponge Fractals Fabricated from Metallodielectric Cu/polyester Media. Science of Advanced Materials, Vol.1,No.2 (December 2009), pp.175-181.

Kosaka, H.; Kawashima, T., Tomita, A., Notomi, M., Tamamura, T., Sato, T. & Kawakami, S. (1999). Photonic Crystals for Micro Wave Circuits Using Wavelength-dependent Angular Beam Steering. Applied Physics Letter, Vol.74, No.8, (March 1999), pp. 1370-1378.

Kutteruf, M. R.; Brown, C. M., Iwaki, L. K., Campbell, M. B., Korter, T. M. & Heilweil, M. J. (2003). Terahertz Spectroscopy of Short-chain Polypeptides, Chem. Phys. Lett., Vol.375, (December 2003), pp.337-343.

Miyamoto, Y.; Kanaoka. H. & Kirihara, S. (2008). Terahertz Wave Localization at a Three-dimensional Ceramic Fractal Cavity in Photonic Crystals, Journal of Applied Physics, Vol.103, (May 2008), pp.103106-1-5.

Noda, S. (2000). Three-dimensional Photonic Crystals Operating at Optical Wavelength Region. Physica B, Vol.279, No.1-3, (April 2000), pp.142-149.

Noda, S.; Yamamoto, N., Kobayashi, H., Okano, M. and Tomoda, K., (1999). Optical Properties of Three-dimensional Photonic Crystals Based on III-V Semiconductors at Infrared to Near-infrared Wavelengths. Applied Physics Letters, Vol.75, No.16, (August 1999), pp.905-907,.

Ohtaka, K. (1979). Energy bang of photons and low energy photon diffraction. Physical Review B, Vol.19, No.10, (May 1979) pp.5057-5067.

Oyama, Y.; Zhen, L., Tanabe, T. & Kagaya, M. (2009). Sub-Terahertz Imaging of Defects in Building Blocks, NDT&E International, Vol.42, (January 2009), pp,28-33.

Oyama, Y.; Zhen. L., Tanabe, T. & Kagaya, M. (2008). Sub-Terahertz Imaging of Defects in Building Blocks. NDT&E International, Vol.42, No.1, (January 2008), pp.28-33.

Soukoulis, C. (1996). Photonic Band Gap Materials, Kluwer Academic Publsher, Netherlands.

Takano, H.; Song, B., Asano, T. & Noda, S. (2005). Highly Efficient in-Plane Channel Drop Filter in a Two-Dimensional Heterophotonic Crystal. Applied Physics Letters, Vol.86, No.24, (June, 2005), pp.241101-1-3.

Temelkuran, B.; Bayindir, M., Ozbay, E., Biswas, R., Sigalas, M., Tuttle, G. & Ho, K. (2000). Photonic Crystal-based Resonant Antenna with Very High Directivity. Journal of Applied Physics, Vol.87, No.1, (January 2000), pp.603-605.

Van Exter, M.; Fattinger, C. & Grischkowsky, D. (2002). Terahertz Time-domain Spectroscopy of Water Vapor. Optics Letters, Vol.14, Iss.20, (October 1989), pp.1128-1130.

Vos, W.; Sprik, R., Blaaderen, A., Imhof, A., Lagendijk, A. & Wegdam, G. (1996). Strong Effects of Photonic Band Structures on the Diffusion of Colloidal Crystals. Physical Review B, Vol.53, No.24, (February 1996), pp.16231-16235.

Wallace, V. P.; Fitzgerald, A. J., Shankar, S., Flanagan, N., et al. (2004). Terahertz Pulsed Imaging of Basal Cell Carcinoma ex Vivo and in Vivo. Br. J. Dermatol, Vol.151 (August 2004), pp.424–32.

Wallace, V.; Fitzgerald, A., Shankar, S. & Flanagan, N. (2004). Terahertz Pulsed Imaging of Basal Cell Carcinoma ex Vivo and in Vivo. The British Journal of Dermatology, Vol.151, No.2, (August, 2004), pp.424-432.

Woodward, R.; Wallace, V., Arnone, D., Linfield, E. & Pepper, M. (2003). Terahertz Pulsed Imaging of Skin Cancer in the Time and Frequency Domain. Journal of Biological Physics, Vol.29, No.2-3, (June 2003), pp.257-259.

Yablonovitch, E. (1987). Inhabited Spontaneous Emission in Solid-state Physics and Electronics. Physical Review Letter, Vol.58, No.20, (May 1987), pp. 2059-2062.

Yamaguchi, M.; Miyamaru, F., Yamamoto. K. , Tani, M. & Hangyo, M. (2005). Terahertz Absorption Apectra of L-, D-, and DL-alanine and Their Application to Determination of Enantiometric Composition. Appl. Phys. Lett, Vol. 86, (April 2005), pp.053903.

# 10

# Evaluation of Nanometer Cutting Tool Edge for Nanofabrication

Yuki Shimizu,
Takemi Asai and Wei Gao
*Tohoku University,
Japan*

## 1. Introduction

Demands for the nanofabrication of the precision surfaces, which have three- dimensional (3D) structures to add specific functions, are continuously increasing. For the nanofabrication on machining tools such as an ultra-precision turning, an ultra-precision diamond cutting tool is generally employed because of its sharp edge and durability. Figure 1(a) shows the illustrations of a chip of the diamond cutting tool. The diamond chip is mounted on the tip of the cutting tool. Figure 1(b) and (c) show a top view from the rake-face side and a side view of the chip, respectively. The chip has a certain nose (corner) radius $R$ at its edge top as shown in Fig. 1(b). The rake angle, radius of the cutting edge, wedge angle and clearance angle are determined as shown in Fig. 1(c).

In the nanofabrication, a quality of the fabricated surface will be dominated by the accuracy of the cutting edge profile. Although the design of the cutting edge profile has some tolerances for its shape, the allowed tolerances for ultra-precision machining are so tight that the highly-accurate measurement technique would be required for assuring the accuracy of the cutting edge profile.

Moreover, a wear of the tool edge should also be taken into consideration for the nanofabrication. Basically, the amount of the wear is very small because the diamond is a hard material. However, a slight wear of the tool edge cannot be avoided even though target materials are the non-steel alloys; namely, the cutting edge profile could change every short periods of the use. The slight change of the cutting edge profile could affect the accuracy of the fabricated surfaces. Therefore, the cutting edge profile should be checked periodically. Generally, the accurate measurement of the cutting edge profile is a time-consuming task. That is why, not only the accuracy but also the easiness of the measurement setting will be required for the nanometer-level evaluation of the cutting edge profile.

Based on such a background described above, in this chapter, the authors describe a method to evaluate the nanometer cutting tool edge by using AFM, including not only a measurement theory but also experimental setup and some results of the tool edge measurement.

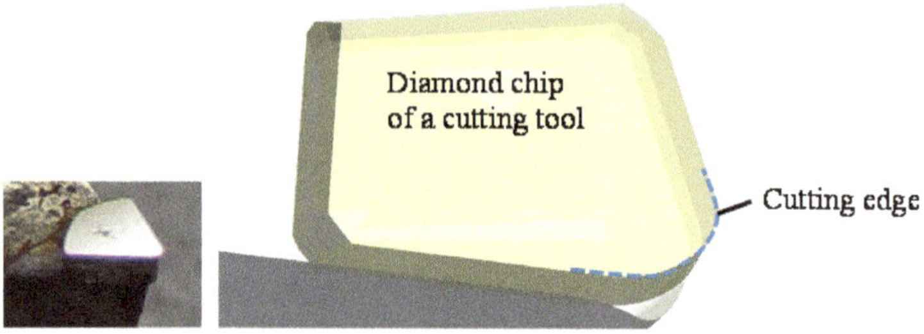


(a) Bird view

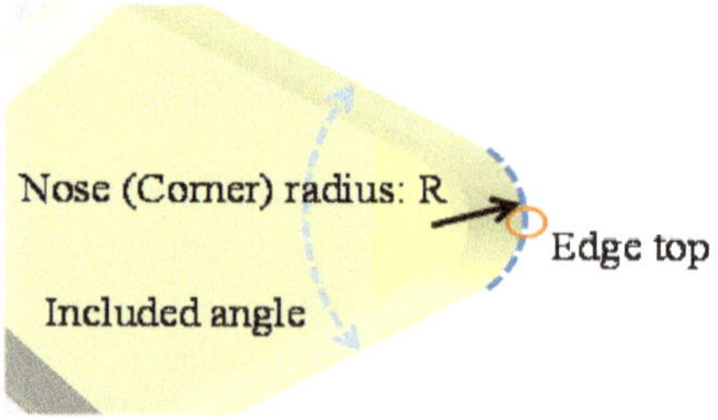


(b) Top view (from the race face side)

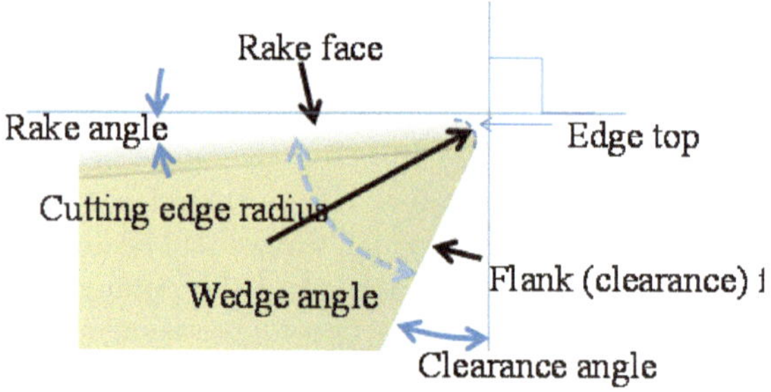


(c) Side view

Fig. 1. Illustrations of a diamond cutting tool chip

## 2. A method for compensation of the AFM tip geometry

### 2.1 Cutting edge profile measurement by AFM

There are several methods to evaluate the cutting edge profile. In most of the cases, scanning electron microscopes (SEMs) are used because of its ease. However, the SEMs require vacuum chambers and vacuum pumps, resulting in bulky systems. On the other hand, the Atomic force microscope (AFM), which will provide a high accuracy and a high resolution in both vertical and lateral directions, could be used for the evaluation of the cutting edge profile until the measurement system would be set properly (Gao et al., 2006). The compact AFM system for the measurement of the cutting edge profile of the diamond tool was previously proposed (Gao, 2010). In the system, because of the simplicity, the contact mode AFM was selected among several measurement mode of the AFM. A unique method was

also proposed to align the AFM probe with the cutting edge of the ultra-precision diamond cutting tool. Figure 2 shows a schematic of the cutting edge profile measurement by AFM. In the alignment method, a pair of laser beam and photodetector was employed to align tool edge and cantilever of the AFM. By using the compact AFM system with the unique alignment method, the cutting edge profile could successfully be carried out on-machine.
The cutting edge profiles measured by the AFM are affected by the tip shape of the AFM probe. For nanofabrication, the influence of the shape of the AFM probe would not be negligible in some cases. In this section, (a) the influence of the shape of the AFM probe on the measured cutting edge profiles, and (b) how to eliminate the influence of the shape of the AFM probe will be described. Furthermore, the evaluation method how to (c) measure the tip shape of the AFM probe will also be described.

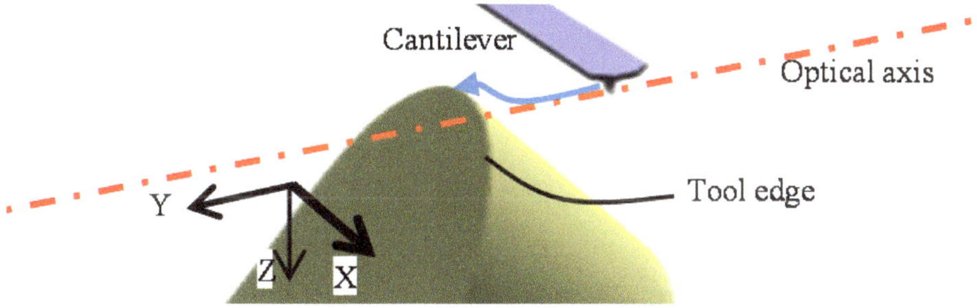


Fig. 2. Schematic of the cutting edge profile measurement by AFM

### 2.2 Influence of the AFM tip geometry on profile measurement

When we employ a contact-type sensor to carry out form accuracy measurements, the evaluated form data always contain a certain form error caused by the roundness or the tip shape of the styluses employed in the measurement setups. In case of the measurement of the cutting edge profile with an AFM, the evaluated cutting edge radius includes the roundness of the AFM tip. Figure 3 shows schematics of the paths of the AFM tip during the scanning over a surface with a step for the imaging. In the contact-mode measurement, AFM tip tries to trace the target surface. Basically, the AFM tip traces the target surface accurately when scanning over the relatively flat sections. On the other hand, the tip cannot trace the target surface accurately when scanning over a step-shaped surface because of its tip radius. When the shape of AFM tip is defined as $t\,(x)$, the reflected tip shape $p\,(x)$ can be described as

$$p(x) = -t\,(-x) \quad (1)$$

When the sample surface is defined as $f\,(x)$, the measured profile $g\,(x)$ can be described by a following equation (Villarrubla, 1997), while the Z-directional servo control of the AFM is properly performed.

$$g(x) = \max_{x'}\left[f(x') - t(x' - x)\right] \quad (2)$$

Namely, the series of the coordinate $(x, g(x))$ can be acquired by searching $x'$, which gives the maximum value of $f(x') - t(x'-x)$, at each $x$. As shown in Eq. (2), the acquired image profile $g(x)$ contains the information of the convex-shaped AFM tip $t(x)$. Such a phenomenon, in which the measured profile contains a certain influence of the measurement stylus, is called as "convolution" often used in the optical imaging. Furthermore, the operation reconstructing the original image from the measured image, which contains the information of the measurement stylus, is called "deconvolution". In case of the AFM measurement, the deconvolution would be possible if the shape of AFM tip $t(x)$ is known in advance, as shown in Fig. 3.

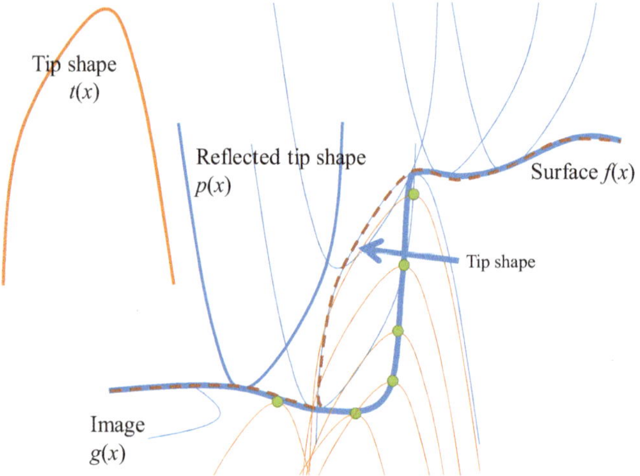


Fig. 3. Conceptual illustration of the imaging

Figure 4 shows a conceptual illustration of the reconstruction of the target surface from the measured results. Although the reconstructed surface sometimes does not represent the actual surface because there still remains some area where the AFM tip cannot contact, the reconstructed surface represents the target surface more accurately. The more precise the AFM tip shape is known, the more accurate the target surface profile can be measured. Although the discussion shown above was carried by using two-dimensional profiles in Figures 3 and 4, the same discussion could be applied to the three-dimensional profiles obtained by AFM. It should be noted that the discussion above would be effective if there is no damage or wear on the AFM tip during the measurements.

There are some methods to observe or measure the roundness at the apex of the AFM tip. One of the most popular methods is to use scanning electron microscope (SEM). The SEM is the primary method to observe the sharp profile such as AFM tips (Bloo et al., 1999), and can be used to estimate the size of the tip apex. There are some other techniques to obtaining the shapes of tip apex. A tip characterizer, which has well-known shapes on its surface, is also used to estimate the AFM tip shape; namely, the shape of the tips will become known if there are well known structures to be imaged by the AFM. Moreover, the blind tip estimation, which is based on the theory of morphology, is another approach (Villarrubia, 1994; Dongmo et al., 2000). Because the convex of the image obtained by an AFM could have information about the tip, the tip shape can be estimated by correcting the AFM information. Furthermore, the implementation of the blind tip estimation has been explained in detail (Villarrubia, 1994). Some implementation is available on-line (Nečas &

Klapetek, n.d.). That is now modified to be applied to the tip with overhang (Tian et al., 2008).
In the following sections, we will discuss (A) the observation of the shape of AFM tip by SEM and (B) the estimation of the shape of AFM tip by tip characterizer.

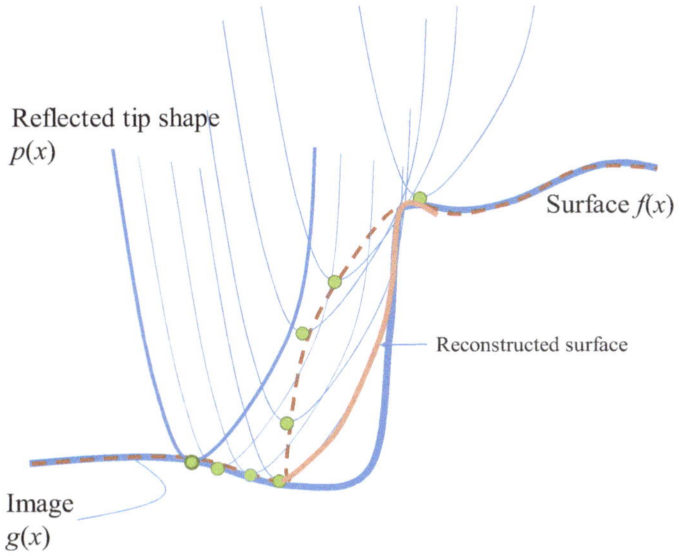


Fig. 4. Conceptual illustration of the reconstruction of the surface

### 2.3 Observation of the shape of AFM tip by SEM

Figures 5 show examples of the images of AFM tip and its apex measured by SEM. Figure 5(a) shows a whole shape of the AFM probe. The AFM probe has a conical-shaped sharp tip on the edge of the probe. Figure 5(b) is the magnified image of the same AFM tip shown in Figure 5(a). Although the shape of the AFM tip is not clear in the magnified images, the tip radius could be estimated around 30 nm from the SEM images. In fact, the AFM tip was said to be less than 20 nm in the specification sheet of the AFM probe, which roughly agrees with the measured tip radius.
It is not so difficult to acquire SEM image of the magnified AFM tip. Therefore, the SEM is one of the most popular methods among the various ones. However, it should be noted that the SEM observation itself could sometimes affect the shape of AFM tip. Figure 5(c) shows another SEM image of the same tip shown in Fig. 5(b). The SEM image in Fig. 5(c) was captured just a few minutes after the image in Fig. 5(b) was captured. The tip apex seemed to become dull from the original shape, which is considered to be due to a discharge or contamination on the AFM tip. It is well known that the SEM observation could easily damage or contaminate the very sharp tip without enough experiences (Egerton et al., 2004; Taniguchi et al., 1997). These results shown in Figs. 5 indicate that the radius of the AFM tip described here could be estimated slightly larger than which really was. However, it can be said that the SEM is one of common methods to evaluate the shape of AFM tip because of its ease.

(a) Image of the tip (b) Image of the tip apex magnified from (a)

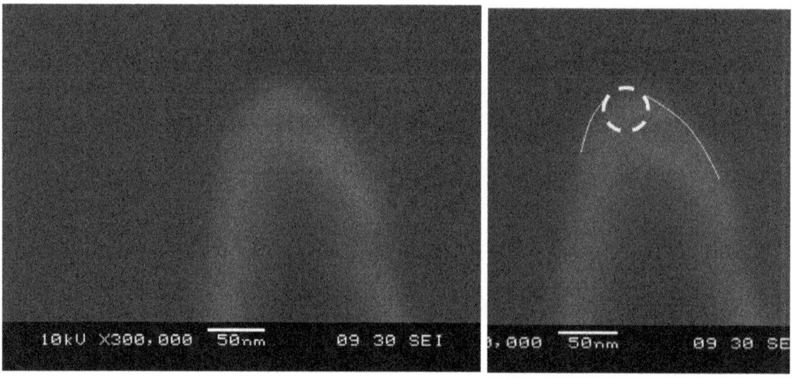


(c) Image of the tip apex re-measured a few minutes after the image (b) was captured

Fig. 5. SEM images of an AFM-tip, which were taken from the same AFM tip

### 2.4 Estimation of the shape of AFM tip by tip characterizer

Using a tip characterizer, which has well-known profile on its surface, is also used to estimate the shape of the AFM tip. A tip characterizer named TGG1, which has a very sharp edge, was used as an example. Figure 6 shows a schematic image of the profile on TGG1 surface. Several triangular steps were fabricated on a silicon wafer with a certain pitch, and the angles of each sharp edge were designed to be 70 degrees. Table 1 summarizes a detailed specification of TGG1. The radius of the sharp edge was fabricated to be less than 10 nm, and the pattern pitch was also controlled to be 3.00 μm with a tolerance of ± 0.05 μm.

Figure 7 shows a bird view of the profile of the tip characterizer measured by AFM. Figure 8 shows a cross-section image of the profile of the tip characterizer in Fig. 7, extracted at a certain X-axis position. To estimate the tip radius, a circle was fitted on the top of the measured profile in Fig. 8 by using least square method. The radius of the fitted circle shown in Fig. 8 was slightly less than 30 nm. Actually, the value of the radius of the fitted circle could vary from 25 to 30 nm when the fitted region was selected arbitrary. Assuming that the edge radius of TGG1 is less than or exactly 10 nm referring to Table 1, the tip radius of the AFM probe could be from 15 to 30 nm, according to Eq. (1). On the other hand, the tip radius of the AFM probe should be less than 20 nm if we believe the datasheet of the AFM probe. However, considering the stability of the measurement instrument used here, it can be said that the AFM tip radius estimated by the tip characterizer roughly agrees with the value indicated in the datasheet.

It should be noted that the definition of edge radius may strongly affect to both the radius in datasheet and the radius measured by experiments; here we assume that there is an AFM tip, which tip width is 40 nm, with dulled apex shape as shown in Fig. 9. Without any attention to the apex shape of the AFM tip, we might determine the tip radius is less than 20 nm even though the real tip radius of the AFM probe should be less than 20 nm in such a case; namely, the apex shape is locally flattened as shown in Fig. 9. Moreover, the region of the tip apex could affect to the determination of the tip radius. The definition of cutting edge radius will be discussed in the following section.

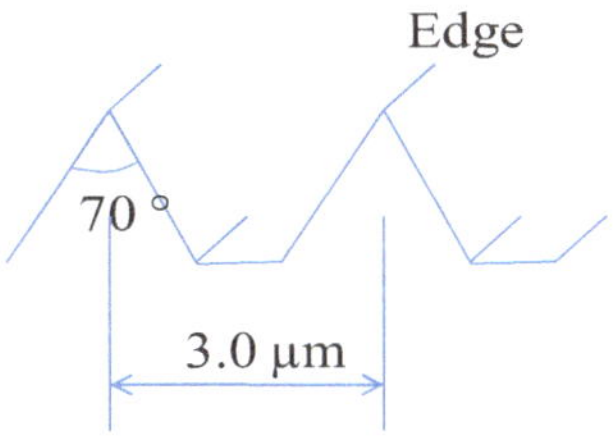


Fig. 6. Schematic of the structure on TGG1 surface

| | |
|---|---|
| Structure | Grating formed on Si wafer |
| Pattern type | One-dimensional array of triangular steps (in X- or Y-direction) having precise linear and angular sizes |
| Edge angle | 70 degrees |
| Edge radius | ≤ 10 nm |
| Period | 3.00 ± 0.05 μm |

Table 1. Specification of the tip characterizer (TGG1)

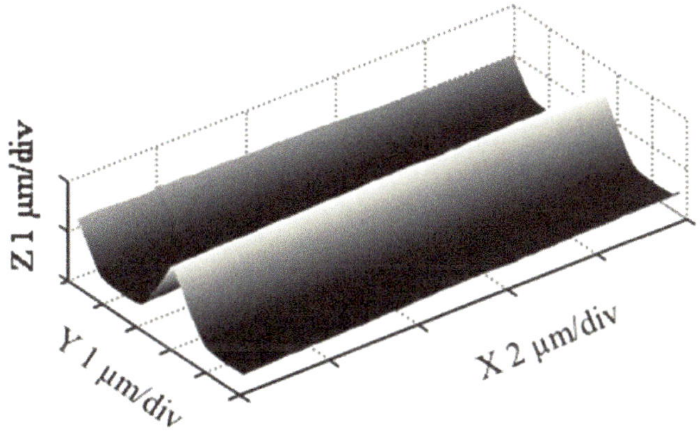


Fig. 7. Schematic of a tip characterizer "TGG1"

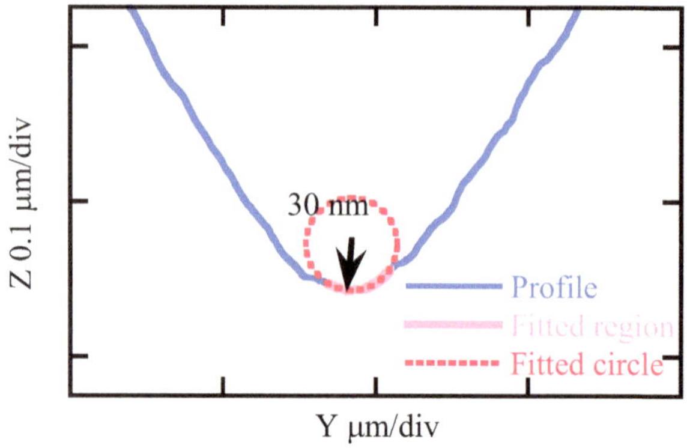


Fig. 8. Cross-section image of TGG1 in Fig. 7, at a certain X-axis position

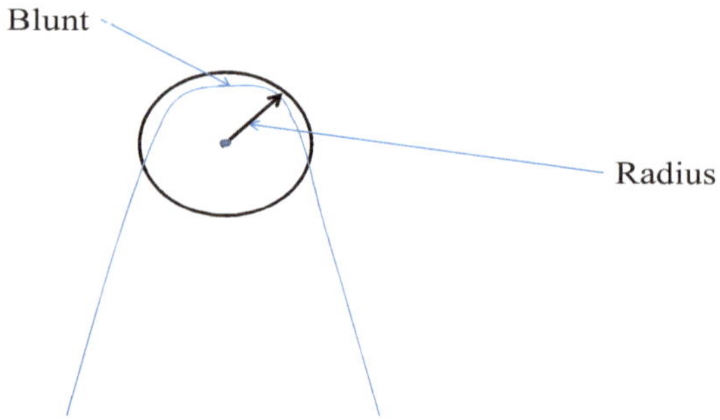


Fig. 9. Cross-sectional image of TGG1 in Fig. 5

## 3. An evaluation method for cutting edge radius

The value of the cutting edge radius is one of the most important parameters for the ultra-precision diamond cutting tool. There are some examples to obtain the cutting edge radius; some of them just fit a circle and employ its radius as the cutting edge radius, as shown in previous section (see Fig. 5(c)). In most cases, the definition of the cutting edge radius is not clear. Although methods shown in previous section will provide us pretty good estimations for the cutting edge radius, some concerns will still remain; where does the curvature starts?, which part of the curvature should we take as a radius?, and so on. Without a suitable method, we cannot define the cutting edge radius in a good repeatability. Therefore, many discussions on how to define the edge radius have been done, and some definitions have been proposed and utilized so far. In fact, however, we'll have to choose a proper method case by case, while taking both measurement accuracy and repeatability into consideration.

Based on such a background, in this section, the methods on how to handle the data of the cutting edge profile will be discussed. Because the cutting edge radius is named as "radius", a circle is usually fitted in many cases. Figure 10 shows a schematic of an ideal cutting edge profile. In Fig. 10, the cutting edge radius and edge angle were set to be 40 nm and 90 degrees, respectively. When the cutting edge radius shown in Fig. 10 is fitted by using the least squares method, the radius should be evaluated to be 40 nm because the applied profile was ideal and any noise component was not added onto the profile in Fig. 10. However, the situation becomes slightly different when the noise component is added, just like as a real case. Figure 11 shows a profile with a noise component. The noise component was added to the ideal profile shown in Fig. 10. The noise component is considered to affect the curve fitting for the cutting edge radius.

Therefore, a simulation was carried out to evaluate the influence of the noise component on the curve fitting. For the simulation, 100 lines of cutting edge profiles with noise component were numerically prepared in a personal computer. The noise components, which have the normal distribution with the standard deviation of 2 nm, were generated by using random number generator. Then the circle fitting was carried out for numerically prepared each cutting edge profile.

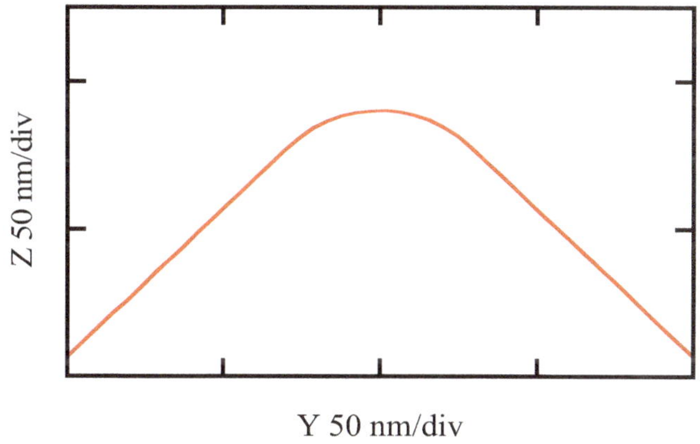


Fig. 10. An artificial edge profile having edge radius of 40 nm

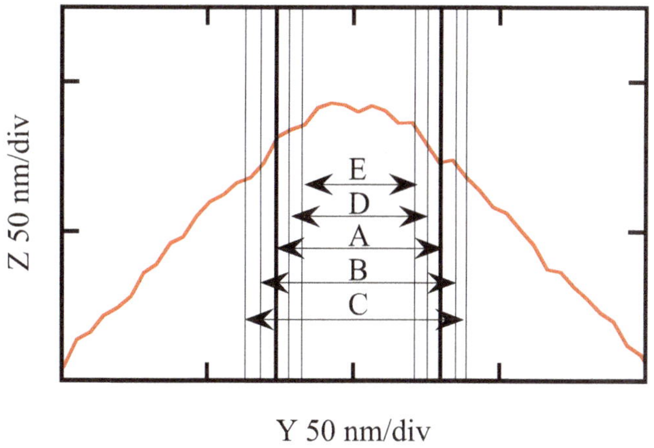


Fig. 11. An artificial edge profile having edge radius of 40 nm with the artificial noise with standard deviation of 2 nm

When we carry out the circle fitting, a data region for the fitting will also affect the calculated cutting edge radius. Therefore, the data region was set to A, B, C, D and E as shown in Fig. 11, and the calculated cutting edge radius were compared.

Table 2 summarizes the average value and the standard deviation of the cutting edge radius calculated at each region from A to E for each 100 lines of cutting edge profiles. Focusing on the result at region A, we can find out that the average of the calculated radius is about 30 nm, with standard deviation of about 3 nm. At each region, the average values of the calculated radius were smaller than 40 nm, which is the radius of the ideal profile. As can be seen in Table 2, the wider data region resulted in larger cutting edge radius. On the other hand, results in Table 2 also tell us that the wider data region also resulted in the smaller standard deviation of the evaluated radius.

| Mark | Evaluated radius (nm) | Standard deviation (± nm) |
|---|---|---|
| E | 15 | 4.2 |
| D | 22 | 4.6 |
| A | 30 | 3.4 |
| B | 35 | 2.8 |
| C | 38 | 2.1 |

Table 2. Fitted circles size at each region (A, B, C, D and E are shown in Fig. 11)

In the case of the simulation shown above, the evaluated radius was considered to be estimated most properly when the region was selected as slightly larger than region C. However, no one can tell how wide we should set the data region in the actual measurement condition. If the evaluation range is set to be larger, a scale of the measurement not only data size but also measurement setup would become large. Anyway, the most important thing here is that the cutting edge radius could be evaluated to be very small or large, on purpose. Local disturbance of the profile due to noise components could cause the misevaluation. Therefore, the cutting edge radius should be evaluated without intentional operations as far as possible.

By the way, generally, the amount of the noise was defined as one sixth of the noise level of the instrument. If the value of standard deviation is determined as $\sigma$, about 99 % of the whole sample is expected to be included in the range of $\pm 3\sigma$ from the average. Of course, this assumption owes to the normal distribution such as Gaussian distribution. It should be noted that the resolution in Y-direction was about 4.6 nm due to quantization of the digital data, which would affect the results in Table 2.

There was an effort to define the cutting edge radius from the measured profile (Asai et al., 1999a; Asai et al., 1999b). A conic curve was fitted to the measurement result and the cutting edge radius was evaluated. There is another reason for selecting a conic curve instead of a circle; the set of conic curves includes ellipses, parabola and hyperbola. The circles are included by the set of ellipses. Therefore, conic curves are more general compared with circles. In case of the dulled edges, the profile is likely to be as shown in Fig. 12. In this case, an ellipse is considered to be the best choice to describe the tip part of this profile. The evaluation by using the circle might take small value compared with the local curvature radius of the edge top, depending on the region to be evaluated. Moreover, the status of the tool whether it is okay or already broken / worn off is important when we want to know the edge radius. However, it is not shown clearly if we use a circle for the fitting. Although there are other definitions proposed in some researches, of course, the edge is considered to be one of the regions marked as A and B in Fig. 12, depending on the case.

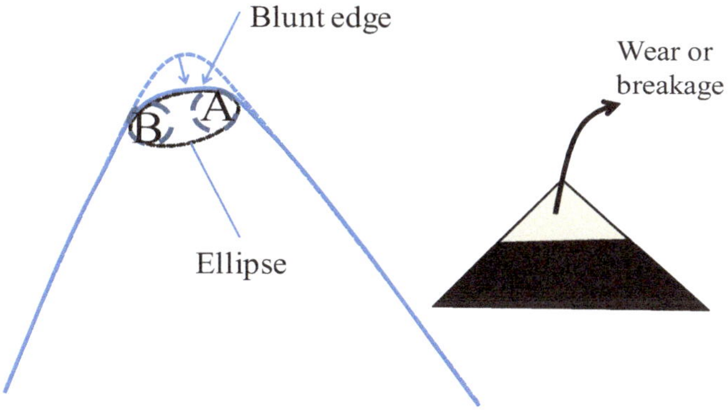


Fig. 12. An example of the blunt edge

According to the references, the actual operations can be described as follows (Asai et al., 1999a; Asai et al., 1999b). Figure 13 shows a schematic of a fitted curve function $F(x)$, which consists of two lines and one curve. Let ($x_i$, $z_i$) be one of the coordinate of the measured data point. The evaluation function $\varepsilon$ can be described as follows:

$$\varepsilon(p, Q_{SX}, Q_{TX}) = \frac{1}{n}\sum\left\{(F(x_i) - z_i)^2\right\} \tag{3}$$

where $n$ is a number of the measurement points. If there are two crossing straight lines SO and OT as shown in Figure 13, the X- and Y-coordinates ($x$, $y$) of the conic curves which are tangent to SO and OT at the points S and T, respectively, can be expressed as follows.

$$x(t,p) = \frac{X(t,p)}{W(t,p)} \tag{4}$$

$$y(t,p) = \frac{Y(t,p)}{W(t,p)} \tag{5}$$

while

$$X(t,p) = \left\{(1-p)Q_{SX} - 2pQ_{OX} + (1-p)Q_{TX}\right\}t^2 - 2\left\{(1-p)Q_{SX} - pQ_{OX}\right\}t + (1-p)Q_{SX} \tag{6}$$

$$Y(t,p) = \left\{(1-p)Q_{SY} - 2pQ_{OY} + (1-p)Q_{TY}\right\}t^2 - 2\left\{(1-p)Q_{SY} - pQ_{OY}\right\}t + (1-p)Q_{SY} \tag{7}$$

$$W(t,p) = 2(1-2p)t^2 - 2(1-2p)t + (1-p) \tag{8}$$

where $t$ and $p$ are the parameters. When $t$ = 0 and $t$ = 1, these equations represent the coordinate of the position S and T, respectively. If the parameter $p$ is set to $0 < p < 0.5$, $p = 0.5$

and $0.5 < p < 1$, the fitted conic curve by Eqs. (4)-(8) could be ellipse, parabola and hyperbola, respectively. If the parameter $p$ is getting smaller and smaller, $a/b$, the ratio of the major axis $a$ to the minor axis $b$ of the ellipse would be getting larger and larger.

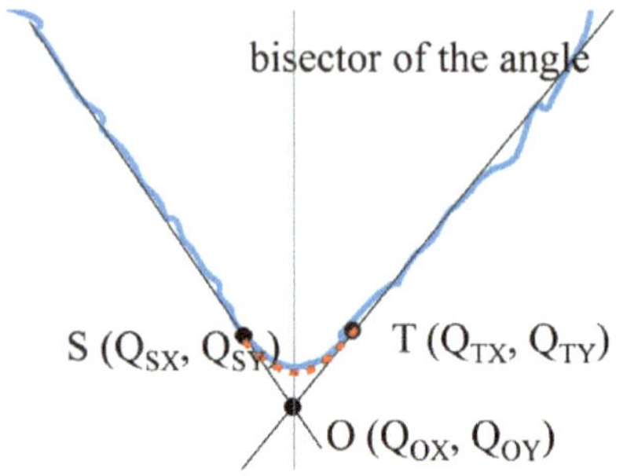


Fig. 13. Conceptual illustration of this first fitting

By using Fig. 13, we will explain how to carry out the fitting. Firstly, select the region to include an edge to be evaluated. Then fit the two lines by using the least squares method, while considering $Q_{SX}$ and $Q_{TX,}$ which are the X- coordinates of the Point S and T, respectively. $Q_{SX}$ and $Q_{TX}$ will be set in this order. It should be noted that not only parameter $p$ but also the coordinates of points S and T would be varied to search the best fitting.
Then find parameters $p$, $Q_{SX}$, $Q_{TX}$ which minimize the parameter $\varepsilon$ in Eq. (3) so that the best set of the parameters $p$, $Q_{SX}$ and $Q_{TX,}$ describing the actual profile, could be obtained. For the minimization, the following methods could be applied;

- Multidirectional search method (Sourceforge.net, n.d.)
- Nelder-Mead method (Sourceforge.net, n.d.)
- Genetic algorithm (Wall, n.d.)

After the fitting operations, draw a line which equally divides the angle between lines TO and SO as shown in Fig. 13. The curvature radius, at the point where the bisector line and the fitted curve, could be determined as one for the cross-sectional profile. However, in practice, the minimization algorithms can only find a local minimum and cannot always find a global minimum. That is a rule of the numerical minimization processes. Besides, a profile sometimes looks as if they do have a dulled edge by accidental noise peaks as shown in Fig. 14 (b).
Fig. 15 shows a histogram of the edge radii, which were calculated by using the previously-prepared profile data in simulation. Because of these problems, the histogram of the evaluation looks like Fig. 15.
Comparing the result in Fig. 15 with the results in Table 2, it seems that the evaluation by fitting a circle looks better because of its small standard deviation. Actually, it is very interesting that the evaluation results with circle fitting do not have large deviation even if there is a profile as shown in Fig. 14(b). Having a large deviation with the fitting method shown above seemed to fits the lines more properly to a raw data. It does not mean that the evaluation of the edge radius is not possible at all, though. When we watch the histogram shown in Fig. 16 carefully, we can find out a peak around the value to be evaluated (40 nm in this case). The value to be obtained is about the peak.

There is a clear difference between a profile shown in Fig. 14(a) and (b). Therefore, a certain criterion is required to tell which evaluation is an outlier. Figure 16 shows the histogram of the edge evaluation as a function of parameter $p$ in Eqs. (4)-(8). The parameter $p$ is the main factor of the radius evaluation if the angle bisector is set properly. Therefore, in order to correct that there is a lower limit in the edge radius evaluation but not upper limit, the practical limit was applied to $p$. Assuming that the standard deviation of $p$ is $\sigma_p$ and the average value of $p$ is $\mu_p$, the lower limit of $p$ can be defined as

$$p > \mu_p - \sigma_p \tag{9}$$

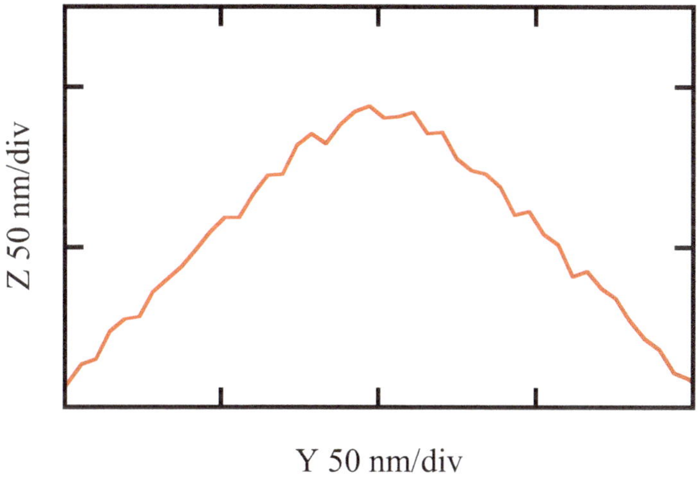


(a) A profile

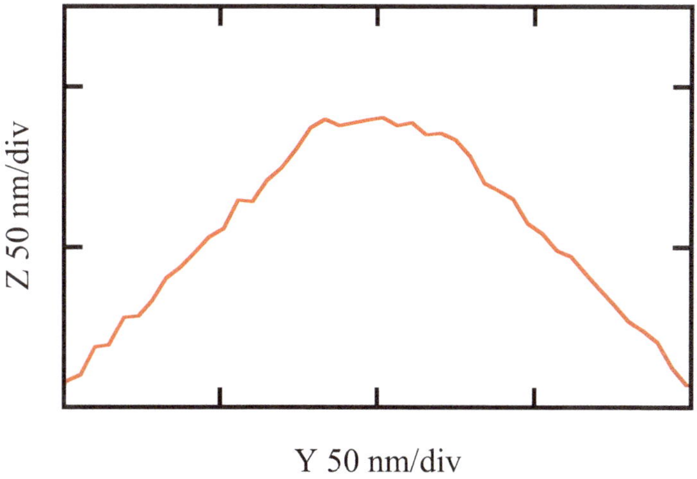


(b) A profile looked blunter than the profile in (a).

Fig. 14. Examples of the pseudo profiles.

The evaluation that does not satisfy this limitation will be treated as an outlier. If the limitation in Eq. (9) is applied to the histogram shown in Fig. 16, the histogram of edge radius evaluation can be re-drawn as shown in Fig. 17. There still seemed to be two peaks around 0.040 and 0.065 μm, however, the average of the evaluations is 43 nm, which is closer to the actual radius designed above. It should be noted that the limitation from upper side like as $p < \mu_p + \sigma_p$ could also be required if the histogram of $p$ has peak in upper side of the average, although the limitation from lower side is only mentioned above.

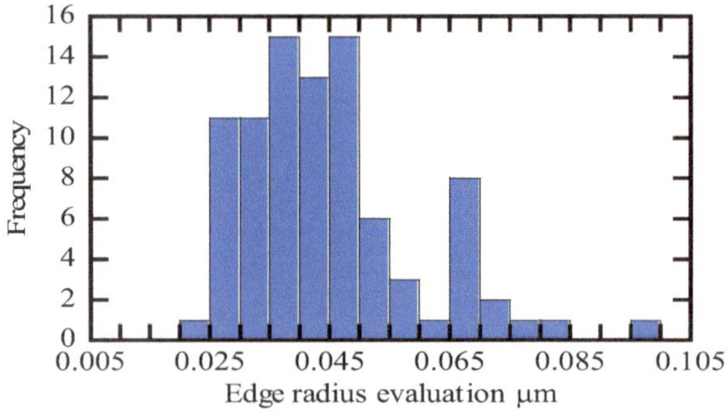


Fig. 15. Histogram of the edge radius evaluation

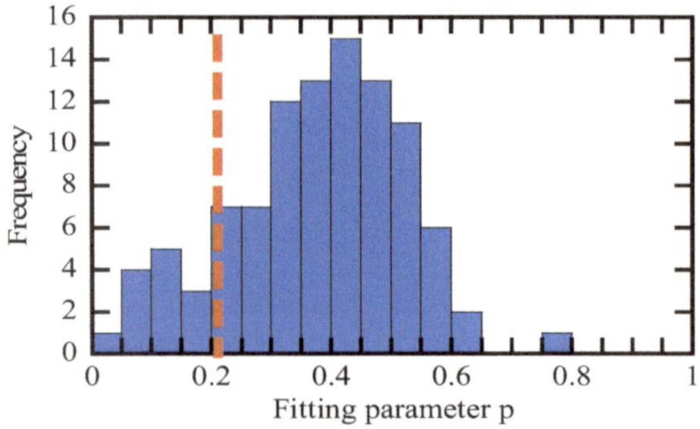


Fig. 16. Histogram of the edge radius evaluation as a function of fitting parameter $p$

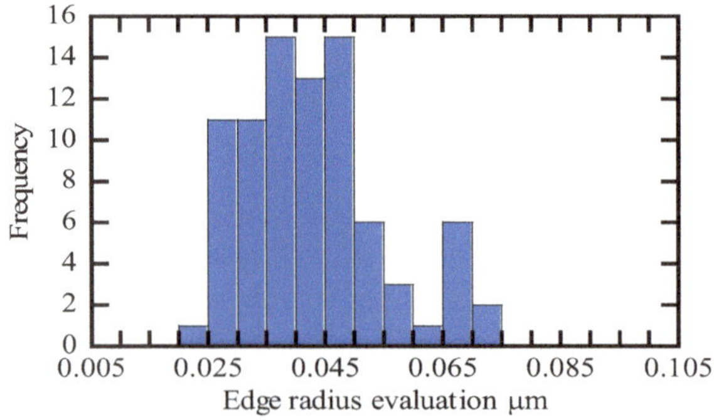


Fig. 17. Histogram of the edge radius evaluation when $p$ is limited

Such tests have been carried out in some cases, while the Edge radius to be evaluated was changed from 10 nm to 60 nm. If the radius to be evaluated is smaller than about 30 nm, the parameter $p$ in Eqs. (3)-(7) has a possibility of taking a value about 0. However, in this evaluation method, if the parameter $p$ takes 0 or 1, it is not possible to evaluate the actual radius, because the evaluated radius becomes $\infty$ or 0, in that case. Therefore, if the fitting parameter $p$ becomes 0 or 1, the evaluations were removed from further calculations. The evaluations results were summarized as shown in Table 3. The term "preprocess" in Table 3 means the removal of outliers. It should be noted that the the edge with edge radius of 10 nm could not be evaluated properly when the point interval is about 4.6 nm, as mentioned in the previous section, because there should obviously be a lack of measure point.
If the parameter $p$ always takes small values, that mean the actual edge is really dulled like Fig. 14(a). Therefore, if $p$ is very small, there is a case that the edge should be considered separately; namely, only one of the edges A and B in Fig. 12 should be taken into consideration depending on the operator's decision, the region could be reselected to include one of two. It is thought that the former evaluation is more sensitive to the wear.
It should be noted that the X-direction represents vertical direction in the explanations shown above. In the case of the instrument mentioned in the previous section, the X and Y are exchanged. Therefore, the definition is switched to agree with that from here.

| Edge radiusnm | Evaluation (preprocessed) nm | Standard deviation (preprocessed) nm (±) |
|---|---|---|
| 10 | - | - |
| 20 | 20 | 9 |
| 30 | 29 | 10 |
| 40 | 43 | 12 |
| 50 | 48 | 11 |
| 60 | 59 | 10 |

Table 3. Evaluations results for each edge radius data (The edge radius to be evaluated is changed from 10 nm to 60 nm)

In practice, the region where the function $F(y)$ is fitted would be found in image to image, and line to line. Therefore, the region to be evaluated is selected like as follows, automatically.
At first, obtain the 2nd order of derivative along $y$ of the cross section image. In order to obtain the 2nd order of derivative along $y$, a "2nd order of least squares filter" is utilized (Savitzky & Golay, 1964). The phrase "2nd order of least squares filter" means the repeating of the quadratic function fitting operations about the $y_i$. Fig. 18 shows the simple explanation of the operation. In Fig. 18, a series of the marks "+" is the filtered data, which is more likely to preserve the height of the peaks than moving average. To carry out this filtering, a moving average type convolute function was selected.
Figure 19 shows the profile and the obtained 2nd order of derivative. The window was set to about 200 nm. From the point where the 2nd order of derivative takes maximum value to the point of the half of the maximum value are used as criterion. The lengths ($L_1 + L_1'$), and points A, E, F and D were defined as shown in Fig. 19. If there are two peaks in the 2nd

order of the derivative, that means the edge or the AFM-tip is dulled. In such case, the window would be enlarged for the peak of the 2nd of derivative to take one obvious peak at first, aiming at the same effect as moving average to flat the sharp peaks.

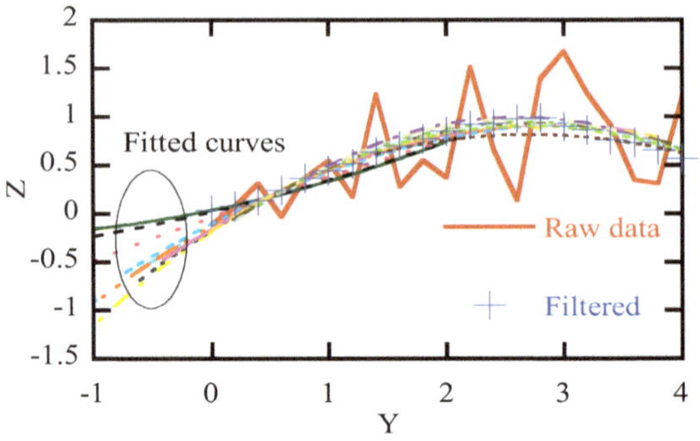


Fig. 18. Least squares filter used in finding the edge

Therefore, as a result, the lines AB and CD in Fig.19 would be fitted using least squares method to the data in the region between point A to point E, and point F to point D. A conic curve would be fitted in the region between point A to point B in the condition that the fitted conic curve tangent to the two lines, at first. As mentioned above, the point E and F would be replaced by the points S and T in the fitting process, respectively. Moreover, the line AB and line CD would also be replaced depending on the points S and T, respectively. The curvature radius of the cross point of the bisector of two straight lines AB, CD and fitted conic curve would be called cutting edge radius, as previously mentioned.

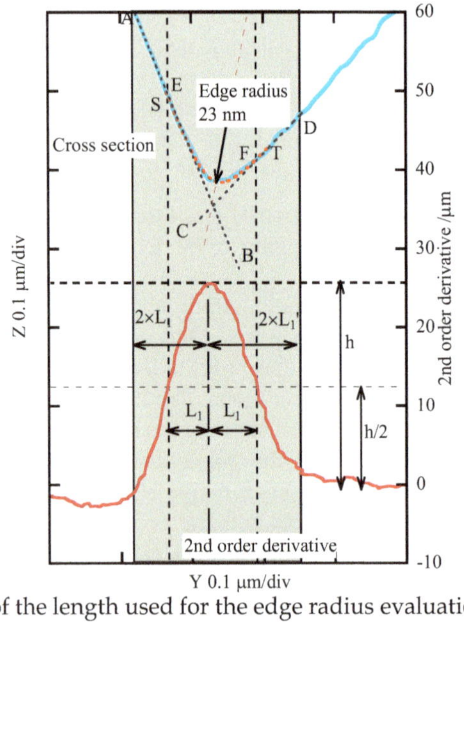


Fig. 19. Determination of the length used for the edge radius evaluation

# 4. Experiments

## 4.1 Measurement of a tool with nose radius of 0.2 mm

Some measurements of the diamond cutting tools were performed using a 3D cutting edge profile measuring instrument (Gao, 2010). The examples of the measurement results on the "real" profile of the cutting tools, in which the influences of the probe shape are eliminated, will be described. The measurements of the tools with nose radius of 0.2 mm, and 1.5 µm were performed. The results showed the clear edge between the rake face and clearance face, which could be used in the evaluation of the profile tolerance and the sharpness.

(a) Reconstructed image

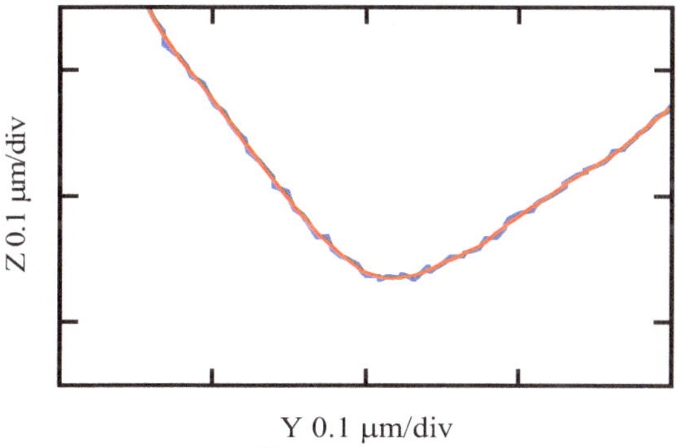


(b) A cross section

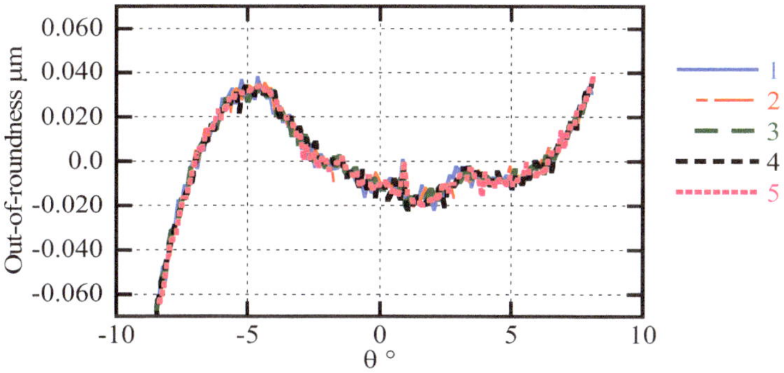


(c) Out-of-contour

Fig. 20. Measurement of a tool having nose radius of 0.2 mm

Although the scan range of the PZT stages of AFM used in the measurement setup limits the measurable region, it is possible to measure the tool having nose radius of 0.2 mm. Figs. 20 show examples of the measurement. Figure 20(a) shows a three-dimensional AFM image of the measured tool tip profile, and Fig. 20(b) shows its cross section image. The tool tip profile was measured five times, and both the averaged profile and non-averaged profile were imposed in Fig. 20(b). Figure 20(c) shows out-of-roundness obtained from Fig. 20(b). The cutting edge radius was evaluated to be 60 nm with standard deviation of 23 nm in the five-time measurement. The reason why the standard deviation became large is the large interval of the lines, which might make the evaluation of the local cutting edge radius impossible. The region to be evaluated is so small that the nose radius could not be evaluated with a repeatability on the order of nanometers (ranging from 0.1878 to 0.1880 mm / 5 times), and the out-of-roundness seemed to be evaluated with repeatability up to 10 nm. That is surprisingly good repeatability compared with the result shown above. The Z-direction positioning is a key to evaluate the tool having large nose radius because the movement of the AFM-tip on the sample surface in X-direction could not be easily translated to the Z-direction like cross talks in the case of the tools having large noses.

### 4.2 Measurement of a tool with very small nose radius

Measurement a tool with round nose radius of 1.5 μm was performed. Figure 21 shows the evaluated cutting edge sharpness for the results of a measurement of a tool with nose radius of about 1.5 μm. The sectional profile of the cutting edge top in the Y-Z plane was plotted. In Fig. 21, the results of repeated 3 times measurement were imposed. By using the method described in the previous section, the edge sharpness was evaluated to be approximately 39 nm with a standard deviation of 9 nm. It should be noted that the influence of the tip radius of the AFM cantilever had not been removed from the result.

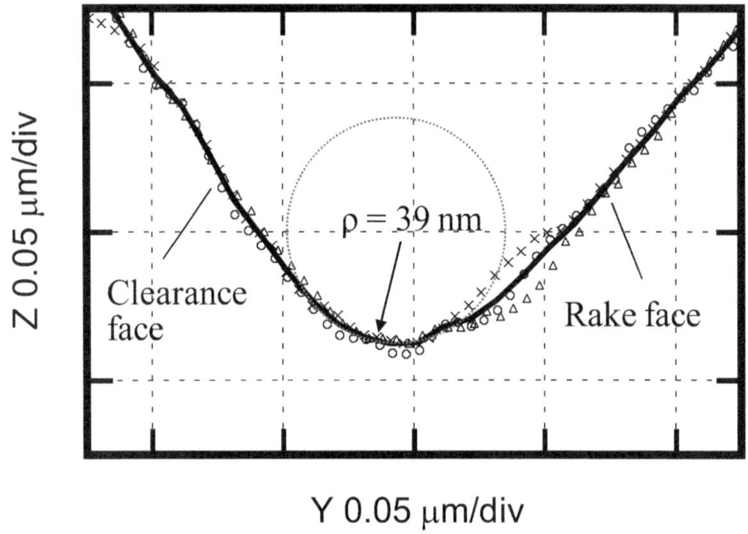


Fig. 21. Evaluated cutting edge sharpness for the results of a measurement of a tool with nose radius of about 1.5 μm.

## 5. Conclusions

Measurements of the diamond cutting tools were performed using an AFM-based 3D cutting edge profile measuring instrument. The cutting edge profiles measured by AFM are affected by the shape of AFM probe. For nanofabrication, the influence of the shape of the AFM probe would not be negligible in some cases. That is why, in the second section, (a) the influence of the shape of the AFM probe on the measured cutting edge profiles, and (b) how to eliminate the influence of the shape of the AFM probe were described. Further, the evaluation method how to measure the shape of AFM probe and apply the acquired shape of the AFM probe to the measured cutting edge profile will also be described.

The value of the cutting edge radii is one of the most important parameters for the ultra-precision diamond cutting tool. There are some examples to obtain the cutting edge radii; some of them just put a circle and show the evaluated value. In most of the cases, the definition of the edge radius is not clear. Without a suitable method, we cannot define the edge radii of the cutting tools with a good repeatability. Throughout much discussion on how to define the edge radii, some definitions have been proposed and utilized. However, we'll have to choose a proper method case by case, while taking the measurement accuracy and the repeatability into consideration. In the third section, the methods how to handle the data of the cutting edge profile by the proposed method in the second section was discussed.

Moreover, the measurements of the tools with nose radius of 0.2 mm and 1.5 μm were performed. The characteristic defects of the size of about 20 nm could be detected in each measurement. The results show the clear edge between the rake face and clearance face, which could be used in the evaluation of the profile tolerance or the sharpness. And in the cases of the tools with nose radii of micrometers, the repeatability of nose radius evaluation and the out-of-contour are good. That is about ±10 nm level. Even in other cases, if only the out-of-roundness is concerned, not the radius of the nose, the order of the repeatability is on that level. Like them, by using the constructed instrument the evaluation of the cutting edge radii and the edge contour encompassing the nose radii and out-of-roundness are possible. This instrument was found to be useful for the evaluation of not only the diamond tool but also the tactile stylus tip.

## 6. References

Asai, S., Taniguchi, Y., Horio, K., Kasai, T., & Kobayashi, A. (1990). Measuring and Analysys on Cutting Edge Radius of Single Point Diamond Tools using Newly Developed Scanning Electron Microscope (SEM), *Journal of JSPE*, Vol. 56, No. 7, pp. 145-150. (in Japanese)

Asai, S., Taniguchi, Y., Horio, K., Kasai, T., & Kobayashi, A. (1990). Measuring the Very Small Cutting-Edge Radius for a Diamond Tool Using a New Kind of SEM Having Two Detectors, *CIRP Annals*, Vol. 39, No. 1, pp. 85-88.

Bloo, M. L., Haitjema, H., & Pril, W.O. (1999). Deformation and wear of pyramidal, silicon-nitride AFM tips scanning micrometre-size features in contact mode. *Measurement*, Vol. 25, pp. 203-2111.

Dongmo, L. S., Villarrubia, J. S., Jones, S. N., Renegar, T. B., Postek, M. T., & Song, J. F. (2000). Experimental test of blind tip reconstruction for scanning probe microscopy, *Ultramicroscopy*, Vol. 85, pp. 141-153.

Egerton, R. F., Li, P., & Malac, M. (2004). Radiation damage in the TEM and SEM, *Micron*, Vol. 35, pp. 399-409.

Gao, W. (2010). Precision Nanometrology -Sensors and Measurement Systems for Nanomanufacturing-, *Springer*, ISBN 978-1-84996-253-7, London

Gao, W., Motoki, T., & Kiyono, S. (2006). Nanometer edge profile measurement of diamond cutting tools by atomic force microscope with optical alignment sensor, *Precision Engineering*, Vol. 30, pp. 396-405.

Nečas, D., Klapetek, P. (July 2011). Gwyddion - Free SPM (AFM, SNOM/NSOM, STM, MFM, ...) data analysis software Users Guide, In: *Gwyddion user guide*, 05.07.2011, Available from http://gwyddion.net/documentation/user-guide-en/

Savitzky, A., & Golay, M. J. E. (1964). Smoothing and Differentiation of Data by Simplified Least Squares Procedures, *Anal. Chem.*, Vol. 36, No. 8, pp. 1627-1639.

Sourceforge.net (July 2011). Documentation: GNU Octave and the Octave-Forge packages, In: *Octave-Forge*, 05.07.2011, Available from http://octave.sourceforge.net/optim/function/fmins.html

Taniguchi, J., Miyamoto, I., Ohno, N., & Honda, S. (1997). Utilizing of hydrocarbon contamination for prevention of the surface charge-up at electron-beam assisted chemical etching, *Nuclear Instruments and Methods in Physics Research B*, Vol. 121, pp. 507-509.

Tian, F., Qian, X., Villarrubia, J. S. (2008). Blind estimation of general tip shape in AFM imaging, Ultramicroscopy, Vol. 109, pp. 44-53.

Villarrubla, J. S. (1997). Algorithms for Scanned Probe Microscope Image Simulation, Surface Reconstruction, and Tip Estimation. *Journal of Research of the National Institute of Standards and Technology*, Vol.102, pp. 425-454.

Villarrubia, J. S. (1994). Morphological estimation of tip geometry for scanned probe microscopy. *Surface Science*, Vol. 321, pp. 287-300.

Wall M. (July 2011). GAlib: Matthew's Genetic Algorithms Library, In: *GAlib*, 05.07.2011, Available from http://lancet.mit.edu/ga/

# 11

# Advantages of a Programmed Surface Designed by Organic Monolayers

Naoto Shirahata
*National Institute for Materials Science (NIMS), Sengen, Tsukuba, Ibaraki, Japan*

## 1. Introduction

Increasing attention has been paid to the control over surface chemical property of bare metals, semiconductors, ceramics, and polymers in the development of nanoscience and nanotechnology. The appropriate functionalization has a potential to lead to the appearance of new function. Organic monolayers, i.e., self-assembled monolayers (SAMs), are simple monomolecular system, convenient, and flexible to tailor the substrate's surface chemical and physical properties including wetting property, conductivity, and thermal, chemical, photochemical stabilities, etc. In the following molecular diagram of organic monolayers, its molecular anatomy and characteristics are highlighted as shown in Figure 1. Such monolayer structures are formed by chemisorption of organic constituents onto the specific solid substrate through liquid or gas phase. A headgroup of the organic constituent reacts to the outermost atoms on the substrate to give a chemical linkage. The chemisorbates organize epitaxially to form crystalline monomolecular film structures. As shown in the molecular

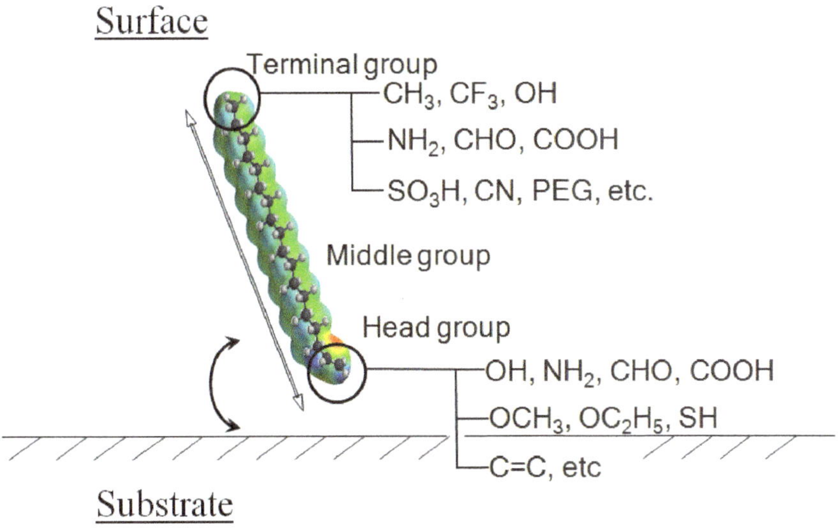


Fig. 1. Schematic diagram of organic monolayers formed on solid substrates

diagram, there is a number of headgroups which can recognize the outermost chemical groups of substrates, resulting in a chemical bond. In contrast, there is no limitation in the selection of terminal functional groups. A wide variety of functionalities or reactivities are applicable. A lateral molecular space, which is controlled by hydrophobic molecular interaction between hydrocarbon chains, allows the control of the film crystallinity and electron transferring behavior between the terminal group and substrate surface. Thus, programmed surface can be designed by careful control of monolayer attachment. Since the finding of siloxane SAM structure in 1980, a number of research efforts have been performed. In this review article, organic monolayer systems on polymer and silicon substrates are focused, and their potential applications is also demonstrated.

## 2. Chemical modification of a polymer surface

### 2.1 Motivation

Recently, the deposition of inorganic materials, e.g., ceramics, metals, and semiconductors, has interestingly attracted attention in the development of mechanically-flexible device applications including OLED, flexible flat panel and wearable displays, electronic papers, biomedical tools. These technological developments take advantage of polymer's excellent properties including mechanical flexibility, lightweight, ease-of-design and coloring, low-cost and good impact resistance. In order to fabricate such flexible device applications, it is a key to control chemical properties of polymer's surface that induce the nucleus formation of inorganic crystals, and accelerates subsequent crystalline growth or film growth. Organosilane SAMs have been frequently employed to modify surface chemically properties of inorganic substrates [1-5]. The resultant surfaces play important roles in molecular recognition events. For example, some of functional groups induce the nucleus formation of inorganic crystals, and accelerates subsequent crystalline growth or film growth. This section describes the preparation of well-ordered SAM structures on polymer sheet, and then the deposition of metal and ceramics films on the SAM-covered polymer sheets through solution processes.

### 2.2 SAM formation

When silane molecules are treated directly on as-received polymer sheet, they cannot form a monomolecular structure. Similar can be seen even on photochemically treated polymer sheet. In other words, when silane precursors are directly treated on surface-activated polymer substrates, well-ordered SAM formation cannot effectively proceed, since the polar-functional groups formed on a polymer surface through surface modification methods using plasma or UV light are generally inhomogeneous and randomly distributed [5]. Such a heterogeneous surface cannot provide adequate support for the preparation of a well-ordered SAM. As an example, Table 1 summarizes the contact angles and surface free energies of polyimide surfaces covered with $NH_2$- and $CF_3$-terminated SAMs, and their values are compared with those of SAM-covered silicon substrates used as standards. The formation of the SAMs is performed on photooxidized polyimide and silica-covered silicon substrates, respectively. The details of the SAM preparation method are described in elsewhere [5], but the surface energies of each sample are calculated using the following equation:

$$\gamma_L(1+\cos\theta_S{}^L) = 2[(\gamma_s{}^d \bullet \gamma_L{}^d)^{1/2} + (\gamma_s{}^P \bullet \gamma_L{}^P)^{1/2} + (\gamma_s{}^h \bullet \gamma_L{}^h)^{1/2}]$$

where $\gamma_L$ is the liquid's surface energy, and $\theta_S^L$ is its contact angle on the solid surface. The superscripts d, p, and h indicate the dispersion, polar and hydrogen bonding forces, respectively. Since the $\gamma_L$, $\gamma_L^d$, $\gamma_L^P$, and $\gamma_L^h$ values of the three liquids are known (i.e., for water, $\gamma_L$ = 72.8 mN/m, $\gamma_L^d$= 29.1 mN/m, $\gamma_L^P$ = 1.3 mN/m, and $\gamma_L^h$ = 42.4 mN/m; for methylene iodide, $\gamma_L$ = 50.8 mN/m, $\gamma_L^d$= 46.8 mN/m, $\gamma_L^P$ = 4.0 mN/m, and $\gamma_L^h$ = 0 mN/m; and for *n*-decane, $\gamma_L$ = 23.9 mN/m, $\gamma_L^d$= 23.9 mN/m, $\gamma_L^P$ = 0 mN/m, and $\gamma_L^h$ = 0 mN/m), the surface free energy ($\gamma_S$) of the sample is estimated to be the sum of the calculated $\gamma_S^d$, $\gamma_S^P$, and $\gamma_S^h$ values.

| | SAM on $SiO_2/Si$ | SAM on Polyimide ($PI_{ox}$) |
|---|---|---|
| FAS-SAM: $CF_3(CF_2)_7(CH_2)_2Si(OCH_3)_3$ | | |
| | $\theta_c$ = 115° | $\theta_c$ = 107° |
| | $\gamma_s$ = 11.9 mN/m | $\gamma_s$ = 27.4 mN/m |
| | $\gamma_s^d$ = 11.7 mN/m | $\gamma_s^d$ = 13.2 mN/m |
| | $\gamma_s^P$ = 0.1 mN/m | $\gamma_s^P$ = 14.2 mN/m |
| | $\gamma_s^h$ = 0.1 mN/m | $\gamma_s^h$ = 0.0 mN/m |
| AHAPS-SAM: $H_2N(CH_2)_6NH(CH_2)_3Si(OCH_3)_3$ | | |
| | $\theta_c$ = 60.0° | $\theta_c$ = 69.6° |
| | $\gamma_s$ = 61.9 mN/m | $\gamma_s$ = 45.2 mN/m |
| | $\gamma_s^d$ = 23.9 mN/m | $\gamma_s^d$ = 23.9 mN/m |
| | $\gamma_s^P$ = 26.2 mN/m | $\gamma_s^P$ = 13.1 mN/m |
| | $\gamma_s^h$ = 11.8 mN/m | $\gamma_s^h$ = 8.2 mN/m |

Table 1. Surface free energies of SAM-covered photooxidized polyimide (PI), i.e., $PI_{ox}$, and are compared with those of standard, i.e., SAM-covered $SiO_2/Si$.

Let's see the impact direct formation of FAS-SAM onto $PI_{ox}$, i.e., photooxidized polyimide, will have its intermolecular structure. The water contact angle of the FAS/$PI_{ox}$ surface is somewhat smaller than that of the standard sample, that is, FAS-SAM/$SiO_2$/Si, but high hydrophobic character can be seen even on it. However, this small difference in surface hydrophobicity influences strongly on the crystal growth behavior of metal oxide film, as described later. The reason why small difference in wetting property occurs between the surfaces can be well understood by comparison of surface free energy of the sample with that of the standard. As shown in Table 1, the $\gamma_S$ value of the FAS/$PI_{ox}$ substrate (27.4 mN/m) is twice larger than that of a FAS/$SiO_2$/Si substrate (11.9 mN/m). This indicates that the molecular density of the FAS-SAM attached to the $PI_{ox}$ surface is considerably low compared to the reference substrate. In particular, the $\gamma_S^P$ value of the FAS/$PI_{ox}$ substrate (14.2 mN/m) is much larger than that of the FAS/$SiO_2$/Si substrate (0.1 mN/m). This clearly demonstrates that the $PI_{ox}$ surface underneath the FAS layer is probably exposed to air to some extent. It can be concluded that the degree of surface coverage by the FAS molecules differ greatly between the $PI_{ox}$ and $SiO_2$/Si substrates. In the sample covered with an amine group, its static-water contact angle is estimated about 69.6° which is slightly larger than that of the reference sample. This small increase can be also discussed comparing surface free energies between sample and reference. The $\gamma_S$ value (45.2 mN/m) is approximately three-fourths of that calculated for the standard, and , is slightly larger than that of a surface terminated with $CH_2$ groups (36 mN/m). Both $\gamma_S^d$ and $\gamma_S^h$ values are consistent with those of standard, but the $\gamma_S^p$ value of the AHAPS/$PI_{ox}$ substrate (13.1 mN/m) is almost half of those obtained for the AHAPS/$SiO_2$/Si substrates (26.2 mN/m). These experimental results suggest that the degree of surface molecular coverage is much less than those on the reference surfaces. As a result, such heterogeneous surfaces cannot provide adequate support for the preparation of well-ordered SAMs. Chemical templates consisting of such disordered SAMs are considered to be unfavorable for the deposition of inorganic materials on polymer substrates. Furthermore, after a relatively short time, such disordered moieties will lose their functionalities, resulting in a decrease of surface activity on the modified polymeric surface. Thus, to avoid hydrophobic recovery, SAM fabrication must be conducted immediately after surface photooxidation. Otherwise it is difficult to form covalent bonds between the SAM and the polymer surface, resulting in weak adhesion. Thus, in spite of polymer substrates offering the attractive advantages described above, such obstacles may prevent the preparation of high quality SAMs.

To overcome this shortcoming, this section demonstrates the formation of a well-ordered SAMs on PI sheets through the chemisorption of organosilane molecules at the solid/vapor interface. PI has been utilized primarily as a substrate for printed circuit boards due to its excellent thermal stability, mechanical strength and insulating properties. In the present study, in order to give silica-like properties to the PI surface, we fabricated an ultrathin buffer layer on the substrates using a either one of precursor vapors of tetraethoxysilane (TEOS) or 1,3,5,7-tetramethylcyclotetrasiloxane (TMCTS) prior to forming the SAM. Let's focus on the effects of this buffer layer on the quality of the obtained SAMs. Well-ordered SAM structures are formed on PI surfaces according to the process illustrated in Figure 2. In the first step Figure 8a), a commercial PI sheet is photochemically hydrophilized to give a $PI_{ox}$ surface. Due to this photoirradiation, the PI surface becomes completely hydrophilic with its water-contact angle dropping from about 71° to 5° or less. In the next step (Figure 2b), either one of organosilane molecules is chemisorbed onto the $PI_{ox}$ surface. The $PI_{ox}$ sheet is placed together with a glass cup filled with 0.2 $cm^3$ TMCTS (or TEOS) into a 65 $cm^3$ $Teflon_{TM}$ container in a dry $N_2$ atmosphere with less than 10% relative humidity. The container is sealed with a cap and heated in an oven maintained at 80°C for 3 h. Next, in order to convert the chemisorbed TMCTS (or TEOS) molecule into an ultrathin $SiO_2$ layer, the sample is again irradiated for 30 min in a manner similar to the PI surface modification mentioned above. Finally, a SAM is formed on the $SiO_2$-covered $PI_{ox}$ surface through a vapor phase (Figure 2e). The thickness of the $SiO_2$ nanolayer formed on the $PI_{ox}$ substrate is estimated from a cross-sectional image acquired by HR-TEM, as shown in Figure 3. As clearly indicated by the arrows, an extremely thin layer is found to form on the $PI_{ox}$ substrate, as a black contrasted image. Two important things can be seen in this cross-sectional view. One is that the ultrathin layer uniformly deposits over the entire surface of the $PI_{ox}$. There are no defects such as a void and a pinhole. The other is an actual thickness of this layer which is estimated from this image to be only 1 nm or less. As confirmed by XPS, the binding energy of the Si2p spectrum for this sample is located at 103.6 eV, the value of which is consistent with that of a Si substrate covered with native oxide ($SiO_2$/Si, 103.6 eV). Therefore, the nanolayer undoubtedly formed on the $PI_{ox}$ substrate. Due to this $SiO_2$ nanolayer coating, the surface chemical properties of the PI substrate are expected to be the same as those of $SiO_2$/Si substrates, as evidenced in Table 2. In the case of the

SAM/nanolayer/$PI_{ox}$ sample, a highly oriented SAM is expected to have grown similarly on the substrate. Unlike SAM sample without $SiO_2$ nanolayer (see Table 1), the $\gamma_S$ values of the SAM-covered $PI_{ox}$ samples (12.4 mN/m for FAS and 62.8 mN/m for AHAPS) are almost consistent with those of the standard substrate (11.9 mN/m for FAS and 61.9 mN/m for AHAPS). Specifically, the SAM formation on the nanolayer leads to the $\gamma_S^P$ which nearly corresponds to the value on the reference sample. As a result, this nanolayer opens SAM chemistry even on polymer sheet [5,6].

| | **SAM on $SiO_2$/Si** | **SAM/Nanolayer/$PI_{ox}$** |
|---|---|---|
| FAS-SAM: $CF_3(CF_2)_7(CH_2)_2Si(OCH_3)_3$ | | |
| $\theta_c$ | = 115° | = 115° |
| $\gamma_S$ | = 11.9 mN/m | = 12.4 mN/m |
| $\gamma_S^d$ | = 11.7 mN/m | = 12.2 mN/m |
| $\gamma_S^p$ | = 0.1 mN/m | = 0.1 mN/m |
| $\gamma_S^h$ | = 0.1 mN/m | = 0.1 mN/m |
| AHAPS-SAM: $H_2N(CH_2)_6NH(CH_2)_3Si(OCH_3)_3$ | | |
| $\theta_c$ | = 60.0° | = 61.4° |
| $\gamma_S$ | = 61.9 mN/m | = 62.8 mN/m |
| $\gamma_S^d$ | = 23.9 mN/m | = 23.9 mN/m |
| $\gamma_S^p$ | = 26.2 mN/m | = 28.5 mN/m |
| $\gamma_S^h$ | = 11.8 mN/m | = 10.4 mN/m |

Table 2. Surface free energies of SAMs on $SiO_2$ nanolayer/$PI_{ox}$ substrates, and are compared with those of standard, i.e., SAM-covered $SiO_2$/Si.

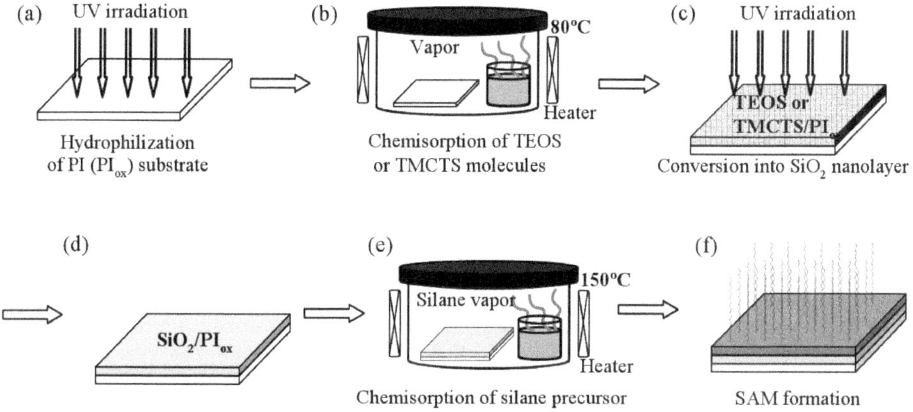


Fig. 2. Schematic diagram for SAM formation on $PI_{ox}$ substrate with a 1 nm thick $SiO_2$ buffer layer.

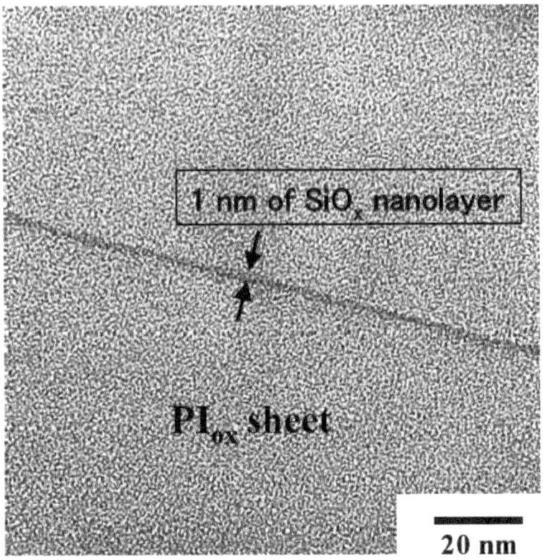


Fig. 3. A cross-sectional HR-TEM image of TEOS-covered $PI_{ox}$ substrate after photooxidation. The TEOS layer is converted into 1 nm thick silica layer as a consequence of the photooxidation with vacuum UV light at $10^3$ Pa.

### 2.3 Attachment of ceramics film on polyimide

This subsection describes the potentials as chemical templates of SAMs formed on nanolayers in the deposition of metal and ceramics from liquid phase. To achieve a better understanding of the potentials, SAM-covered samples with and without silica nanolayer are micropatterned by photoirradiation with UV lamp through photomasks.

Among the various types of reactive functional groups, amino groups are of particular interest owing to their chemical reactivities. Palladium colloid is particularly important as a catalyst for the fabrication of metal micropatterns on non-conductive substrates like a polymer. Figure 4 illustrates a deposition process of copper microlines on polymer sheets with and without a silica nanoalyer. Each of SAM-covered $PI_{ox}$ sheets are photolithographically micropatterned (Figure 4a). The micropatterned sample substrates are activated for 20~30 min by immersion in a Pd(II) solution containing $PdCl_2$ and HCl (adjusted to pH 5.0 by adding aq-NaOH solution), followed by gentle rinsing with Milli-Q water (Figure 4c). These activated substrates are then immediately immersed in a commercial plating bath containing formaldehyde as a reducing agent (Figure 4d). The pH value of this solution is 12.5. As a result, copper microlines are deposited on the AHAPS chemical templates, as shown in Figure 4e. The comparison of LFM images (a) with (b) in Figure 5 shows the significant difference in a quality of SAMs formed on $PI_{ox}$ with (b) and without (a) nanolayer. In these LFM images, bright and dark areas correspond to photoirradiated and unirradiated regions, respectively. As can be seen in Figure 5b, the 10-μm-wide micropatterns are clearly imaged through the friction force contrast between the photoirradiated and masked regions of the AHAPS/nanolayer/$PI_{ox}$ substrate. The photoirradiated regions exhibit stronger friction forces than the unirradiated AHAPS-covered regions. Due to UV irradiation, an AHAPS-SAM is removed through oxidation with UV-activated oxygen species. Accordingly, in the UV-irradiated regions of the sample, the bottom part of the AHAPS layer, which consisted of a 0.2~0.27 nm-thick $SiO_2$ monolayer, remains on the ONS/$PI_{ox}$ substrate surface. This $SiO_2$ monolayer-covered nanolayer/$PI_{ox}$ surface is highly hydrophilic with a water-contact angle of 5° or less. It therefore adheres more strongly to the Si probe surface, which is also covered with a hydrophilic native $SiO_2$ layer. The friction force contrast on the UV-irradiated regions is thus greater than that on the AHAPS-covered areas. Although it can be expected that this $SiO_2$ monolayer would similarly remain on the UV-irradiated regions of the AHAPS/$PI_{ox}$ substrate, the friction force contrast in its LFM image (Figure 5a) is relatively small compared to that of the AHAPS/nanolayer/$PI_{ox}$ substrate. In addition, several dark areas can be observed, even in the $SiO_2$ monolayer-covered regions. This reveals that areas of weak friction exist on the

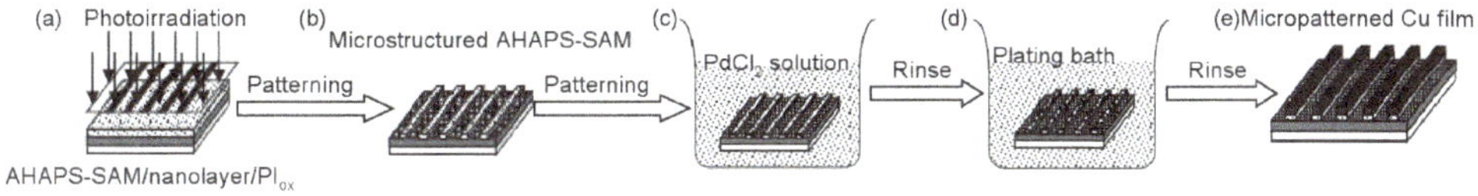


Fig. 4. Schematic illustration showing the fabrication of copper microlines on $PI_{ox}$ substrate.

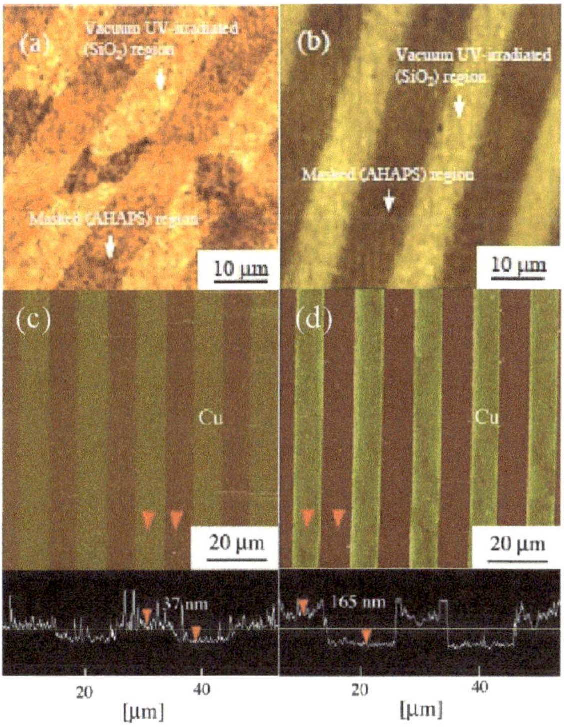


Fig. 5. LFM images of micropatterned $PI_{ox}$ substrates covered with (a) AHAPS-SAM and (b) AHAPS-SAM/nanolayer, and AFM images and their cross-sectional views of copper microlines area-selectively electroless plated on the micropatterned (c) AHAPS-SAM/PIox and (d) AHAPS-SAM/nanolayer/$PI_{ox}$ substrates.

$SiO_2$ monolayer-covered regions. These are believed to be due to incomplete termination with AHAPS molecules. As mentioned in the previous section, since the packing density of the AHAPS molecules on the $PI_{ox}$ substrate is considered to be low, parts of the $PI_{ox}$ substrate surface which does not react with AHAPS molecule probably remain even after its SAM formation. However, the polar-functional groups terminating such a surface are unstable and gradually lose their polarity due to hydrophobic recovery, resulting in a decrease in hydrophilicity of the $PI_{ox}$ surface. The dark regions are considered to be hydrophobic areas regenerated by this hydrophobic recovery.

Copper films are deposited on both micropatterned samples, since both the amount of Pd catalyst adsorbed and the subsequent growth of Cu through electroless plating are thought to be related to the termination states of the $NH_2$ groups in the AHAPS molecules. Typical AFM images of the electroless-plated copper microlines and their schematic cross-sections on both micropatterned AHAPS/$PI_{ox}$ and AHAPS/nanolayer/$PI_{ox}$ substrates are shown in Figures 5c and d, respectively. In both images, continuous film-like deposits of Cu can be seen on the masked regions, while the surrounding photoirradiated regions remain free of deposits. 10-μm-wide Cu microlines are successfully formed in a 10×10 $\mu m^2$ area on both substrates. The film thicknesses as estimated from these images are about 37 and 165 nm for Figures 5c and d, respectively. Accordingly, the growth rate of each is calculated to be about 3.7 and 16.5 nm/min, respectively, because of the deposition time of a 10 min in both cases. It is clear that the electroless Cu plating proceeded more efficiently on the AHAPS/nanolayer/$PI_{ox}$ substrate. Its Cu growth rate is roughly consistent with that observed on an AHAPS/$SiO_2$/Si substrate (20 nm/min or less). In spite of the fact that immersion time is identical, the growth rate on the AHAPS/$PI_{ox}$ substrate without the silica buffer layer is only about one-fifth of that on the AHAPS/nanolayer/$PI_{ox}$ substrate, offering further proof that the degree of surface coverage by Pd catalyst particles is much higher on the AHAPS/nanolayer/$PI_{ox}$ substrate than on the AHAPS/$PI_{ox}$ substrate. As with the plating rate, the pattern resolution also depends on the quality of the AHAPS-SAM. As clearly seen in the cross-section of the Cu microlines on the AHAPS/nanolayer/$PI_{ox}$ substrate (Figure 5d), the edge regions are relatively sharp. On the contrary, the microlines' edge sharpness is low on the AHAPS/$PI_{ox}$ substrate (Figure 5c). These results agree well with the LFM results described above. It is thus concluded that the silica interlayer proves effective in fabricating a well-defined AHAPS-SAM even on a polymer substrate, in a manner similar to that seen on a $SiO_2$/Si substrate. In addition, the micropatterned substrate prepared by the present approach serves as an ideal template for spatially defined electroless Cu plating on nonconductive polymer substrate.

Next, the role of a high quality chemical template for depositing ceramics film, i.e., tin oxide, is demonstrated. The FAS-covered samples with and without silica nanolayer are photolithographycally micropatterned as shown in Figure 6a. Each patterned samples is immersed into a 1 M HCl solution containing 0.05 mol/l of $SnCl_2{\cdot}2H_2O$ (see Figure 6b). After immersion, the substrates are immediately washed with Milli-Q water several times, blown dry with $N_2$ gas, and subsequently sonicated for 0.5~2.0 h in absolute toluene (Figure 6c) to remove the deposits from FAS-covered regions. We named this "solution lithography" [1]. As a result, a microstructured tin oxide film is obtained (Figure 6d). Typical optical micrographs of the films deposited on both the micropatterned FAS/nanolayer/$PI_{ox}$ and FAS/$PI_{ox}$ substrates (deposition time of 24 h; sonication time of 2 h) are shown in Figures 7a and 7b, respectively. As can be seen in Figure 7a, tin oxide deposits with continuous film-like features remain on the photoirradiated regions, while the

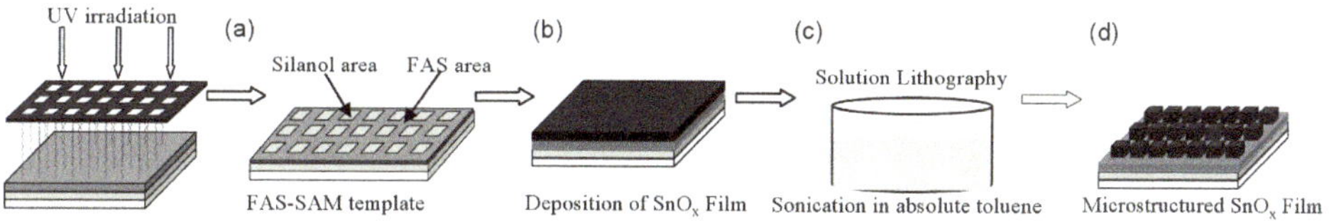


Fig. 6. Schematic illustration showing the microfabrication process of metal oxide film on $PI_{ox}$ substrate.

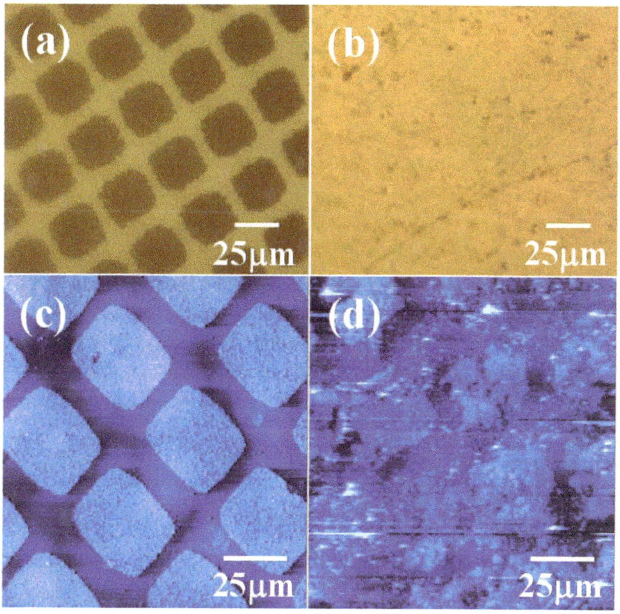


Fig. 7. Optical (a and b) and DFM images (c and d) of tin oxide film microstructured on two different substrates: (a), (c) on FAS/nanolayer/$PI_{ox}$ substrate, while (b), (d) on FAS/$PI_{ox}$ substrate, directly.

surrounding regions, which are not photoirradiated and are thus FAS-covered, are free of deposits. This clearly indicates that the deposits grown on the UV-irradiated regions stay tightly fix to the surface, while the deposits on the FAS-covered regions have been completely eliminated. On the other hand, no microstructured features can be seen on the micropatterned FAS/$PI_{ox}$ substrate shown in Figure 7b. Both sample surfaces are observed in detail by DFM. As clearly seen in the DFM image in Figure 7c, highly resolved micropatterns composed of tin oxide film grow site-selectively on the photoirradiated regions of the micropatterned FAS/nanolayer/$PI_{ox}$ substrate. In contrast, it is difficult to achieve clear micropatterns on the FAS/$PI_{ox}$ substrate, as shown by Figure 7d. This sample has been treated in absolute toluene for 30 min. As can be clearly seen, the tin oxide remains deposited relatively evenly on the FAS-covered regions and there is marked distortion of the deposits on the UV-irradiated regions. When sonication is prolonged up to 2 h (image not shown), the micropattern becomes hardly recognizable by DFM. This significant difference in surface morphology between the samples must be considered in terms of the surface chemical properties of the UV-irradiated and the unirradiated regions on both samples. However, since it is difficult to determine the actual surface properties of these regions selectivity, both FAS/nanolayer/$PI_{ox}$ and FAS/$PI_{ox}$ substrates before and after VUV irradiation without micropatterning are characterized.

Figure 8 shows the variation in the ζ-potential with the pH of UV-irradiated FAS/nanolayer/$PI_{ox}$, (i.e., SML (silica monolayer)/ONS/$PI_{ox}$) and FAS/$PI_{ox}$, (i.e., SML/$PI_{ox}$) substrates (indicated by the open and solid circles, respectively). As a control experiment, an observation of a $SiO_2$/Si substrate is also performed in the same manner (indicated by the solid triangles). Over the pH range from 3 to 11, the $SiO_2$/Si surface shows negative ζ-potentials from ca. -22 to -82 mV due to the partial ionization of surface silanol (Si-OH) groups. The isoelectric point (IEP) of this surface is estimated to pH 2.0. It is noteworthy that the ζ-potential vs pH curves for the SML/ONS/$PI_{ox}$ and $SiO_2$/Si substrates are nearly identical in shape and magnitude over the entire pH range. Thus, it can be concluded that the UV-irradiated FAS/nanolayer/$PI_{ox}$ surface shows chemical properties very similar to the $SiO_2$/Si substrate, and that the oxide layers (i.e., the SML and nanolayer) has covered the PI substrate almost completely. On the other hand, the UV-irradiated FAS/$PI_{ox}$ surface is found to 15-39% less charged than the SML/nanolayer/$PI_{ox}$ and $SiO_2$/Si substrates. In addition, the IEP of this sample, observed to be around pH 3.5, is higher than those of the SML/nanolayer/$PI_{ox}$ and $SiO_2$/Si substrates. This clearly indicates that the SML insufficiently covered the $PI_{ox}$ surface. Since the density of the FAS molecules is considered to have been low, a very small part of the hydrophilic $PI_{ox}$ surface is probably exposed to air. As a result, the molecules at the bottom of the FAS comprising SML are loosely packed. This marked difference in surface properties must account for both the final morphology and the adhesion properties of the deposits on the UV-irradiated regions.

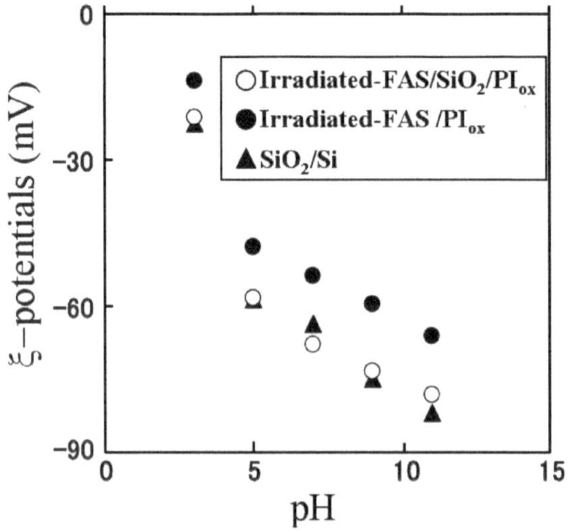


Fig. 8. Comparison of ζ-potentials as a function of pH for an irradiated FAS/nanolayer/$PI_{ox}$ (open circles), an irradiated FAS/$PI_{ox}$ (solid circles), and a silicon covered with native oxide (solid triangles), respectively. SML means siloxane monolayer.

SAM formation technique demonstrated here offers the advantage that, by selecting an organosilane precursor, a well-ordered SAM with a wide variety of terminal-end groups can be fabricated on a mechanically-flexible polymer sheet (not only PI) without any marked change in surface morphology or bulk properties. Thus I expect this technique to be readily and widely applicable to novel polymer surface modification based on organosilane SAM chemistry. In addition, patterned SAM-covered polymer substrates could be applied as microtemplates for the site-selective immobilization of various kinds of materials including biomolecules, semiconductors and other functional molecules to fabricate novel polymer-based flexible devices.

## 3. Parallel detection of biomolecular recognition event on a microarray of monolayers

Direct-writing using molecular ink plays a leading role in the development of bottom-up nanofabrication, due to its unparalleled, lithography-free properties, that allow precise positioning of the ink on a predefined surface site to obtain a high-resolution pattern on a substrate. To date, a number of direct-writing methods including microcontact printing, dip-pen nanolithography (DPN), microfluidic devices, nanoscale pipetting, and inkjet printing have been developed. A combination with either one of the molecular inks including SAMs, oligonucleotides, proteins, carbohydrates, nanoparticles, and polymers leads to the fabrication of its fine pattern. Many works in SAM chemistry have used the high resolution SAM patterns in particular, as chemical templates that serve as prospective candidates to miniaturize and simplify traditional laboratory techniques, bioassays, cell culture experiments, and multifunctional gas-sensing array. Achieving these goals requires the production of a chemical template in which different varieties of SAMs are precisely positioned on predefined surface sites on a given substrate. Thus far, a SAM system using a combination of thiol-ink and gold-substrate, i.e., ink-paper chemistry, has already satisfied this requirement. Actually, Mirkin and co-workers produced a microarray of four different alkanethiol SAMs by DPN [7]. The chemical functionalization of the gold surface allows the use of surface plasmon resonance and an oxidation-reduction potentiometer to detect the chemical or biomolecular interaction on thiol-template. However, its high-conductivity and optical inflexibility in the visible region has often limited the enviable versatility of SAM microarrays. This section thus describes the novel methodology to produce SAM microarrays on other substrates including glass, oxidized silicon, OH-terminated diamond, metal oxides, and polymers. The transparency in the visible region of glass is particularly essential to study biomolecular interactions under the microscope. Moreover, using a substrate diamond having both transparency and conductivity facilitates the electrochemical detection of specific biomolecular events.

The lithography-free method for oxide-based microarrays starts with the transfer of an alkoxysilane ink to the predefined surface position on an oxidized substrate under ambient conditions, as shown in Figure 9. A wide variety of pens including syringe and AFM tips can be lined up as candidates for painting with molecular inks. To verify the methodological potential, three different types of trialkoxysilanes, that is, triethoxysilylundecanal (TESUD), N-(6-aminohexyl)-3-aminopropyltrimethoxysilane (AHAPS), and octadecyltrimethoxysilane (OTS), are selected as model molecular inks. As shown in Figure 9, these organosilane inks are respectively transferred from the tips of syringes onto the predefined surface positions. At each deposit, silanization occurs during hydrolysis of the alkoxy groups by physisorbed

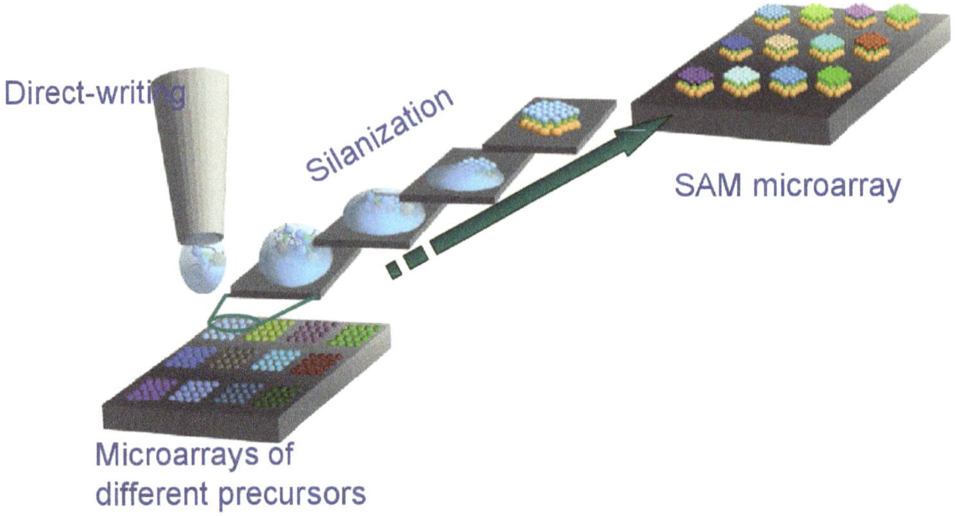


Fig. 9. Schematic illustration showing ink-paper chemistry on oxide substrate to fabricate a siloxane-based SAM microarray.

water on the hydrated surface, followed by a condensation reaction to give an interfacial covalent linkage between the silane molecule atom and the substrate. All of the analyses shown in Table 3 suggest that the alkoxysilane molecules form their one-molecular thick structures by the SAM formation process. The measurement of the water-contact angle is still a facile and useful method for estimating SAM formation. A high value of the water-contact angle for the OTS-treated sample is comparable to that of OTS-SAM prepared by chemical vapor deposition. A correspondence in the surface wetting property is observed between the AHAPS-treated sample and amine-terminated SAMs. The prepared OTS- and AHAPS-treated samples are then analyzed using atomic force microscopy (AFM) and ellipsometry. Although the ellipsometrical thickness of the AHAPS-treated sample is fairly in good agreement with that of AHAPS-SAM structure, the OTS-treated film is slightly thinner than its ideal thickness, that is, 25–28 Å. As expected in the case of small water-contact angle hysteresis $\Delta\theta$, which is defined as $\Delta\theta = \theta_a - \theta_r$, both the samples' surface roughnesses measured by AFM at angstrom level are small, resulting in the successful formation of SAMs with or without less defective surfaces. On the other hand, the water-contact angle of TESUD-treated samples is substantially lower than that of the non-orientated TESUD-SAMs (e.g., $\theta_{a/r}$ = 78.1°/76.2°). In order to elucidate the lower water contact-angles for this TESUD-treated sample, attenuated total reflection Fourier transform infrared (ATR-FTIR) spectroscopy, which is a convenient and well-established method of examining a SAM in surface chemical composition and intermolecular packing structure, is used. The frequencies of antisymmetric and symmetric methylene stretching modes are important indicators for exploring in-plane structures, including molecular ordering of alkane chains. The appearance of the $\nu_a(CH_2)$ at 2925–2930 $cm^{-1}$ and the $\nu_s(CH_2)$ at 2855–2858 $cm^{-1}$ suggests that the hydrocarbon chains in a SAM structure are disordered, resulting in a liquid alkane or an amorphous film. In contrast, shifts from their vibrational peak positions to the frequencies of $\nu_a(CH_2)$ and $\nu_s(CH_2)$ centered at 2917 $cm^{-1}$ and 2950 $cm^{-1}$ indicate the formation of a highly crystalline ordered SAM structure due to the strong hydrophobic interactions between adjacent alkyl chains. On the basis of this common monomolecular packing theory, the methylene stretching vibrations observed at 2916 $cm^{-1}$ for $\nu_a(CH_2)$ and at 2849 $cm^{-1}$ for $\nu_s(CH_2)$ in the sample, indicate that the internal hydrocarbon chains are completely ordered to produce the solid TESUD-SAM structure. Due to this well-ordering, the TESUD-SAM has a higher hydrophilic surface property, and is slightly thicker than the disordered structures ($d_{ellips}$ = 1.2–1.4 nm). The frequencies of $\nu_a(CH_2)$ and $\nu_s(CH_2)$ for the OTS-SAM are respectively observed at 2922 $cm^{-1}$ and 2852 $cm^{-1}$, which suggests that this monolayer is relatively ordered when compared with its amorphous film. In Table 3, the ATR-FTIR measurement indicates that the molecular structure of the AHAPS-treated sample is akin to a non-crystalline film, but its molecular thickness and surface wetting properties correspond approximately to the reported values ($d_{ellips}$ = 1.1−1.3 nm, $\theta_{a/r}$ = 62.0±3°/58.0±2°). Due to the limited early reports about AHAPS-SAM, further comparison is restricted, but the measurements shown in Table 1 yield conclusive evidences about the formation of a single-molecule thick structure of AHAPS.

| Reactants | Water contact-angle $\theta_{a/r}$ (deg.) | Thickness $d_{ellips}$ (nm) | asymmetric stretch $\nu_a^{p\text{-}pol}$ ($cm^{-1}$) | $\nu_a^{s\text{-}pol}$ ($cm^{-1}$) |
|---|---|---|---|---|
| OTS $CH_3(CH_2)_{17}Si(OCH_3)_3$ | 108°/97° | 1.96 ± 0.1 | 2922 | 2852 |
| TESUD $CHO(CH_2)_{10}Si(OC_2H_5)_3$ | 69°/62° | 1.44 ± 0.1 | 2916 | 2849 |
| AHAPS $NH_2(CH_2)_6NH(CH_2)_3Si(OCH_3)_3$ | 58°/50° | 1.29 ± 0.2 | 2925 | 2855 |

Table 3. Surface characterization of OTS-, TESUD-, and AHAPS-treated films.

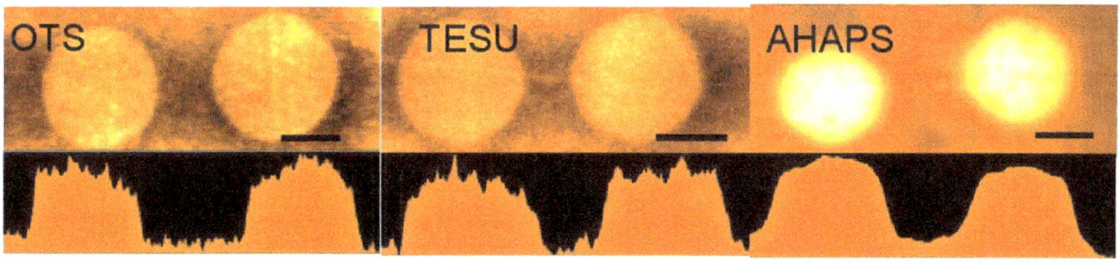


Fig. 10. KPFM and cross-sectional images of the microstructured OTS-, TESU-, and AHAPS-SAMs formed on $SiO_2$/Si substrate

As shown in Figure 10, SAMs' structures shown in Table 3 are microfabricated, and are analyzed using Kelvin probe force microscopy (KPFM), which promotes a better understanding of the difference in chemical properties between the SAM−covered and −uncovered surfaces. As expected, the coplanar microstructures of OTS/$SiO_2$, TESUD/$SiO_2$, and AHAPS/$SiO_2$ are clearly imaged through the difference in surface potential between the

SAM-covered and -uncovered regions, respectively. In the images, the round-shaped features, as well as their frictional contrast images are highly resolved. Besides, the surface potential of the surrounding $SiO_2$ surfaces is consistently lower in magnitude than those of every SAM-covered region. Depth distribution profiles acquired from those KPFM images show potential difference between the SAM-modified and -unmodified surfaces of about 6 mV (for TESUD/$SiO_2$), 18 mV (for ODS/$SiO_2$) and 37 mV (for AHAPS/$SiO_2$), respectively. A surface potential distribution between the SAM-covered and –uncovered regions is given by following equation [8-10]:

$$\Delta V_{SAM\text{-}SiOH} = \mu / A\varepsilon_{SAM}\varepsilon_0$$

where $\mu$ is the actual molecular-dipole moment normal to the substrate, $\varepsilon_{SAM}$ and $\varepsilon_0$ are the permittivity of the SAM and the free space, respectively, and A is the area occupied by the molecule. Accordingly, a comparison of the theoretical dipole moment between two different surfaces leads to the identification of the terminal functional-group of the SAM, in addition to its molecular packing density. In order to calculate the dipole moments of the SAMs' precursors used here, their simplified molecular models are assumed, as shown in Figure 11. In the model SAMs, the molecules lie perpendicular to the substrate, and the terminal silicon atoms in their head-groups are replaced with hydrogen atoms. Since the directional vector of a dipole moment points from the positive charge source to the negative charge source, the three types of SAM-covered surfaces shown in KPFM images must have a positive potential contrast when compared with the surrounding $SiO_2$ surface. In Figure 11, the theoretical dipole moment of the AHAPS-model is 29.91 Debye, oriented to the substrate with a 0° of tilt angle to its alkyl chain. The amorphous-like molecular structure of AHAPS-SAM does not allow the estimation of its tilt-angle by ATR-FTIR, but the direction of its theoretical dipole moment is consistently toward the substrate, regardless of the actual molecular tilt angle. The dipole moment of the OTS-model is equal to 0.85 Debye, and is tilted at 150.0° from the surface normal. This indicates that the OTS-SAM has a positive surface potential higher than the surrounding $SiO_2$ surface. On the other hand, the dipole moment of the TESUD-model is tilted at 90° to the surface normal. The component of its dipole moment vertical to the substrate is thus almost zero. In order to explain the surface potential in contrast to TESUD/$SiO_2$, the most probable dipole moment is determined as follows. Prior to the calculation of the molecular dipole moment, ATR-FTIR is used to estimate the tilt angle of the TESUD-SAM by measuring its dichroic ratio, given the strong dependence between the molecular orientation and the direction of dipole moment. The dichroic ratio, which is defined as the ratio of peak-intensity of the s-polarized absorbance, $A^{s\text{-}pol}$, to p-polarized absorbance, $A^{p\text{-}pol}$, gives information on the molecular tilt angle of a SAM. The dichroic ratio of the asymmetric methylene stretch observed in the TESUD-SAM has a value of 1.05, which is in fairly good agreement with those of the ordered OTS-SAMs prepared by Sagiv (D = 1.03 and 1.09) and Ulman (D = 1.04) [11,12]. Using the IR dichroism data, a tilt angle, $\theta_{tilt}$, is estimated to be 27°, with respect to the surface normal. When the TESUD-SAM is tilted at 27° with respect to the surface normal, the vertical component of the dipole moment is equal to 1.05 Debye oriented towards the substrate. As a result, the positively-charged surface potential of the TESUS-SAM on the $SiO_2$ matrix is verified. These KPFM observations indicate that the microscale features of SAM surfaces shows chemical properties very similar to OTS-, TESUD-, and AHAPS-SAMs, respectively.

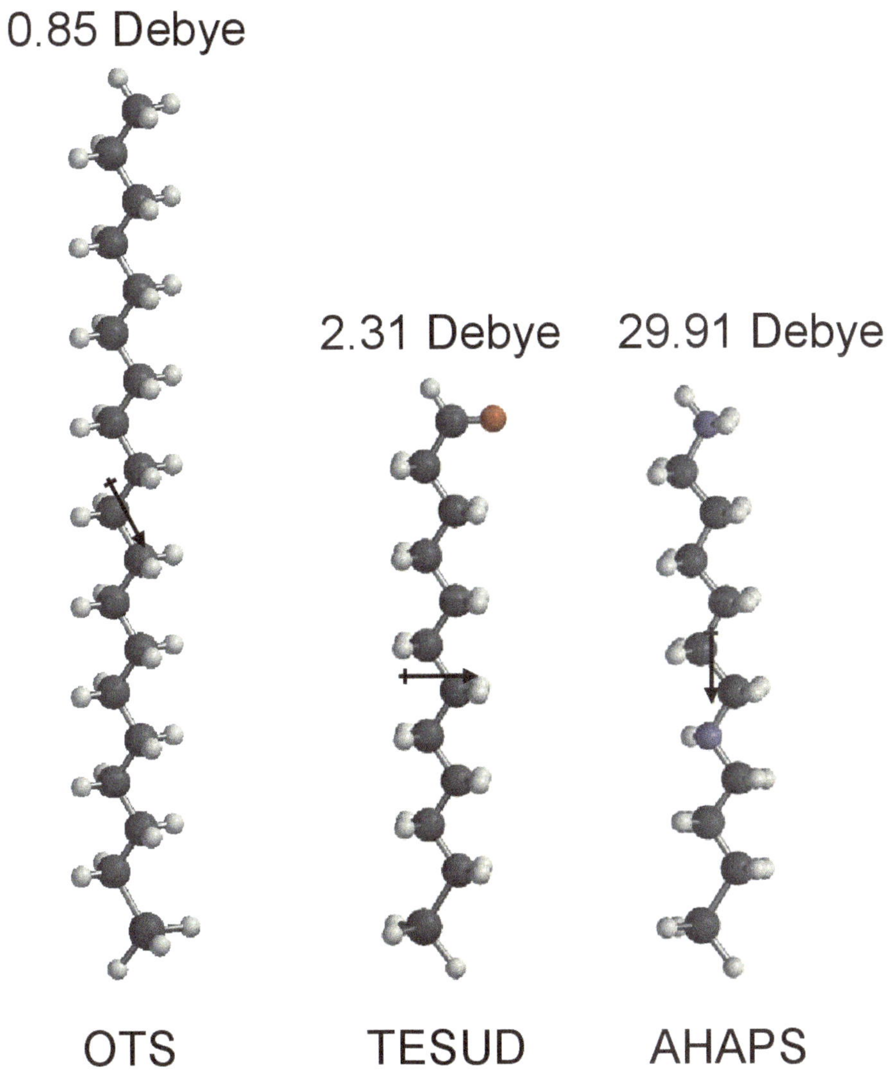


Fig. 11. Molecular structures and dipole moments of the model molecules for OTS, TESUD, and AHAPS-SAMs.

A SAM microarray is produced by combining the non-lithographical method and two different alkoxysilane inks. In a manner similar to that shown in Figure 9, two different molecular inks, TESUD and AHAPS, are respectively positioned under ambient conditions to create a microarray of two silane inks, as shown in Figure 12a. This sample is then treated in a manner very similar to the above-mentioned SAM formation process. Figure 12b shows a KPFM image of the resultant sample. In this surface potential image, well-defined microstructures composed of round-shaped 20 μm features are observed in contrast to the $SiO_2$ matrix. The surface potential contrasts between SAM-covered and -uncovered regions are +6 mV for TESUD/$SiO_2$ and +38 mV for AHAPS/$SiO_2$, respectively. The magnitude of each surface potential contrast approximately corresponds to the values observed in the single microarrays of TESUD- and AHAPS-SAMs in Figure 10. As a result, the successful fabrication of the SAM microarray, in which two different types of SAMs are positioned on each predefined surface site, is confirmed. Next, the potentiality of this SAM microarray as single-molecule thick platform for protein array is examined using protein G, which is an immunoglobulin-binding protein. An amine in protein G reacts with the outermost aldehyde group of the TESUD region on the microarray to give corresponding schiff base, which is reduced to the secondary amine by sodium cyanoborohydride. Throughout this reaction, protein G is covalently immobilized onto the TESUD-SAM surface. Due to the use of protein primarily labeled with Alexa 488, its presence on the sample surface is observed under a fluorescence microscope. As shown in Figure 12c, a strong fluorescence signal is detected with area-selectively on the TESUD-covered region. This successful detection, even after strong surface-cleaning, implies that protein G is not simply physisorbed, but is covalently bound to the aldehyde-terminated SAM surface. In contrast, the dark areas in Figure 12c correspond to the regions covered with AHAPS-SAM. This indicates that the AHAPS-SAM surface strongly prevents even nonspecific adsorption of protein G from the contacting buffer when compared with that on $SiO_2$ matrix. Accordingly, the fluorescence signal from the Alexa-labeled protein is the highest in magnitude on the TESUD-SAM region of the microarray.

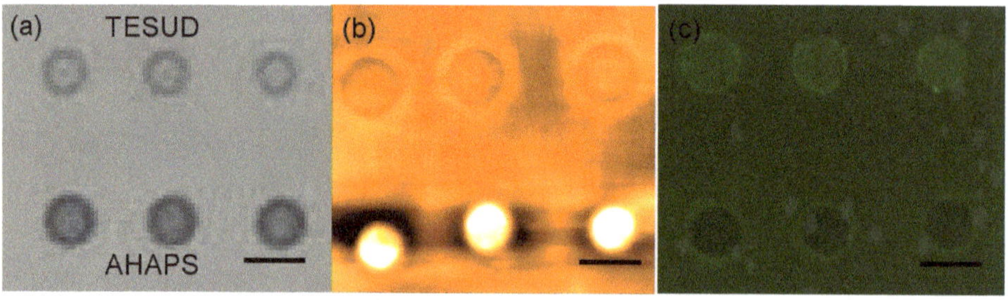


Fig. 12. (a) Optical image of a microarray having droplets of AHAPS and TESUD molecules, (b) KPFM image of TESUD/AHAPS-SAMs microarray, (c) Fluorescence microscopic image of TESUD/AHAPS-SAMs microarray after treated with Alexa-labeled Protein G. Scale bar: 20 μm.

As demonstrated here, the use of SAM microarray allows the parallel detection of different protein-behaviors on one platform surface under same experimental conditions. Although the two different kinds of SAMs are microstructured in this case, further increase in varieties will allows the parallel detection of many host-guest molecular interactions. Thus, such parallel observation of different variations in chemical reaction and biomolecular interaction on the same platform, where a number of different functional SAMs are integrated at predefined positions, will be able to lead to the achievement of facile and fast medical screening, or miniaturizations of traditional laboratory techniques, bioassays, and cell culture experiments.

## 4. Highly luminescent silicon nanoparticles

### 4.1 Background

Si is one of the most common chemical elements; it can be found everywhere and makes up 28% of Earth's crust by mass. Bulk Si wafers, which provide a platform for large-scale integrated circuits in modern microelectronics, are essential in consumer electronics, but have not attracted interest in the fields of optics and optoelectronics. This is because of its indirect bandgap nature due to which the minimal-energy state in the conduction band differs in a k-vector of the Brillouin zone from the maximal-energy state in the valence band. Therefore, in 1990, the discovery of red light emission from porous Si with a quantum yield (QY) of 1%, which was 10,000 times higher than that of bulk Si, created a sensation. As a result, many researchers recognized other features of Si that are useful in functional applications, including illumination sources. Free-standing nanoparticles provide much higher QYs than do porous Si. The emission color can be tuned in an astonishingly wide range from near-UV through visible to near-IR wavelengths, but the direct relationship between size and optical properties is still poorly understood, and thus inhibiting reproducible synthesis of luminescent Si nanoparticles with distinct emission spectra. In this section, I provide a brief overview of recent progress in this field, and highlight some of the most pertinent parameters to improve luminescence quality from Si nanoparticles. To achieve the color tunability, interfacial configuration formed between monolayers and nanoparticles is one of keys [14]. The difference in the interfacial chemical property affects on the optical absorption and emission properties of Si nanoparticles. Perhaps, the most obvious model for this demonstration is a colloidal form of nanoparticles. Furthermore, a high chemical affinity of Si has the potential to produce unlimited variations in the derivatives of the nanoparticles that can be mounted on optical and optoelectronic applications.

### 4.2 Synthesis of Si nanoparticles

A hydrogen-terminated silicon surface, which is habitually used as in semiconductor processing, is used as a basal plane in a chemical modification, as demonstrated in this study. Many methods have been developed for H-terminated silicon nanocrystal synthesis, as listed in Table 4. Among these methods, the most widely used ones are the mechanical fracturing of H-terminated silicon wafers and the gas-phase decomposition of silane or organosilane gas. Mechanical fracturing is further classified into two types: electrochemical anodization and pulsed laser ablation. Electrochemical etching, which is a well-known method for porous silicon production, is normally performed by anodic oxidation of H-terminated silicon wafers with a mixture of hydrofluoric acid (HF) and hydrogen peroxide. Therefore, nanocrystals can be obtained from the anodically etched wafer, i.e. porous silicon. Although their size distributions are broad, a relatively large quantity of luminescent

| Methods | References |
|---|---|
| Electrochemical anodization | 15—19 |
| Pulsed laser ablation | 20—23 |
| Gas-phase decomposition | 24—30 |
| Liquid phase solution synthesis | 31—33 |

Table 4. A list of representative methods for the preparation of silicon nanoparticles

crystals can be produced. On the other hand, pulsed laser ablation, generates extremely narrow size distribution of silicon nanocrystals. In a typical experimental setting, an H-terminated wafer of silicon is placed on a rotating target holder in a vacuum chamber, and then is ablated at the focal point of a laser beam. The ablation generates large quantities of vapors of elemental silicon. The vapors then assemble to form nanostructured configurations, such as nanoparticles and/or nanoparticles. The successful preparation of such nanosized configurations requires careful control in the experimental process. A wide variety of lasers, such as YAG, $CO_2$, $N_2$, KrF and ArF, have been hitherto employed for the ablation. An inert gas such as helium, introduced into the chamber, prevents unfavourable reactions with small amounts of oxygen that exists even in vacuum conditions. The background gas pressure influences the particle size distribution. In addition, the laser irradiation density affects the particle sizes and their distribution. The laser ablation method utilizes the homogeneous vaporization and subsequent rapid condensation to achieve high monodispersity. Generally speaking the drawback of the use of laser ablation method is lower yield of the nanoparticles. Since the early investigation by Brus [24], the gas-phase decomposition of silane or organosilane molecules has been studied. In general, such silane-based gases together with helium and hydrogen are introduced into the reaction field within the vacuum chamber, and are pyrolytically decomposed at high temperature, typically 950°C, which results in a silicon nanoparticle. In addition to $CO_2$ lasers, microwave plasmas are also effective for similar thermal decomposition. Both methods, i.e., mechanical fracturing and gas-phase decomposition, are technically practicable for producing large amounts of silicon nanoparticles; however, subsequent chemical etching caused by the acidic system, composed of $HNO_3$/HF or HF, is definitely required for the complete passivation of outermost silicon layer with hydrogen atom. The first report about successful liquid phase solution synthesis was achieved by Hearth [31]. Since then, several methods for producing H-terminated silicon nanoparticles have been reported, as listed in Table 1. Hydrogen reduction of silicon halides, i.e. $SiX_4$ where X = Cl, Br or I, by metal hydrides such as lithium aluminium hydride ($LiAlH_4$, LAH) is the most common synthesis route. Moreover, the reductive reaction is basically performed in a non-polar aliphatic solvent with complete absence of oxygen and water molecules in an inert atmosphere. The LAH reduces Si(IV) to Si(0). Next, the neutral Si(0) assembles to form nanoparticles with a passivation layer of Si-H (318–351 kJ/mol), as schematically illustrated in Figure 13. In the hydride reduction process, the additional use of an inverse micelle such as tetraoctylammonium bromide leads to remarkable improvement in size monodispersity because the nanoscale gauge formed by the inverse micelle controls the final crystal size. Unlike mechanical fracturing and gas-phase decomposition, hydride reduction enables complete passivation of the crystalline surface. Therefore, this method does not require subsequent HF treatment. The use of a metal silicide promises large quantity synthesis of silicon nanoparticles even by a liquid phase solution process. For example, Kauzlarich and co-workers successfully produced more than a decamilligram of silicon nanoparticles by the following chemical reaction [33]:

$$NaSi + NH_4Br \rightarrow Si/H + NaBr + NH_3 \quad (1)$$

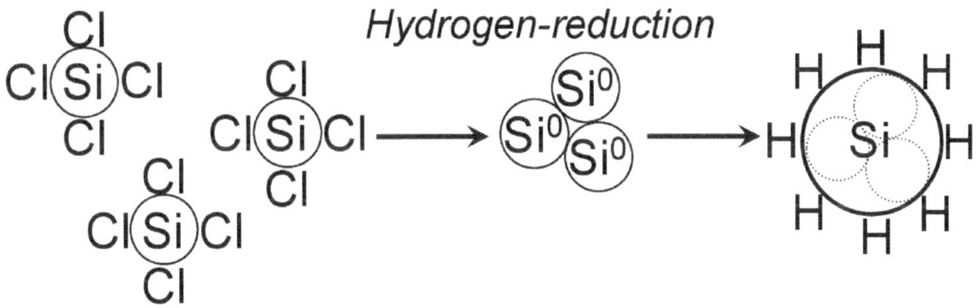


Fig. 13. Hydrogen reduction process for H-terminated silicon nanoparticles.

Identification of the chemical bonding states of a nanoparticle's surface has been frequently performed by Fourier transform infrared spectroscopy (FT-IR). Absorption at around 2100 $cm^{-1}$ indicates the presence of the Si-$H_x$ stretch. On the other hand, the relatively broad peak observed at 1000–1100 $cm^{-1}$ can be attributed to the Si-O-Si linkage, which indicates that the silicon surface is oxidised. More than 99% of the outermost silicon atoms can be hydrogen terminated; however, complete absence of oxygen is a difficult-to-achieve goal on a nanoparticle. As a clear-cut example, some samples were prepared by laser pyrolysis of gaseous silane, and then were etched with a mixture of HF (48 wt%) and $HNO_3$ (69 wt%). In a FT-IR spectrum of the sample, strong peaks were visible at around 1087 $cm^{-1}$ and 2105 $cm^{-1}$. In contrast, acidic treatment with HF/$HNO_3$ (10/1, v/v) was expected to result in non-oxidised surfaces. However, the presence of oxygen, i.e. Si-O-Si, was seen in FT-IR spectrum, although its intensity in absorbance was relatively weak when compared with the first model sample. Therefore, the etching chemistry on a nanoparticle is considerably different from that on a wafer. In fact, the etching rate of native oxide on a wafer is thousands times larger than that of oxidized nanoparticle.

In current microelectronics, halide-terminated surfaces are less used when compared to H-terminated ones; however, they are extremely useful as the starting stages in wet chemistry on silicon nanoparticles. Due to their relatively high reactivity, it is imperative to treat the halogenated surfaces in an inert condition. Following the first demonstration by Heath [31], a series of solution syntheses that use the combination of Zintl salts (KSi, NaSi and $Mg_2Si$) and $SiCl_4$ have been developed by Kauzlarich and co-workers. In the case of KSi, Si nanoparticles can be obtained through the following reaction:

$$4n\text{KSi} + n\text{SiCl}_4 \rightarrow \text{Si nanoparticle} + 4n\text{KCl} \quad (2)$$

The synthesis utilising Zintl salts presumably produces a chloride-terminated silicon surface, which is very sensitive to oxygen and moisture. For the evaluation of the intra-particle structure FT-IR, solid-state Nuclear Magnetic Resonance (NMR) and High-Resolution Transmission Electron Microscopy (HR-TEM) has been used. A combination of HR-TEM and selected-area diffraction (SAD) has showed the presence of crystalline silicon in a product synthesized by the reaction of Zintl salts with $SiCl_4$; however, macroscopic observation based on $^{29}$Si MAS-NMR revealed that the product was composed of a mixture of crystalline and amorphous phases. Unlike X-ray powder diffraction, NMR allows acquisition of structural information even from a small quantity of fluid specimen. In many cases, the Zintl salts are generally synthesized via the reaction of explosively reactive alkali metals with silicon at high temperatures of more than 600°C. Recent technical advancements in wet chemistry have achieved room temperature solution synthesis of Cl-capped silicon nanoparticles using alkali naphthalides as reducing agents. The synthesis of alkali naphthalides is very simple. In particular, alkali naphthalides can be easily prepared at room temperature by the reaction of alkali metals, such as lithium or sodium, with naphthalide. As this reaction requires a moisture-free environment, the use of a dehydrated solvent, such as THF, glyme or diglyme, is necessary. If the reaction is correctly processed, the resultant solution turns dark green in colour. Unlike the reductant sodium naphthalide, it was reported that lithium naphthalide gives only amorphous nanoparticle.

Bromine-terminated silicon nanoparticles can be synthesised by the oxidation of magnesium silicide with bromine, according to the following equation.

$$n\text{Mg}_2\text{Si} + 2.5n\text{Br}_2 \rightarrow \text{Si}_n\text{Br} + 2n\text{MgBr}_2 \quad (3)$$

The reaction with $Br_2$ requires careful selection of the solvents. Molecular bromine is a strong oxidant capable to decompose ether solvents such as glyme. One possible solution is the use of inert solvents, such as alkanes, for bromine. A halogenated surface shows high reactivity to various functional molecules, such as alcohol, amine, aldehyde, alkyllithium and Grignard reagents, as demonstrated in the following sections.

Structural analysis is performed by combining HR-TEM and either one of XRD, XPS and Raman spectroscopy. It has been widely recognized that the energy position of Si2p spectrum in XPS profile is far from the known position for Si-Si bond, i.e., 99.5 eV because of the presence of charge trapping which arises from monolayer-shells covering NCs. Raman spectrum, which is a powerful macroscopic observation tool, is sensitive to the size and crystallinity of nanoparticles. The relationship between size and Raman peak maxima is summarized in Figure 14. Raman spectrum for crystalline bulk Si exhibits a single distinct peak at 520 $cm^{-1}$ whereas that of amorphous Si is composed of a broad peak at around 476-480 $cm^{-1}$. Interestingly, a blueshift in the Raman spectrum appears with decreasing physical size, as shown in Figure 14. In particular, a relatively broad Raman peak appears at about 503 $cm^{-1}$, whereas vibrational features are absent in the 510–517 $cm^{-1}$ region and at around 476 $cm^{-1}$, suggesting that the majority of the product is constructed of crystalline structures ≤2.5 nm [35]. This was consistent with HR-TEM study [35]. Raman spectroscopy is one of powerful tools to perform macroscopic observation, while microscopic observation requires HR-TEM. According to the careful observation with HR-TEM, a true atomic arrangement is revealed, as shown in Figure 15. These nanoparticles were prepared at room temperature by sodium biphenylide reduction of $SiCl_4$ in the presence of surfactant, i.e., dimethyldioctyl ammonium bromide (DMDOAB) [35]. In every high-resolution image, periodic arrangements of channel structures surrounded by Si tetrahedrons are clearly visible. The nanoparticles having diameter d > 1.5 nm have diamond cubic lattice structures. The same is true for 1.5-nm nanoparticles; thus, a 1.5-nm nanoparticle is large enough to form a diamond cubic lattice. Interestingly, a 1.1-nm nanoparticle, unlike other nanoparticles, cannot have a diamond cubic lattice. At 1.1 nm, about 30 atoms are required to produce a diamond cubic lattice. Although a complementary simulation study is needed to determine the real atomic geometry, the minimum size that allows the formation of a diamond cubic lattice may be found to lie in the 1.2–1.5-nm size regime.

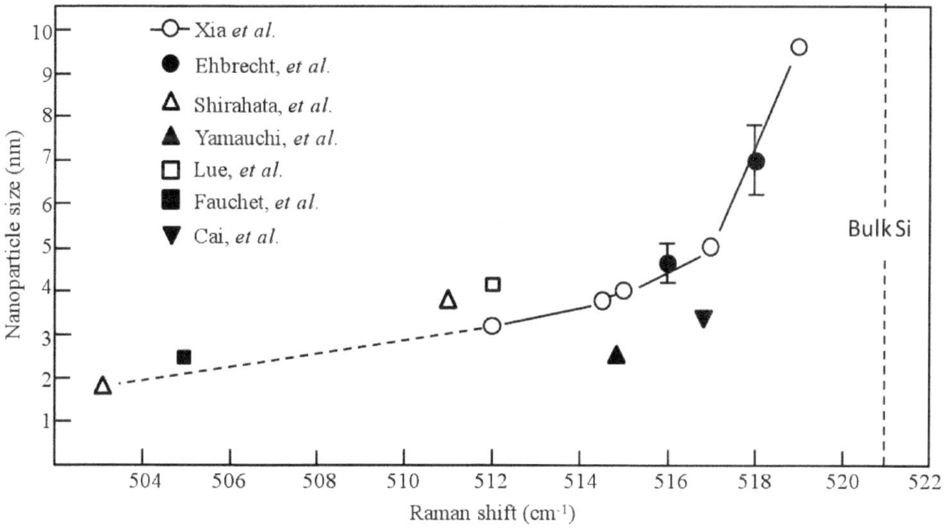


Fig. 14. Comparison of our results with the Raman resonances published by others (refs. 34-41).

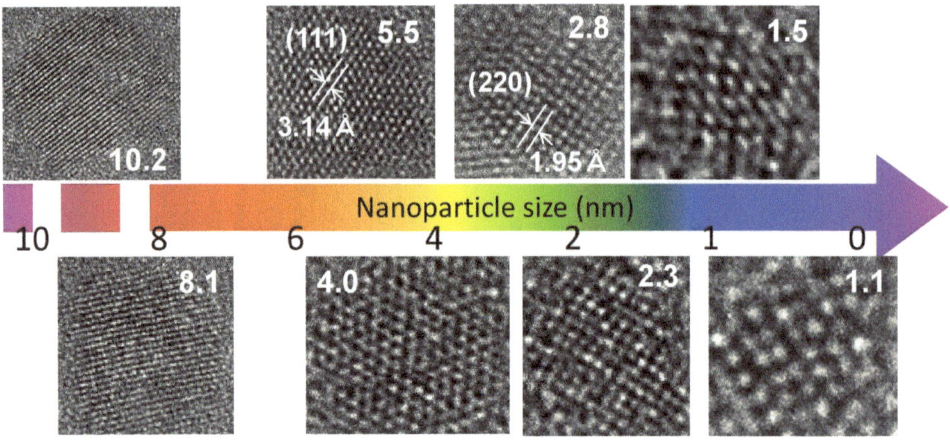


Fig. 15. HR-TEM images showing lattice fringes of the different size of Si nanoparticles synthesized by inverse micelle method [35].

### 4.3 Surface modification of nanoparticles

An elemental Si exhibits a high chemical affinity for carbon, oxygen, nitrogen and sulfur. The first three elements form covalent bonds with Si and provide the differences in polarity. On the other hand, sulfar can interact specifically with Si as well as gold, yielding a S-Si linkage. Such a range of options in mode of binding creates variations in the organic derivatives of Si nanoparticles hybridized at the molecular level. As shown in Figure 16, various chemical routes are available for modifying the surface of Si nanoparticles as in to the case with wafer chemistry [14,43]. Interestingly, a molecular attachment to the surface of nanoparticles yields the change in physical nature from solid products to viscous liquid, because the organics possibly dominate the physical nature in such nanoparticle systems [42]. Therefore, surface modified nanoparticles are easily soluble in any solvents without severe sonication. The most comprehensive and accessible are the use of luminescent nanoparticles as fluorophors for biomolecular imaging and monitoring. To give water-solubility, the end groups of monolayer-capped nanoparticles are modified into either one of hydrophilic groups including amine, carboxyl, aldehyde, ester, oligonucleotides, and carbohydrate [14]. As shown in Figure 16a, water-soluble luminescent Si nanoparticles exhibit the superior stability against photobleaching when compared to conventional organic dyes. Furthermore, the presence of surface monolayers gives nanoparticles an exceptional stability over a wide pH range (pH=1-13) and high temperatures (120°C), and influences on a cytotoxicity of NCs [44,45]. Therefore, their use allows the prolonged monitoring of specific biomolecular interaction events and a parallel detection of different molecular recognitions. A recent investigation for bioimaging is demonstrated in Figure 16b [46]. In this fluorescence observation, a two-photon microscopic image of microphage cell when excited with a 780 nm of laser is presented. As clearly seen in inter-comparison experiment of (c) and (d), the presence of the luminescent nanoparticles in the cytoplasm of the incubated cell is visible, indicating the endocytic mode of internalization. In order to avoid the photodamage, the use of low photon energy as an excitation source is suitable for applications in dense tissue imaging. These fluorescence images verify the successful in-vitro imaging using a multiphoton microscopy. Thus, organic derivatives of Si nanoparticles have a great potential to be used in a wide variety of optical and optoelectronic applications.

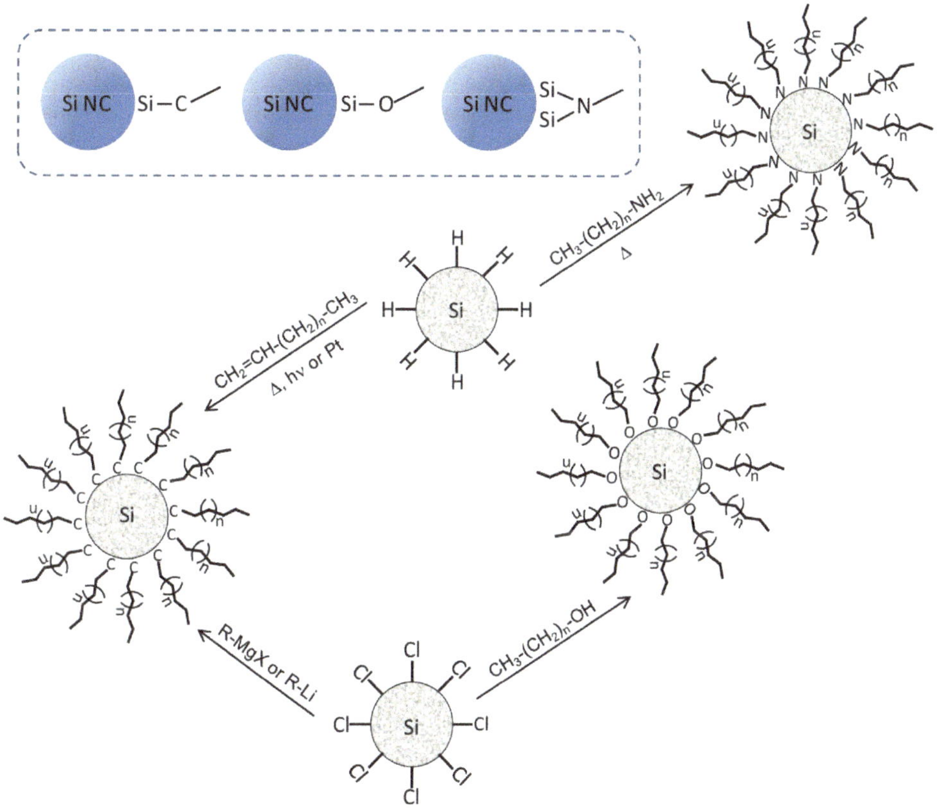


Fig. 16. Representative modification routes for monolayer attachment to the surfaces of hydrogen- and chloride-terminated silicon nanoparticles.

## 5. Summary

In this article, some of programmed surfaces are demonstrated as models using organic monolayers. Organic films including self-assembled monolayers, which works as a prototypical chemical template and linkers, plays an important role in the development of bioinspired inorganic materials. Due to SAMs' high ability to assemble onto the surfaces of any geometry, size, and component, their employments allow a general and highly flexible industrial uses. SAMs enable the control of various surface functions including wettability, electrochemistry, magnetism, friction, biofunctionality, and others. The notable advance of top-down lithography approach will be able to produce fine structures in tens of nanoscale resolution, but real nanostructures (or parts) must require to be produced by self-assembly processing. Silicon is one of the few nontoxic, abundant, and environmentally friendly elements, and it has dominated the microelectronics industry owing to its impressive performance as an electronic material. Therefore, the optical use of Si could rapidly gain popularity for its ability to help build a sustainable society by turning indirect bandgap semiconductors like Si and Ge into efficient light emitters. Another great attraction of luminescent Si is the potential ability to tune the emission photon energy over a very wide range from 1.1 eV (for bulk) to 4.0 eV, if high efficiencies of light emissions at each wavelength in the ranging from near-IR to near-UV region are achieved. Furthermore, Si exhibits high chemical affinity for covalent linkages with carbon, oxygen, and nitrogen, which can produce a variety of organic derivatives hybridized at the molecular level. A well-designed ligand working as a molecular wire is expected to decrease a potential energy barrier between nanoparticles and monolayers, leading to the efficient electronic conduction. In addition to the use as fluorescent material for molecular labeling, the chemical compatibility could generate further vitality in its optical use even for fragile items which have direct contact with our skin, such as textile fabrics, cosmetics and paints. Rigorous control of size distribution, surface/interface chemistry and structural phase of Si nanoparticles provides a readily accessible range of color with high purity, and direct grafting of molecular chains on the nanoparticles creates unlimited variations in sophisticated architectures in which Si islands are positioned in a molecular networking matrix.

## 6. References


[1] N. Shirahata, W. Shin, N. Murayama, A. Hozumi, Y. Yokogawa, T. Kameyama, Y. Masuda K. Koumoto, Adv. Funct. Mater. 14 (2004) 580.

[2] N. Shirahata, W. S. Seo, T. Kinoshita, T. Yonezawa, A. Hozumi, Y. Yokogawa, T. Kameyama, Y. Masuda, K. Koumoto, Langmuir 20 (2004) 8942.

[3] A. Hozumi, M. Inagaki, N. Shirahata, Surf. Sci. 600 (2006) 4044.

[4] Y. Yamanoi, N. Shirahata, T. Yonezawa, N. Terasaki, N. Yamamoto, Y. Matsui, K. Nishio, H. Masuda, Y. Ikuhara, H. Nishihara, Chem. Eur. J. 12 (2005) 314.

[5] N. Shirahata, A. Hozumi, Chem. Mater. 17 (2005) 20.

[6] A. Hozumi, S. Asakura, A. Fuwa, N. Shirahata, T. Kameyama, Langmuir 21 (2005) 8234.

[7] A. Ivanisevic, K. V. McCumber, C. A. Mirkin, J. Am. Chem. Soc. 124 (2002) 11997.

[8] V. Palermo, M. Palma, P. Samorì, Adv. Mater. 18 (2006) 145.

[9] A. Liscio, V. Palermo, D. Gentilini, F. Nolde, K. Müllen, P. Samorì, Adv. Funct. Mater. 16 (2006) 1407.

[10] K. Hayashi, N. Saito, H. Sugimura, P. Takai, N. Nakagiri, Ultramicroscopy 91 (2002) 151.

[11] N. Tillman, A. Ulman, J. S. Schidkraut, T. L. Penner, J. Am. Chem. Soc. 110 (1998) 6136.

[12] R. Maoz, J. Sagiv, J. Colloid Interf. Sci. 100 (1984) 465.

[13] N. Shirahata, J. Nakanishi, Y. Echikawa, A. Hozumi, Y. Masuda, S. Ito, Y. Sakka, Adv. Funct, Mater. 18 (2008) 3049.
[14] N. Shirahata, Phys. Chem. Chem. Phys. (Perspective Article) 13 (2011) 7284.
[15] J. L. Heinrich, C. L. Curtis, G. M. Credo, M. J. Sailor, K. L. Kavanagh, Science 255 (1992) 66.
[16] R. A. Bley, S. M. Kauzlarich, J. E. Davis, H. W. H. Lee, Chem. Mater. 8 (1996) 1881.
[17] M. H. Nayfeh, N. Barry, J. Therrien, O. Akcakir, E. Gratton, G. Belomoin, Appl. Phys. Lett. 78 (2001) 1131.
[18] L. Wang, V. Reipa and J. Blasic, Bioconjugate Chem. 15 (2004) 409.
[19] J. Choi, N. S. Wang and V. Reipa, Langmuir 23 (2007) 3388 .
[20] T. Seto, Y. Kawakami, N. Suzuki, M. Hirasawa and N. Aya, Nano Lett. 1 (2001) 315.
[21] E. Werwa, A. A. Seraphin, L. A. Chiu, Chuxin Zhou and K. D. Kolenbrander, Appl. Phys. Lett. 64 (1994) 1821.
[22] T. Makimura, Y. Kunii and K. Murakami, Jpn. J. Appl. Phys. 35 (1996) 4780.
[23] L. Patrone, D. Nelson, V. I. Safarov, M. Sentis, W. Marine and S. Giorgio, Appl. Phys. Lett. 87 (2000) 3829.
[24] W. L. Wilson, P. J. Szajowski and L. E. Brus, Science 262 (1993) 1242.
[25] M. L. Ostraat, J. W. De Baluwe, M. L. Green, L. D. Bell, M. L. Brongersma, J. Casperson, R. C. Flagan and H. A. Atwater, Appl. Phys. Lett. 79 (2001) 433.
[26] L. E. Brus, P. J. Szajowski, W. L. Wilson, T. D. Harris, S. Schuppler, P. H. Citrin, J. Am. Chem. Soc. 117 (1995) 2915.
[27] G. Ledoux, J. Gong and F. Huisken, Appl. Phys. Lett. 79 (2001) 4028.
[28] X. Li, Y. He, S. S. Talukdar and M. T. Swihart, Langmuir 19 (2003) 8490.
[29] F. Hua, F. Erogbogbo, M. T. Swihart and E. Ruckenstein, Langmuir 22 (2006) 4363.
[30] H. Takagi, H. Ogawa, Y. Yamazaki, A. Ishizaki, T. Nakagiri, Appl. Phys. Lett. 56 (1990) 2379.
[31] J. R. Heath, Science 258 (1992) 1131.
[32] J. P. Wilcoxon and G. A. Samara, Appl. Phys. Lett. 74 (1999) 3164.
[33] D. Neiner, H. W. Chiu and S. M. Kauzlarich, J. Am. Chem. Soc. 128 (2006) 11016.
[34] N. Shirahata, M. R. Linford, S. Furumi, L. Pei, Y. Sakka, R. J. Gates and M. C. Asplund, Chem. Commun. (2009) 4684.
[35] N. Shirahata, T. Tsuruoka, T. Hasegawa and Y. Sakka, Small 6 (2010) 915.
[36] H. Xia, Y. L. He, L. C. Wang, W. Zhang, X. N. Liu, X. K. Zhang, D. Feng and H. E. Jackson, Appl. Phys. Lett. 78 (1995) 6705.
[37] M. Ehbrecht, B. Kohn, F. Huisken, M. A. Laguna and V. Paillard, PRB 1997, 56, 6958
[38] P. Shen, N. Uesawa, S. Inasawa, Y. Yamaguchi, JMC 2010, 20, 1669
[39] S. -K. Ma and J. T. Lue, Thin Solid Films 1997, 304, 353.
[40] J. von Behren, et al., Solid State Commun. 1998, 105, 317.
[41] W. Cai, et al., JAP 2008, 104, 023516
[42] N. Shirahata, S. Furumi and Y. Sakka, 311 J. Cryst. Growth (2009) 634.
[43] N. Shirahata, A. Hozumi, T. Yoenzawa, Chem. Rec. 5 (2005) 145.
[44] M. Rosso-Vasic, E. Spruijt, Z. Popović, K. Overgaag, B. Van Lagen, B. Grandidier, D. Vanmaekelbergh, D. Domínguez-Gutiérrez, L. De Cola and H. Zuilhof, J. Mater. Chem. 19 (2009) 5926.

[45] L. Ruizendaal, S. Bhattacharjee, K. Ournazari, M. Rosso-Vasic, L. H. J. De Haan, G. M. Alink, A. T. M. Marcelis and H. Zuilhof, Nanotoxicology 3 (2009) 339.
[46] C. Tu, X. Ma, P. Pantazis, S. M. Kauzlarich and A. Y. Louie, J. Am. Chem. Soc. 132 (2010) 2016.

# 12

# Nanocomposites Preparation Method Based on Bubbles Explosion and Nanocomposites Capability Evaluation Method Base on Fractal Theory and TEM Image

Peng Nie and Kai-feng Zhang
*Shenyang Aerospace University,*
*China*

## 1. Introduction

Spontaneous agglomeration tends to occur between nanometer scale particles for their high surface free energy. The size of particle agglomerates granules is always beyond nanometer scale. The spontaneous agglomeration of nanometer scale particles and high viscosity of epoxy resin solution made it difficult to disperse nanomaterials into polymer as the configuration of nanometer scale particles. How to achieve nanomaterials' symmetrical decentralization in epoxy resin by the configuration of nanometer scale particles is one of the "choke points" of polymer nanocomposite preparation. To solve the problem, lots of researching works have been developed by experts and many inaugurated results be achieved. Presently, there are many methods applied in the polymer nanocomposite preparation, such as intercalation polymerization, solgel, in-situ polymerization, ultrasonic method, blending, chemical dispersant method and so on.

In this chapter, a nanocomposites preparation system based on diminutive bubbles explosion is designed. By applying this preparation system, we can disperse inorganic nanometer material into polymers as the configuration of nanometer scale particles. Compare to traditional dispersion methods, the proposed method tends to obtain ultrahigh stretching rate and increase the freedom degrees of nanoparticles dispersion. As the particles aggregates serve as nucleation agents, and the bubbles expand on the surface of the aggregates, the nanoparticles would disperse more effectively into polymer without dispersant .

At present, there are many researchs and reports about preparation and performance evaluation of nanocomposites. Methods usually used to evaluate uniformity of nano-particles are X-ray diffraction, transmission and scanning electron microscope. Detecting thermodynamics performance of nanocomposites is an indirect method for uniformity evaluation. Although these methods aforementioned are explicit, they can't satisfy the routine and fast examination request for production and application departments. And the application and promotion of nanocomposites has been affected to some extent. So, the quantitative evaluation method for size distribution and space distribution of nano-particles is essential for preparation and research of new nanocomposites.

To rightly evaluate the size distribution and space distribution of nano-particles, the exact Mathematical model must be established and the picking-up method for corresponding parameter should be found. Nanocomposites belong to the category of ultra fine powder. Researchs show that the size distribution of ultra fine powder has fractal characterization. Many irregular discrete objects in nature distribute in space as fractal. So, fractal geometry provide a good mathematics frame for this kind of complicated system. It should be regard as a irregular dispersion process to disperse nanomaterials into polymer. The TEM images of nanocomposites can be directly obtained by transmission electron microscope. In this chapter, we will propose a evaluation method for size distribution and space distribution of dispersed phase in nanocomposites based on fractal theory and TEM image. Applying the method to describe distribution of filling particles in the polymer composites, the results show that it can exactly evaluate the size distribution and space distribution of nano-particles.

## 2. Nanocomposites preparation method based on bubbles explosion

To obtain suspended dispersion of nano-particles in liquid without any dispersant, a novel epoxy resin/$SiO_2$ nanocomposites preparation method based on diminutive bubbles explosion is presented. And, corresponding nanocomposites preparation system was designed. The preparation system applies compressed gas as transmission medium to carry nanomaterials into epoxy resin solution. The compressed gas with nanomaterials turns into diminutive bubbles distributing in epoxy resin/$SiO_2$. The great pressure difference between inner and outer-bubbles led to bubbles inflation and explosion. During the bubble inflation, bubble oscillation may generate. The stretching rate may reach $10^6 s^{-1}$, which favors more homogeneous dispersion of nano-particles. During the bubbles explosion the released energy and the explosion shock waves disperse the nano-particles into epoxy resin solution. By using the preparation system, epoxy resin/$SiO_2$ nanocomposites were prepared. The $SiO_2$ dispersed into epoxy as the configuration of 15nm - 30nm particles.

### 2.1 Preparation method

In preparing polymer nanocomposites, laminar flow mixing of the nanomaterials is brought about by the shearing and stretching of polymer solution. In the preparation method based on diminutive bubble explosion, the key variant s about nanomaterials dispersion are the shearing rate and stretching rate.

#### 2.1.1 Stretching dispersion effect

According to Tadmor and Gogos's dumbbell model, under the stretching rate ε, in terms of granule size(r) and the polymer viscosity(μ) the maximum dispersion force exerted is given by

$$F_{Max}=6\pi\mu\varepsilon r^2 \tag{1}$$

To separate the agglomerated nanomaterials granules, Fmax must be larger than the interaction force between nanomaterials granules $F_M$. The interaction forces between nano-particles are mainly due to surface interactions. The action distance range of attractive forces between nano-particles agglomerated granules is about 50 nm. The intermolecular interaction between two granules with radius r at spacing h may be approximated to

$$F_M = \frac{Hr}{24h^2} \tag{2}$$

where H is the Hamaker constant. Then, we got the minimum granule radius that can be separated under stretching rate ε, it can be expressed as

$$r_{min} = \frac{H}{144\pi\mu\varepsilon h^2} \tag{3}$$

Eq. (3) demonst rates that stretching is an effective means of separating agglomerated granules. Nano-particles can be dispersed into polymer with enough high stretching rate ε.

**2.1.2 Diminutive bubble explosion dispersion method**

When the diminutive bubble inflation in the liquid polymer, the liquid polymer around the bubble is subject to stretching, the stretching rate is

$$\varepsilon = \frac{1}{r_0}\frac{dr}{dt} \tag{4}$$

where $r_0$ is the initial radius of the compressed bubble. According to Zaremba-DeWitt model, we got the diminutive bubble inflation model, shown as Fig. 1.

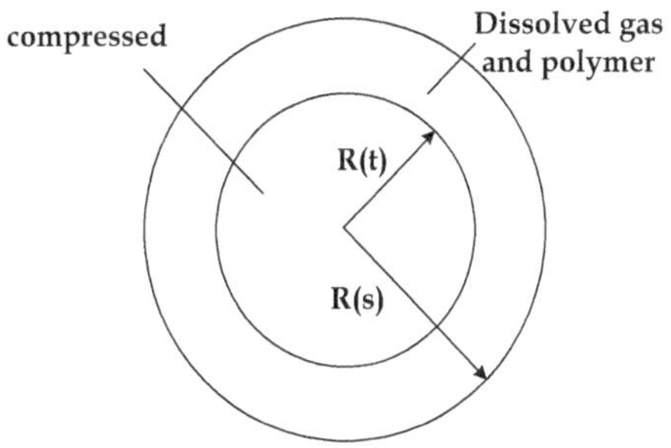


Fig. 1. Bubble and polymer shell

During the bubble inflation, the liquid polymer surrounding the bubble subject to high stretching rate. When the bubble expands, viscoelastic deformation will occur in the surrounding liquid polymer. As the expansion process is very fast, a certain amount of the elastic potential energy is stored in the surrounding liquid polymer macromolecules because there is no time for energy release. Following bubble expansion, the gas pressure in the bubble decreases monotonically, whereas the elastic potential energy stored in surrounding polymer macromolecules increases. When the elastic storing force of the liquid polymer surrounding the bubble becomes larger than the bubble expansion force, the bubble will be compressed. In the bubble compressing process, the elastic potential energy decreases, whereas the bubble pressure rises, and when the bubble pressure reaches a certain value, the bubble starts to expand again. The aforementioned process repeated under a certain conditions, bubble oscillation occurs. Considering the continuity and momentum equations as well as the force balance on bubble wall and the liquid boundary remote from the bubble wall, the compressed bubble inflation may be described as

$$\ddot{R}=\frac{P_g}{R_\rho}-\frac{2\sigma}{\rho R^2}-\frac{P_\infty}{\rho R}-\frac{3}{2}\frac{\dot{R}^2}{R\rho}-\frac{12\eta_0}{R\rho\lambda_0}\int_0^t e^{(s-t/\lambda)}\frac{R^2(s)\dot{R}(s)}{R^3(t)-R^3(s)}\ln\left|\frac{R(t)}{R(s)}\right|ds \tag{5}$$

where $P_g$ is the gas pressure in the bubble. $P_\infty$is thepressure of polymer solution remote from the bubble surface. $\sigma$ is the surface tension of the bubble. R is the bubble radius. $\eta_0$ is the zero-shear viscosity. $\rho$ is the density of polymer solution. $\lambda$ is the relaxation time of the polymer.

According to Fick's law of diffusion, the relationship between $P_g$ and R can be shown as

$$P_g=[\frac{6\rho^2 D(C_0-C_w)^2 R}{\rho_g R^3-\rho_{g0}R_0^3}-\frac{3\dot{R}\rho_g}{R}]\frac{R_g T}{M} \tag{6}$$

where $\rho_g$ is the density of gas in the bubble. $\rho_{g0}$ is theinitial density of the gas in the bubble. D is the diffusion coefficient. $C_0$ is the initial concent ration of dissolved gas in the polymer solution. $C_w$ is the concent ration of dissolved gas in the polymer solution on the bubble surface wall. T is the temperature. M is the relative molecular mass of the gas. Using Eq. (1) and Eq. (2), we can get the bubble inflation process, shown as Fig. 2 ($P_g$ = 0.4052MPa, $P_\infty$=0.1013MPa, $\lambda$ = 0.9s, $\rho$ = 0.89g/cm$^3$, $\sigma$ =2.97 ×10$^{-2}$ N/m).

The bubble oscillatory frequency in Fig. 2 reached the order of $10^7$Hz. In the very early period of bubble inflation, the initial bubble radius is small. Let the initial bubble radius equal 1μm, the stretching rate reaches the order of $10^6$s$^{-1}$.

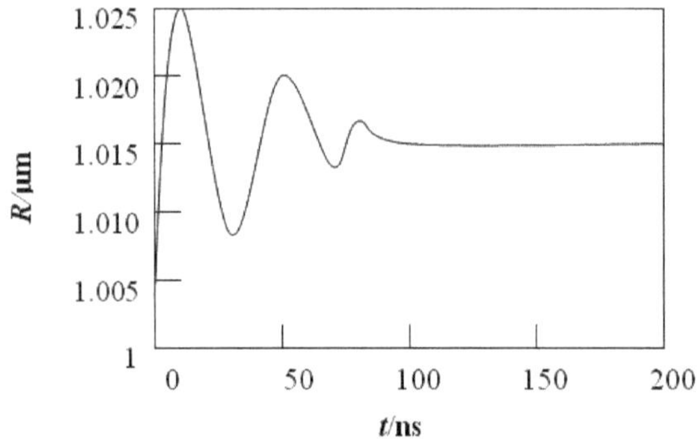


Fig. 2. Variation of bubble radius in the process of bubble inflation

According to Eq. (3), the minimum granule radius that can be separated under the bubble oscillation shown in Fig. 2 is 6 nm.

On the other hand, in the anaphase of diminutive bubble inflation, bubble instabilities may occur and eventually cumulative jet penetrate the bubbles, just as Fig. 3 shows. The cumulative jet leads to micro-droplets formation and subsequent evaporation. The micro-droplets are spatially homogeneously distributed inside the gas-nanomaterials mixture. The mixture process made nanomaterials in the bubbles enwrapped with liquid polymer micro-droplet s and evaporation, which ameliorates the interphase binding state between nanomaterials and liquid polymer. What's more, the bubble oscillation frequency reached ultrasonic frequency, ultrasonic cavitations may appear. This can also promote nano-particles dispersion.

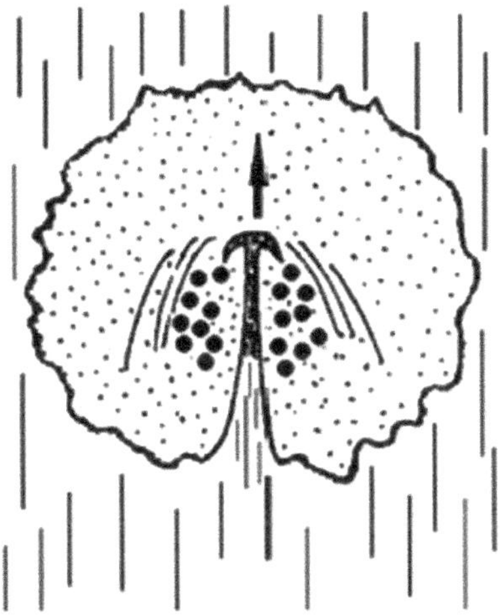

Fig. 3. Cumulative jet

### 2.2 Preparation system

The polymer nanocomposites preparation system includes compressed gas preparation subsystem, gas-nanomaterials mixture subsystem and polymer nanocomposites preparation subsystem (shown as Fig. 4).

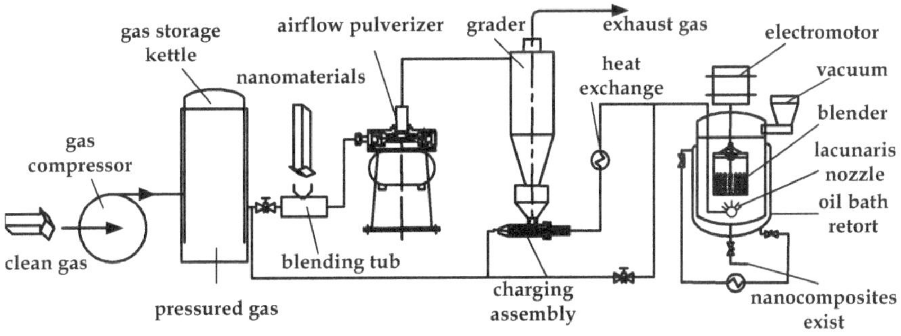


Fig. 4. Sketch of the nanocomposites preparation system

#### 2.2.1 Compressed gas preparation subsystem

The compressed gas preparation subsystem is made of gas compressor, gas storage kettle and some accessories, which can supply compressed gas for nanocomposites preparation.

#### 2.2.2 Gas2nanomaterials mixture subsystem

The gas-nanomaterials mixture subsystem is made of blending tube, siphon and heat exchanger. In the gas-nanomaterials mixture process, nanomaterials were commixed with compressed gas. That leads to gas-nanometer scale particles mixture jet formation. A blending tube was designed to commix nanomaterials into compressed gas (Fig. 5).

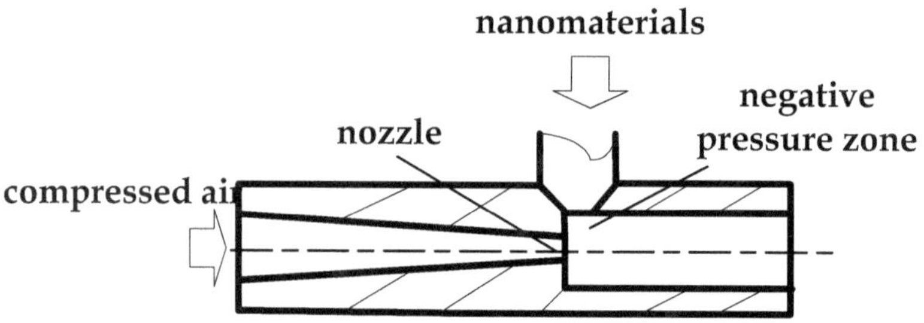


Fig. 5. Blending tube

The compressed gas from gas storage kettle turns into high2velocity jet when it rushes through the small nozzle in blending tube. Because of the high-velocity gas flow's entrainment effect, a negative pressure zone is formed around the nanomaterials entrance. As the pressure of negative zone is much lower than atmospheric pressure, $SiO_2$ powder will is absorbed into the blending tube, where the $SiO_2$ powder is blended with the gas flow by the way of momentum exchange and forms $SiO_2$ powder-gas admixture. When the admixture rushes through the siphon (whose inner diameter is 4 mm) between blending tube and lacunaris nozzle, the impact and shear caused by the high-velocity gas flow rubbing and impacting with pipe wall lead to $SiO_2$ powder conglomerations' being elementarily destroyed to smaller particles. The heat exchanger increases the temperature of the gas-nanomaterials so high that the liquid polymer near the lacunaris nozzle will not be cooled.

### 2.2.3 Polymer nanocomposites preparation subsystem

The polymer nanocomposite preparation subsystem is made of oil bath retort, cylinder blender, electromotor, vacuum pump and lacunaris nozzle. The oil bath retort makes the retort inner temperature high enough to keep polymer in liquid state. The gas-nanomaterials jet f rom lacunaris nozzle rushes into oil bath retort. As the gas-nanomaterials jet is an under-expanding jet, it will turns into a mass of diminutive bubbles distributing in the liquid. The diminutive bubbles with compressed gas and nanomaterials are closed inside working as nanomaterials dispersion energy cell and minisize nanometer scale particles storage cell.

The vacuum pump keeps the retort in low pressure. As the pressure in the bubbles is much higher than the retort inner pressure, bubble expansion occurs in the liquid. During the bubble expansion, the polymer solution around the bubble is subject to stretching with high stretching rate. The high stretching rate will disperse nanomaterials agglomeration granules effectively.

Cylinder blender includes a cylinder with a lot of holes and a parabola impeller. The high speed electromotor drives the impeller to rotate with high speed. When the impeller is rotating, there will be a pressure difference between the polymer solution in the cylinder and those out of it, just like impeller pump do. The difference of pressure drives the polymer solution to rush through the holes in the cylinder. That will do strong shear and stretching action on the admixture liquid. The impeller can drive the polymer solution rotating at high speed. Inflated bubble may collapse in this process. The energy released by bubble collapse or explosion may disperse the nano-particles into polymer ulteriorly.

### 2.3 Experiment

Raw material : epoxy resin E-20, produced by XIN GCHEN chemical new material Co. Ltd. Intenerating point: 71°C, the epoxy value: 0.20ep/100g, hydrolyzed hydrogen: 0.03%, volatile: 0.3%. Solidifying reagent: T31. A. R. acetone, produced by Harbin XINCHUN Chemist Plant.

Nanomaterials: nanometer $SiO_2$ (the diameter is about 20 nm). The nanometer $SiO_2$ is shown as Fig. 6.

Fig. 6. Nano-$SiO_2$ TEM image

Dissolving 500g epoxy resin into 1500mL A. R. acetone, then we configure them into solution. The epoxy resin/acetone solution's density (denoted as ρ) is 0.93g/cm$^3$, moving viscosity at 20°C(denoted as v) is $7.75\times10^{-6}$m$^2$/s, and dynamical viscosity (denoted as μ) is $7.21\times10^{-3}$Pa · s, surface tension (denoted as σ) is $2.97\times10^{-2}$N/m. The contact angle of epoxy resin/acetone solution and $SiO_2$ at room temperature is measured by contact angle analyzer, which is less than 90°. The epoxy resin/acetone solution and $SiO_2$ is soakage. That's to say, the surface energy of $SiO_2$ and epoxy resin/acetone solution is almost at the same order. As the epoxy resin/acetone solution and $SiO_2$ is soakage, the $SiO_2$ particles will accrete on the bubble wall during the process of bubble expansion and raising. The energy released by the bubbles explosion will disperse the nanometer scale particles into liquid polymer.

The pressure of compressed gas is 0.4MPa, the flux of pressured air (denoted with Q) is 44 L/min. The diameter of lacunaris nozzle is 4mm. We spout 10g nanometer $SiO_2$ into the solution from the jet gun. The epoxy resin/acetone solution is milled at the speed of 3000 r/min. After spouting, we make it mill for 20min, then put away the admixture of epoxy resin and $SiO_2$ for 12h. Then, we take out 20mL of the admixture of epoxy resin and $SiO_2$. By adding 5mL T31 into the 20mL solution, it will solidify. When epoxy is solidified, we make the sample slice from it and observe it under TEM.

According to Eqs. (3), (5), (6), the minimum granule radius that can be separate under the bubble oscillation under the condition $P_g$=0.4052MPa, $P_\infty$=0.1013MPa, λ=0.9s, ρ=0.89g/cm$^3$, σ=$2.97\times10^{-2}$N/m is about 6 nm. Then, the $SiO_2$ may be dispersed into epoxy as the configuration of nanometer scale particles.

### 2.4 Results and discussion

Effective distribution and dispersion is necessary for the preparation of polymer/nano-particles composites. The diminutive bubbles explosion dispersion method is a dispersion mixing method and has significant effect on dispersion agglomerated nano-particles. A CM-12 transmission electron microscope (made by Philip Co. Ltd) was used to observe the E-20/ $SiO_2$ nanocomposites prepared by the proposed method. The TEM image was shown as Fig. 7. The TEM images show the nano-$SiO_2$ equably dispersed in polymer by the configuration of 15nm-30nm particles.

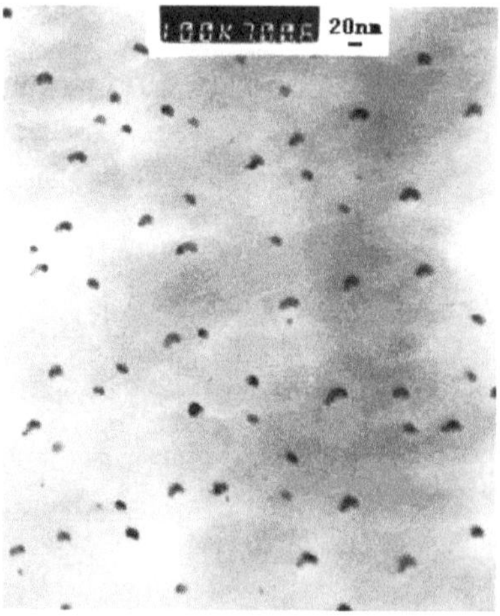


Fig. 7. E-20/ $SiO_2$ nanocomposites TEM image

Compared to traditional dispersion methods, the diminutive bubble explosion dispersion method is more effective in dispersion of nano-particles aggregates in polymer without dispersant because the particles aggregates serve as nucleation agents, and the bubbles expand on the surface of the aggregates. During the bubble inflation, the polymer surrounding the inflated bubble is subject to stretch in both longitudinal direction and latitudinal direction over the surface of inflated bubble. The bidirectional stretching increases the freedom degrees of nano-particles dispersion, which favors more homogeneous dispersion. The diminutive bubble explosion dispersion method may obtain ultrahigh stretching rates of around $(10^5\text{-}10^6)s^{-1}$, which is superior to the shearing dispersion effects obtained with screw extruders that have shearing rate about $(10^3\text{-}10^4)s^{-1}$.

According to the numerical simulation, the initial radii of diminutive bubbles have close relation to the stretch rate and bubble oscillation frequency in bubble inflation. The smaller initial radii of the bubble will lead to the higher st retch rate and oscillation frequency. High stretch rate and oscillation frequency favor more microcosmic homogeneous dispersion.

## 3. Nanocomposites capability evaluation method base on fractal theory and TEM image

Size distribution and space distribution of dispersed phase in nanocomposites is one of the key factors that effect material performance, and it's also an important indicator for evaluating the effect of nano-particles dispersion.

Size distribution of dispersed phase in nanocomposites means the range of particle diameter dimension. Size distribution of dispersed phase reflected the broken up effect of the agglomerated nanomaterials granules. The wider the range of size distribution of dispersed phase is, the worse the broken up effect of the agglomerated nanomaterials granules is. Wider range means that different degree of dispersed phase agglomeration phenomena exist. On the contrary, the narrower the range of size distribution of dispersed phase is, and the size accords with the demands of required nanometer scale, the better the broken up effect of the agglomerated nanomaterials granules is. Narrower range means that dispersed phase agglomeration is light.

Space distribution of dispersed phase in nanocomposites means the uniformity of nano-patricles dispersed into polymer and the uniformity of the number of nano-particles in unit volume polymer. Symmetrical dispersed phase can make the performance of nanocomposites good consistency. Contrarily, asymmetrical dispersed phase will result in stress concentration of nanocomposites and the performance of nanocomposites bad consistency.

### 3.1 Digital processing of nanocomposites TEM images

Because nanocomposites TEM images usually have the characteristic that their background brightness isn't homogeneous and gray scale difference between dispersed phase boundary and background is very small, we need to do some digital processing to TEM image for extracting the number of image elements in dispersed phase.

#### 3.1.1 Balancing partition gray scale

Divide TEM image into n×n grids. Then, calculate gray scale of every image element according to Eq. (7) for balancing gray scale of the whole image whose gray scale average value will turn into 128.

$$g_{n1}=g_a-(m_1-128) \tag{7}$$

where $g_{n1}$ is new gray scale of a image element. $g_a$ is original gray scale of that image element. $m_1$ is gray scale average value of the grid which the image element belong to.

#### 3.1.2 Adjusting brightness of every image element

Make every image element as a center. Choose a grid include k×k image elements. Then, calculate gray scale value of every image element according to Eq. (8) for adjusting brightness so that every image element in the whole image will has almost the same brightness.

$$g_{n2}=g_a-(m_2-128) \tag{8}$$

where $g_{n2}$ is new gray scale of a image element. $g_a$ is original gray scale of that image element. $m_2$ is gray scale average value of the grid whose center is the image element.

#### 3.1.3 Superimposing image

Superimpose the two images which received separately from the two steps aforementioned according to Eq. (9), for increaseing gray scale difference between dispersed phase and background.

$$g_{n3} = g_{n1} - (g_{n2} - 255) \tag{9}$$

where $g_{n3}$ is gray scale of a image element after superimposing. If $g_{n3}<0$, set $g_{n3}=0$.

### 3.1.4 Thresholding

After the processing aforementioned, transform the image into binary image through assigning gray value 0 to image elements that belong to dispersed phase and gray value 255 to other image elements.

## 3.2 Theory of sand box method and gyration radius method

We can measure the number of particles in a ball with its radius ε. If particles are distributed on a plane, the number M(ε) of particles on the cross section of a ball (radius ε) can be expressed as $M(\varepsilon) \propto \varepsilon^2$. If particles are distributed in space in a ball (radius ε), the number M(ε) of particles can be expressed as $M(\varepsilon) \propto \varepsilon^3$. In a similar way, if the distribution of particles has the characteristic of self-similarity and the fractal dimension is D, we can get the expression that

$$M(\varepsilon) \propto \varepsilon^{D} \tag{10}$$

where M(ε) is the number of dispersed phase particles in a box(radius ε). D is fractal dimension. Here, we finish the definition of sand box method. On the base of sand box method theory, we can continue to derive gyration radius method.

Define gyration radius $R_g$, it can be expressed as

$$R_g(Z) = \left(\frac{1}{Z}\sum_{i-1}^{Z} r_i^2\right)^{\frac{1}{2}} \tag{11}$$

where $r_i$ is the distance between particle i and the center of fractal mass. Z is the number of particles in the box whose center is O and radius is $\mathbf{R_g}$.

According to Eq. (10), if make $\mathbf{R_g}$ as the radius of a box, the number of particles in the box can be expressed as

$$Z \propto R_g^D \tag{12}$$

Eq. (12) can be transformed into another form that

$$R_g \propto Z^{\frac{1}{D}} \tag{13}$$

## 3.3 Size distribution of dispersed phase in nanocomposites

After transform TEM image into binary image, we need to calculate the number of image elements in every dispersed phase and make eight neighbourhoods mark to every dispersed phase in binary image. The number of image elements with the same value mark is defied as the area ($S_k$) of a corresponding dispersed phase (k). Then, the equivalent radius ($R_k$) can be expressed as

$$R_k = \sqrt{\frac{S_k}{\pi}} \tag{14}$$

Determine the centers of dispersed phases according to centroid method. Then, mark every center with the equivalent radius value of corresponding dispersed phase and other image elements with 0. Then, the TEM image turned into dot matrix image. The dot matrix image can be reduced to a row matrix, in which the first element is 0 and the second to the h+1 elements are the equivalent radius values arranged from small to small. h is the sizefraction of disperse phase equivalent radius. The first element is defined as the center of the row matrix.

The dispersed phases in nanocomposites can be regarded as fractals making up of dispersed objects and have the characteristic of self-similarity. According to Eq. (10), the size function of dispersed phase can be expressed as

$$Q(r) \propto r^{p} \tag{15}$$

where Q(r) is the number of dispersed phase in a circle whose radius value is r. p is the parameter of size distribution of dispersed phase.

Then, the number of elements ($N_1$) that the distances between the center of the row matrix and them are smaller than step-length (m) can be given by

$$N_1 = K_1 m^{p} \tag{16}$$

where $K_1$ is constant.

Set different step-length (m=1, 2, 3…), calculate $N_1$ according to Eq. (15), then we can get a series of datum ($\ln(m)$,$\ln(N_1)$), in which ln( ) is natural logarithm. Then, we can transform Eq. (16) to another form that

$$\ln(N_1) = p\ln(m) + \ln(K_1) \tag{17}$$

Obviously, we can draw a line with the points ($\ln(m)$,$\ln(N_1)$) in double-log coordinate. The linear slope is just the parameter p. Because $K_1$ is a constant, $\ln(K_1)$ can not affects slope p.

Fig. 8. Nano-composites $SiO_2$ /acrylic esters by shearin(TEM image(left), binary image (right))

Fig. 9. Nano-composites $SiO_2$ /acrylic esters by high speed ball milling(TEM image(left), binary image(right))

Fig. 10. E-20/$SiO_2$ nanocomposites by inflated bubble(TEM image(left), binary image(right))

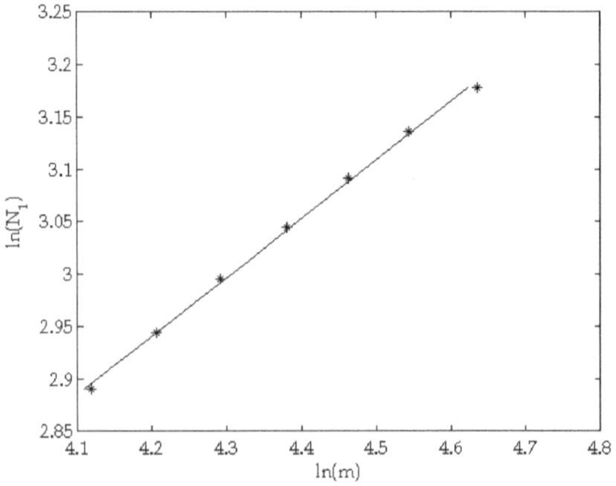


Fig. 11. Fitting curve of figure 8 with dimension $p_1$

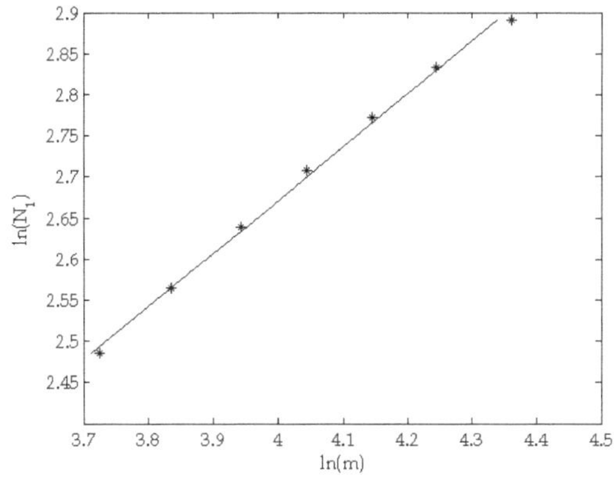


Fig. 12. Fitting curve of figure 9 with dimension $p_2$

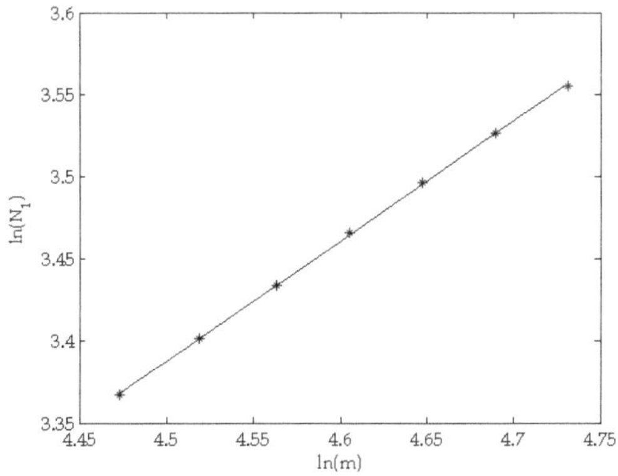


Fig. 13. Fitting curve of figure 10 with dimension $p_3$

Fig. 8, 9, 10 are TEM images and binary images of three kinds of composite. Fig. 11, 12, 13 are fitting curves correspondingly drawn according to Eq. (17). $p_1$, $p_2$, $p_3$ are their slopes as well as parameters of size distribution of dispersed phase. After calculating, we can know that $p_1$=1.806, $p_2$ =1.793, $p_3$=1.382 and $p_1$> $p_2$> $p_3$. According to Fig. 8, 9, 10, we can easily discover that their size distribution of dispersed phase status has relatively obvious difference. In Fig. 8, dispersed phase agglomeration phenomena is the most serious, and the range of size distribution is the widest. In Fig. 10, dispersed phase agglomeration phenomena is the slightest, and the range of size distribution is the narrowest. It can be shown that the smaller the value of p is, the narrower the range of size distribution is, and parameter p can be used to described the fractal characteristic of size distribution of dispersed phase.

### 3.4 Space distribution of dispersed phase in nanocomposites

After transform TEM image into binary image, determine the center of TEM image and the centers of dispersed phases according to centroid method. Then, mark every centers of dispersed phases with 1 and other image elements with 0. Then, the TEM image turned into dot matrix image.

According to Eq. (11), we can get the equation that

$$R_g = (\frac{1}{N_2}\sum_{i-1}^{N_2} r_i^2)^{\frac{1}{2}} \tag{18}$$

where $r_i$ is the distance between the two centers of dispersed phase i and TEM image. $N_2$ is the number of dispersed phases in the circle whose center is O and radius is $R_g$.
According to Eq. (13), the space distribution uniformity function of dispersed phase can be expressed as

$$R_g \propto N_2^{\frac{1}{q}} \tag{19}$$

where q is the parameter of space distribution uniformity of dispersed phase.
Set a constant $K_2$, we can get Eq. (20) that

$$R_g = K_2 N_2^{\frac{1}{q}} \tag{20}$$

Draw a circle with the center O and the radius $m_i$ (i=1, 2, 3,...), then calculate the number (N2) of dispersed phase and radius ri corresponding to every dispersed phase in the circle. Then, we can calculate Rg according to Eq. (18). Choose different radius mi, we can get a series of datum (ln(N2),ln(Rg)).
Eq. (20) can be transformed to another form that

$$\ln(R_g) = \frac{1}{q}\ln(N_2) + \ln(K_2) \tag{21}$$

Obviously, we can draw a line with the points ($\ln(N_2)$,$\ln(R_g)$) in double-log coordinate. The reciprocal of the linear slope is just the parameter q. Because K2 is a constant, ln(K2) can not affects slope parameter q.
Fig. 14, 15, 16 are fitting curves correspondingly drawn according to Eq. (17) corresponding to Fig. 8, 9, 10 respectively. q1, q2, q3 are the reciprocal values of the linear slopes as well as parameters of space distribution uniformity of dispersed phase respectively. After calculating, we can know that q1=1.3604, q2 =1.535, q3=1.991 and q1< q2< q3. According to Fig. 8, 9, 10, we can easily discover that their space distribution of dispersed phase status has relatively obvious difference. In Fig. 8, the distribution uniformity is the worst. In Fig. 10, the distribution uniformity is the best. Our research shown that the ideal fractal dimension is 2 for the fractal whose particle are dispersed on a plane. It can be shown that the closer the value of q is to 2, the better the the distribution uniformity is, and parameter q can be used to described the fractal characteristic of space distribution uniformity of dispersed phase.

## 4. Conclusion

The epoxy resin/$SiO_2$ nanocomposites preparation method based on diminutive bubbles explosion can evenly disperse $SiO_2$ particles into epoxy resin as the configuration of nanoscale particles. At the very start of bubble explosion , the bubble will expand and lead to high stretching rate in the polymer surrounding the bubble. The stretching rate may reach $10^6 s^{-1}$, which favors more homogeneous dispersion of nano-particles. Diminutive bubbles explosion method can be used for the dispersion of nanometer scale particles.

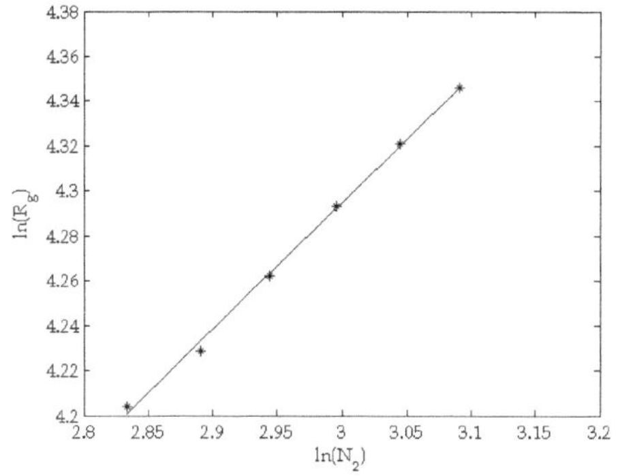


Fig. 14. Fitting curve of figure 8 with dimension $q_1$

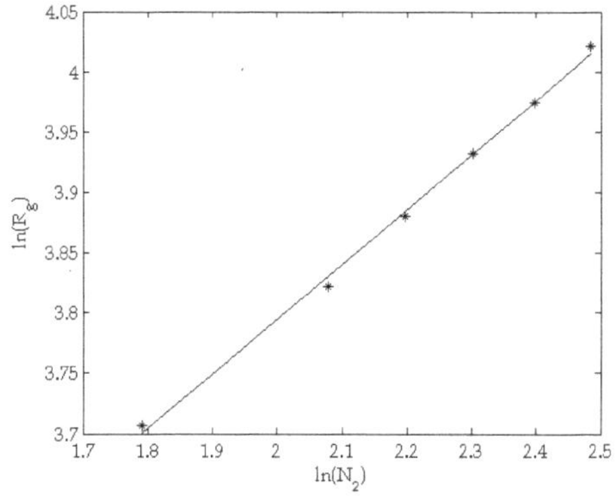


Fig. 15. Fitting curve of figure 9 with dimension $q_2$

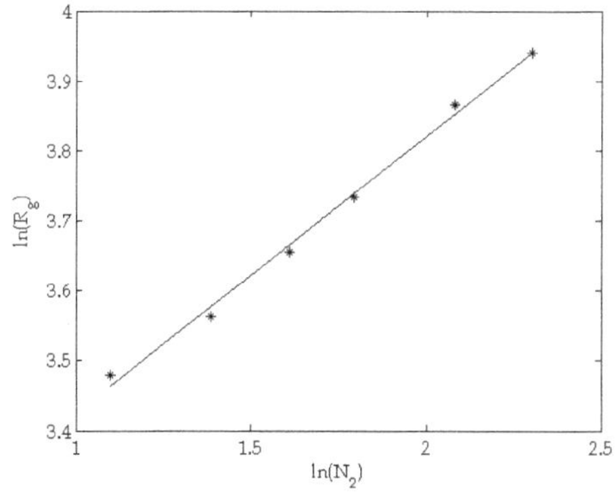


Fig. 16. Fitting curve of figure 10 with dimension $q_3$

A nanocomposites preparation system was designed. By using the preparation system, epoxy resin/$SiO_2$ nanocomposites were prepared. The $SiO_2$ dispersed into epoxy as the configuration of 15nm - 30nm particles.

A quantitative description method was proposed to evaluate size distribution and space distribution of dispersed phase in nanocomposites based on fractal theory and TEM image of nanocomposites. The models of size distribution and space distribution of dispersed phase were established using fractal dimension of natural fractal. After TEM image processing, parameters of size distribution and space distribution uniformity of dispersed phase were extracted. Then, lines were fitted according to linear equations correspondingly and it is obvious that the parameters and the values of linear slope have corresponding relations. Using the parameters to describe distribution of dispersed particles in the nanocomposites, the results show that it can exactly evaluate the size distribution and space distribution of nano-particles.

## 5. References

Cohen-Addad, J P. & Schmit, C. Nuclear magnetic resonance approach to fractal chain structure in molten polymers and gels: Characterization method of the spin-system response. *Polymer*, Vol 29, No. 5, (1988), pp. 883 - 893

Zhang, Shiwei. & Yang, Naiwei. Nanometer particles agglomeration in gas flows. *Vacuum Science and Tecnology*, Vol 21, No. 2, (2001),pp.87 - 90 (in Chinese)

Schmidt, D. Shah, D. & Giannelis, E P. New advances in polymer/layered silicate nanocomposites. *Current Opinion in Solid State & Materials Science*, Vol 6, No.3 (2002),pp. 205 - 212

Warrier, K G K. & Anilkumar, G M. Densification of mullite-SiC nanocomposite sol-gel precursors by pressureless sintering. *Materials Chemistry and Physics*, Vol 67, (2001) ,pp. 263 - 266

Jiao, Ningning. & Wang, Jianming. Advance in polymer nanocomposites. *Petrochemical Technologu & Ap2plication*, Vol 19, No. 2, (2001), pp. 57 - 61 (in Chinese)

Li, B., Xie, Y. & Huang, J., et al. Sonochemical synthesis of silver, copper and lead selenides. *Ultrasonics Sono-chemistry*, Vol 6, No. 4, (1999), pp. 217 - 220

Pramanik, M., Srivastava, S. & Samantaray, B K., et al. Rubber-clay nanocomposite by solution blending. *Journal of Applied Polymer Science*, Vol 87, No. 14, (2003), pp. 2216 - 2220

Lu, Yunfeng., Lu, Mengcheng. & Huang, Jinman., et al. Functional nanocomposites prepared by self-assembly and polymerization of diacetylene surfactants and silicic acid . *Journal of the American Chemical Society*, Vol 125, No. 5, (2003), pp. 1269 - 1277

Meng, Qingyun. & Wu, Daming. Oscillatory behavior gas bubbles in a melted viscoelastic polymer. *Journal of Beijing University of Chemical Technology*, Vol 30, No. 4, (2003), pp. 71 - 72 (in Chinese)

Hee, Ju Yoo. & Han, C D. Oscillatory behavior of a gas bubble growing in viscoelastic liquid. *AIChE*,Vol 28, No. 6, (1982), pp.1002 - 1008

Enomoto, H. & LernerM, M. Synthesis of polymer/1T-$TaS_2$ layered nanocomposites. *Materials Research Bulletin*, Vol 37, No. 8, (2002), pp. 1499 – 1507

Carotenuto, G. Nicolais, L. & Kuang, X. Preparation of PMMA-$SiO_2$ nanocompositeswith high homogeneity. *Applied Composite Materials*, Vol 2, No. 6, (1995), pp. 385 – 393

Akelah, A., Salahuddin, N. & Hiltner, A., et al. Morphological hierarchy of butadienea crylonitrile/montmorillonite nanocomposite. *NanostructuredMaterials*, Vol 4, No. 8, (1994), pp. 965 – 978

Du, T. & Ilegbusi, O J. Synthesis and morph-ologyical characterization on PVP/ZnO nano hybridfilms. *Journal ofMaterials Science*, Vol 39, No. 19, (2004), pp. 6105 – 6109

Xu, Tao., Lei, Hua. & Yu, Jie., et al. The particle fractal distribution in pp/talc materials and its relation with impact properties. *PolymerMaterials Science & Engineering*, Vol 18, No.1, (2002), pp. 135 -139 (in Chinese)

# 13

# Nanofabrication of Particle Assemblies and Colloidal Crystal Patterns

Yoshitake Masuda
*National Institute of Advanced Industrial Science and Technology (AIST), Anagahora, Shimoshidami, Moriyama-ku, Nagoya, Japan*

## 1. Introduction

Nano/micro periodic structures have attracted much attention as next-generation devices[1,2] such as photonic crystals[3-5] in which the refractive index changes periodically to show a photonic band gap. Various scientific and engineering applications, such as control of spontaneous emission, zero-threshold lasing, sharp bending of light, and so on, are expected to become possible by using the photonic band gap and the artificially introduced defect states and/or light-emitters. The structures were prepared by semiconductor nanofabrication techniques such as lithography and etching processes[6,7], advanced wafer-fusion technique[8], lithographic layer-by-layer approach[9], holographic lithography[10], advanced silicon microelectromechanical systems[11], glancing angle deposition[12] or auto cloning technique[13], and theoretical studies were performed to estimate the properties of the structures. These studies confirmed the high potential of nano/micro periodic structures as future devices.

However, a simple process which requires a short time for fabrication, low energy and less amount of material needs to be developed to enable mass production. Additionally, the processes of patterning the structures need to integrate various elements for application to commercial devices. The regularity and feature edge acuity of periodic structures should also be improved in order to enhance the performance.

Nano/micro periodic structures can be prepared with short fabrication time and low energy by self-assembly of mono-dispersed particles in which particles and air (wall and air for inverse opal) are arranged periodically[14,15]. Self-assembly and patterning of nano/micro particles have attracted much attention recently[1,2,16,17]. Micropatterns of close-packed particle assemblies with high arrangement accuracy have been realized by using templates such as micromolds[18], grooves[19], cylindrical holes[20] or trenches[20]. However, substrates having micromolds or grooves are necessary in these processes, and feature edge acuity and regularity need to be improved further in order to fabricate various complicated structures for photonic devices.

In this section, nano/micropatterns of colloidal crystals were fabricated self-assembly. They were realized with several patterning methods in static solution systems[21-23] or in drying processes[24,25] without using molds or grooves. Microstructures constructed from particles such as micropatterns of particle layers, narrow particle wires, arrays of particle wires and so on were prepared under moderate conditions using self-assembled monolayers (SAMs).

Additionally, 2D patterning of colloidal crystals were realized with two solution method. The method has the advantage of both the static solution system and drying process. It offers excellent self-assembling performance for the fabrication of nano/micro periodic structures to be applied to next-generation devices.

## 2. Patterning of colloidal crystals in liquids[21-23]

### 2.1 Patterning of colloidal crystals in liquids

Novel processes to realize low-dimensional arrangement of $SiO_2$ particles were proposed[21-23]. Particle wires and a pattern of the close-packed particle monolayer were fabricated in the solution at room temperature. SAMs (self-assembled monolayers) were formed on Si substrates and modified to be suitable for templates of precise arrangement. Particles were arranged precisely in the desired positions in the solution using well controlled electrostatic interactions and chemical bond formation between particles and substrates.

### 2.2 SAM Preparation for patterning of coloidal crystals

Octadecyltrichlorosilane (OTS)-SAM was prepared by immersing the Si substrate in an anhydrous toluene solution containing 1 vol% OTS for 5 min under a $N_2$ atmosphere. SAMs were exposed for 2 h to UV light (184.9 nm) through a photomask. The UV-irradiated regions became hydrophilic due to Si-OH group formation, while the non-irradiated part remained unchanged, i.e., it was composed of hydrophobic octadecyl groups, which gave rise to patterned OTS-SAM. To check successful film formation and functional group change, water drop contact angles were measured for irradiated and non-irradiated surfaces. Initially deposited OTS-SAM had a water contact angle of 96°, while the UV-irradiated SAM surface was saturated (contact angle $< 5°$).

The patterned OTS-SAM was immersed in a toluene solution containing 1 vol% APTS (aminopropyltriethoxysilane) in air for 1 h. APTS molecules combined to silanol groups of SAM and hence, octadecyl / amino groups pattered SAM was fabricated. OTS-SAM exhibited a water contact angle of 96°, while that of the amino surface was 28°. These observations indicated successful fabrication of octadecyl / amino-groups patterned SAM.

### 2.3 Surface modification of $SiO_2$ particles

Silica particles (1μmφ, HIPRESICA UF, UNK, Ltd.) were immersed in a dicyclohexyl and sonicated for 10 min under a $N_2$ atmosphere for good dispersion. 1 vol% of trichlorocyanoethylsilane (TCES) was added to the dicyclohexyl solution under a $N_2$ atmosphere, and the solution was stirred gently for 30 min in order to chemisorb TCES onto the $SiO_2$ particle surfaces. $SiO_2$ particles with TCES were centrifuged several times to remove unreacted TCES using dicyclohexyl.

The $SiO_2$ particles with TCES were further dispersed in a tetrahydrofuran solution containing potassium tert-butoxide (t-BuOK) and 18-crown 6-ether for 48 h under an ambient atmosphere to oxidize the CN-groups to carboxyl groups. The solution was centrifuged several times using distilled water to remove t-BuOK, 18-crown 6-ether, and a tetrahydrofuran. $SiO_2$ particles modified with carboxyl groups were thus obtained.

$SiO_2$ particles covered by silanol groups or carboxyl groups were arranged selectively in silanol regions or amino regions of SAM using interactions between particles and SAMs. Zeta potentials of $SiO_2$ particles that have silanol groups and $SiO_2$ particles modified by carboxyl groups were measured (Zetasizer 3000HSA, Malvern Instruments Ltd.) as shown in Fig. 1. Zeta potentials measured in aqueous solutions (pH=7.0) for the surface of silicon substrate covered with silanol groups, phenyl groups (PTCS) and amino groups (APTS) are -38.23 mV, +0.63 mV and +22.0 mV, respectively.

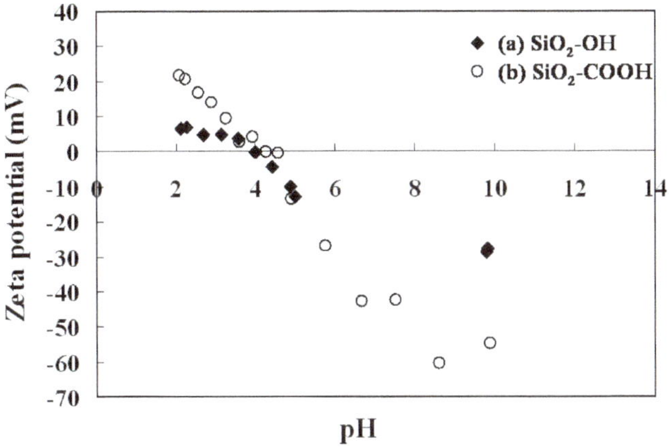




Fig. 1. Zeta potential of (a) $SiO_2$ particles and (b) $SiO_2$ particles modified with carboxyl groups.

## 2.4 Fabrication of particle wires employing selective arrangement process

OTS-SAM was modified by a diamond tip to form a line of silanol groups of approximately 0.5 µm width (Fig. 2). The diamond tip was contacted to OTS-SAM surface lightly and traced with low contact pressure in order to modify the SAM surface. The surface modified by a diamond tip, i.e., the white area in Figure 2 corresponds to silanol groups, showed low contact angle (<5°). This modified region was shown to be white compared with OTS-SAM region in a scanning electron micrograph (SEM; S-3000N, Hitachi, Ltd.). Octadecyl groups were broken mechanically by contact pressure with the diamond tip, and they possibly changed into silanol groups. The diamond tip was used to avoid contamination from a metal tip and the influence of a chemical reaction between the tip and the SAM.

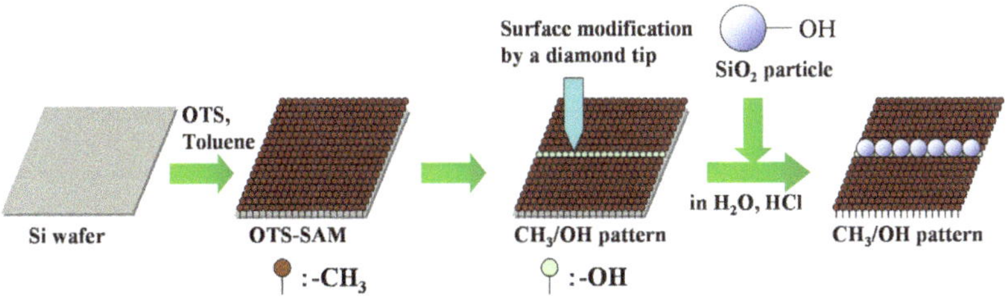




Fig. 2. Conceptual process for fabrication of a particle wire on a patterned SAM modified by a diamond tip.

Patterned SAMs were immersed in the aqueous solution containing $SiO_2$ particles and a hydrochloric acid as a condenser, rinsed in water, and were observed by a SEM. $SiO_2$ particles were observed on lines of silanol groups selectively indicating particles were successfully arranged well (Fig. 3(a)). Because particles were not easily removed by sonication, it was judged that iloxane bonds had been formed by condensation of silanol groups between particles and a SAM.

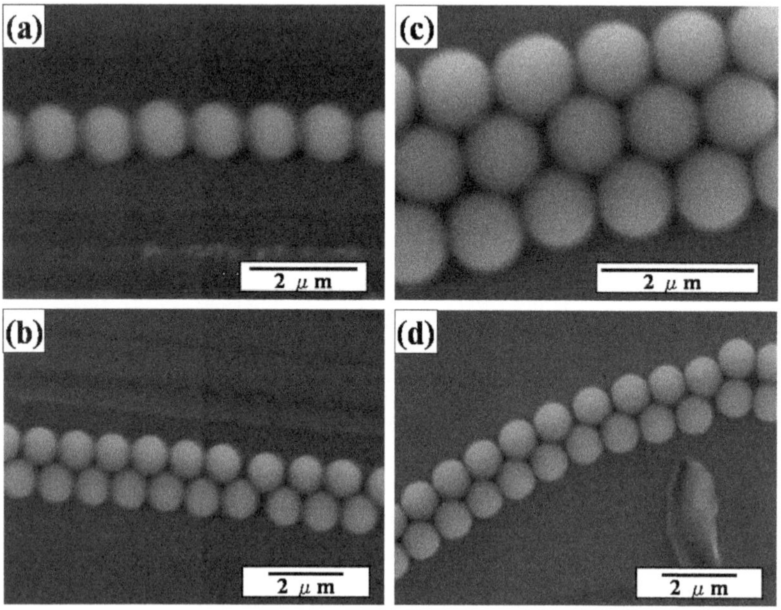


Reprinted with permission from Ref. [23], Masuda, Y., Itoh, M., Yonezawa, T. and Koumoto, K., 2002, *Langmuir*, 18, 4155. Copyright @ American Chemical Society

Fig. 3. SEM micrographs of (a) a single particle wire, (b) a double particle wire, (c) a triple particle wire and (d) a curved double particle wire

It is clearly seen that the accuracy of particle arrangement has been improved compared to our former experiments[21,22]. A double particle wire and a triple particle wire were likewise fabricated on wide silanol groups regions with about 1.4 μm and 2.2 μm in width, respectively (Fig. 3(b), (c)). The double particle wire that has a triangular lattice also demonstrates a high arrangement accuracy, though there is a defect in arrangement between seventh particle from left and eighth particle. Additionally, a curved double particle wire was fabricated on curved region of silanol groups. Curved double particle wires have not been reported previously, and they may have useful applications for an optical waveguide. Accuracy of particle arrangement was evaluated from Fig. 3(a). Center position ( $(x_i, y_i)$ μm ) of each particle was plotted to estimate the standard deviation (Fig. 4). The bottom left corner of Fig. 3(a) was set to be the origin of the x-y coordinate. The approximated straight line( $f(x)$ ) and its slope ( $\theta$ ) are represented as follows.

$$f(x) = 0.0061x + 2.9609 \tag{1}$$

$$\cos\theta = 0.9940 \tag{2}$$

Standard deviation from the approximated straight line is described by the expression,

$$\text{S (standard deviation)} = \frac{\left[\sum_i \{\cos\theta \cdot (f(x_i) - y_i)\}^2\right]^{\frac{1}{2}}}{n-1} \tag{3}$$

where $n$ is the number of particles ( $n = 6$ ). S = 0.0126 was obtained. The accuracy of particle arrangement in Fig. 3(b) and (c) was estimated by the same manner. Standard deviation of seven particles from left in an upper particle line and an bottom particle line in Fig. 3(b) were estimated to be S = $5.66 \times 10^{-3}$ and S = $3.84 \times 10^{-3}$. And standard deviation of an upper particle line, a middle particle line and a bottom particle line in Fig. 3(c) were estimated to be S = $8.11 \times 10^{-4}$, $8.27 \times 10^{-3}$ and $2.30 \times 10^{-2}$, respectively.

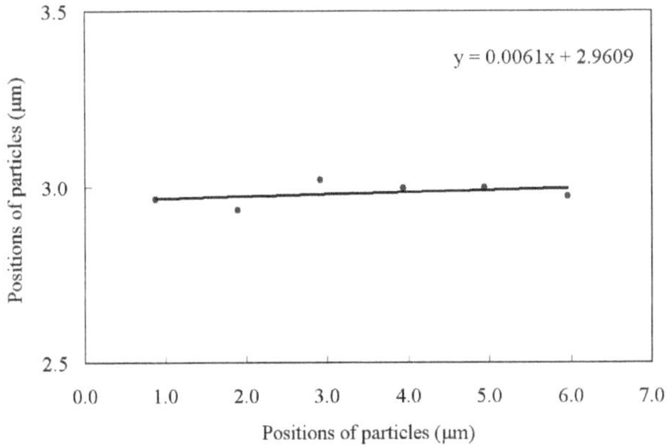


Reprinted with permission from Ref. [23], Masuda, Y., Itoh, M., Yonezawa, T. and Koumoto, K., 2002, *Langmuir*, 18, 4155. Copyright @ American Chemical Society

Fig. 4. Positions of particles in Fig. 3 (a) showing the accuracy of particle arrangement.

### 2.5 Precise arrangement of particles on small-area silanol sites modified by AFM lithograph

OTS-SAM was modified to silanol groups by an AFM (atomic force microscope, Nanoscope E, Digital Instruments) to control position of arrangement accurately (Fig. 5). A source measure unit (SMU Model 236, Keithley) was installed in the AFM in order to control the electric current passing through the probe and a SAM. The SAM was biased positively, and the AFM probe was scanned with constant current mode (50 nA), and the scanned area was used as a template for arrangement. Scanning area (100 nm×100 nm) was set smaller than the diameter of the particles (500 nmφ) to facilitate precise arrangement ofparticles.

$SiO_2$ particles (500 nmφ, powder, Admatechs Co., Ltd., SO-E2) modified with carboxyl groups were sonicated for 10 min in tetrahydrofuran or dichloromethane, and this solution was refrigerated to -20 °C for 1 h. N,N'-dicyclohexylcarbodiimide was added to this solution as a condenser to form ester bonds between carboxyl groups of $SiO_2$ particles and silanol

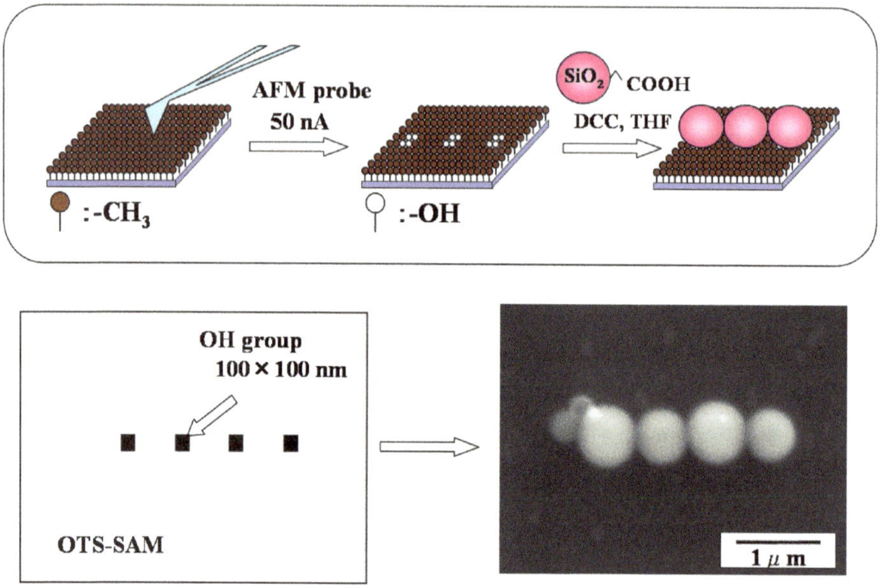


Reprinted with permission from Ref. [23], Masuda, Y., Itoh, M., Yonezawa, T. and Koumoto, K., 2002, *Langmuir*, 18, 4155. Copyright @ American Chemical Society

Fig. 5. Conceptual process and SEM micrograph of particle arrangement on a patterned SAM modified by AFM lithography.

groups of a SAM. Modified OTS-SAM was then immersed in this solution for 2 h. The temperature of the solution was increased slowly to 25 ℃ and kept for 2 h. After having been rinsed in water, a SAM was observed by a SEM. Particles were arranged in silanol regions and line of particle was fabricated (Fig. 5). Two-dimensional arrangement with required features can easily be realized with this technique, though it takes a long time to modify a SAM with an AFM probe. Particles weren't removed easily from a SAM by sonication, indicating that ester bonds were formed by condensation. The accuracy of particle arrangement in Fig. 5 was estimated to be $S = 1.17 \times 10^{-2}$. This might be decreased by decreasing the dimension of each silanol region.

In order to verify the formation of ester bonds between carboxyl groups and silanol groups, bromopropionic acid, whose molecule has a carboxyl group at one end and a bromo group at the other, was reacted with silanol groups of a Si substrate using the same reaction scheme as used to attach $SiO_2$ particles to silanol groups. After having been sonicated in acetone for 5 min, the substrate surface was analyzed by X-ray photoelectron spectroscopy (XPS; ESCALAB 210, VG Scientific Ltd., $1\text{-}3 \times 10^{-7}$ Pa, measurement area; 3 mmx4 mm). The X-ray source (MgK*a*, 1253.6 eV) was operated at 15 kV and 18 mA. The spectrum corresponding to Br 3d binding energy centering at 74.35 eV was observed. Although the observed binding energy is higher than that of KBr, this chemical shift must have been caused by carbon atoms neighboring bromine atoms. Since bromo groups can't react directly with silanol groups under the present conditions, the XPS result firmly indicates that carboxyl groups of bromopropionic acid reacted with silanol groups to form possible ester bonds.

## 2.6 Patterning of close-packed particle monolayers

$SiO_2$ particles modified with carboxyl groups were dispersed in water. Octadecyl / amino-groups patterned SAM was then immersed in the solution for several minutes (Fig. 6). The substrate was rinsed with water and observed with a SEM. $SiO_2$ particles were observed in silanol regions selectively forming a close-packed mono-particle layer (Fig. 7 (a)). Boundaries between the mono-particle layer and otcadecyl region is clearly observable, and a few particles are observed in octadecyl region. $SiO_2$ particles modified with carboxyl groups are charged negative, and amino groups of SAM are charged positive in water. Accordingly, particles are attracted to amino groups and form a mono-particle layer. Particles in the solution did not adhere to the mono-particle layer, since both the particles and the mono-particle layer have negative charges and repel each other. Particles were also deposited randomly in some areas (Fig. 7 (b)), and this suggests that it is difficult to obtain the pattern of the close-packed particle monolayer in a large area.

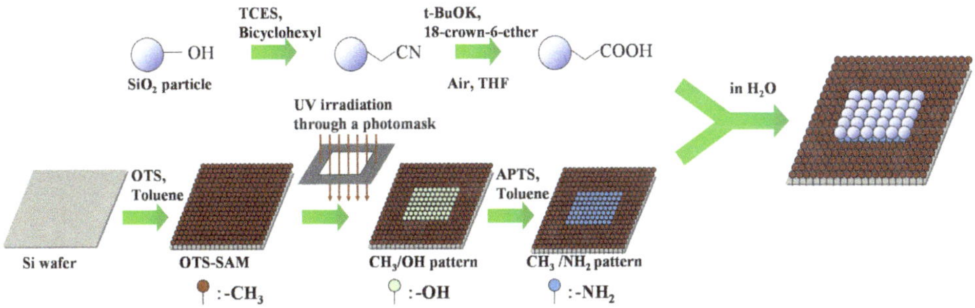




Fig. 6. Conceptual process for patterning of the close-packed particle monolayer.

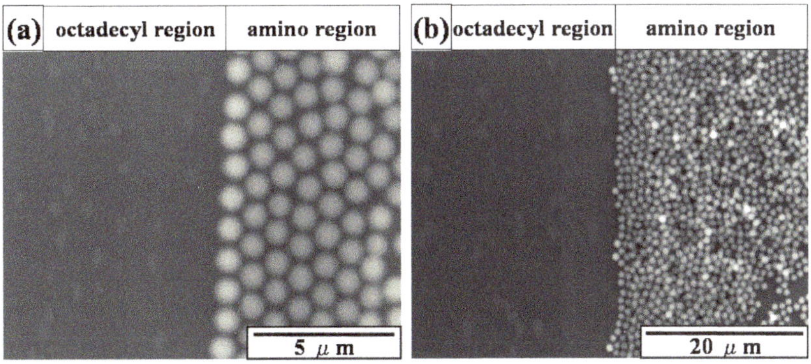




Fig. 7. SEM micrographs of (a) a close-packed particle monolayer of $SiO_2$ particles modified with carboxyl groups formed in the amino region and (b) randomly deposited particles in the amino region.

Additionally, octadecyl / amino-groups patterned SAM was immersed in the solution containing non-modified $SiO_2$ particles for several minutes (Fig. 8). While $SiO_2$ particles were observed in silanol region predominantly, the feature edge acuity of the pattern was lower than that of the pattern in which $SiO_2$ particles modified with carboxyl groups was used. This demonstrates applicability of surface modification of $SiO_2$ particles with carboxyl groups. Furthermore, octadecyl / silanol-groups patterned SAMs were immersed in solutions containing $SiO_2$ particles modified with carboxyl groups or non-modified $SiO_2$ particles for several minutes, respectively. Particles were not adhered to either octadecyl groups or silanol groups. This means that the difference in surface potential between $SiO_2$ particles modified with carboxyl groups and amino groups of SAM accelerate the adhesion of particles to the amino groups. Particles were attracted and adhered to amino groups predominantly by electrostatic interactions between particles and SAMs and chemical bonds weren't formed because pure water was used as a solution with no condensation agent. Surface of $SiO_2$ particles modified with carboxyl groups and amino SAM must have changed into $-COO^-$ and $-NH_3^+$, respectively, in water to attract each other.

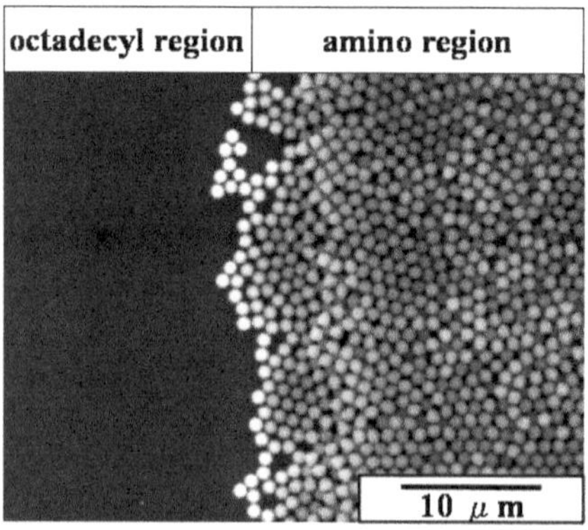


Reprinted with permission from Ref. [23], Masuda, Y., Itoh, M., Yonezawa, T. and Koumoto, K., 2002, *Langmuir*, 18, 4155. Copyright @ American Chemical Society

Fig. 8. SEM micrograph of randomly deposited $SiO_2$ particles in the amino region.

## 3. Drying patterning of colloidal crystals and 2D arrays

### 3.1 Self-assembly patterning of colloidal crystals by drying patterning[24]

Particle wires were fabricated through self-assembly on hydrophilic regions of SAMs (self-assembled monolayer)[24]. An SAM of octadecyltrichlorosilane was formed on a silicon substrate and modified by UV irradiation to create a pattern of hydrophobic octadecyl and hydrophilic silanol groups. Ethanol or water containing particles (550 nmφ or 800 nmφ) was dropped onto a patterned SAM. The solution was separated into two droplets with a liquid bridge between the droplets along the hydrophilic regions of a patterned SAM. The droplets and the liquid bridge were used as a mold for fabrication of a two-dimensional pattern of colloid crystals. Particle wire was formed between two droplets and colloid crystals such as an opal structure were formed at both ends of the particle wire after drying the solution. The particle wires constructed from a close-packed structure or non-close-packed structure, i.e. square lattice, were fabricated through self-assembly at room temperature using this method.

The UV-irradiated regions became hydrophilic due to the formation of Si-OH groups, while the non-irradiated part remained unchanged, i.e. it was composed of hydrophobic octadecyl groups, which gave rise to patterned OTS-SAM. To confirm successful film formation and functional group change, water drop contact angles were measured for irradiated and non-irradiated surfaces. Initially deposited OTS-SAM had a water contact angle of 96°, while the UV-irradiated SAM surface was saturated (contact angle $<5°$). This observation indicated successful fabrication of SAM patterned with octadecyl/silanol groups (Fig. 9). Polystyrene particles in water (150 μl) (550 nmϕ particle or 820 nmϕ carboxylated particle, 10% wt, dispersed in water, Seradyn Co., Ltd.) were further dispersed in ethanol (3 ml) or water (3 ml), and poured onto a patterned OTS-SAM. The contact angles of the ethanol solution or water solution measured 10-20° or 96° on the OTS-SAM, respectively, while they were saturated (contact angle <5°) on silanol groups. The droplets were observed to separate into two drops and a bridge of solution was formed on a silanol line. The droplets and the liquid bridge were used as a mold for fabrication of a two-dimensional pattern of colloid crystals. After evaporation of the solution, substrates were observed by a scanning electron microscope (SEM; S-3000N, Hitachi, Ltd.).

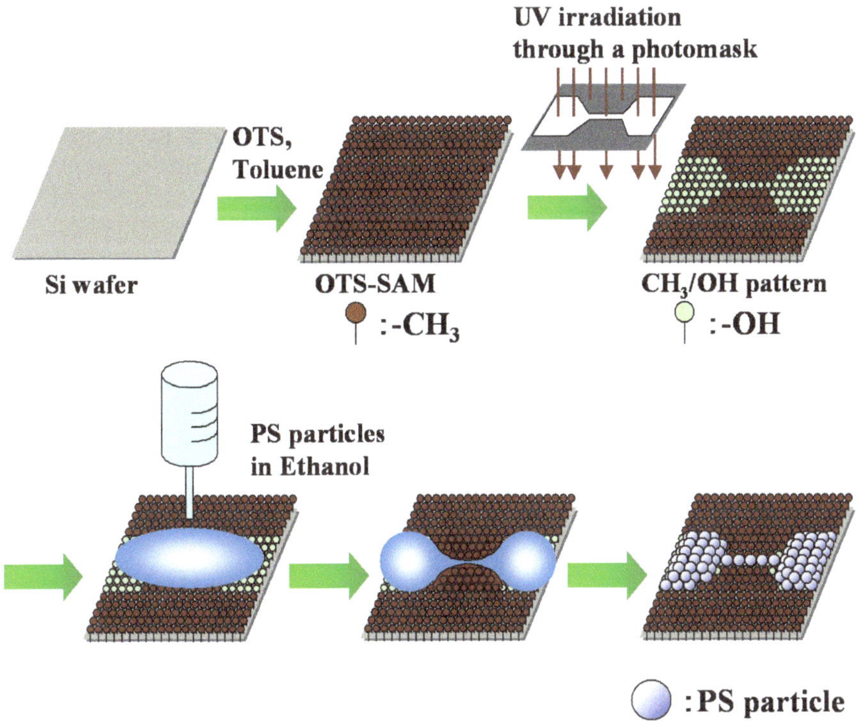


Reprinted with permission from Ref. [24], Masuda, Y., Tomimoto, K. and Koumoto, K., 2003, *Langmuir*, 19, 5179. Copyright @ American Chemical Society

Fig. 9. Conceptual process for fabrication of a particle wire using a patterned SAM and a liquid bridge.

The dispersibility of particles in the solution is very important for particle assembly and high dispersibility is necessary to form a close-packed structure. The zeta potentials of particles dispersed in the solutions were examined by direct measurement of electrophoretic mobility using an electrophoretic light scattering spectrometer (Zetasizer 3000HS, Malvern Instruments Co., Ltd.). The zeta potentials of polystyrene particles (550 nmϕ) in water, carboxylated particles (820 nmϕ) in water, polystyrene particles in ethanol and carboxylated particles in ethanol were determined to be -38.3 mV, -50.2 mV, -53.9 mV and -44.0 mV, respectively. Surface modification by carboxyl groups decreased the negative zeta potential in both solutions. Furthermore, particles in the ethanol had slightly low negative zeta potentials compared to those in water, i.e. the particles were slightly well-dispersed compared to those in water.

Polystyrene particles in water were poured onto a patterned OTS-SAM (Fig. 9), and observed after evaporation of the water. The water at the liquid bridge evaporated in about 24 h to form a particle wire, and droplets at the two ends completely evaporated in about 48 h. In this manner, particle wires constructed from a close-packed structure, i.e. triangular lattice, were produced from the water solution (Fig. 10a). The middle of a particle wire was narrower than its end (Fig. 10b, c). The width of the particle wire does not depend on the width of the silanol line, but rather on the interfacial tensions between solution and substrate, solution and atmosphere, and atmosphere and substrate. The silanol line was not used to decide the width of the particle wire, but rather the position of the liquid bridge and particle wire. Close-packed structures were also formed on large silanol regions (Fig. 10d, e). The right-hand area of Fig. 10d can be regarded as the {100} plane of the fcc structure and the left-hand area can be regarded as the {111} plane of the fcc structure or the {0001} plane of the hcp structure. The close-packed structure was thus considered to be an fcc structure.

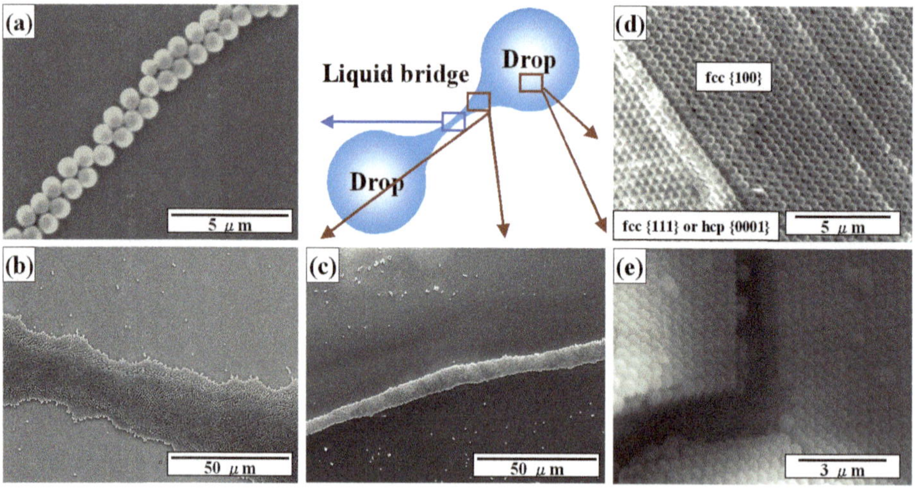




Fig. 10. SEM micrographs of particle structures fabricated from aqueous solution containing micro particles using a liquid bridge. (a) - (c) particle wires constructed from triangular lattice (close-packed structure) and (d) - (e) a close-packed 3D structure.

Figure 11 shows particle wires and 3D structures fabricated from ethanol solution containing polystyrene particles. The ethanol at the liquid bridge evaporated in about 1 min to form a particle wire, and droplets at the two ends evaporated in about 20 min. The liquid bridge of ethanol evaporated faster than that of water for several reasons. The saturated vapor pressure of ethanol (59 mmHg (0.078 atm) at 25°C) is higher than that of water (24 mmHg (0.031 atm) at 25°C), explaining the difference in the evaporation rate of the two droplets. The ratio in evaporation rate of the ethanol liquid bridge to the ethanol droplets is higher than that of the water liquid bridge to the water droplets. This can be explained as follows: Water has high surface tension ($71.8\times10^{-3}$ N/m at 25°C) compared with ethanol ($22.0\times10^{-3}$ N/m at 25°C). Ethanol existed along patterned hydrophilic regions with small meniscus at the angle between droplets and a liquid bridge. However, water formed large meniscus at the angle between droplets and a liquid bridge, causing a wide line width of water liquid bridge compared with ethanol on our patterned surfaces. Additionally, the water liquid bridge was higher than that of ethanol due to high surface tension. These made the cross-section area of water larger than that of ethanol. The thick liquid bridge evaporated slowly because of its large volume and low vapor pressure calculated from the Kelvin equation in which the smaller convex liquid surface gives rise to higher internal pressure and faster evaporation rate. Furthermore, the solution at the droplets flowed into a liquid bridge and this further complicated the evaporation mechanism.

Fig. 11. SEM micrographs of particle structures fabricated from ethanol solution containing micro particles using a liquid bridge. (a), (b) particle wires constructed from square lattice, (c) particles deposited on edge of silanol line and (d), (e) close-packed 3D structures.

A narrow particle wire was formed at the center of the liquid bridge (Fig. 11a, b), and wide wires were formed at the edges of the liquid bridge (Fig.11c) along the silanol line. The particle wire in Fig. 11a is not a close-packed structure and is constructed from a square lattice, which is a metastable phase compared with a close-packed structure. High dispersibility of particles is necessary to form a close-packed structure in the solution.

However, the particles dispersed well in ethanol and the dispersibility of particles in ethanol is similar to that in water. This shows that the non-close-packed structure was caused not only by the influence of dispersibility but also by many other factors. Movement and rearrangement of deposited particles is necessary to construct a close-packed structure. However, the ethanol evaporated quickly and suppressed the movement of particles by liquid bridge force. Additionally, adhesion between particles and a substrate, and cohesion between particles probably caused moderate suppression of the rearrangement of particles. Factors such as evaporation rate, interaction force between particles, and interaction force between particles and a substrate were important in the packing process. Close-packed 3D structures were also formed on large silanol regions (Fig. 11d, e), and they contained many defects (Fig. 11e). The ethanol evaporated so quickly that the particles did not rearrange well to form a close-packed structure during evaporation of ethanol. This is one of the factors of forming a loosely packed structure. To directly evaluate the effect of the evaporation rate, a similar experiment using ethanol was conducted in a small airtight container with small pinholes to allow the ethanol to evaporate slowly. The size and number of pin holes were adjusted for ethanol at the liquid bridge to evaporate in about 24 h. The particle wire constructed from a close-packed structure was formed after about 24 h, and droplets at both ends were dried after about 48 h. The close-packed 3D structures were formed in large silanol regions. The number of defects was smaller than that formed from the ethanol solution with a shorter time and was similar to that formed from water. The humidity in the container was close to 100% and the saturated vapor pressure of ethanol was 59 mmHg (0.078 atm) at 25°C. This showed that particles dispersed well in ethanol and the interaction between particles and a substrate was sufficiently weak to produce a close-packed structure in the drying process of 24 h. Additionally, the results showed that not only the interaction force between particles and that between particles and a substrate, but also the evaporation rate needs to be controlled to fabricate particle wires.

Accuracy of particle arrangement was evaluated from Fig. 11(a) as calculated in recent work[23]. The center position (($x_i$, $y_i$) μm) of each particle in an upper layer was plotted to estimate the standard deviation. The bottom left corner of Fig. 11(a) was set to be the origin of the x-y coordinate.

The approximated straight line ($f(x)$) and its slope (θ) are represented as follows.

$$f(x) = 1.3965x - 5.3344 \tag{4}$$

$$\cos\theta = 0.5822 \tag{5}$$

Standard deviation from the approximated straight line is described by the expression,

$$\text{S (standard deviation)} = \frac{\left[\sum_i \{\cos\theta \cdot (f(x_i) - y_i)\}^2\right]^{1/2}}{n-1} \tag{6}$$

where $n$ is the number of particles ($n = 19$). The accuracy of the particle arrangement in Fig. 11(a) was estimated to be S = $1.63\times10^{-3}$. This is lower than that of the particle arrangement obtained in our previous work[23].

Particle wires were fabricated on hydrophilic regions of a patterned SAM. Ethanol or water containing particles was separated into two droplets with a liquid bridge between the droplets along hydrophilic regions of a patterned SAM. Particle wires constructed from a close-packed structure or non-close-packed structure were then formed through self-assembly between two droplets after drying of the solution.

### 3.2 Self-assembly 2DArray of colloidal crystal wires by drying patterning[25,26]

An orderly array of particle wires constructed from a close-packed colloidal crystal were fabricated without preparation of patterned templates[25,26]. A substrate was immersed vertically into a $SiO_2$ colloidal solution, and the liquid surface moved downward upon evaporation of solution. Particles formed a mono/multi-particle layer, which was cut by the periodic drop-off of solution. The orderly array of particle wires was successfully fabricated, showing the suitability of the self-assembly process for the fabrication of nano/micro structures constructed from nano/micro particles or blocks. The mechanism of the assembly process and control of thickness, width and interval of particle wires were further discussed. Moreover, an array of particle wires constructed not from close-packed fcc (or hcp) structure but from two kinds of particles was realized to fabricate an array of particle wires with NaCl structure by this self-assembly process.

Octadecyltrichlorosilane (OTS)-SAM was prepared by immersing Si substrate in an anhydrous toluene solution containing 1 %vol OTS for 5 min under an $N_2$ atmosphere. The contact angles of the ethanol solution or water solution measured 10-20° or 96° on the OTS-SAM, respectively. The water solution showed a contact angle of 10-20° on a silicon wafer that was kept in air.

The OTS-SAM was immersed in ethanol solution (80 ml) containing $SiO_2$ particles (1000 nm$\varphi$, 10 mg). The bottom of the solution was heated at 70°C and a condenser tube was kept at the top of the solution to cool it. The temperature difference between the top and bottom of the solution was controlled so as to stir and move particles by convection. The surface of the solution moved on the OTS-SAM surface upon evaporation of the ethanol. Particles began to assemble at the surface of the solution (Fig. 12(a)) and the particle layer was fabricated by movement of the solution surface (Fig. 12(b)). Further evaporation of the solution caused separation of the particle layer and solution surface (Fig. 12(c)) because particles were not sufficiently supplied from the solution. The liquid surface then dropped off and the particle layer separated from the solution surface (Fig. 12(d)). The next particle layer was formed by the same procedure (Fig. 12(e)). Consequently, separated particle wires, i.e. an array of particle wires, were successfully fabricated by our newly developed method (Fig. 12(f), 12(a-d)).

After having been immersed in the solution, which evaporates quickly, the substrates were observed using a scanning electron microscope (SEM; S-3000N, Hitachi, Ltd.), an optical microscope (BX51WI Microscope, Olympus Optical Co., Ltd.) with a digital camera (DP50, 5.8 megapixels, Olympus Optical Co., Ltd.) and a computer for capturing data, and a digital video camera recorder (DCR-TRV 50, Sony Corporation) with optical magnifying glass.

The width and interval of particle wires fabricated from the ethanol solution (80 ml) containing $SiO_2$ particles (10 mg) at 70°C were shown to be about 150 $\mu$m and 200 $\mu$m, respectively (Fig. 13(a-d)). Particle wires were constructed from a close-packed particle structure and their upper side showed high feature edge acuity (Fig. 13(c)). The array of particles finished suddenly as shown on the bottom side of the particle wires (Fig. 13(c-d)). These observations suggest that particle wires were formed from the upper side and were cut by the drop-off of solution, and are consistent with the procedure shown in Fig. 12.

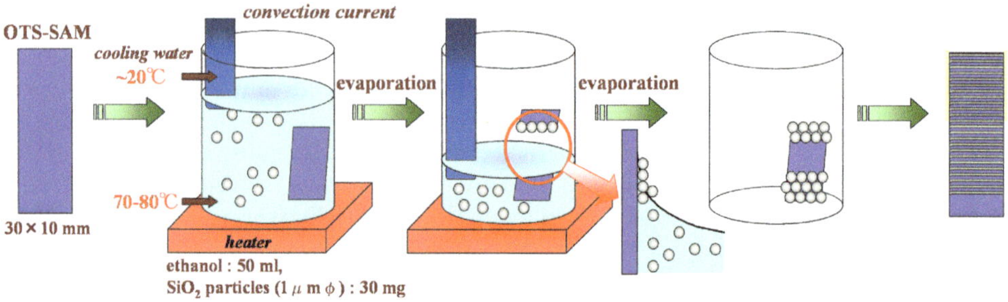


Reprinted with permission from Ref.[26], Masuda, Y., Itoh, T., Itoh, M. and Koumoto, K., 2004, *Langmuir*, 20, 5588. Copyright @ American Chemical Society

Fig. 12. Schematic for self-assembly process to fabricate an orderly array of particle wires constructed from a close-packed structure.

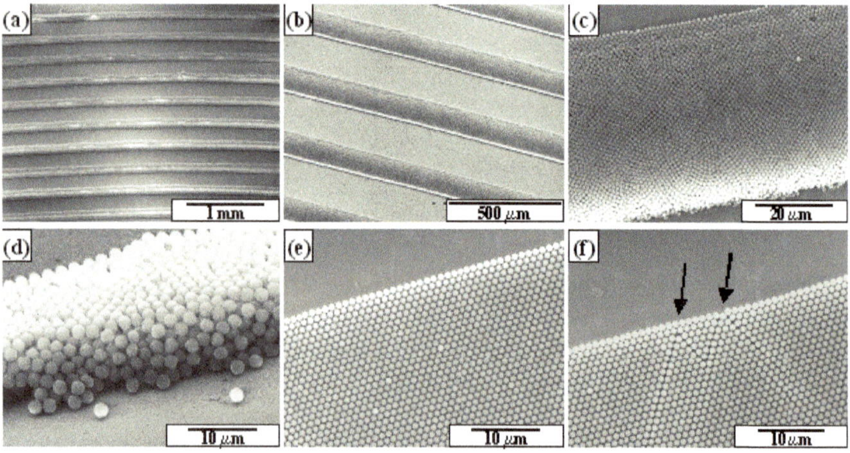


Reprinted with permission from Ref.[26], Masuda, Y., Itoh, T., Itoh, M. and Koumoto, K., 2004, *Langmuir*, 20, 5588. Copyright @ American Chemical Society

Fig. 13. SEM micrographs of array of particle wires constructed from (a-d) multiparticle layer or (e-f) monoparticle layer, (d) bottom side of a particle wire, (f) a monoparticle wire having two defects on the upper side.

Particle wires were formed on the OTS-SAM from the ethanol solution (80 ml) containing a small amount of $SiO_2$ particles (1000 nmφ, 1 mg) to fabricate thin wires constructed from a mono-particle layer (Fig. 13(e-f)). Particles were slowly supplied to particle wires from the solution. The number of particle layers was shown to be controlled by the change of particle number in the solution. The upper side of the particle wire in Fig. 13(f) has two defects, shown by arrows. The influence on the arrangement below the defects shows that particles were constructed from the upper side to form a close-packed structure. In addition, the disorder disappeared at the middle of the particle wire (Fig. 13(f)), which shows that the formation process of particle wire has self-recovery ability.

Figure 14 shows the formation process of particle wire. The upper particle line was formed (Fig. 14(a)) first and particles were supplied gradually from the solution to form a close-packed structure (Fig. 14(b)) to prepare a wide mono-particle wire (Fig. 14(c)). High dispersibility of particles in the solution and the effective meniscus force allowed us to prepare a highly-ordered close-packed structure.

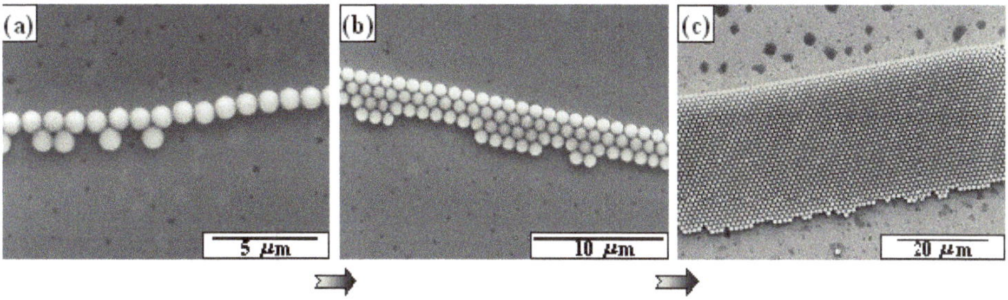


Reprinted with permission from Ref.[26], Masuda, Y., Itoh, T., Itoh, M. and Koumoto, K., 2004, *Langmuir*, 20, 5588. Copyright @ American Chemical Society

Fig. 14. SEM micrographs of the formation process of a particle wire.

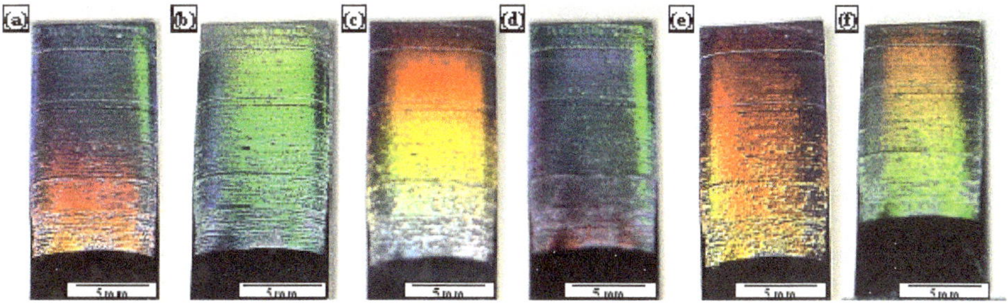


Reprinted with permission from Ref.[26], Masuda, Y., Itoh, T., Itoh, M. and Koumoto, K., 2004, *Langmuir*, 20, 5588. Copyright @ American Chemical Society

Fig. 15. Photographs of particle wires taken from different directions. (a) front view, (b-f) slightly cross shot of the same sample.

Particle wires showed iridescent diffraction (Fig. 15) caused by the high regularity of the particle array shown in Fig. 13. Diffracted wave number was changed drastically by the diffraction angle.

A close-packed structure was easily obtained by the use of meniscus force compared to the site-selective deposition in the solution[21-23]. A two-dimensional ordered array can be fabricated without the preparation of a template although templates are required for the liquid mold method[24].

Control of the interval of particle wires was achieved by change in solution temperature (Fig. 16, 17). The interval of particle wire fabricated from the ethanol solution (80 ml) containing $SiO_2$ particles (20 mg) at 70°C (Fig. 16(a)) was about three times that of particle wires fabricated at 60°C (Fig. 16(b)) or 50°C (Fig. 16(c)) (Fig. 17). The same tendency was confirmed in the dilute solution system (80 ml) containing $SiO_2$ particles (10 mg) (Fig. 17). Wide interval was caused from high descent speed of solution surface at high temperature. The regularity of particle wires at 70°C was much higher than that at 60°C or 50°C in which the movement of particles caused by convection and the descent speed of solution surface was low compared to that at 70°C.

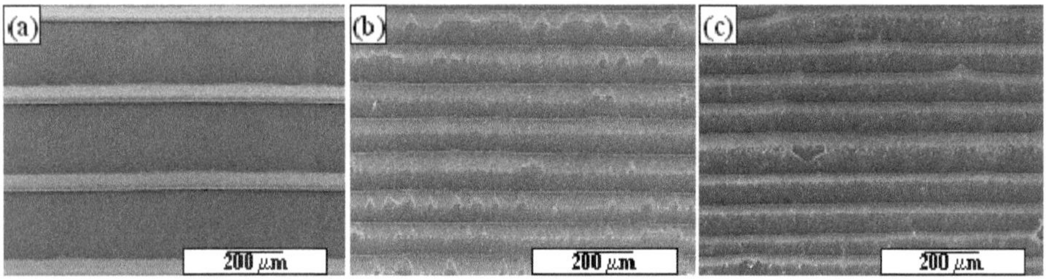


Reprinted with permission from Ref.[26], Masuda, Y., Itoh, T., Itoh, M. and Koumoto, K., 2004, *Langmuir*, 20, 5588. Copyright @ American Chemical Society

Fig. 16. SEM micrographs of array of particle wires constructed at (a) 70 °C, (b) 60 °C or (c) 50 °C.

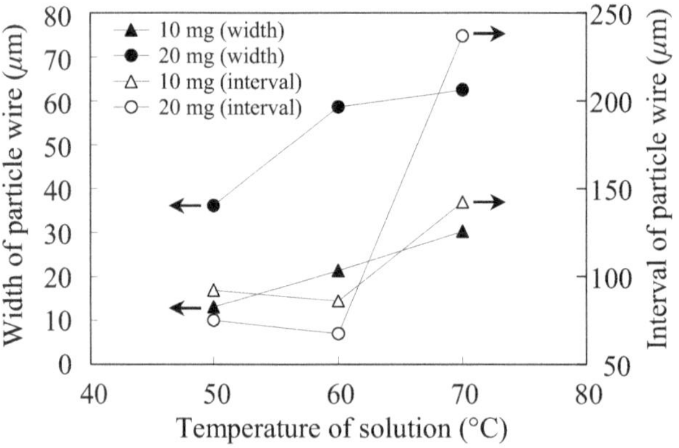


Reprinted with permission from Ref.[26], Masuda, Y., Itoh, T., Itoh, M. and Koumoto, K., 2004, *Langmuir*, 20, 5588. Copyright @ American Chemical Society

Fig. 17. Width and interval of particle wires as functions of temperature of solution.

The width of particle wires was also controlled by change in solution temperature or particle concentration in the solution (Fig. 17). The width of particle wires was increased by the increase of solution temperature in both particle concentrations (10 mg or 20 mg), and the width at high particle concentration (20 mg) was about two times that at low particle concentration (10 mg). High descent speed of solution surface and high movement of particles caused by convection were brought about at high temperature. These factors probably cause the change in width; however, further precise control and clarification of mechanism are necessary. Descent speed of solution surface would be controlled by control of the lift speed of a substrate instead of control of the evaporation speed. Further improvement of the process is required for application to future devices.

Self-assembly of particle wires was realized using water instead of ethanol without an SAM (Fig. 18). The particle wires also showed iridescent diffraction and the diffracted wave number was changed drastically by the diffraction angle. A particle layer was reported to form on the whole area of OTS-SAM by the use of water, which evaporates slowly compared with ethanol. Fast movement of the solution surface caused by quick evaporation or lift of substrate is probably necessary for drop-off of the solution surface, which allows fabrication of separated particle wires. The condenser tube, which is used to cool the solution surface, was removed to evaporate water more quickly. This makes the descent speed of the water surface fast enough to separate the particle wires. Additionally, silicon wafer, which was kept in air to show a contact angle of 10-20° to water, was used instead of OTS-SAM. Because the contact angle of the water on the OTS-SAM (96°) was higher than that of ethanol (10-20°) the shape of the water surface was not suitable for making an array of particle wires. These improvements allowed us to fabricate an array of particle wires using water without an SAM. This process is environmentally friendly compared with the process using ethanol with OTS-SAM; however, the regularity and the feature edge acuity of particle wires were low compared to the process shown in Fig. 13 and thus further improvement is required in this system.

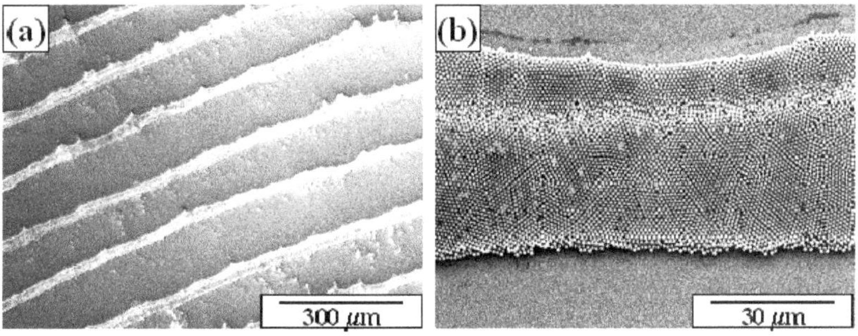


Reprinted with permission from Ref.[26], Masuda, Y., Itoh, T., Itoh, M. and Koumoto, K., 2004, *Langmuir*, 20, 5588. Copyright @ American Chemical Society

Fig. 18. SEM micrographs of array of particle wires fabricated from water solution.

Array of particle wires constructed not from fcc (or hcp), but from NaCl structure (rock salt structure) in which each ion is 6-coordinate and has a local octahedral geometry, was realized by the use of two kinds of particles ($SiO_2$ modified with carboxyl groups: 2000 nm$\varphi$, $SiO_2$ modified with amino groups: 1000 nm$\varphi$; ratio of particle radii is 0.5). $SiO_2$ particles 2000 nm$\varphi$ in diameter and $SiO_2$ particles 1000 nm$\varphi$ in diameter were modified to have carboxyl groups or amino groups on their surfaces, respectively. The zeta potential of $SiO_2$ particles modified with carboxyl groups was measured to be -40 mV in an aqueous solution at pH 7.0[8] by direct measurement of electrophoretic mobility using an electrophoretic light-scattering spectrometer (Zetasizer 3000HS, Malvern Instruments Co., Ltd.). On the other hand, APTS (3-Aminopropyltriethoxysilane)-SAM was prepared by the immersion of silicon wafer in an anhydrous toluene solution containing 1 vol% APTS for 1 h in air to measure the zeta potential instead of $SiO_2$ particles modified with amino groups. Amino groups of APTS-SAM were measured to be +22.0 mV in aqueous solutions (pH = 7.0) by an

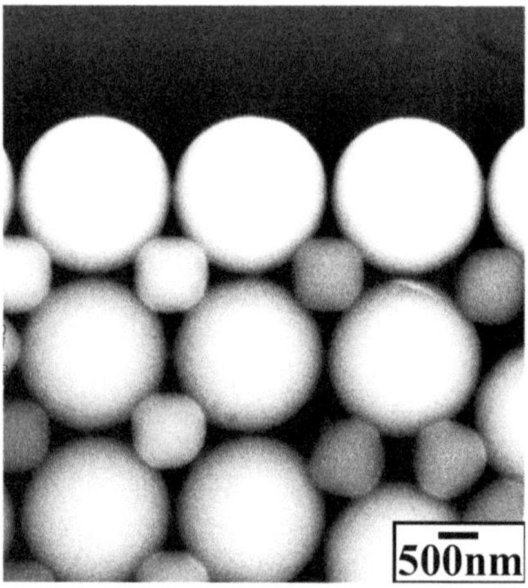


Reprinted with permission from Ref.[26], Masuda, Y., Itoh, T., Itoh, M. and Koumoto, K., 2004, *Langmuir*, 20, 5588. Copyright @ American Chemical Society

Fig. 19. SEM micrograph of array of particle wires constructed from NaCl structure.

electrophoretic light-scattering spectrophotometer (ELS-8000, Otsuka Electronics Co., Ltd.). Electrostatic interaction between $SiO_2$ particles modified with carboxyl groups and $SiO_2$ particles modified with amino groups was utilized for self-assembly of these particles to form ionic crystal such as NaCl structure. Although the ratio of ion radii is 0.611 (= $r^+(Na^+)/r^-(Cl^-)$) in NaCl, cations can contact to anions in the range of 0.414 - 0.732 in ratio of ion radii to form NaCl structure. Figure 19 shows the upper side of the particle wire and is the same arrangement as (100) face of NaCl structure. Each particle can be 6-coordinate when the same particle layer stacks on this layer with a slide of half lattice constant (sum of each particle radii). Small spherical particles (cations in NaCl structure) are shown as square or triangular in SEM micrographs because electrons from the SEM electron gun flowed from large particles into small particles to show the contacted area as white. Two kinds of dispersed particles would be adhered on the substrate and rearranged well to form 2-D layer. Particles were also deposited randomly in some areas, and this suggests that it is difficult to obtain NaCl structure in a large area. In addition, NaCl structure was prepared from two kinds of $SiO_2$ particles without surface modification ($SiO_2$: 2000 nm$\varphi$, $SiO_2$: 1000 nm$\varphi$; ratio of particle radii is 0.5) though the regularity of particle array was slightly lower than that prepared from $SiO_2$ particles with surface modification. NaCl-type arrangement of bimodal $SiO_2$ particles can be formed thermodynamically without electrostatic interaction because (100) face of NaCl structure constructed from large particles ($x$ nm$\varphi$) and small particles $\left(\left(\sqrt{2}-1\right)\chi\text{nm}\varphi\right)$ was close-packed structure. This process should be improved to prepare NaCl structure without the use of surface modification of particles to apply for various kinds of particles. Additionally, the interaction between particles, substrates and liquid and the behavior of particles should be controlled well to form particle assembly having high regularity. Particles should be dispersed well in the solution to avoid aggregation, and attractive interaction between particles and a substrate should be decreased enough to accelerate rearrangement of particles on the substrate. Appropriate movement of particles accelerate formation of particles assembly, however, excess convection suppresses assembly of particles. This process has many factors should be improved to fabricate particle assembly having desired structure and high regularity. The fabrication of an array of particle wire constructed from NaCl structure shows high ability of self-assembly process to realize future devices such as photonic crystal in which various structures are required to effectively utilize PBG.

Particle wires constructed from a close-packed multi-particle layer or mono-particle layer were fabricated without patterned templates. The array shows iridescent diffraction caused by high regularity of the particle array. The mechanism of the process was discussed and the control of the thickness, width and interval of particle wires was realized by change of solution temperature and concentration of colloidal solution. Furthermore, an array of particle wires constructed from an ionic crystal such as NaCl structure was fabricated using electrostatic interaction between particles showing positive zeta potential and particles showing negative zeta potential. This shows the suitability of the self-assembly process for creating future devices such as photonic crystals.

## 3. Patterning of colloidal crystals and spherical assemblies by two-solution method

### 3.1 Self-assembly patterning of colloidal crystals by two-solution method[27]

Desired patterns of colloidal crystals having high feature edge acuity and high regularity were fabricated by two-solution method[27]. A micropattern of colloidal methanol prepared on a self-assembled monolayer in hexane was used as a mold for particle patterning, and slow dissolution of methanol into hexane caused shrinkage of molds to form micropatterns of close-packed $SiO_2$ particle assemblies. This result is a step toward the realization of nano/micro periodic structures for next-generation photonic devices by a self-assembly process.

Silicon substrate was immersed into toluene solution containing 1 vol % octadecyltrichlorosilane (OTS) molecules under nitrogen atmosphere for 5 min to prepare a hydrophobic OTS-SAM. OTS-SAM was irradiated by ultraviolet light (PL21-200, SEN Lights Co., 18 mW/cm$^2$, distance from a lamp 30 mm, 24°C, humidity 73 %, air flow 0.52 m$^3$/min, 100 V, 320 W) through a photomask for 10 min. UV irradiation modified hydrophobic octadecyl groups to hydrophilic silanol groups forming a pattern of octadecyl regions and silanol regions. Patterned OTS-SAM having hydrophobic octadecyl regions and hydrophilic silanol regions was used as a template for patterning of colloidal solution.

$SiO_2$ particles (1 $\mu$m in diameter) (0.002 - 0.2 mg) were thoroughly dispersed in methanol (20 $\mu$l) and dropped on a patterned OTS-SAM (Fig. 20). The solution was lightly repelled by hydrophobic regions and mainly exists on hydrophilic silanol regions. The substrate was then immersed into hexane and carefully swung to remove the residual solution. The solution was repelled well by octadecyl regions in hexane. The contact angle of the methanol solution on OTS-SAM was confirmed to increase from 51.6° in air to 129.5° in hexane (Fig. 20), indicating that the methanol solution tends to exist on silanol regions selectively.

Methanol was then gradually dissolved into hexane to shrink the colloidal solution mold containing particles. The shrinkage of the mold increased the concentration of particles in the solution. The particles then attracted each other by meniscus force to form a close-packed fcc (face-centered cubic) or hcp (hexagonal closest packing) structure during the drying process of methanol. Close-packed structures were thus formed on hydrophilic silanol regions selectively (Fig. 21, A to H).

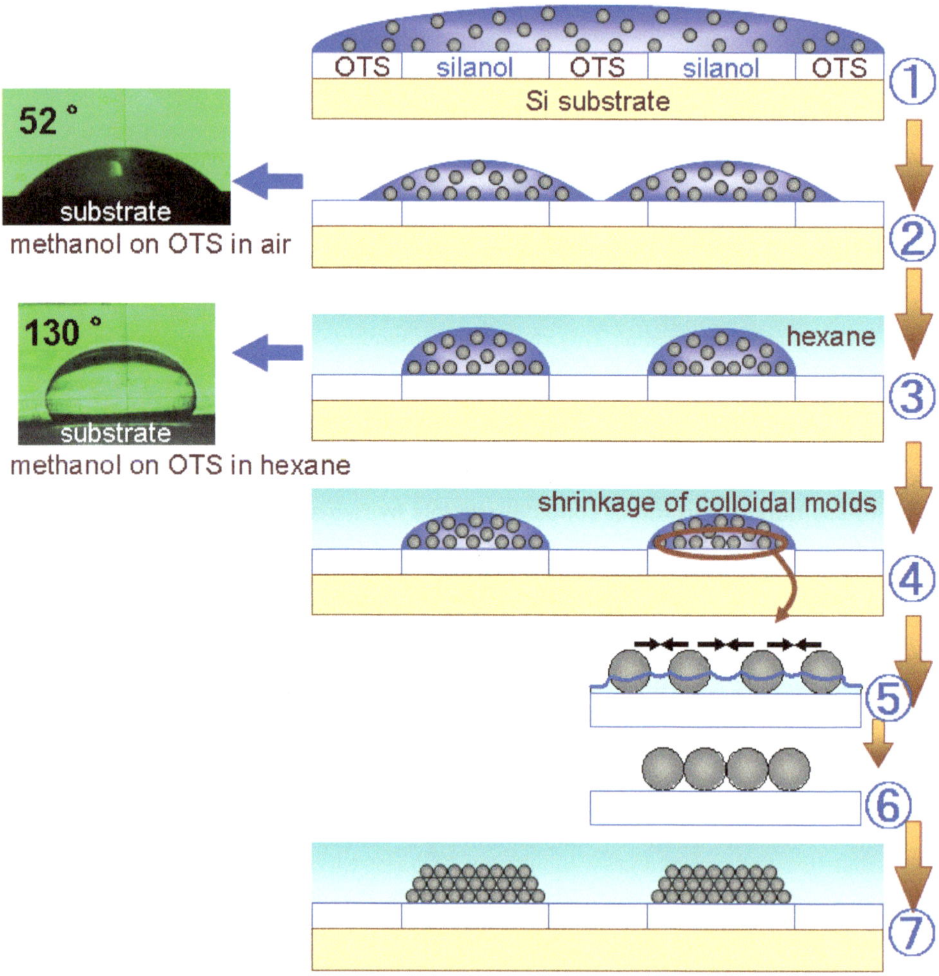


Reprinted with permission from Ref. [27], Masuda, Y., Itoh, T. and Koumoto, K., 2005, *Langmuir*, 21, 4478. Copyright @ American Chemical Society

Fig. 20. Conceptual process for two-solution self-assembly method to fabricate patterned colloidal photonic crystals.

2D micropatterns of multi particle-layers (Fig.21, A and B), partially double particle-layers (Fig. 21, C and D), or single particle-layers (Fig. 21, E to H) were successfully fabricated by changing the particle concentration in the solution and solution volume per unit of hydrophilic area ($SiO_2$/methanol0.2 mg/20 $\mu$l for Fig. 21, A and B, 0.02 mg/20 $\mu$l for Fig. 21, C and D, 0.002 mg/20 $\mu$l for Fig. 21, E to H). 2D patterns of colloidal crystals with high regularity in particle assembly have not been prepared by our processes previously reported[21-25]. The feature edge acuity of patterns and regularity in particle assembly[21-25]

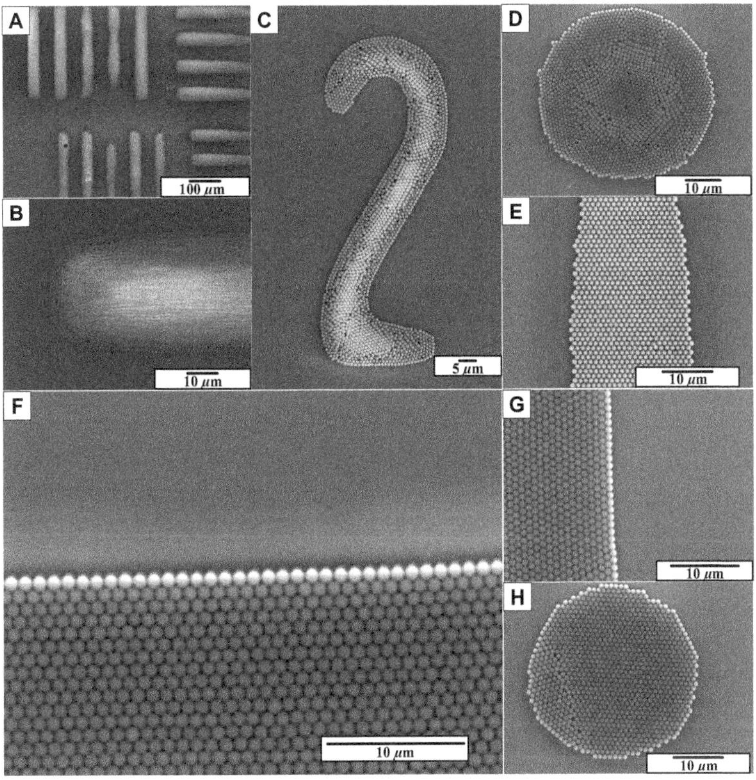




Fig. 21. SEM micrographs of patterned colloidal photonic crystals constructed from (A-B) multi particle-layers, (C-D) partially double particle-layers or (E-H) single particle-layers. Image (B) is a magnified area of (A).

presented here are clearly higher than those previously reported[21-25]. Figure 4-2B shows a magnified area of the patterns constructed from thick particle-layers (Fig. 21A). The edge of patterns (Fig. 21, A to H) shows high feature edge acuity due to close-packing induced by meniscus force in the drying process. The core area of the particle circle (Fig. 21D, Fig. 22) was a double particle-layer of close-packed hexagonal lattice, i.e., the arrangement of fcc{111}, and the outer shell of the circle was a single particle-layer of hexagonal lattice (fcc{111}). The boundary area of these two flat terraces, i.e., inner shell, was constructed from a nested structure of square lattice, i.e., the arrangement of fcc{100}, to form a gentle slope between the core double-layer and outer shell single-layer. The lattice constant of the square lattice at the inner shell increased gradually with distance from the core of the circle to form a gentle slope. The difference in height was caused by the assembling process and the shape of the liquid mold in which the center is higher than the outside. The particle arrangement in the patterns constructed from a single particle-layer (Fig. 21, E to H) were assigned to the arrangement in fcc{111} which is a close-packed structure. The border line of particle layer in Fig. 21E showed different shape with that in Fig. 21F, G because the border

line of particle layer in Fig. 21F, G was fcc <1, 1, 0> and that in Fig. 21E was fcc <1/2, 1/2, 1> which is orthogonal to fcc <1, 1, 0>. There were far fewer defects in the particle patterns than in our former processes[21-25] because of the effective meniscus force. The standard deviation for the edge of the pattern constructed from a single particle-layer (Fig. 21F) was calculated in the same manner reported previously[24]. The center position (($x_i$, $y_i$) $\mu$m) of each particle at the edge was plotted to estimate the standard deviation. The particle at the far left in the edge line of Fig. 21F was set to be the origin of the x-y coordinate.
The approximated straight line ($f(x)$) and its slope ($\theta$) are represented as follows.

$$f(x) = 0.0273x - 0.09 \tag{7}$$

$$\cos\theta = 0.9996 \tag{8}$$

The standard deviation from the approximated straight line is given by the expression,

$$\text{S (standard deviation)} = \sqrt{U} = \frac{\left[\sum_i \{\cos\theta \cdot (f(x_i) - y_i)\}^2\right]^{1/2}}{n-1} \tag{9}$$

where $n$ is the number of particles ($n$ = 35). Unbiased variance (U) was used because the number of particles ($n$) is smaller than universe. The accuracy of the particle arrangement in Fig. 21F was estimated to be S = 8.75 × $10^{-3}$. This is lower than that of the pattern constructed from a single particle-layer prepared in the solution using chemical reactions (S = 3.89 × $10^{-2}$, Fig. 7 (a) in previous report[23]). The standard deviation for the edge of the pattern constructed from a single particle-layer (Fig. 21H) was calculated in the same manner.Distance ($r_i$) from center of a circle (($x_o$, $y_o$) $\mu$m) to each particle (($x_i$, $y_i$) $\mu$m) at the edge and its average ($\bar{r}$) are presented as follows.

$$r_i = \left((x_i - x_o)^2 + (y_i - y_o)^2\right)^{1/2} \tag{10}$$

$$\bar{r} = \frac{\sum_{i=1}^{n} \left((x_i - x_o)^2 + (y_i - y_o)^2\right)^{1/2}}{n} \tag{11}$$

$x_o$= 21.9 $\mu$m, $y_o$=17.7 $\mu$m, $\bar{r} = 16.53\ \mu$m
The standard deviation from the approximated circle is given by the expression,

$$\text{S (standard deviation)} = \sqrt{V} = \left[\frac{\sum_{i=1}^{n} \left\{\left((x_i - x_o)^2 + (y_i - y_o)^2\right)^{1/2} - \bar{r}\right\}^2}{n}\right]^{1/2} \tag{12}$$

where $n$ is the number of particles ($n$ = 112). Variance (V) was used instead of unbiased variance because the number of particles ($n$) is the same as universe, i.e., all particles at the edge. The accuracy of the particle arrangement in Fig. 21H was estimated to be S = $3.89 \times 10^1$. This is higher than that in Fig. 21F because a perfect circle can't be constructed from a small number of particles which were packed in hexagonal arrangement.

The assembly process can be assumed from the details of structures and defects. Particle circles (Fig. 21, D and H) showed no defects at their core, implying that particles were probably assembled from the core of the particle circle and not from the outer shell. The particle circle (Fig. 21D, Fig. 22) would be formed not layer-by-layer, and the upper layer at

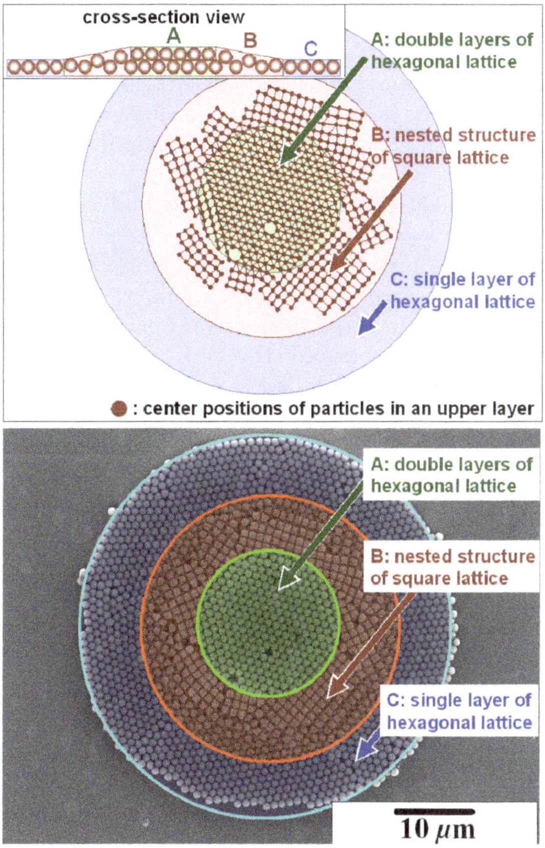


Reprinted with permission from Ref. [27], Masuda, Y., Itoh, T. and Koumoto, K., 2005, *Langmuir*, 21, 4478. Copyright @ American Chemical Society

Fig. 22. Particle arrangement view (top) prepared from SEM micrograph (bottom) of patterned colloidal photonic crystals constructed from partially double particle-layers. Center positions of particles in an upper layer show hexagonal or square lattice arrangement regions. The cross-sectional view shows the difference in height caused from the difference in particle arrangement modes. The three colored areas show regions of (A) double layers of hexagonal lattice, (B) nested structure of square lattice and (C) single layer of hexagonal lattice.

the core was also formed before the outer shell was assembled, since the first layer at the boundary area was not a close-packed assembly. A close-packed hexagonal lattice would be formed in the lower layer as shown in the single particle circle (Fig. 21H) in the case particles were assembled layer-by-layer. Other particle patterns (Fig. 21, A to C) would be assembled from the core area in the same manner.

Methanol solution containing particles was dropped onto a patterned OTS-SAM and dried in air without immersion into hexane for comparison. However, a particle pattern only with a low feature edge acuity was obtained. This indicates that the contact angle of methanol solution on OTS-SAM, i.e., the shape of the methanol solution mold, is important for particle patterning. Additionally, $SiO_2$ particles were dispersed in water instead of methanol, and a patterned OTS-SAM covered with the water colloidal solution was immersed into hexane, but the particle pattern was not obtained because the water solution cannot form a micropattern of the mold on hydrophilic regions selectively due to its high contact angle. These results suggest that the combination of methanol, hexane, OTS and silanol allowed us to fabricate micropatterns of particle assembly having high feature edge acuity and high regularity.

This two-solution self-assembly method successfully produced precise 2D patterns of colloidal photonic crystals, i.e., close-packed fcc or hcp particle assemblies having a photonic band gap, with short fabrication period and low energy. Various complicated patterns of colloidal crystals having high feature edge acuity and high regularity in particle assembly were fabricated, and the number of particle stacking layers, i.e., thickness of colloidal crystals, was demonstrated to be controlled. This process has the advantage of the drying process in which meniscus force can be utilized for close-packed particle assembly, and also the advantage of the static solution process in which particle assembling processes can be controlled well. The low dissolution rate of methanol into hexane and the appropriate contact angle of methanol on a substrate achieved by the combination of methanol, hexane, OTS and silanol allowed us to utilize the solution mold for particle patterning. Additionally, high dispersibility of particles and high repulsion force between particles and substrate suppressed the aggregation of particles, and the meniscus force between particles was effectively utilized to form a close-packed structure. The quality of SAM, shape and shrinkage rate of solution molds also greatly influence the feature edge acuity and the regularity of the particle assembly. The newly developed method achieved much higher regularity in particle assembly and feature edge acuity of the pattern than those previously reported[21-25]. The result is a step toward the realization of nano/micro periodic structures for next-generation photonic devices by the self-assembly process.

### 3.2 Self-assembly patterning of spherical colloidal crystals by two-solution method[28]

Micropatterns of spherical particle assemblies were fabricated by two-solution method[28]. Hydrophilic regions of a patterned self-assembled monolayer were covered with methanol solution containing $SiO_2$ particles and immersed in decalin to control the shape of droplets and gradually dissolve the methanol into decalin. Interfacing of methanol/decalin and shrinkage of methanol droplets were utilized to obtain meniscus force to form spherical particle assemblies; additionally, its static solution system allowed precise control of the conditions. Particles were assembled to form spherical shapes on hydrophilic regions of an SAM and consequently, micropatterns of spherical particle assemblies were successfully fabricated through self-assembly. This patterned two-solution process has the advantages of both a drying process having meniscus force and a static solution process having high controllability.

An Si wafer (p-type Si [100], NK Platz Co., Ltd.) was sonicated in water, ethanol or acetone for 10 min, respectively, and exposed for 2 h to UV light (184.9 nm) (low-pressure mercury lamp, NL-UV253, Nippon Laser & Electronics Lab.) to clean the surface. The OTS-SAM or HFDTS-SAM[29] were prepared by immersing the Si substrate in an anhydrous toluene(Aldrich Chemical Co., Inc.) solution containing 1 vol% OTS(Acros Organics) or HFDTS(Lancaster synthesis Ltd.) for 5 min under an $N_2$ atmosphere (Fig. 23). The substrate with the SAM was baked at 120°C for 5 min to remove residual solvent and promote chemisorption of the SAM.

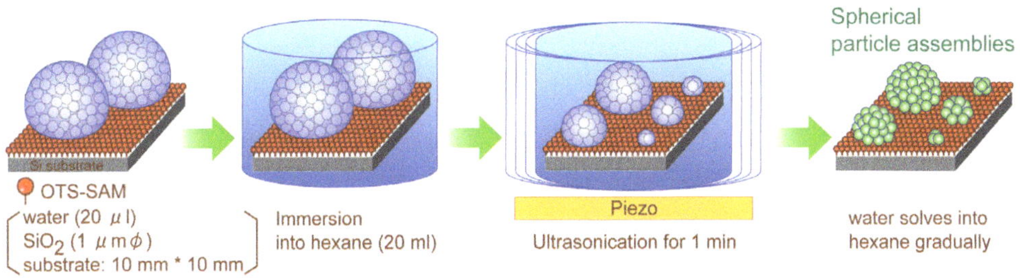


Reprinted with permission from Ref. [28], Masuda, Y., Itoh, T. and Koumoto, K., 2005, *Adv. Mater.*, 17, 841. Copyright @ Wiley-VCH Verlag GmbH & Co.

Fig. 23. Conceptual process for fabrication of spherical particle assemblies.

The HFDTS-SAM on the silicon substrate was exposed for 2 h to UV light through a photomask such as a mesh for transmission electron microscopy (Okenshoji Co., Ltd.) to be used as a template for micropatterning of spherical particle assemblies. UV-irradiated regions became hydrophilic due to silanol group formation, while the non-irradiated part remained unchanged. Formation of the SAMs and the modification to silanol groups by UV irradiation were verified using the water drop contact angle ($\theta$w). The initially deposited OTS-SAM or HFDTS-SAM showed a water contact angle of 105° or 112°, but the UV-irradiated surface of SAM was wetted completely (contact angle < 5°).

$SiO_2$ particles (1.13 $\mu$m$\varphi$, Hipresica UF N3N, CV: 3.57 %, specific gravity: 1.8 ± 0.1 g/cm$^3$, Ube-Nitto Kasei Co., Ltd.) (0.2 mg, 1.5 × $10^{27}$ particles) were thoroughly dispersed in water (20 $\mu$l) and dropped on a hydrophobic OTS(octadecyltrichlorosilane)-SAM (Fig. 23). The OTS-SAM with droplets was then immersed in hexane (20 ml, solubility of hexane in water at 20°C: 0.0013 g / 100 ml, specific gravity: 0.7) and ultrasonicated for 1 min. Large water droplets containing $SiO_2$ particles were separated into many small emulsions that kept them spherical on hydrophobic OTS-SAM. Water in the emulsions was gradually extracted to hexane to reduce the size of emulsions forming spherical particle assemblies[30]. After having been immersed for 12 h, spherical particle assemblies with different diameters were observed on OTS-SAM; it was also observed that the assemblies were constructed from various numbers of particles such as 3, 5, 6, 8 or many particles (Fig. 24, 25, 26). Quantities of 3, 5, 6 or 8 particles were assembled into triangular, pyramidal, octahedral or decahedral particle clusters, respectively. The number of particles in spheres can be controlled by the change of emulsion size or particle concentration in water[30]. The spherical shape of particle assemblies was caused by the high contact angle of water emulsion on hydrophobic OTS-SAM in hexane. Consequently, various sizes of spherical particle assemblies can be prepared using this emulsion process.

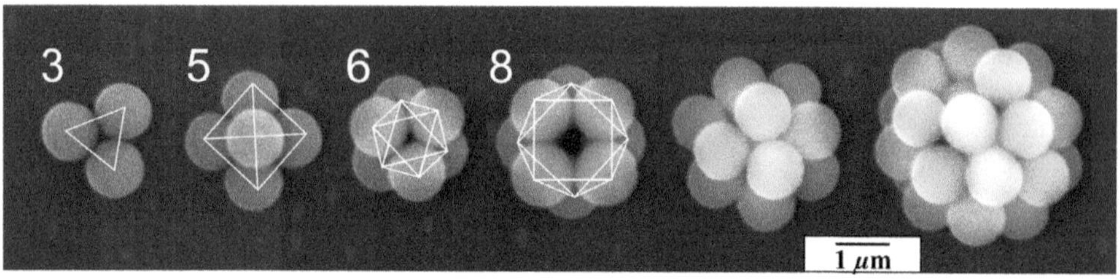


Reprinted with permission from Ref. [28], Masuda, Y., Itoh, T. and Koumoto, K., 2005, *Adv. Mater.*, 17, 841. Copyright @ Wiley-VCH Verlag GmbH & Co.

Fig. 24. SEM micrographs of spherical particle assemblies with different diameters. Particle assemblies constructed from a small number of particles such as 3, 5, 6, 8 and so on.

Reprinted with permission from Ref. [28], Masuda, Y., Itoh, T. and Koumoto, K., 2005, *Adv. Mater.*, 17, 841. Copyright @ Wiley-VCH Verlag GmbH & Co.

Fig. 25. Spherical particle assembly and tilted images.

The particles would be strongly bound to the droplet interfaces by surface tension in emulsions[31]. In the formation process of small clusters of particles (Fig. 24), the water in emulsions dissolved into hexane to reduce droplet size and this restricted the area in which particles could exist[30]. Particles touched together by the reduction of emulsion size and formed a spherically packed assembly. Deformation of the interface then led to the rearrangement that formed close-packed particle assemblies. The clusters were formed using emulsions not dispersed in the solution but adsorbed on the flat substrate, and thus some of the clusters showed imperfect symmetry such as pyramidal, octahedral or decahedral, which have a large flat face touching the substrate.

Additionally, large spherical particle assemblies (for instance ~57 $\mu m\varphi$ in Fig. 26) were prepared from large methanol emulsions (~100–300 $\mu m\varphi$) in decalin (decahydronaphthalene) without the use of ultrasonication, which makes emulsions smaller. Many linear disclinations, i.e., grain boundaries, were formed on the surface to reduce elastic strain energy because a close-packed triangular particle lattice composed of a particle surrounded by 6 particles in plane cannot cover a spherical surface[32]. Linear disclinations were composed of an alternative arrangement of the particle surrounded by 5 particles in plane (red, this particle can be assumed to have a charge of +1) and the particle surrounded by 7 particles in plane (yellow, charge -1). Both ends of a linear disclination were particles surrounded by 5 particles in plane to make the total charge of each linear disclination +1 as observed in previous report[32]. The minimum number of linear disclinations can be calculated for the large sphere (~57 $\mu m\varphi$) (Fig. 26) to be 22 (*N*: minimum number of linear disclinations $N = 12 + 0.41 r/d = 22$, *r*: radius of sphere ~28.5 $\mu$m, *d*: mean particle spacing 1.13 $\mu$m) from geometric calculations[32,33]. The surface area in a purple circle can be also calculated to be about 10.5 % of the total surface area of the sphere from the formula (surface area ratio: $S_2/S_1$, surface area of the sphere: $S_1 = 4\pi r^2$, surface area in a purple circle: $S_2 = 2\pi r(r - r_1)$, radius of the purple circle shown in Fig. 26 $\sqrt{r^2 - r_1^2} = 17.5\mu m$). The minimum number of linear disclinations on the surface area in a circle can therefore be estimated to about 2.3(*n*: minimum number of linear disclinations in a purple circle $n = N \cdot S_2/S_1$). However, many linear disclinations were observed in an SEM micrograph (Fig. 26 (Right)). It is suggested that our assembly method can be further optimized to reduce linear disclinations.

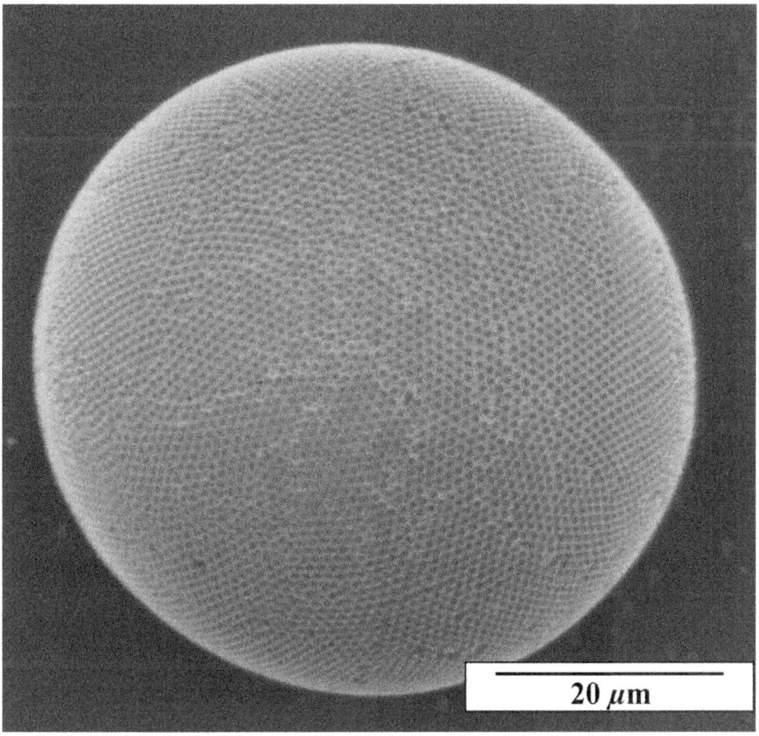


Fig. 26. (Continued)

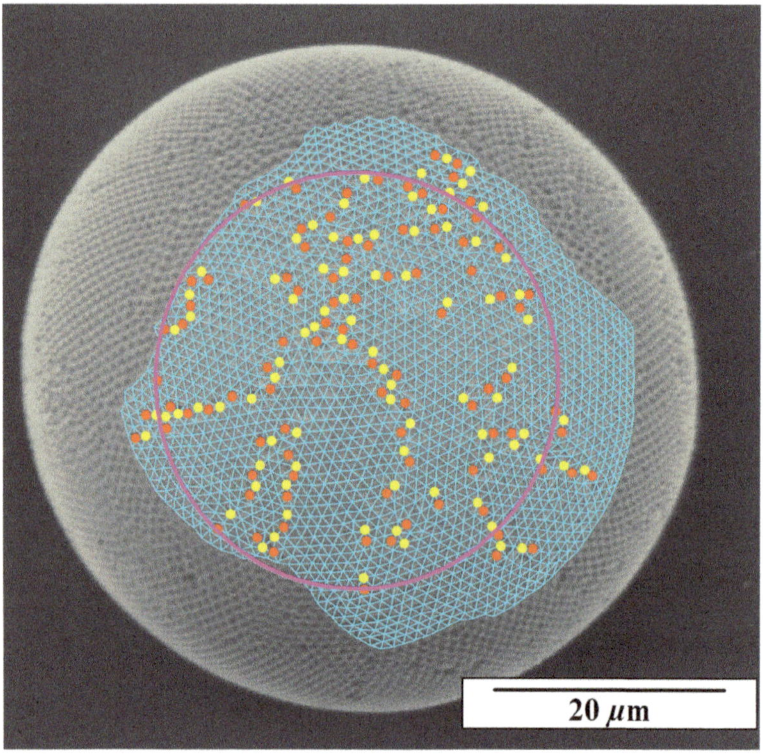




Fig. 26. Top: A large spherical particle assembly. Bottom: Linear disclinations on a large spherical particle assembly.

Furthermore, the upper sides of spherical particle assemblies were removed using a manipulator installed in an optical microscope (BX51WI Microscope, Olympus Optical Co., Ltd.) to evaluate the packing structure. Particle assemblies have a densely packed structure (Fig. 27). Particles would be completely rearranged to form a densely packed structure during the reduction of emulsion size due to the high dispersibility of particles in emulsions. $SiO_2$ particles (1.13 $\mu m\varphi$, 1 g/l) were thoroughly dispersed in methanol solution (10 $\mu$l) and dropped onto a patterned HFDTS(heptadecafluoro-1,1,2,2-tetrahydrodecyltrichlorosilane)-SAM[29] having hydrophobic HFDTS-SAM regions and hydrophilic silanol regions, photopatterned using a mesh for transmission electron microscopy as a photomask (Fig. 28). The patterned SAM covered with the solution was then carefully immersed in decalin so as not to remove the solution because the density of methanol (0.79) is lower than that of decalin (0.88) causing methanol to float on decalin. The patterned SAM was then gently vibrated to remove additional methanol solution and assist the movement of droplets to silanol regions. The methanol solution was selectively contacted on hydrophilic regions to form a micropattern of the solution, which became clearer after immersion for a few hours. The methanol solution containing particles formed a spherical shape because of the surface

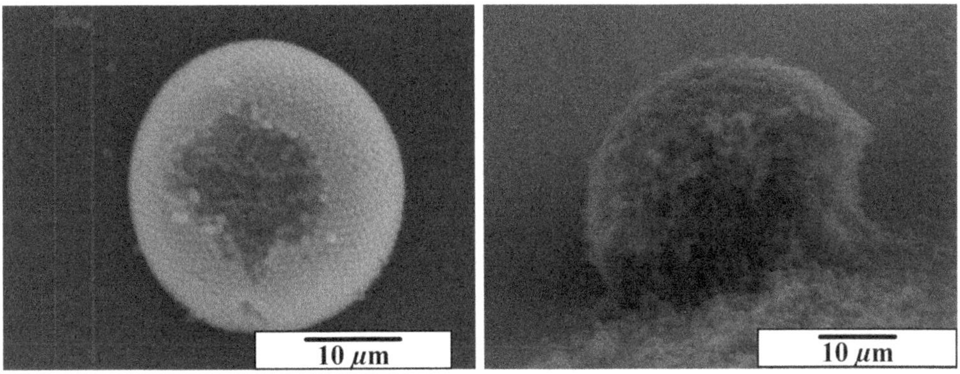


Reprinted with permission from Ref. [28], Masuda, Y., Itoh, T. and Koumoto, K., 2005, *Adv. Mater.*, 17, 841. Copyright @ Wiley-VCH Verlag GmbH & Co.

Fig. 27. SEM micrographs showing the inside of spherical particle assemblies formed from methanol emulsions in decalin.

interaction between methanol, decalin and surface of a SAM and the buoyant force of methanol in decalin. Methanol in emulsions was gradually dissolved into exterior decalin phase to form particle assemblies. After having been immersed for 12 h, particle assemblies having a dome shape were formed at the center of each silanol region (Fig. 29). The distance between the centers of each spherical particle assembly was same to the distance between holes of a mesh. The diameter of spherical particle assemblies was about 18 $\mu m\varphi$. It was smaller than that of hydrophilic regions (about 100$\mu$m) due to low particle concentration in methanol and the shape of methanol droplets on a substrate. This indicates the arrangement regularity of spherical particle assemblies can be improved more by the use of the photomask having small holes to decide positions of each droplet precisely. Some extra particles, i.e., noise particles, were also deposited on hydrophobic regions. The process should be further optimized to control many factors such as volume of methanol solution on a substrate, quality of a SAM or aggregation of particles in the solution to avoid noise particles. Particle assemblies were shown to have a densely packed structure by destructive inspection using a manipulator. Consequently, the dot array of spherical particle assemblies was successfully fabricated by this self-assembly process.

Micropatterns of particle assemblies were fabricated without the use of a template, having microstructures such as molds or grooves. A two-dimensional array of spherical particle assemblies was fabricated by self-assembly with this method. Interfacing between two solutions and shrinkage of droplets were utilized to obtain meniscus force to form spherical particle assemblies, and additionally, its static solution system allowed precise control of the conditions. These showed the high ability of self-assembly processes to prepare microstructures constructed from colloidal crystals. Further investigations of the solution-solution, solution-SAMs and solution-particles interfaces and the behavior of particles and solutions would allow us to develop this two-solution system to prepare desirable particle assembly structures.

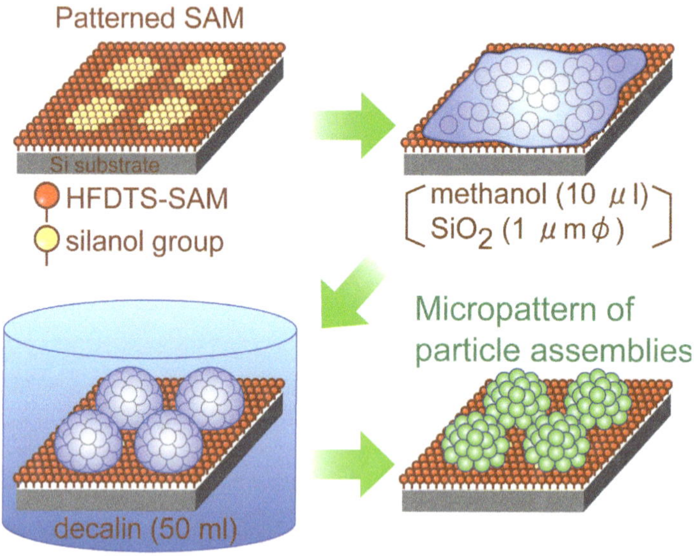




Fig. 28. Conceptual process of micropattern of spherical particle assemblies.

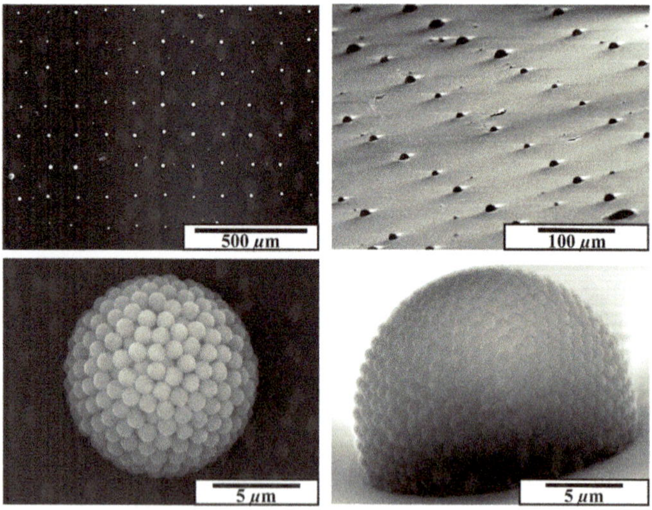




Fig. 29. SEM micrographs of micropattern of spherical particle assemblies. Top left: Micropattern of spherical particle assemblies. Bottom left: Magnified area of top left. Top right: Tilted micropattern of spherical particle assemblies. Bottom right: Magnified area of top right.

## 4. Summary

Self-assembly patterning of colloidal crystals was realized by liquid phase patternig, drying patterning and two-solution method. Micropatterns of colloidal crystals were fabricated under environmentally friendly conditions using self-assembled monolayers. They can be applied to photonic crystals, dye-sensitized solar cell, molecular sensors, gas sensors, nano-particle devices, etc. These showed high ability of self-assembled monolayers and self-assembly processes.

## 5. References


[1] Masuda, Y.; Koumoto, K. *J. Soc. Inor. Mater. Jpn.* 2000, *7*, 4-12.
[2] Masuda, Y.; Tachibana, K.; Itoh, M.; Koumoto, K. *Materials Integration* 2001, *14*, 37-44.
[3] Yablonovitch, E. *Phys. Rev. Lett.* 1987, *58*, 2059-2062.
[4] John, S. *Phys. Rev. Lett.* 1987, *58*, 2486-2489.
[5] Joannopoulos, J. D.; Meade, R. D.; Winn, J. N. *Princeton Press, Princeton, New Jersey* 1995.
[6] Chow, E.; Lin, S. Y.; Johnson, S. G.; Villeneuve, P. R.; Joannopoulos, J. D.; Wendt, J. R.; Vawter, G. A.; Zubrzycki, W.; Hou, H.; Alleman, A. *Nature* 2000, *407*, 983-986.
[7] Colombelli, R.; Srinivasan, K.; Troccoli, M.; Painter, O.; F. Gmachl, C.; Tennant, D. M.; Sergent, A. M.; Sivco, D. L.; Cho, A. Y.; Capasso, F. *Science* 2003, *302*, 1374-1377.
[8] Noda, S.; Tomoda, K.; Yamamoto, N.; Chutinan, A. *Science* 2000, *289*, 604-606.
[9] M. Qi; E. Lidorikis; P. T. Rakich; S. G. Johnson; J. D. Joannopoulos; E. P. Ippen; H. I. Smith *Nature* 2004, *429*, 538-542.
[10] M. Campbell; D. N. Sharp; M. T. Harrison; R. G. Denning; A. J. TurberReld *Nature* 2000, *404*.
[11] J. G. Fleming; S. Y. Lin *Optics Letters* 1999, *24*, 49-51.
[12] S. R. Kennedy; M. J. Brett; O. Toader; S. John *Nano Letters* 2002, *2*, 59-62.
[13] Kuramochi, E.; Notomi, M.; Kawashima, T.; Takahashi, J.; Takahashi, C.; Tamamura, T.; Kawakami, S. *Optical and Quantum Electronics* 2002, *34*, 53-61.
[14] Vlasov, Y. A.; Bo, X. Z.; Sturm, J. C.; Norris, D. J. *Nature* 2001, *414*, 289-293.
[15] A. Blanco; E. Chomski; S. Grabtchak; M. Ibisate; S. John; S. W. Leonard; C. Lopez; F. Meseguer; H. Miguez; J. P. Mondia; G. A. Ozin; O. Toade; H. M. van Driel *Nature* 2000, *405*, 437-440.
[16] Xia, N.; Gates, B.; Yin, Y.; Lu, Y. *Adv. Mater.* 2000, *12*, 693-713.
[17] Lopez, C. *Adv. Mater.* 2003, *15*, 1679-1704.
[18] Kim, E.; Xia, Y.; Whitesides, G. M. *Adv. Mater.* 1996, *8*, 245-247.
[19] Ozin, G. A.; Yang, S. M. *Adv. Func. Mater.* 2001, *11*, 95-104.
[20] Xia, Y. N.; Yin, Y. D.; Lu, Y.; McLellan, J. *Adv. Func. Mater.* 2003, *13*, 907-918.
[21] Masuda, Y.; Seo, W. S.; Koumoto, K. *Thin Solid Films* 2001, *382*, 183-189.
[22] Masuda, Y.; Seo, W. S.; Koumoto, K. *Jpn. J. Appl. Phys.* 2000, *39*, 4596-4600.
[23] Masuda, Y.; Itoh, M.; Yonezawa, T.; Koumoto, K. *Langmuir* 2002, *18*, 4155-4159.
[24] Masuda, Y.; Tomimoto, K.; Koumoto, K. *Langmuir* 2003, *19*, 5179-5183.
[25] Masuda, Y.; Itoh, M.; Koumoto, K. *Chem. Lett.* 2003, *32*, 1016-1017.
[26] Masuda, Y.; Itoh, T.; Itoh, M.; Koumoto, K. *Langmuir* 2004, *20*, 5588-5592.
[27] Masuda, Y.; Itoh, T.; Koumoto, K. *Langmuir* 2005, *21*, 4478-4481.
[28] Masuda, Y.; Itoh, T.; Koumoto, K. *Adv. Mater.* 2005, *17*, 841-845.
[29] Masuda, Y.; Sugiyama, T.; Koumoto, K. *J. Mater. Chem.* 2002, *12*, 2643-2647.

[30] Manoharan, V. N.; Elsesser, M. T.; Pine, D. J. *Science* 2003, *301*, 483-487.
[31] Levine, S.; Bowen, B. D.; Partridge, S. J. *Colloids and Surfaces* 1989, *38*, 325-343.
[32] Bausch, A. R.; Bowick, M. J.; Cacciouto, A.; Dinsmore, A. D.; Hsu, M. F.; Nelson, D. R.; Nikolaides, M. G.; Travesset, A.; Weitz, D. A. *Science* 2003, *299*, 1716-1718.
[33] Bowick, M. J.; Nelson, D. R.; Travesset, A. *Phys. Rev. B* 2000, *62*, 8738-8751.

# Permissions

The contributors of this book come from diverse backgrounds, making this book a truly international effort. This book will bring forth new frontiers with its revolutionizing research information and detailed analysis of the nascent developments around the world.

We would like to thank Dr. Yoshitake Masuda, for lending his expertise to make the book truly unique. He has played a crucial role in the development of this book. Without his invaluable contribution this book wouldn't have been possible. He has made vital efforts to compile up to date information on the varied aspects of this subject to make this book a valuable addition to the collection of many professionals and students.

This book was conceptualized with the vision of imparting up-to-date information and advanced data in this field. To ensure the same, a matchless editorial board was set up. Every individual on the board went through rigorous rounds of assessment to prove their worth. After which they invested a large part of their time researching and compiling the most relevant data for our readers. Conferences and sessions were held from time to time between the editorial board and the contributing authors to present the data in the most comprehensible form. The editorial team has worked tirelessly to provide valuable and valid information to help people across the globe.

Every chapter published in this book has been scrutinized by our experts. Their significance has been extensively debated. The topics covered herein carry significant findings which will fuel the growth of the discipline. They may even be implemented as practical applications or may be referred to as a beginning point for another development. Chapters in this book were first published by InTech; hereby published with permission under the Creative Commons Attribution License or equivalent.

The editorial board has been involved in producing this book since its inception. They have spent rigorous hours researching and exploring the diverse topics which have resulted in the successful publishing of this book. They have passed on their knowledge of decades through this book. To expedite this challenging task, the publisher supported the team at every step. A small team of assistant editors was also appointed to further simplify the editing procedure and attain best results for the readers.

Our editorial team has been hand-picked from every corner of the world. Their multi-ethnicity adds dynamic inputs to the discussions which result in innovative outcomes. These outcomes are then further discussed with the researchers and contributors who give their valuable feedback and opinion regarding the same. The feedback is then collaborated with the researches and they are edited in a comprehensive manner to aid the understanding of the subject.

Apart from the editorial board, the designing team has also invested a significant amount of their time in understanding the subject and creating the most relevant covers. They scrutinized every image to scout for the most suitable representation of the subject and create an appropriate cover for the book.

The publishing team has been involved in this book since its early stages. They were actively engaged in every process, be it collecting the data, connecting with the contributors or procuring relevant information. The team has been an ardent support to the editorial, designing and production team. Their endless efforts to recruit the best for this project, has resulted in the accomplishment of this book. They are a veteran in the field of academics and their pool of knowledge is as vast as their experience in printing. Their expertise and guidance has proved useful at every step. Their uncompromising quality standards have made this book an exceptional effort. Their encouragement from time to time has been an inspiration for everyone.

The publisher and the editorial board hope that this book will prove to be a valuable piece of knowledge for researchers, students, practitioners and scholars across the globe.

# List of Contributors

**Susmit Kumar, Shilpi Karmakar and Ross Rinaldi**
National Nanotechnology Laboratory, Istituto Nanoscienze-CNR, Italy

**Alessandro Bramanti**
STMicroelectronics Srl, Italy

**Giuseppe Maruccio**
National Nanotechnology Laboratory, Istituto Nanoscienze-CNR, Italy
Department of Physics, University of Salento, Lecce, Italy

**Serban F. Peteu**
The National Institute for R&D in Chemistry and Petrochemistry, Bucharest, Romania
Department of Chemical Engineering and Materials Science, Michigan State University, East Lansing, Michigan, USA

**Sabine Szunerits**
Institut de Recherche Interdisciplinaire, Universite of Lille 1, Villeneuve d'Ascq, France

**Alina Vasilescu**
International Centre for Biodynamics, Romania

**Wolfgang Knoll**
Austrian Institute of Technology AIT, Vienna, Austria

**Yoshitake Masuda**
National Institute of Advanced Industrial Science and Technology (AIST), Anagahora, Shimoshidami, Moriyama-ku, Nagoya, Japan

**Yanfeng Gao**
State Key Laboratory of High Performance Ceramics and Superfine Microstructure, Shanghai institute of Ceramics, Chinese Academy of Sciences, Shanghai, China
Research Center for Industrial Ceramics, Shanghai institute of Ceramics, Chinese Academy of Sciences, Shanghai, China
Xinjiang Key Laboratory of Electronic Information Materials and Devices, Xinjiang Technical Institute of Physics & Chemistry, Chinese Academy of Sciences, Beijing Road, Urumqi, Xinjiang, China

**Litao Kang and Zhang Chen**
State Key Laboratory of High Performance Ceramics and Superfine Microstructure, Shanghai institute of Ceramics, Chinese Academy of Sciences, Shanghai, China

**Hongjie Luo**
Research Center for Industrial Ceramics, Shanghai institute of Ceramics, Chinese Academy of Sciences, Shanghai, China

**Minoru Osada and Takayoshi Sasaki**
International Center for Materials Nanoarchitectonics (MANA), National Institute for Materials Science (NIMS), Tsukuba, Ibaraki, Japan

**Ie-Hong Hong**
Department of Electrophysics, National Chiayi University, Chiayi, Taiwan

**Yongqi Fu**
School of Physical Electronics, University of Electronic Science and Technology of China, China

**Fengzhou Fang and Zongwei Xu**
Key Laboratory of Precision Measuring Technology & Instruments, Centre of Micro Nano Manufacturing Technology, Tianjin University, China

**Soshu Kirihara**
Osaka University, Japan

**Yuki Shimizu**
Takemi Asai and Wei Gao Tohoku University, Japan

**Naoto Shirahata**
National Institute for Materials Science (NIMS), Sengen, Tsukuba, Ibaraki, Japan

**Peng Nie and Kai-feng Zhang**
Shenyang Aerospace University, China

**Yoshitake Masuda**
National Institute of Advanced Industrial Science and Technology (AIST),
Anagahora, Shimoshidami, Moriyama-ku, Nagoya,
Japan